PREPARATIVE METHODS OF POLYMER CHEMISTRY

Third Edition

PREPARATIVE METHODS OF POLYMER CHEMISTRY

Third Edition

WAYNE R. SORENSON
WILFRED SWEENY
TOD W. CAMPBELL

A JOHN WILEY & SONS, INC., PUBLICATION

New York • Chichester • Weinheim • Brisbane • Singapore • Toronto

This book is printed on acid-free paper.♾

Published simultaneously in Canada.

For ordering and customer service, call 1-800-CALL-WILEY.

Library of Congress Cataloging-in-Publication Data:

Sorenson, Wayne R. (Wayne Richard), 1926-
Preparative methods of polymer chemistry/Wayne R. Sorenson, Fred Sweeny, Tod W. Campbell.—3rd ed.
p. cm.
Includes index.
ISBN 0-471-58992-6 (cloth: alk.paper)
1. Polymers. 2. Polymerization. I. Sweeny, Fred (Wilfred) II. Campbell, Tod W. III. Title.

QD281.P6 S58 2001
547′.28—dc21 00-047989

Printed in the United States of America.

10 9 8 7 6 5 4 3 2 1

We'd like to dedicate this book to the memory of our former friend and colleague, Tod Campbell, who died in 1968. Tod participated in the conception of the first edition of this book and to the contents of the first two editions. His zest for organic chemistry, polymer chemistry and for life at large will, we expect, remain evident in this third edition, since these elements of his life have stayed in the memories of those of us who knew him well.

CONTENTS

PREFACE

Polymer chemistry has grown rapidly in the last 40 years, and world production of polymeric materials, including plastics, rubber, paint and adhesives, to more than 100 million tons per year. Key advances include polymers with strength and stiffness four to five times that of steel on a weight basis, electrical conductivity approaching that of copper, and broad structural functionality at low density in the area of composites.

Since the last edition of this book (1968), we have witnessed tremendous advances in the chemistry of polymer synthesis from new intermediates and new fabrication techniques (e.g., gel spinning combined with extreme polymer chain extension, air-gap fiber spinning, and solid-state extrusion). These improvements have not occurred by chance but from development of a deeper understanding of the principles involved in preparing and processing polymers.

The intent of this book is to provide the organic chemist with tried and true methods of making specific polymers. Its intended audience are students, both undergraduate and graduate, and practicing chemists whose work involves or portends the need *(1)* to synthesize polymers and *(2)* to characterize them. The latter, as we approach the subject here, involves those first steps required to ascertain the basic properties of the polymer under study. It should be possible to gain enough insight to guide the experimenter in making course corrections in synthetic strategy or perhaps even in continuing the project. We also provide a chapter titled "Advanced Processing," which goes more deeply into the physical and chemical aspects of polymer processing. The reason for this is that in today's world of polymer research, especially in industry, no researcher is likely to want, or to be allowed, to confine his or her work solely to synthesis. The wishes of the individual and the prevailing forces of the work place are likely to put the researcher into the processing environment to follow-up the laboratory work on a polymer problem.

Each chapter on synthesis includes a discussion of the background and principles relating to polymers of that general type. This should enable the chemist to apply the synthesis technology intelligently and to make desired changes without loss of molecular weight or functionality. Regarding experimental strategy, the chemist should ask questions such as: What is my goal? Do I understand the mechanism of the reaction? What do I have to do to ensure the purity of my reactants? What are the best conditions for the reaction? Are any catalysts suggested from related simple organic reactions? What possible side-reactions do I have to be alert to? How do I minimize these? Are the reactants and polymer stable under the chosen conditions? Do I need a solvent? Might the solvent react? Is it a good solvent for the polymer product? Are moisture or oxygen deleterious to the reactants? What do the structural aspects of the polymer suggest about processing problems and conditions?

The book includes polymers and technologies from previous editions, because they encompass many important commercial polymers. Updated technology has been particularly expanded and includes new polymer types, processing technologies and characterization methods. We apologize in advance for any omissions.

Regarding polymer science, it is fair to say that the discipline is at a stage at which polymer application is being stressed at the expense of seeking new knowledge and new concepts. Advances may be expected in the so called speciality polymers (i.e., polymers designed for a particular end use or function). Specific examples include engineering plastics with superior mechanical properties and thermal resistance, liquid-crystalline polyarylates, polymer alloys, conductive polymers, advanced composites for aerospace use from high-modulus carbon fiber, and toughened thermosets or thermoplastic polyimides. To optimize these high-performance materials requires precise control of the polymer structure at the molecular, macromolecular and supramolecular levels. In the instances of polymers from olefins and diene, emphasis is on new catalysts which can more precisely control properties, less so on variations in monomer structures.

Regarding fibers, close to theoretical tensile modulus has been achieved in some polymers by efficient chain alignment, but fiber-breaking strength remains at less than 10% of theoretical, based on C–C single chain bond breakage. Reducing defect level and increasing the percentage of load-bearing chains will be required for further improvements. Advances should also be expected in the electrical and optical properties of polymers and their use in the rapidly expanding electronic and telecommunication industries.

We wish to note here and throughout this edition a general statement about the safe handling of chemicals and the safe execution of laboratory procedures. While this will be understood by, and will have become an article of belief to, chemical professionals and students of chemistry and chemical engineering, it is wiser, we believe, to be repetitious than omissive in the following: All properly trained chemists and chemical engineers understand that many of the materials they work with, often routinely, are inherently hazardous to some degree and under some circumstances, and that some are much more hazardous than others. Information on real and potential hazards is now abundant and every chemical professional, student and technician is obligated to be informed on current safety knowledge and practice regarding the material and equipment to be used. The point cannot be over stressed, whether the subject is polymerization or any other field of chemistry. We have tried to identify the most evident hazards to the reader, but leave it to the readers to use the full range of information available about the materials and methods described here, and their own laboratory equipment, to ensure their own and their co-workers' safety.

PREFACE TO THE SECOND EDITION

Since the first edition of this book appeared in 1961, polymer chemists have made major advances in many areas of new polymer chemistry, with a resultant increase in polymer types, polymer forming reactions, and catalytic techniques to polymerize species hitherto considered not polymerizable. This remarkable growth is indicated by the considerable growth in size of this second edition. Major advances have been made, particularly in the vinyl field, which have necessitated an almost complete revision of chapter 4. Most of the other chapters have been markedly changed by addition of new material and deletion of that which is less appropriate. Noteworthy additions to chapter 3 are an expanded section on Block Condensation Elastomers, and a section on High Temperature Polymers.

We have retained our original philosphy of keeping background discussion to a minimum. However, we have also attempted to provide a framework of commentary to highlight the significance of certain individual syntheses and place each section in the proper perspective. We have tried as before to give ample reference to the specific texts and articles that can supply detailed discussion on most polymer categories.

We again wish to thank our respective Managements for permission to write this second edition, and to our many friends throughout the world who provided advice, criticism, and encouragement. Among the latter, we wish especially to thank W. K. Wilkinson, E. J. Frazza, and J. Fath. A special group of our colleagues have gone so far as to provide us with entire preparations or supplementary detail which we have used. Our indebtedness to these individuals will be acknowledged at the appropriate place in the text.

Finally, for typing we wish to thank Virginia Mize, Dolores Cline, Suzeahn Massey, Rose Quincy, and Mary Pack.

WAYNE R. SORENSON
TOD W. CAMPBELL

January 1968

PREFACE TO THE FIRST EDITION

The purpose of this book is to provide a reference work containing detailed procedures for the synthesis and handling of a wide variety of polymer types. Although such information is readily available for most low-molecular-weight organic compounds, in the case of polymers it is scattered over a large number of more or less (usually less) accessible journals and patents. We have therefore assembled here the procedures involved in the laboratory preparation of most of the known classes of polymers. In cases where the original literature did not provide sufficient detail, supplementary information is given based on our own experience and that of our colleagues. In many instances this has involved a careful checking of the preparation and the procedures may therefore be used with confidence in their operability.

Since the preparation of useful high-molecular-weight polymers required that considerable attention be paid to the purification of monomers, solvents, and other reaction intermediates, as well as the choice of suitable equipment and reaction conditions, such information is provided in the different procedures wherever necessary.

As indicated in chapter 1 (Introduction) it is also our hope that the book might be of use as a supplementary text for a laboratory course in polymer chemistry or an advanced organic synthesis course.

As is always the case, we are indebted to our colleagues, without whose unselfish contribution of time and knowledge this book would not have been possible. We would like to offer special thanks to Professor C. S. Marvel of the University of Illinois and Professor H. Mark of Brooklyn Polytechnic Institute, for their unfailing enthusiasm and encouragement. We are also indebted to Warren Watanabe of Rohm and Haas, J. P. Schroeder, John Wynstra and R. K. Walton of the Union Carbide Plastics Company, and C. G. Overberger of Brooklyn Polytechnic Institute, for most helpful reviews of various portions of the book.

Of our colleagues at the du Pont Company, we would like to thank Fred Billmeyer, Roger A. Hines, Alfred C. Webber, and Neil Keiser, all of the Polychemicals Department, L. P. Hubbuch of the Fabrics and Finishes Department, and A. C. Stevenson and C. M. Barringer of the Elastomers Department for most helpful comments and suggestions.

We would like to thank the management of our own Textile Fibers Department for permission to write this book. To the following of our colleagues in the Textile Fibers Department, our gratitude for encouragement, technical assistance, and many trenchant comments: Jim Van Oot, Sid Maerov, Norton Higgins, Emerson Wittbecker, Fred Sweeny, John Schaefgen, Vic Shashoua, Bill Statton, Frank Moody, Tom Mackey, Helen Anderson, Fran Cramer, Bob Taylor, Al Goodman, Herman Marder, Wayne Hill, Paul Morgan, Ray Tietz, Lup Jung, Ralph Beaman, and Hal Bonner.

Thanks are due to Dan Sauers, Walt Brown, and Norm Van Hove for help in preparing the illustrations.

Finally, our special thanks to Tena Evlom, Mary Joan Reese, Hilda Smith, and Jean Iannarone, who waded through reams of illegible scribblings, stacks of incoherent records, and numerous revisions to produce the final manuscript.

WAYNE R. SORENSON
TOD W. CAMPBELL

1

POLYMERIZATION AND CHARACTERIZATION OF POLYMERS

1.1. TYPES OF POLYMERIZATIONS AND MOLECULAR WEIGHT REQUIREMENTS

In this book, polymers are considered as high molecular weight materials linear or branched structures, usually with defined repeating structural units. The utility and strength of polymers depends on having a sufficiently high molecular weight. The latter usually corresponds to >5000 molecular weight and/or >500 Å in chain length.

The objective of polymer synthesis should be to achieve the maximum molecular weight that can be readily processed to provide optimum properties. Maximum tensile properties depend on *(1)* chain length; *(2)* chain alignment, i.e., maximum number of load-bearing chains; *(3)* intermolecular forces; *(4)* crystallinity; and *(5)* low level of defects. These will be discussed further in chapter 3.

There are basically two major polymerization methods used to convert small molecules to polymers (the small molecules are called monomers although this is applied correctly only when the small molecule can be converted to polymer without loss of any product molecule).

The polymerization methods are termed step growth, or condensation polymerization and chain growth, or addition, polymerization.

Step growth polymerization is characterized by elimination of a small molecule. Examples are as follows:

$$NH_2—R—NH_2 + HOOC—R'—COOH \longrightarrow \left[NH—R—NH—\overset{O}{\overset{\|}{C}}—R'—\overset{O}{\overset{\|}{C}} \right]_n + H_2O \quad \text{polyamide}$$

$$HO—R—OH + HOOC—R'—COOH \longrightarrow \left[O—R—O—\overset{O}{\overset{\|}{C}}—R'—\overset{O}{\overset{\|}{C}} \right]_n + H_2O \quad \text{polyester}$$

$$Cl—R—Cl + Na—S—Na \longrightarrow [S—R]_n + 2NaCl \quad \text{polysulfides}$$

In chain growth polymerization, a small molecule is not eliminated. Examples are as follows:

$$CH_2{=}CHR \longrightarrow [CH_2—CHR]_n \quad \text{vinyl or acrylate polymer}$$

$$R—N{=}C{=}O \longrightarrow \left[\underset{R}{\underset{|}{N}}—\overset{O}{\overset{\|}{C}} \right]_n \quad \text{polyisocyanate or 1-nylon}$$

Comparison of both types of polymers are given in Table 1.1(1).

If we first consider step growth polymerization, the requirements for high molecular weight polymers are stoichometric balance of intermediates, high degree of monomer purity, and high yield reaction with lack of side reactions (little to no reaction with solvent; no decarboxylation). Polymer molecular weight

Table 1.1. Distinctions in Polymerization Categories

Step Growth	Chain Growth
One reaction type responsible polymerization, propagation, termination reactions	Several reactions with different rates for initiation, propagation, chain transfer, termination
Slow random chain growth; any species present can react; monomer(s) quickly consumed.	High MW[a] chains form quickly, independently of conversion
MW rises steadily throughout until end groups are used up or immobilized. Initial reactants rapidly depleted	Monomer(s) slowly consumed to form high MW in a rapid reaction
Broad calculable MW distribution as a function of degree of polymerization.	Broad MW distribution coexists with depleting monomer
Conversions always 100%	Conversions usually limited to 70–90%

[a]Molecular weight.

and degree of polymerization are of considerable importance. At the beginning of the polymerization, we have a certain concentration of monomer units, N_0 (mol/g). As the polymerization proceeds, N_0 is depleted to a value N after a specific time period.

If we define the extent of polymerization by a quantity p, then $(1 - p)$ is the fraction of moles that have not polymerized. Then

$$p = (N_0 - N)/N_0 \qquad \text{hence} \qquad N = N_0(1 - p)$$

and

$$(1 - p) = N/N_0 = T/T_0$$

where T_0 and T represent the number of end groups at the beginning and end of the polymerization. The number average degree of polymerization $\overline{\mathrm{DP}}$ defines the concentration of chains in the polymer and

$$\overline{\mathrm{DP}} = N_0/N = T_0/T = 1/(1 - p).$$

The number average molecular weight $\overline{M}_n$ is $\overline{\mathrm{DP}} \times W_r$ (where W_r is the molecular weight of the repeat unit).

The effect of the extent of polymerization on the degree of polymerization of the product can be exemplified as follows:

$$\overline{\mathrm{DP}} = 1/(1 - p)$$

If a reaction proceeds to 90% completion, then

$$\overline{\mathrm{DP}} = 1/(1 - 0.9) = 10$$

The product is a decamer, on average, and is too low in molecular weight to provide useful mechanical properties.

Usually a $\overline{\mathrm{DP}} > 50$ is required for useful properties. Hence if $50 = 1/(1 - p)$, $p = 0.98$. Thus at least 98% conversion of monomers to polymer is usually required for useful mechanical properties. This conclusion applies only to step growth polymerizations. Useful products are often obtained from chain growth polymerizations at relatively low conversions.

As mentioned earlier, effects of monomer purity on $\overline{\mathrm{DP}}$ are profound. If we consider the effects of a monofunctional impurity in a typical AB polyamide polymerization, the $\overline{\mathrm{DP}}$ is given by the expression

$$\begin{aligned}\overline{\mathrm{DP}} &= (1 - N/N_0)/[(1 - p) + N/N_0] \\ &= (1 + r)/[(1 - p) + r] = (1 + r)/[(1 + r) - p]\end{aligned}$$

where $N/N_0 = r =$ (mol of impurity)/(mol of monomer). For example, with a 2% molar impurity

$$\overline{\mathrm{DP}} = (1 + 2/98)/[(1 - 0.98) + 2/98] = 25 \quad \text{(vs. 50 without the impurity)}$$

For AA–BB polymerizations the expression is

$$\overline{\mathrm{DP}} = (1 + r)/[2r(1 - p) + (1 - r)] = (1 + r)/[(1 + r) - 2rp]$$

where $r = (\mathrm{COOH})/\mathrm{NH_2}$. With 2% excess of the AA intermediate; i.e., $r = 100/102 = 0.98$.

$$\overline{\mathrm{DP}} = (1 + 0.98)/[(1 + 0.98) - 2(0.98 \times 0.98)]$$
$$= 33 \quad \text{(vs. 50 for molar balanced intermediates).}$$

In the specific case of polymers like 6-6 nylon, where ends imbalance can occur from loss of diamine, cyclization, etc., $\overline{M}_n$ can be calculated from the expression:

$$\overline{M}_n = 2 \text{ million/total number of ends}$$
$$\text{(including stabilized ends, determined as parts per million)}$$

1.2. MOLECULAR WEIGHT DISTRIBUTION AND AVERAGES

In linear condensation polymers with a given average degree of polymerization (P_n) the chains do not all have the same length, but exist in a broad distribution as described by Flory (2) in an equation where the mole or number average fraction containing x repeat units is given by:

$$n_x = p^{x-1}(1 - p)$$

and the weight fraction by:

$$w_x = xp^{x-1}(1 - p)^2$$

where n_x is the mole or number fraction of chains containing x repeat units, w_x is the weight fraction containing x repeat units, and p is the degree of polymerization. Representative curves of these distributions are shown in Figure 1.1. At any degree of polymerization, there is more monomer than any other species on a mole fraction basis (2, 3).

We have mentioned molecular weight without specific definition and should clarify it. If polymer molecules were exactly the same size (termed monodisperse), then molecular weight could easily be defined. However we usually have a broad distribution of chain lengths, hence a range of molecular weights. Thus the molecular weight we calculate is an average of the species. Two averaging procedures are commonly used to define molecular weight. The first, called number average molecular weight $\overline{M}_n$, is obtained by summing the product of mole fraction of each species and its molecular weight. This is equivalent to dividing the total

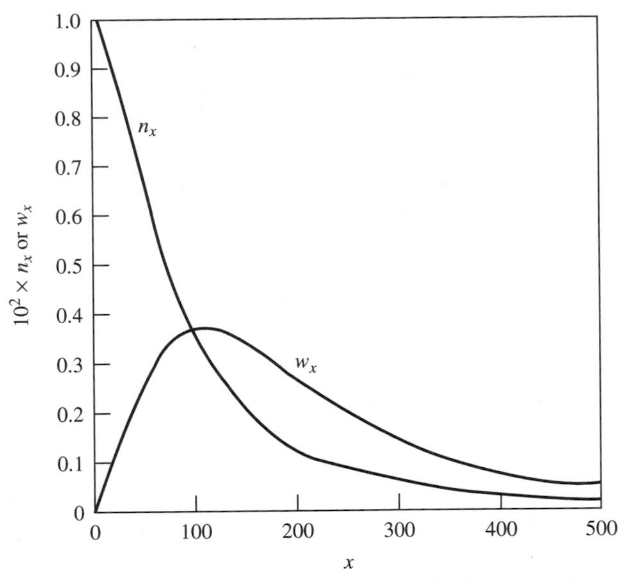

Figure 1.1. Most probable molecular weight distribution for linear condensation polymers. Variation of mole or weight fraction of x-mol with x for $p = 0.94$, $\overline{P}_m = 100$, $\overline{P}_w = 199$.

weight of the sample by the number of molecules contained in it. Thus

$$\overline{M}_n = \sum_{i=0}^{\infty} \left[N_i \Big/ \sum N_i \right] M_i = \sum [P_{n,i} M_i]$$

where

$$M_i = \text{molecular weight of any mole fraction } N$$

The number average degree of polymerization is given by $\overline{DP}_n = \sum_{x=1}^{\infty} x n_x$, where $n_x =$ the mole or number fraction of chains containing repeat units.

The second type of average molecular weight is called the weight average ($\overline{M}_w$). For this average, the weight fraction of each species times its molecular weight is summed. Thus

$$\overline{M}_w = \sum_{i=0}^{\infty} \left[W_i \Big/ \sum W_i \right] M_i = \sum [P_{w,i} M_i] = \sum [N_i M_i^2] \Big/ \sum [N_i M_i]$$

The weight average degree of polymerization, where the weighting factor in the average is the weight fraction of chains with a given chain length x_n is given by

$$\overline{DP}_w = \sum_{x=1}^{\infty} x w_x = \sum_{x=1}^{\infty} [x^2 n_x] \Big/ \sum_{x=1}^{\infty} [x n_x]$$

At high degrees of polymerization and normal molecular weight distributions, $\overline{DP}_w/\overline{DP}_n = 2$. This can be deduced from $\overline{DP}_n = 1/(1-p)$, and $\overline{DP}_w = (1+p)/(1-p)$, where $p =$ the extent of the polymerization. Thus

$$\overline{DP}_w/\overline{DP}_n = (1+p)/(1-p) \times (1-p)/1 = (1-p) = 2$$

when $p = 1$. In terms of the extent of the polymerization reaction (p) then

$$\overline{M}_n = M_0/(1-p)$$
$$\overline{M}_w = M_0(1+p)/(1-p)$$

where $M_0 =$ polymer unit molecular weight.

A third measurement of molecular weight $\overline{M}_z = [\sum_{l=0}^{\infty} N_i M_i^3]/[\sum_{l=0}^{\infty} N_i M_i]$ stresses the effect of any high molecular weight tail fraction and its impact on the viscosities of molten polymer at low rates of deformation. Ward (4) provides a tabular example of the disproportionately large contribution of large molecules on average molecular weights involving the higher moments of the distribution (Table 1.2). Four types of molecules are selected from a hypothetical polymer sample, with N_i and M_i being listed. M_n and M_z are listed in columns three and four. All molecular weights contribute significantly to the calculation for M_n, whereas the lighter molecules may be neglected in the sum leading to M_z. It is noted that 10 molecules of 1 million g/mol contribute more to the third moment sum than do 100 molecules of 400,000 molecular weight units. The point to be stressed is that the presence of a very small number of very large molecules can have an enormous effect on any physical or chemical property of a polymer that is a function of the higher moments of the molecular weight distribution. Figure 1.2 shows the importance of polymer chain lengths on the various molecular weight averages.

One common way of measuring number average molecular weight is based on ends determination, e.g., in the case of nylons, determining the amine and carboxyl ends plus an estimate of "stabilized" ends formed from added end capper, such as acetic acid, or from decarboxylation. Thus $\overline{M}_n =$ 2 million/SF (total ends including stabilized ends). Ends are measured in parts per million grams.

Table 1.2. Relative Contribution of Different Molecular Weights to the Average Values

N_i, mol	$M_i/10^3$ g/mol	$N_i M_i/10^3$	$N_i M_i^3/10^3$
10	10	10^2	10^4
100	100	10^4	10^8
100	400	4×10^4	6.4×10^9
10	1000	10^4	10^{10}

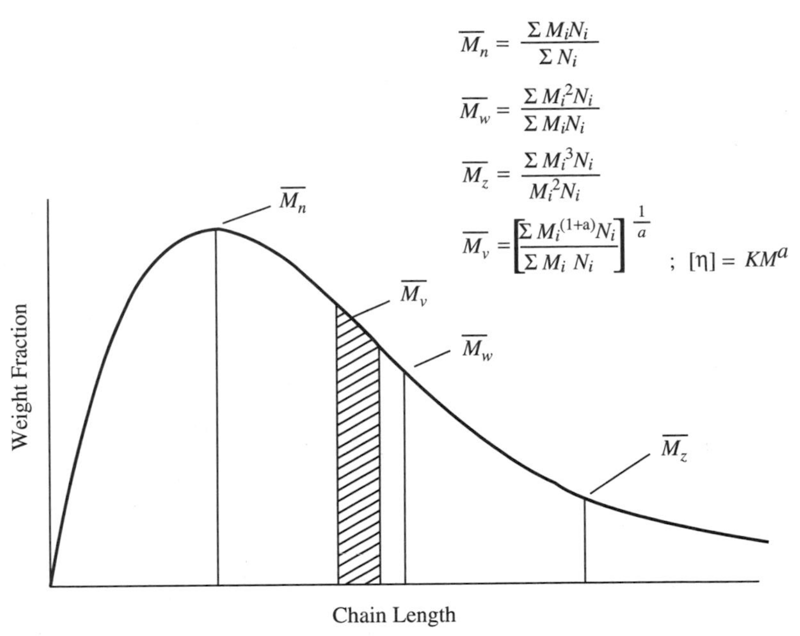

$\overline{M_n} = \frac{\Sigma M_i N_i}{\Sigma N_i}$
$\overline{M_w} = \frac{\Sigma M_i^2 N_i}{\Sigma M_i N_i}$
$\overline{M_z} = \frac{\Sigma M_i^3 N_i}{M_i^2 N_i}$
$\overline{M_v} = \left[\frac{\Sigma M_i^{(1+a)} N_i}{\Sigma M_i N_i}\right]^{\frac{1}{a}}$; $[\eta] = KM^a$
$\overline{M_n}$
$\overline{M_v}$
$\overline{M_w}$
$\overline{M_z}$
Weight Fraction
Chain Length

Figure 1.2. Molecular weight averages.

Probably the most commonly used laboratory methods of assessing polymer molecular weight are solution viscosity and size exclusion chromatography. These will be discussed in more detail in chapter 2.

Intrinsic viscosity $[\eta]$ (see chapter 2) can be related to viscosity average molecular weight by the semiempirical Mark–Houwink–Sakurada equation:

$$[\eta] = KM_v^a \sim KM_w^a$$

K and a are constants for each specific linear polymer, depending on solvent and temperature. Polymer shape information can be derived from the value of a. For a random coil polymer in a theta solvent (theta conditions describe solvent, polymer, and temperature at a point where the polymer assumes its "unperturbed" average shape. At infinite molecular weight the polymer would be on the verge of precipitating). $[\eta]$ varies as V_h/M, where V_h is the volume of solvated molecule, and M is the molecular weight. But V_h varies as $M^{3/2}$. Therefore, $[\eta]$ varies as $M^{1/2}$. Therefore, $a = 0.5$ for a random coil. For a better solvent than a theta solvent $a > 0.5$. For stiff-rod polymers, $a = 1 - 2$, i.e., the rod sweeps out a larger volume than a random coil. For hard spheres, V_h varies as M; hence $[\eta] = KM^{\circ}$.

Significant modification in the K and a values occur when the polymer is branched. Lower values of $[\eta]$ are obtained from branched polymers vs. linear polymers at equivalent molecular weights. Some values of a in the Mark–Houwink–Sakurada equation are shown in Table 1.3.

Table 1.3. Values for a Range of Polymers

Polymer	Solvent	α	Comment
Polystyrene	Toluene	0.69	Coil
Poly(methylmethacrylate)	Chloroform	0.79	Coil
Poly(*p*-aminobenzoic acid)	H_2SO_4	1.1–1.6	Stiff rod
Poly(*p*-phenyleneterephthalamide)	H_2SO_4	1.09	Semistiff

Chain growth polymerization incorporates a range of ring-opening polymerizations in which relief of strain is the driving force. Many of the polymerizations relate to changing the sp^2 bonding of a double bond to sp^3 of a single bond. $H_2C{=}CHR \longrightarrow CH_2{-}CHR{-}$.

The double bond is usually activated by an electron-withdrawing or -donating R group or by complexing of the double bond with a metal ligand. The electronic effect of the attachment dictates the mechanism of polymerization. Cationic (or carbocationic when C^+ is the active center) polymerization occurs most readily with monomers having electron-donating groups adjacent to the double bond, e.g., in isobutene or vinyl ethers.

Anionic polymerization occurs most readily with monomers having electron-withdrawing groups adjacent to the double bond, e.g., in methyl methacrylate or acrylonitrile. Monomers with intermediate donating and, to a certain extent, electron-withdrawing groups, polymerize via a free-radical mechanism.

The mechanism of polymerization consists of three distinct steps: initiation, propagation, termination. In the initiation step an initiator molecule is thermally decomposed or undergoes a chemical reaction to generate an active species, (which can be a free radical, a cation, anion, or complex), which then initiates polymerization by adding to or complexing with the monomer's (usually carbon-carbon) double bond. The reaction occurs in such a manner that a new free-radical, cation, anion, or complex is generated. The initial monomer becomes the first repeat unit in the incipient polymer chain. In the propagation step, the newly generated active species adds another monomer in the same manner as the initiation step. This procedure is repeated over and over again until the final step in the process (termination) occurs. In this step, the growing chain terminates through coupling with another growing chain; by reaction with another species in the polymerization mixture (e.g., transfer of the active site to another molecule) or by spontaneous decomposition of the active site by disproportionation etc. The following are characteristics of this type of polymerization:

- Once initiation occurs, the polymer chain forms quickly.
- The concentration of the active species is low. Hence the polymerization mixture consists primarily of newly formed polymer and unreacted monomer.

- Because the carbon-carbon double bonds in the monomer are in effect converted to two single bonds in the polymer, energy is released, making the polymerization exothermic and often requiring cooling.
- Chain reactions normally afford polymers of high molecular weights (10^4 to 10^7).
- Free-radical and carbocationic polymers are often branched, resulting from abstraction of hydrogen from the backbone main chain and generation of a radical or cation at that site.

Two or more different monomers are often employed in chain-reaction polymerization to yield polymer containing the corresponding repeat units. Such a process is called copolymerization and the product is a copolymer. It should be pointed out that chain reactions are not limited to C=C groups but include others such N=C; C=O; P=N etc. In anionic polymerization (and in some carbocationic or free-radical polymerizations) termination can be minimized and "living polymers" result. In such cases $\overline{M}_w/\overline{M}_n$ approaches 1.

Free-radical polymers rarely can be held in the living state at ambient temperatures and proceed to a termination reaction. They are invariably branched. Step-growth types of polymers are far less likely to be branched. Free-radical polymerization is initiated by radicals from decomposition, typically, of peroxides or azo compounds or by light. Unlike ionic polymerization, radical polymerization is not sensitive to water but is to oxygen. If polymerizations are run in water, care should be taken to ensure that the water is free of dissolved air, i.e., the water reaction medium should be boiled before use to remove oxygen and then cooled under nitrogen.

Many types of free-radical initiators are used for vinyl polymerizations, chiefly organic peroxides, water-soluble inorganic peroxides (commonly called redox initiators) and aliphatic azo compounds. Choice depends on the polymerization system and reaction conditions. Polymer termination occurs by two growing chain radicals combining, by disproportionation, or by radical transfer.

$$-CH_2-C^*HR + {}^*CHR-CH_2- \longrightarrow -CH_2-CHR-CHR-CH_2- \quad \text{combination}$$

$$-CH_2-C^*HR + {}^*CHR-CH_2- \longrightarrow -CH_2-CHR + -CH{=}CHR \quad \text{disproportionation}$$

$$-CH_2-C^*HR + R'SH \longrightarrow -CH_2-CHR + R'S^* \quad \text{transfer}$$

In ionic systems, termination obviously does not occur by combination but usually by transfer or reaction with adventitious species such as water. Ionic-chain polymerizations usually require more stringent conditions than free-radical polymerizations. The formation of ions of sufficient stability to propagate to high molecular weight usually requires low temperatures (-100°C to 50°C) and polar solvents. One is limited, however, to mildly polar solvents, because polar solvents stabilize ions so well that propagation is slow. Water often destroys the initiator.

Kinetics are complicated due to the simultaneous presence of two types of propagating species: ---(C^+A^-), the ion pair; and ---C^+A^-, the free ions. In carbocationic polymerization, initiation requires a cationic species to react with the π electrons and bond to carbon of the monomer double bond. Proton sources, such as HCl or H_2SO_4 generate low molecular weight products because the counter ion is closely attached to the carbocation (C^+) and hinders polymerization. BF_3 and other Lewis acids provide higher molecular weight products.

Work by Kennedy and Faust (5) has demonstrated that the proper catalyst and solvent system provide "living" carbocationic systems with polymer $\overline{M}_w/\overline{M}_n$ close to 1.

Although thermodynamics assists us in predicting whether a polymerization will proceed, we rely on the kinetic aspects to tell us how fast these reactions may occur (propagation). It is beyond the scope of this book to discuss the kinetics of free-radical polymerization in detail, and further information can be found in Jenkins and Ledwith (6) and Ham (7). In brief, however, if one makes the reasonable assumption that the rate of initiation equals that of termination under steady state conditions, then the rate of polymerization will depend on several constants and will be proportional to the first power of the monomer concentration and to the square root of the initiator concentration. Practically, this means if the initiator concentration is doubled, the rate will increase only by a factor of 1.4. Of course, initiator efficiency is never 100%, and is closer to 50 to 80%.

Chain transfer occurs when the growing polymer chain abstracts an atom (H, Cl, etc.) from another chain, a small molecule, solvent, etc. to terminate the growing chain but produces a new and growing radical at the abstracted center. What this step does is regulate the molecular weight and affects molecular weight distribution. It does not necessarily decrease the rate of polymerization if one assumes that the new radical will again reinitiate more monomer at the same rate of polymerization as before.

The amount of chain transfer depends on the structure of the agent in question and can roughly be related to its reactivity with a small molecule radical species. An aromatic molecule like benzene does not react effectively with small molecule radical species, and indeed the chain transfer constant with growing polymer chain systems is small. Toluene, however, is a fair transfer agent. Other molecules like carbon tetrachloride and particularly mercaptans are more effective transfer agents. One can assess the effectiveness of a chain transfer agent by running polymerizations with various ratios of transfer agent to monomer and by plotting the reciprocal of the number average molecular weight or number average degree of polymerization versus this ratio.

Although we have emphasized chain transfer in radical systems, it should be pointed out that this occurs readily in carbocationic systems; abstraction of hydrogen from the polymer backbone is common, leading to branching. "Living" carbocationic systems developed by Kennedy and Faust (5) minimize chain transfer by close association of the growing cation with a bulky counterion.

In contrast to carbocationic systems, living polymer often occurs in anionic systems. Conditions, monomers, and catalysts can be selected to minimize chain transfer and termination, resulting in $\overline{M}_w/\overline{M}_n$ close to 1. These will be discussed more fully in chapter 9. One last type of polymerization that should be mentioned is that involving metal complexing catalysts. These include Ziegler-Natta and later types of metallic catalysis in olefin polymerizations, polymerization of olefin/CO with palladium catalysts and olefin metathesis with tungsten catalysts. These will also be discussed in chapter 9.

REFERENCES

1. J. K. Stille, *J. Chem. Ed.*, **58**, 862 (1981).
2. P. J. Flory, *Principles of Polymer Chemistry*, Cornell University Press, Ithaca, New York, 1953.
3. J. Zimmerman, *Encyclopedia of Polymer Science and Engineering*, vol. 2, Wiley, New York, 1988.
4. T. C. Ward, *J. Chem. Ed.*, **58**, 868 (1981).
5. J. P. Kennedy and R. Faust., *J. Poly. Sci. A.*, **25**, 1847 (1987).
6. A. Jenkins and A. Ledwith, eds., *Reactivity, Mechanism and Structure in Polymer Chemistry*, Wiley, New York, 1988.
7. G. Ham, ed., *Kinetics and Mechanism of Polymerization*, vol. 1, *Vinyl Polymerization*, Decker, New York, 1967.

2

PREPARATION, FABRICATION, AND CHARACTERIZATION OF POLYMERS

This chapter discusses general methods of preparation, fabrication, and characterization of polymers. Specific methods will be discussed in the chapters that follow. Although the reactions used in the preparations of polymers are usually identical to those used in the synthesis of small molecules, the high molecular weight of the polymeric products and the physical result of chain size and interactions give polymers the properties that set them apart. Consequently, they often require special methods of handling and characterization that are quite different from those used with small molecules.

2.1. TECHNIQUES FOR PREPARING AND HANDLING OF POLYMERS

2.1.1. Use of Constant-Temperature Baths

It is sometimes necessary to heat polymerization reactions for long periods of time at constant temperatures. One convenient way to do this is to use a vapor bath consisting of a large (2 × 15 in.) test tube filled one fourth full with a liquid of a selected boiling point. Several boiling chips are added, the liquid is heated to boiling, and the vessel to be heated is suspended in the vapors. Table 2.1 gives a partial list of materials that are thermally stable and can be used as vapor baths. The preferred sources of heat for the baths are controlled electric heaters to minimize fire in the event of bath breakage. For lower temperatures, electrically heated and controlled water or oil baths are commonly used. Likewise, heating of rotating or shaken vessels is often carried out in a larger container using air warmed by passing over electrically heated coils, or simply in a container filled with temperature-controlled water, when the needed temperature range permits.

Table 2.1. Liquids for Vapor Baths[a]

Compound	BP, °C	Compound	*BP*,°C
Toluene	111	Diethylene glycol	245
Chlorobenzene	133	Diphenyl	255
Cyclohexanol	160	Diphenyl ether	259
o-Dichlorobenzene	179	Methyl naphthyl ether	275
Phenol	181	Dimethyl phthalate	283
m-Cresol	202	Benzophenone	305
Methyl salicylate	222	*o*-Terphenyl	330
n-Decyl alcohol	231	Anthraquinone	380

[a]Other suitable liquids with a wider range of boiling points can be found in sources such as Lange's Handbook of Chemistry.

2.1.2. Purification of Reagents

Purification of reagents is essential in polymer chemistry if high molecular weights are to be achieved. As mentioned in chapter 1, a 2% impurity can cut the molecular weight in half in the instance of step growth reactions. All reagents should be purified by conventional means (crystallization, distillation, etc.) to give a product with a single gas chromatographic peak. NMR should also be considered to characterize the reagent and confirm its purity. Most polymerizations should be run under argon or nitrogen. It should also be pointed out that in nitrogen blanketing of reactions in heated ovens (e.g., thermal stabilities of fibers, film), one must make sure that the circulating fan does not introduce air from the outside. Oxygen level should be monitored in blank experiments. Furthermore, it is essential that samples be shielded from radiant heat from the heater coils; otherwise, specious results will be obtained.

2.1.3. Sealed Tube Reactions

Sealed tubes are often used in the synthesis of both condensation and vinyl polymerizations. A typical tube uses a long-stemmed funnel to add either liquids or finely divided solids without contaminating the upper part of the tube where the molten seal has to be made. The tubes are made of hard glass from heavy glass tubing. Because the tubes must be strain free to withstand high pressures it is recommended that they be purchased commercially or fabricated by a competent glassblower. When sealed, the tube represents a potential bomb and should be protected by a heavy glass or steel-mesh sleeve. Additional protection is afforded by running the heated reaction behind a heavy glass or Lucite shield. Often a lot can be learned (especially about solubility vs. temperature) by rotating clamped sealed tubes in a heavy-walled box with a thick glass front while heating with thermostatically controlled circulating heated air.

2.1.3.A. Vinyl Polymerization in Sealed Tubes and Bottles

Vinyl polymerizations are adversely affected by oxygen, and many vinyl monomers are low boiling; hence it is convenient to seal the polymerization mixture in a glass polymer tube under nitrogen. The tube is then immersed in a constant temperature bath and mechanically shaken or tumbled during the course of the polymerization. For monomers that boil close to room temperature, commercially available heavy-walled bottles may be used instead of tubes. Sealed tubes should not exceed 100 to 200 mL in capacity, must not be filled by more than half, and must be free of residual air. When the glass tubes are sealed with an oxygen torch, care must be taken to ensure that the seal maintains the original thickness of the glass wall. All seals should be annealed to remove any strain in the seal.

In the event of an exothermic polymerization and buildup of high pressure, the polymerization should be run behind adequate safety shielding. After cooling, the tube should be opened behind shielding by heating at the seal to melt the glass to release any pressure, or by scoring the glass with a file just below the seal, then heating at the score with an oxygen torch to induce cracking. Sometimes the top will pop off if the tube is under pressure. Alternatively, commercially available pressure bottles with an appropriate closure may be used when expected pressures are not excessive. For example, butadiene can be safely contained at temperatures up to 60°C. Although sealed bottles are convenient as reaction vessels, one must guard against a cavalier attitude to their potential danger. Recognize that these are potential bombs, and handle them with extreme caution. Protective fabric sleeves must be used over the glass bottle to provide a degree of containment in the event of rupture.

2.1.3.B. Condensation Polymerization in a Sealed Tube

Most high-temperature polycondensation reactions, such as polyamidation and polyesterification, are started in sealed tubes in the absence of air to minimize coloration and degradation.

By way of illustration, a typical polyamide would be prepared as follows (1). A quantity of "nylon salt" is added to the tube through a funnel to about one third full. The tube is then constricted with an oxygen torch, alternately evacuated with an oil pump, flushed with nitrogen several times, and finally sealed under vacuum. The tube is heated either in an oil bath or in a salt bath (several eutectic mixtures can be obtained commercially). For reasons of safety, the glass tube should be contained in a sleeve of glass cloth or in a steel tube, with the open end directed away from the operator. After the heating cycle, the tube containing water and the nonvolatile prepolymer is cooled and opened under nitrogen. It is then equipped with a capillary and side arm for distillation, either fused to the tube or attached with heavy-walled pressure tubing. The nitrogen inlet should extend to the bottom of the polymer tube, below the surface of the prepolymer. If necessary, the tube may be heated under nitrogen to liquify the polymer so that the tip can be introduced into the liquid. The tube is now evacuated, and the nitrogen bubbled in for a prescribed time at a given temperature. The progress of polymerization can be followed by the rate of flow of the nitrogen through the melt and obvious increase in viscosity of the polymer. When the polymerization is considered complete, the tube is removed from the bath and cooled under nitrogen; the contents are removed by carefully breaking the tube using gloves and wearing safety glasses.

2.1.4. Use of High-Speed Home Mixers in Preparation of Polymers

Interfacial polymerization (chapter 4) broadened the synthesis of a range of polymers that could not be made readily other ways, and decreased the time required to making them. This technique involves reaction of an acid chloride in an inert water-immiscible solvent with, for example, an aliphatic diamine or a bisphenol, in water containing an acid acceptor. The polymerization occurs at the interface between the two solvents; hence it is desirable to make the interface as large as possible via high-speed stirring and use of an inert surface-active agent. One of the most convenient ways of doing this on a laboratory scale is to use a home blender (such as a Waring or Oster), preferably modified with a nitrogen inlet screwed into the base to minimize fire hazards when flammable solvents are used.

The aqueous diamine, acid accepter and detergent solution is placed in the jar of the blender, and the motor is turned on. The diacid chloride is run rapidly into the jar in a thin stream and is dispersed throughout the aqueous solution. Polymerizations run this way are complete in a matter of seconds. Numerous examples are found in chapter 4.

2.1.5. Use of Resin Kettles

Probably the most convenient laboratory vessel to use for medium-size polymerization reactions is the resin kettle, equipped with a detachable top with a center stirrer inlet and two to three other openings for condenser, feed port, and thermometer. These are available commercially in a range of sizes and designs. Viscous polymers or solutions can be removed easily by many manipulative techniques.

Often it may be necessary to scale up a polymerization beyond the resin kettle level, and at this point recourse to stirred reactors or autoclaves of several gallons capacity is necessary. For polymerizations under any level of pressure and of any size, stirred or rocking autoclaves not only are functional but are safer than glass containers. The use of such equipment is particularly valuable in polymerizations via coordination catalysts, particularly when it applies to flammable gases such as ethylene or propylene under pressure.

2.1.6. Use of Hypodermic Syringes

In many vinyl polymerizations, particularly those proceeding by ionic mechanisms, it is necessary to add catalyst to a cold anhydrous monomer under completely anhydrous and anaerobic conditions. This can be done conveniently by equipping the polymerization vessel with a serum-type stopper free of sulfur etc., which could act as an inhibitor. The catalyst, or other ingredient, is then injected quantitatively through this closure from a graduated hypodermic syringe.

Besides the transfer of catalyst solutions, syringes can be used for suspensions of solids under the right circumstances using a wide-bore needle. For example, commercial titanium trichloride can be successfully dispensed from suspension in cyclohexane using an 18-gauge or larger needle, using not more than two thirds of the syringe's capacity. The amount of catalyst used can be determined by evaporation of solvent from aliquot of the suspension.

2.1.7. Isolation of Polymers

Polymerizations may yield the desired polymer as a solution, a gel-like mass, an emulsion, a hard solid lump, or as an easily filtered granular suspension.

2.1.7.A. Polymer Solutions

Polymer solutions, after degassing by centrifuging, may be fabricated directly into fibers or films. If it is desired to isolate the polymer, the solution should be added with high-shear stirring into a larger volume of nonsolvent. The precipitated polymer should then be filtered, washed several times with nonsolvent, and then dried by heating in a vacuum oven. If the precipitant is a flammable solvent, electrically heated ovens should be avoided.

2.1.7.B. Emulsions

Vinyl polymerizations are sometimes run in emulsion systems. In such cases, to isolate the polymer it is often necessary to break the emulsion by adding an ionic material such as alum, sodium chloride, or hydrochloric acid or simply by heating. The polymer often precipitates as a curd, like cottage cheese, which is filtered, washed thoroughly with water, and dried. Other nonfilterable suspensions may be separated by centrifugation.

2.1.7.C. Gel-Like Mass

Gel-like masses of precipitated polymer are best soaked or stirred vigorously in a nonsolvent for the polymer to remove the gelling agent and give a filterable solid. Alternatively, the solvent may be evaporated or removed by steam to leave a solid mass.

2.1.7.D. Solid Lump

Polymer obtained as a solid lump may often be fabricated from this form. However, for convenience in handling and extracting possible impurities, it is better to cut the polymer into smaller lumps using a heavy knife and a hammer. The small pieces can then be fed into a Wiley mill and cut to any desired fineness. To avoid polymer fusion, it is preferable to grind the polymer together with dry ice. The polymer should then be vacuum dried to remove moisture.

2.1.8. Dissolving Polymers

Most polymers are soluble in organic media or in strong organic or mineral acids, and it is customary to determine the solubility spectrum of new polymers as they are made. Polymers often dissolve slowly, and solutions are viscous. Therefore, patience is required when assessing the range of solubility. It is important to have the polymer as finely divided as possible and to apply agitation, especially with shearing. Ultrasonic mixing is quite effective. A surprising increase in the rate of polymer dissolution is effected by using a shear-disk stirrer, which consists simply of a steel shaft terminated by a flat steel disk, rotated at high speed. The shear-disk stirrer may be modified by a second disk mounted far enough up the shaft to be above the level of the solution to be stirred. This effectively prevents viscous solutions from climbing the stirrer shaft. Once the solution is obtained, it may be freed of any impurities by dilution and filtration or by pressure filtration through stainless-steel screens.

One specific problem is in isolation of poly(benzobisthiazoles) or poly-(benzobisoxazoles) synthesized in polyphosphoric acid (PPA). The solution viscosity, even at 3% solids is so high that the solution is dough-like. Aqueous isolation yields a polymer that is no longer soluble in the PPA. The preferred mode of handling is to remove the thick dough with a spatula from the reaction kettle under nitrogen onto thin Teflon-aluminum sheets (polymer on the Teflon side) and roll it into cigar-like shapes for subsequent fabrication into fiber or film. The samples should be stored under nitrogen in sealed, dry conditions.

2.2. FABRICATION OF POLYMERS

A complete review of the fabrication of polymers would require extensive treatment. Hence it is intended in this chapter to describe the simplest and most elementary methods of polymer fabrication. In general, polymer fabrication can be classified into melt methods, solution methods and, as a special case, gel spinning methods.

2.2.1. Melt Methods

For melt methods, the polymer is heated until molten, is formed into the desired shape, and is then cooled. The technique is applicable only to polymers that are melt-stable.

2.2.1.A. Melt Pressing of Films

To prepare small pieces of film in the laboratory, a press such as a Carver press is used. The platens of the press are heated electrically to close to the polymer melt temperature, previously determined (discussed below). The polymer is placed in a small pile at the center of a 4 × 4-in. sheet of heavy aluminum foil (copper if the temperature is to be >250°C.). It is covered with another piece of the foil, and the sandwich is placed between the platens of the press. Desirably, the thickness of the film should be predetermined by placing copper shims between the edges of the foil or at least on two opposite sides of the platens. The platens are brought together by the hydraulic-jack, and a pressure of 2000 to 5000 psi is applied for about 30 s. The pressure is released, the foil is removed, and product is cooled in air or water. The two pieces of foil are carefully separated and the film sample removed. It is usually easier to remove the foil from the film rather than film from the foil. Removal of the film is often assisted by prying a razor blade under the edge of the film to make an initial separation and then placing the film/foil in water, where the capillary action will assist further separation. Sometimes if the polymer adheres too well to the foil, one can obtain a good separation by using Teflon-coated foil. However, one should be aware that surface properties (e.g., wettability) can be affected by residual Teflon. As a last resort, if the polymer is unaffected by acid or base, the foil may be dissolved away by one of these. If the film is not clear or completely coalesced, the pressing temperature may be too low. If the film is too thin and shows evidence of excessive flow, or if it contains small gas bubbles and evidence of degradation, the temperature is too high. Much information can be obtained from a film sample. Thus the toughness of the polymer can be determined qualitatively and quantitatively, giving an indication if the polymer has been prepared in useful molecular weight. Drawability, tensile properties, orientation, and crystallinity can all be determined. In most polymer laboratories, film assessment is the route for preliminary screening.

2.2.1.B. Preparation of Molded Objects

The preparation of molded objects, such as bars, circular chips, and cups, may be done either by injection or compression molding. Injection molding is done by an apparatus consisting of a molding chamber, into which molten polymer may be forced. The molding chamber consists of two heavy plates hollowed in the desired shapes, with each compartment connected to the others by channels, through which the molten polymer may flow, and to a channel through which the molten polymer is forced into the mold. The objects so obtained (e.g., small test bars) are all connected and are separated by cutting away the connecting polymer.

Even the simplest such machine represents considerable capital outlay. Machines suitable for production of larger and more complicated objects are quite costly and will be found only in laboratories specializing in this type of work or in commercial injection-molding factories.

Compression molding of simple objects such as bars and chips may be done with simple and readily available equipment. In compression molding, the polymer is placed in the mold, on top of the bottom plate, and the piston inserted. The mold is heated externally with, for example, an electric strip heater to fuse the polymer, and pressure is applied to the piston to compact the polymer. The pressure may be applied by a press, such as an Arbor press. The apparatus is then cooled, the bottom plate and piston removed, and the molded object knocked out (Fig. 2.1).

2.2.1.C. Melt Extrusion of Polymer to Fibers

The easiest method of making short lengths of fiber is to melt the polymer and pull away short lengths of fiber either with a glass rod or with forceps. Fiber so prepared is nonuniform and unsatisfactory for any tensile measurement. The

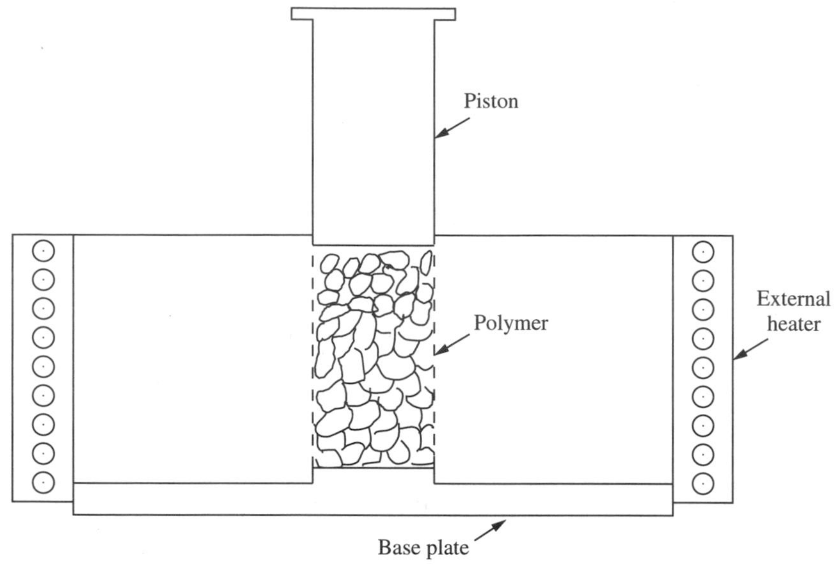

Figure 2.1. Apparatus for compression molding disk.

simplest mechanical method for converting polymer to continuous lengths of fiber is to place the polymer in a heavy-walled, heated steel cylinder equipped with a piston (driven by a hydraulic ram), a spinneret (a small disk drilled with one or more holes), and a motor-driven wind-up (equipped with a cylindrical, removable bobbin) (Fig. 2.2). The polymer must be compacted before placing it in the cylinder of the extruder to minimize entrapment of air, which would cause discoloration, bubbles, and defects in the extruded filament. The dry, finely divided polymer may be compacted to a cylindrical plug in a heated mold equal to or less in diameter than that of the chamber of the extruder. The dried plug is then dropped into the cylinder of the melt-spinning equipment.

A simpler, but less effective, method is to press the polymer to rather thick film, and cut disks from the film with a sharp cork borer of the proper diameter to fit snugly in the barrel of the spinner. A stack of disks 1 to 2 in. in height will produce enough fiber for preliminary evaluation.

2.2.1.D. Plasticized Melt Fabrication

Occasionaly, a polymer shows a very high melt viscosity, or perhaps some instability at the temperature of fusion. In such a case, it is often possible to blend the polymer with a plasticizer, a high-boiling liquid or solid compatible with the polymer. The plasticized polymer can then be melt fabricated at a lower temperature in keeping with the amount of plasticizer added. A number of polymers based on vinyl and vinylidene chloride are fabricated commercially this way, for example Saran fibers. The plasticizer will make the polymer more flexible; and if this is desirable, the plasticizer may be left in the polymer, otherwise it may be extracted with an organic solvent.

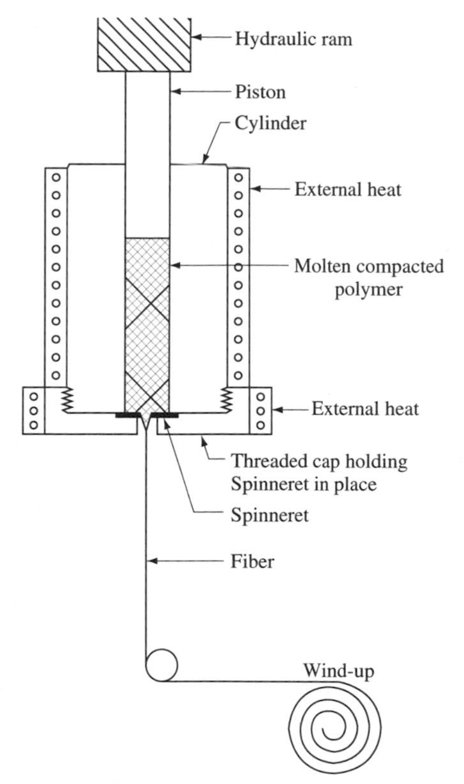

Figure 2.2. A melt spinning apparatus.

2.2.2. Solution Methods

When a polymer cannot be melted or is unstable at its melt temperature, it may be dissolved in a solvent and fabricated to fiber or film. Massive articles, corresponding to molded objects, cannot be made from solution because complete removal of solvent is difficult and form retention is almost impossible.

2.2.2.A. Casting Films

Solutions for casting should be quite viscous to prevent cast films from running, or spreading, over the cast surface. Solutions for casting may be warm (in such a case, the casting surface should also be warm) and concentrations of 10 to 20% are usually satisfactory, although much depends on the polymer's properties. Trial and error will lead to the optimum concentration. To cast a film, the viscous polymer solution is spread onto a glass plate with a casting knife. This is usually an I- bar of brass or stainless steel in which the front edge of the cross-bar is raised to a fixed level (depending on desired film thickness, usually 5 to 30 mil) above the level of the sides. The front edge of the cross-bar is also machined to about a 45° angle to even the spread of the polymer solution during casting. The prepared plate is placed in a forced draft oven to dry, at a temperature well below the boiling point of the solvent to avoid bubbles. More uniform slow removal of the solvent can be accomplished by placing another glass plate, separated by shims, over the film. Pennies or washers make good shims. After drying, the film is stripped from the plate. Sometimes soaking in water assists separation from the plate. A rarely used but interesting technique is to insert the freshly cast film into a dish of another liquid that is a nonsolvent for the polymer and is miscible with the solvent used for dissolving the polymer. Chemical reactions (e.g., deposition of metals by reaction with their salts) can be carried out in this manner. The firmly coalesced film can be removed from the plate, dried, and isolated as above.

2.2.2.B. Spinning Fibers

There are two well-known methods, dry spinning and wet spinning, for converting polymer solutions to fibers. In dry spinning, a viscous polymer solution is forced through a spinneret into a heated gas (air or nitrogen) that evaporates the solvent, leaving a polymeric thread. In wet spinning, the polymer solution is injected into a mixture of solvent and nonsolvent, which coagulates the solution into a thread line. Composition and choice of the nonsolvent are important to avoid obtaining a highly porous fiber. Both methods are commercially important.

In the laboratory, successful dry spinning is difficult to accomplish because of the complexity of the process. Even a simple setup for continuous fiber production with adequate control of variables will cost thousands of dollars. Usually small-scale dry spinning (e.g., 100 to 1000 mL of solution) is performed in a semiworks fitted with such equipment.

More practical on a small scale is wet spinning, although again the nonsolvent, bath temperature, etc. are critical. The choice of precipitating bath is most critical and must be determined experimentally for each situation. Slow gelation of the fiber is desirable otherwise a weak fiber will result. This will be discussed more fully in chapter 3. However, in small-scale wet spinning, a small amount of polymer solution (e.g., in dimethylformamide) may be spun from an ordinary hypodermic syringe into a bath of dimethylformamide diluted with water. The precipitating fiber can be drawn from the tip, through the bath, with a pair of tweezers. For more precise work, a motor driven, metal hypodermic syringe can be used with roller guides to keep the fiber in the bath for a specific length of travel and a motor-controlled cylindrical windup with a traverse guide to spread the fiber uniformly across the wind-up bobbin.

2.3. POLYMER CHARACTERIZATION

The many desirable physical properties associated with polymeric materials, such as strength, elasticity, and viscosity, are a direct consequence of the high molecular weight of the molecules composing the materials. The problem is further complicated by the fact that polymeric materials are not uniform and contain polymer chains with a range of molecular weights. In the following discussion, polymer characterization is treated from the standpoint of the simplest laboratory methods available for estimating the property or value in question. The methods are intended as a minimum framework of characterization that is applicable to most of the polymers prepared in subsequent chapters. In many cases, the methods can be run by the chemist himself or herself. There are others, including many not directly considered here, that are in the realm of the characterization specialist; these, of course, are extremely important but encompass a level of expertise beyond the scope of the present treatment. From a commercial viewpoint, polymer strength, toughness, ability to be fabricated, and transverse and compressive properties are most important. Polymer characterization is indeed a broad topic and covers basic aspects such as elemental analysis and molecular weight through mechanical properties to environmental and end use properties, such as light stability of fibers and burst strength of plastic pipes. A fairly extensive bibliography is provided so that the reader can be initially directed to some of the extensive general and particular treatments of polymer characterization.

2.3.1. Molecular Weight Determination

As has been mentioned before, $\overline{M}_n$ and $\overline{M}_w$ represent average molecular weight measurements of a range of polymer molecules. The polymer molecular weight distribution (or polydispersity) plays an important role in the behavior and properties of polymers. Dispersity is defined by $\overline{M}_w/\overline{M}_n$. For a monodisperse system $\overline{M}_w/M_n = 1$. For melt-equilibrated step growth polymers $\overline{M}_w/\overline{M}_n$ is close to 2. The common polydispersities of various polymer types are listed in Table 2.2.

2.3.1.A. Gel-Permeation Chromatography

Before the development of gel-permeation chromatography (GPC), polymer fractionation was a tedious process; however, this important technique has transformed the fractionation of polymers into a rapid, relatively low cost process, accessible to well-equipped laboratories. Typical instruments in use include the Waters ALC/GPC-244, equipped with a 6000-Å solvent delivery system, a U6K injector, a 440-uv detector, and a R401 differential refractive index detector.

The principle involved is selective permeation of polymer molecules from solution into the pores of a gel of cross-linked polymer (e.g., polystyrene) from

Table 2.2. Polydispersity of Various Polymer Types

Polymer Type	$\overline{M}_w/\overline{M}_n$
Free-radical addition (low conversion)	
Terminated by coupling	1.5
Terminated by disproportionation	2.0
Condensation polymer (100% utilization of monomer)	2.0
Vinyl (high conversion)	2–5
Branched polymers	20–50
Anionic (living type)	1.01–1.05
Hetrogenious catalysis	Broad

which they are subsequently eluted. The smaller molecules penetrate the gel more easily than the larger and hence elute more slowly. The larger molecules are less able to penetrate the pores and hence go through the column faster. Detector systems commonly used for GPC are uv absorption and differential refractometry. Both work by the difference between the sample stream and a reference (either solvent or air). A trace from the GPC gives a plot of concentration vs. eluted volume. Excellent and simply described reviews of the theory and experimental techniques are available (2).

GPC results are relative and depend on calibration of the column system with polymer materials of known molecular weights and narrow distribution, determined by some independent absolute method. Separation occurs on the basis of molecular size or, more correctly, hydrodynamic volume. Hydrodynamic volume is related to molecular weight through the Einstein viscosity law:

$$[\eta] = K(V_h/M)$$

where V_h = hydrodynamic volume, $[\eta]$ = intrinsic viscosity, and M = molecular weight. Then

$$V_h = K'M[\eta]$$

A typical plot of $\log M$ vs. eluted volume $V_e(\sim V_h)$ used for GPC calibration is linear. Using the Mark–Houwink–Sakurada equation for a monodisperse sample, $[\eta] = KM^{\alpha}$; for a monodisperse sample, $M_w \sim M_v$. Multiplying by M and taking the log of both sides gives

$$\log([\eta]M) = \log K + (1 + \alpha) \log M$$

Thus if we know the value of M for the samples and can find the constants α and K for the calibration of polymers under similar conditions of temperature and solvent, we can easily calculate $\log[\eta]M$ and make a universal calibration plot of $\log[\eta]M$ vs. V_e (eluted volume). There are extensive listings of K and α for numerous polymers in a variety of solvents (2).

The gels used are often made of cross-linked polystyrene that is hard and incompressible and able to resist high-pressure failure and permit high flow rates. Pore size is controlled by the extent of cross-linking and solvent/nonsolvents used during polymerization. A typical column consists of 20% polystyrene/divinylbenzene polymer, and 40% pore volume, and 40% interstitial volume. Elution time is inversely proportional to hydrodynamic volume. Key problems in using GPC are selection of appropriate column gels and control polymers that mimic the polymer being evaluated in regard to hydrodynamic size and shape parameters.

2.3.1.B. Other Routes to Molecular Weight Determination

The molecular weight average obtained depends on the method used it. Osmometry gives a number average, whereas light scattering and certain types of ultracentrifugation give weight average molecular weights. The latter two require the use of highly specialized and expensive equipment and sophisticated operational skills. The same can generally be applied to osmometry, and it is fair to say that GPC is by far the best and most widely used system. These methods have the advantage of being absolute methods, in that a molecular weight average may be determined without recourse to any previous measurement. As such, they are the ultimate basis for the use of polymer solution viscosity measurements for the determination of molecular weight (3).

By and large, the simplest and most common laboratory methods that relate to molecular weight are solution methods that compare the viscosities of polymer solutions of known concentration to that of the solvent. A number of viscosity designations have been defined for dilute polymer solutions. For the sake of consistency the most common use is adopted here, with alternative terms given in parenthesis. These are as follows:

Relative viscosity (viscosity ratio)	$\eta_{rel} = t/t_0$
Specific viscosity	$\eta_{sp} = \eta_{rel} - 1$
Reduced specific viscosity	$\eta_{sp}/C = (\eta_{rel} - 1)/C$
Inherent viscosity	$\eta_{inh} = \ln \eta_{rel}/C$ (log viscosity number)
Intrinsic viscosity	$[\eta] = \lim \eta = \lim \eta_{sp}/C$ (limiting viscosity number as conc. $\longrightarrow$ 0)

where t_0 is the flow time through a viscometer of a reference liquid, t is the flow time through the same viscometer of a dilute solution of polymer in the reference liquid, and C is the concentration of polymer solution in grams per 100 mL of solvent. Therefore, the units of intrinsic viscosity are deciliters per gram. The intrinsic viscosity is obtained by extrapolating to zero concentration a plot of either inherent or reduced specific viscosity versus concentration. The above viscosity numbers depend on the solvent, the concentration (except for intrinsic values), and the temperature at which the determinations are made, although the latter usually has little effect within 10 to 15°C in a good solvent for the polymer. A correlation of intrinsic viscosity with molecular weight for linear polymers can be achieved through the empirical equation proposed by Mark–Houwink–Sakurada (4) based on earlier work by Staudinger and Heuer (5).

$$[\eta] = KM^{\alpha}$$

where M is molecular weight and K and α are determined by the intercept and slope, respectively, for a plot of the log of intrinsic viscosity against the log of molecular weight of fractionated samples over a wide range of molecular weight values.

Thus, by determination of intrinsic viscosity, the magnitude of a polymer molecular weight can be estimated by choosing K and α values from known polymer–solvent combination that is as similar as possible to the polymer–solvent system under consideration. Comparison of intrinsic viscosities of two similar polymers measured in the same solvent, one of which is of known molecular weight, can also be used to estimate molecular weight. The qualitative nature of such comparisons must be recognized, however, because minor structural differences may lead to large differences in viscosity.

Viscometric molecular weight determination has been extensively reviewed (6) and a large number of K and α values for various polymer–solvent are given. The *Polymer Handbook* (7) is another excellent source.

When it is not necessary to arrive at a numerical value for the molecular weight of a polymer, it is often convenient simply to relate one given sample of polymer to another by means of inherent viscosity or relative viscosity. If it is known, for instance, that poly(vinyl chloride) of a certain inherent viscosity value (under stated conditions) is required to achieve a desired level of film properties, it may be sufficient characterization to relate newly prepared poly(vinyl chloride) to anticipated properties by determining only the inherent or the reduced viscosity of the new sample. In some cases, relative viscosity may have advantages. In the following procedures for viscosity determinations, percent solutions are given as grams per 100 mL of solvent, e.g., 0.5% is 0.5 g. per 100 mL.

2.3.1.B.1. Procedure for Determination of Relative Viscosity

Relative viscosity can be taken as the ratio of the flow times of a polymer solution and the pure solvent in the same viscometer and at the same temperature. (Strictly, relative viscosity is the ratio of the kinematic viscosities of solution and solvent.) Assuming the same density for dilute solution and solvent and using the same viscometer for solvent and solution, the ratio of efflux time can be taken as relative viscosity, in most cases.

The following method is specifically designed to determine the relative viscosity of dilute (1% or less) solutions. Relative viscosity values generally are used for calculating the intrinsic or inherent viscosity of a polymer. Relative viscosity is easily determined, but its magnitude is more a function of polymer concentration than of molecular weight. Relative viscosity measurements are conveniently made in a thermostatted bath at 30°C. This temperature is chosen so that the bath will always be above room temperature, even during hot weather and thus will not require controlled cooling. For viscometers, Cannon-Fenske 100 to 500 series, Fisher no. 13-616, or an equivalent, machine is useful for these determinations. It is essential that viscometers be kept free of dust, residues, and other foreign matter. The solvent to be used depends on the polymer in question. In general, the solvent should completely dissolve the sample in <30 min. It is desirable that the polymer be dissolved at room temperature, although heating is permissible if no degradation occurs. Degradation can be noted by determining the viscosity at intervals of 1 h or so.

Weigh 125 ± 1 mg of the dry sample, and transfer it quantitatively to a test tube. If some concentration other than 0.5% is run, use a proportionately smaller or larger or smaller sample. Weigh liquid samples directly into a test tube. Accurately measure 25 ± 0.05 mL of the required solvent into the test tube containing the sample. Manual or mechanical stirring will usually be required. Instead of a test tube, a small, stoppered Erlenmeyer flask may be used for making the solution. Here, a magnetic stirrer may be used. Solvent must in no case be allowed to evaporate and alter the concentration in any of the operations that follow. When the sample has completely dissolved, filter about 10 mL without vacuum through a coarse porosity sintered glass filter into a 50-mL flask or other suitable container. While the solution is filtering, add about 10 mL. of the solvent used for dissolving the sample into a 50-mL beaker. Select a viscometer through which the solvent will flow in not less than 100 s and preferably not more than 200 s. Hold the viscometer in a vertical, inverted position and immerse the end of the capillary tube in the solvent. Apply suction to the other (wide) tube until the solvent fills both bulbs and most of the capillary. Remove the beaker and while the viscometer is still inverted, allow the liquid in the capillary tube to drain back until the meniscus just reaches the graduation mark around the working capillary. When the graduation has been reached, quickly invert and wipe the excess liquid off the tube.

Immerse the viscometer in a constant-temperature bath in a vertical position. Allow the viscometer to remain in the bath long enough for its contents to come to the temperature of the bath. Ordinarily, 5 min is sufficient for temperatures between 20° and 40°C; 10 min should be allowed for temperatures outside this range. During the temperature-equilibrating stage, and later while making the measurement, the temperature of the bath must be held constant within ±0.05°C. During the temperature-equilibrating stage, the solvent will drain into the lower reservoir bulb and, while doing so, a bubble may become trapped in the bend at the bottom of the instrument. Remove any such bubble by manipulating the solution with a suction bulb. After the solvent has attained the temperature of the bath, apply pressure to the top of the wide arm (or suction to the capillary) until the liquid has filled the first bulb and is about 1 cm above the mark between the bulbs. Make certain that no bubbles are trapped. Allow the liquid to flow freely and determine the time for the meniscus to pass from the upper to the lower mark. Record this time as the flow time for the solvent. Draw the liquid up again and remeasure the flow time, which should agree with the first flow time within 0.2 s. If it does not, continue until three flow times agreeing within 0.2 s are obtained. When about 10 mL of sample solution filtrate has been collected, immerse the same viscometer, which has been cleaned and dried, in the filtrate and fill as described above. Place in the water bath, allow to come to temperature equlibrium, and determine the flow time as described above. Three values, agreeing within 0.2 s for solution flow times, should be obtained. The relative viscosity is given by: η_{rel} = (solution flow time in seconds/solvent flow time in seconds).

2.3.1.B.2. Procedure for Determination of Inherent Viscosity

Inherent viscosity is calculated from the dilute solution (1% or less) relative viscosity. The relative viscosity of the polymer is determined as described in the preceding section. While the nature of the solvent used and the polymer concentration have an influence on inherent viscosity values, the effect is far less than on relative viscosity. In general, the better the solvent, the higher the observed inherent viscosity for a given polymer. Similarly, the higher the concentration, the lower the observed inherent viscosity. Temperature is important only insofar as it influences solvent power and polymer degradation. Results obtained at 30 and 25°C usually agree within the precision of this method. Results obtained by this method should be precise within 0.04 units absolute, within the range of 0 to 5 inherent viscosity units. The inherent viscosity is calculated as $\eta_{inh} = \ln \eta_{rel}/C$, where C is the concentration of the polymer in grams per 100 mL of solvent (usually, $C = 0.50$); $\ln \eta_{rel}$ is the natural log of the relative viscosity of the dilute (1% or less) polymer solution.

For polymers of limited solubility at 30°C, it is necessary to use a constant-temperature bath held at an elevated temperature. Polyhydrocarbons such as polyethylene are satisfactorily soluble only at 130°C in solvents such as decahydronaphthalene containing 0.2% of an antioxidant (e.g., phenyl β-naphthylamine). The polymer is dissolved in decahydronaphthaline at a concentration of 0.1 g/100 mL instead of the usual 0.5 g, using an ethylene glycol monomethyl ether bath at reflux (125°C) to heat the polymer and solvent. The solution is filtered through a preheated 200-mesh stainless-steel screen into a test tube immersed in the 130 ± 0.1°C bath. Then 10 mL of the solution is transferred by means of a hot pipet (do not use mouth suction) to a suitable viscometer (a Cannon-Fenske series 75) immersed in the bath. After allowing 10 min for temperature equilibrium to be established, the measurements are made and the inherent viscosity is calculated as described above.

2.3.1.B.3. Procedure for Determination of Intrinsic Viscosity

The intrinsic viscosity $[\eta]$ is given by:

$$[\eta] = \lim \eta_{inh} = \lim(\eta_{sp})/C$$

as C approaches zero. Intrinsic viscosity is obtained from plotting inherent viscosity numbers or reduced viscosity numbers versus concentration and extrapolating to zero concentration. The viscosity number at zero concentration is taken as the intrinsic viscosity. Customarily, curves for both inherent and reduced viscosities versus concentration are plotted. The intrinsic viscosity value obtained should be the same in each case. If they are not, the midpoint between them is usually taken. The determination of intrinsic viscosity should be based on a minimum of three inherent or reduced viscosity measurements at different concentrations. Four or five such measurements are preferable. Concentrations should be from 0.1 to 0.5 g per 100 mL solvent. The experimental procedures are the same as those given for η_{sp} and η_{inh} in the preceding sections. Molecular weight estimation by solution viscometry outlined in this section is necessarily brief and nonmathematical and is designed primarily to present the experimental methods involved. The full theory, implications, and limitations of the subject are available elsewhere (3, 6). Additional experimental procedures, as applied to poly(vinyl chloride) and ethylene polymers, for example, are given in ASTM standards (8).

2.3.1.C. End-Group Methods

End-group determinations, when applicable, are useful for molecular weights <25,000. This approach, which gives a number average value, depends on the polymer having a terminal group, or groups, on each chain that can be quantitatively measured. When there are no losses of end groups by side reactions or production of additional end groups by branching, and when all the ends can be accurately determined, a direct measure of the number average molecular weight can be made (3). Molecular weight by end-group determination is much more difficult in the case of free-radical vinyl-type polymerization because of chain termination and chain transfer reactions but may be useful in special cases, such as living anionic polymers for which chain transfer is low.

In condensation polymers, where $-COOH$ or $-NH_2$ groups are present, direct titration is usually an effective means of end-group determination (9–11). The procedures involve the use of a suitable inert solvent for acidic or basic titration and consequently depend on the solubility limits of the polymer. Hydroxyl groups in polyesters have been determined by reaction of the $-OH$ with a reagent that then forms a titratable group, e.g., acetic anhydride followed by titration of the acetic acid after hydrolysis of the excess reagent (12), or succinic anhydride followed by titration of the free carboxyl of the succinic half ester (13). Hydroxyls in polyesters have also been determined through infrared methods (14). The selection of methods to be used for determining a particular group is influenced by the following factors.

2.3.1.C.1. Solvent

The solvent must dissolve the polymer but not interfere with the applicable analytical method. For example, formic acid is an excellent solvent for nylon, but the direct acidimetric titration cannot be carried out on it. Generally speaking, the sample need not be highly soluble in the solvent; 0.1% is ordinarily sufficient.

2.3.1.C.2. Impurities Present

In many cases, the impurities likely to be present in a polymer may be deduced by knowing how it was made. For example, polyamides made by melt polymerization are often free from impurities, whereas those prepared from acid chlorides usually contain salts that are difficult to remove and seriously interfere with many methods. Methods used for purifying the polymer and the solvents with which it has been in contact must also be considered.

2.3.1.C.3. Other Functional Groups Present

Interfering functional groups must be absent. For example, amine groups interfere with the determination of hydroxyl end groups.

2.3.1.C.4. Molecular Weight and Structure of the Polymer

If the molecular weight is greater than about 25,000, end-group methods are, in general, not reliable. The degree of cross-linking and branching are other factors that influence both the solubility of the polymer and the end-group concentration.

2.3.1.D. Other Methods

Both cryoscopic and ebullioscopic methods give number average molecular weights. As the molecular weight of a solute increases, the molar quantity required to give the necessary observable melting point or boiling point change in the solvent becomes greater. Consequently, these methods have generally been restricted to polymers of 30,000 molecular weight or less, although higher values have been determined in some cases.

2.3.2. Crystallinity in Polymers

Polymeric materials may range from highly crystalline to completely amorphous, as assessed by x-ray diffraction assessment. They can also display liquid crystallinity properties (see chapter 6). In addition to x-ray diffraction, crystallinity can be demonstrated and quantified by differential scanning calorimetry (discuss below). Crystallinity is generally observed in polymers when chains have (*1*) structural regularity, (*2*) strong secondary forces between chains (e.g., hydrogen bonding or polar forces), and (*3*) little or no random branching or cross-linking or bulky side groups.

The effects of these structural features are much the same as they are in simple organic compounds but with the important difference that polymer structures often contain both crystalline and amorphous regions. Polymers that are amorphous as made often can be induced to crystallize by treatment with a near solvent or by heating above the glass transition temperature or, in the case of film or fiber, by stretching to produce chain alignment. Crystalline polymers can often be rendered amorphous by rapid cooling from the molten state to below the glass transition temperature. One of the most effective methods of ascertaining the presence and extent of crystallinity in a polymer is x-ray diffraction, which is one of the most informative and useful physical methods of polymer characterization. Although it is beyond the scope of this section to give an in-depth discussion of the theory and practice of x-ray diffraction, some of the basics should be mentioned. When x-ray or other electromagnetic radiation is scattered by an ordered arrangement of atoms, the angle of diffraction can be described by the Bragg equation:

$$n(\lambda) = 2d \sin\theta$$

Where n is the order of diffraction, λ is the wavelength of the radiation, d is the distance between successive layers of scattering points, and θ is the angle of diffraction. The conditions for strong reflections are that the separate rays reflected from successive planes must reinforce one another. This means that differences in path length must correspond to whole number wavelengths. In Figure 2.3., this difference is $CBD = 2d \sin\theta$. Hence the condition for strong reflection is given by the Bragg equation.

The order of the diffraction is defined by the set of crystal planes that diffract the incident radiation. A set of crystal planes can be defined as a geometric arrangement of atoms that are repeated down the crystal lattice by a unit distance. Rearrangement of the Bragg equation gives $d = n(\lambda)/2\sin\theta$, the distance between scattering planes.

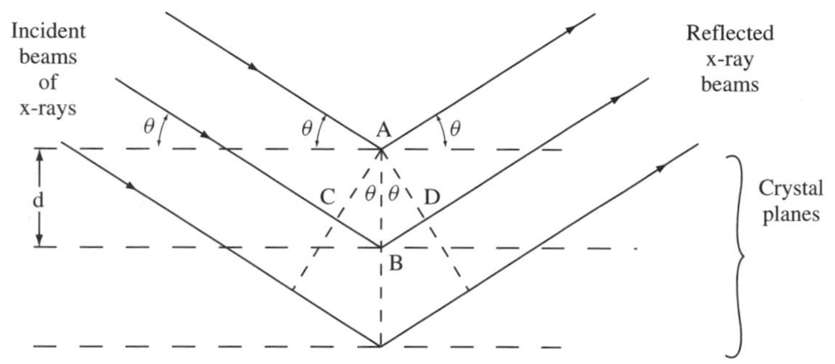

Figure 2.3. X-ray diffraction from crystal planes.

The most common x-ray sources are from copper targets or filters (CuK_{α} and CuK_{β}, in wavelengths of 0.154 and 0.139 nm, respectively). The technique typically used is to expose the polymer sample in a vacuum to a collimated beam of x-rays from the copper source and collect the scattered beam on a flat photographic plate. The unscattered beam is prevented from striking the film by a lead beam stop placed in its pathway. With unoriented samples, a photographic image of concentric circles is obtained. More complex patterns are obtained from oriented samples.

The reflection of x-rays from a sample can be related to the dimension and angles of the unit cell, the smallest repeating unit of the crystal structure. The unit cell is described by the three vectors (a, b, and c) and the three angles between the vectors. All unit cell configurations can be categorized by one of seven crystal systems, shown in Table 2.3.

X-ray measurements are usually classified as wide angle (WAX) or small angle (SAX) scattering. The latter refers to measurements within 1 to 2 degrees of the unscattered beam and the former, to larger angles. WAX provides methods for determining (*1*) the dimensions of the repeating unit of the unit cell, (*2*) the length of the repeating unit along the fiber axis, (*3*) the number of repeat units per turn in helical structures (common in linear polymers) and (*4*) the degree of orientation (i.e., chain alignment). SAX provides information about long-range periodicities and voids in the structure, e.g., the fold length of polymer chains. The unit of length is the nanometer; 1 nm = 10 Å.

Table 2.3. Unit Cell Characterization

Crystal System	Relative Axial Lengths	Angles
Cubic	$a = b = c$	$\alpha = \beta = \gamma = 90°$
Tetragonal	$a = b \neq c$	$\alpha = \beta = \gamma = 90°$
Orthorhombic	$a \neq b \neq c$	$\alpha = \beta = \gamma = 90°$
Rhombohedral	$a = b = c$	$\alpha = \beta = \gamma \neq 90°$
Hexagonal	$a = b \neq c$	$\alpha = \beta = 90°; \gamma = 120°$
Monoclinic	$a \neq b \neq c$	$\alpha = \gamma = 90° \neq \beta$
Triclinic	$a \neq b \neq c$	$\alpha \neq \beta \neq \gamma \neq 90°$

In describing a fiber pattern, one usually begins by marking lines through the center of the pattern parallel and perpendicular to the machine direction of the sample. The parallel line is the meridian and the other, the equator. The spots fall on layer lines roughly parallel to the equator. The equator is designated $L = 0$, and the layer lines on either side are designated $L = 1, 2, 3$, etc. A theoretically perfectly oriented specimen would not display any meridianal reflections. Maltese cross-reflections suggest a helical configuration. Reflections on the nth layer lines are strong evidence of an n-fold structure. A helix with three units per turn will usually produce meridianal reflections on the third, sixth, ninth, etc. layer lines.

For polymers that are crystalline, the crystalline melting point T_m is taken as the temperature at which the last traces of crystallinity disappear under equilibrium conditions. The long heating time required (24 h or more), however, limits the usefulness of this method. The simplest laboratory method (apart from differential scanning claorimity and differential thermal analysis discuss below) is to note the loss of crystalline birefringence in a polymer sample when heated on a hot stage polarizing microscope (described later in this chapter). More detailed descriptions of theory and techniques for structural determination by x-rays are available (2, 15).

2.3.3. Thermal Transitions in Polymers

Two types of transitions (first and second order) are recognized in polymers. A first-order transition is a phase change accompanied by discontinuities in primary thermodynamic properties, such as density and enthalpy. One example is the crystalline melt temperature. This, however, is not as sharp in polymers as it is in small molecules. The crystalline order is lower because polymers consist of a range of molecular weights and exhibit hysteresis effects. Because polymers are composed of a range of molecular weights and contain crystals of different degrees of perfection, a crystalline melting point cannot be precisely defined. The polymer melt temperature T_m, is used to describe the temperature at which the final vestiges of crystallinity in the polymer disappear on heating.

For every amorphous polymer, there is a narrow region in which the polymer changes from a rubber to a glass as the temperature is lowered. This is a second-order transition (because no phase change is involved) and is called the glass-transition temperature (T_g). Other properties, such as heat capacity (calories per gram per degree Celsius), thermal expansion coefficient, and dielectric constant also change markedly over the T_g range. The T_g results from segmental motion due to bond rotation from thermal energy and free volume in the polymeric mass. Below the T_g, the polymer behaves like a glass, the free volume becomes small, and the thermal energy cannot overcome the energy barriers for rotation and translational "jumps" of polymer segments. Glass transation temperature effects are shown only in the amorphous regions. In highly crystalline polymers, it is often difficult to observe T_g effects. As a rule of thumb, T_g is often roughly $2/3T_m$ (in K) (16).

A universal relationship between T_g and the temperature of maximum spherulitic crystal growth rate (T^*) from the melt has been proposed: $T^* = 1.26T_g$ (in K) (17).

2.3.3.A. *Structural Effects on T_g and T_m*

High secondary forces (hydrogen bonding, polar forces etc) and good chain packing increase both T_g and T_m. T_m is more affected by packing, and T_g by secondary forces and chain flexibility. Increased chain rigidity raises both T_m and T_g, but the quantitative effects are not the same. Flexible polymer chains containing groups (e.g., $-CH_2-CH_2-$, $-\overset{\overset{O}{\|}}{C}-O-$, $-SiR_2-O-$, and ethers) glass-transition have low temperatures. Bulky side groups in the polymer unit (e.g., $-CH_2-CHR-$, $-CH_2-CHCl-$, and $-CH_2-CH-Ph-$) often decrease flexibility and raise the T_g. However, 1,1-disubstitution in the polymer unit (e.g., $-CH_2-C-(CH_3)_2-$ and $-CH_2-C-Cl_2-$) often lowers the T_g. Aromatic rings in the polymer backbone stiffen the chains and raise the T_g. T_m increases with degree of polymerization (DP) but reaches a maximum at a $\overline{DP}$ of 20 to 30. T_g is more profoundly affected by changes in $\overline{DP}$ and increases as $\overline{DP}$ increases. This probably results from reduction in free volume in the amorphous regions due to reduction in chain ends and low molecular weight fractions. In copolymers, the T_m is usually lower than that of either homopolymer, except in cases where the copolymer constituents are isomorphous (i.e., capable of crystallizing in the same form). In these cases, the polymer melt temperature does not show a eutectic but increases as the higher melting component increases. Isomorphism in polymers is not common. Block and graph polymers often exhibit two T_g values. The T_g of random copolymers falls in between that of the homopolymers, if the polymer units (or monomers) are similar. Blends of two compatible homopolymers behave similarly. When the two units or monomers differ in polarity, the T_g is depressed.

Overall, copolymerization has a lesser effect on T_g than on T_m, because symmetry is less important in T_g. Regarding T_m, this is a first-order event that can be described by the Gibbs free energy equation:

$$\Delta F = \Delta H_u + T\Delta S_u$$

where ΔH_u = the molar enthalpy of fusion and ΔS_u = the molar entropy of fusion. At the crystalline melting point, $\Delta F = 0$; hence $T = \Delta H_u/\Delta S_u$; ΔS and ΔH are the differences between H and S in the solid and liquid phases.

2.3.3.B. *Determining Crystalline Melt Temperatures*

Probably the most useful and accurate thermal methods for characterizing polymers are differential scanning calorimity (DSC) and differential thermal analysis (DTA) (2). The former differs from the latter in measuring enthalpy rather than temperature change. DSC measures the heat required in a furnace to maintain the same temperature in a sample versus an appropriate reference. A first-order transition or chemical reaction is observed as an abrupt change in

ΔT vs. T. A second-order transition is observed by a change in slope of ΔT vs. T. These methods are the easiest and most accurate ways to determine the glass-transition and melting temperatures of polymers.

A common DSC instrument is the DuPont 990 model. The procedure is quite simple. Samples of polymer (ca. 10 mg) are sealed in a small pan under nitrogen (an empty pan or one containing aluminum oxide is used as the reference). The temperature is raised, usually about 10°C/min, to the desired level with continuous recording of the heat flow vs. temperature. Glass-transition, water/solvent loss, melt temperature, and any other changes are observed as inflections or gross changes in the heat vs. temperature curve. The sample is then cooled, and a second cycle is run. This permits identification of a crystallization temperature and any degree of supercooling.

Typical scans for poly(ethylene terephthalate) are shown in Figure 2.4.

DSC allows one to accurately identify T_g, crystalline phase changes, onset of melting, melt temperature, and enthalpy of melting. It should be pointed out again that highly crystalline polymers often fail to show a T_g, because segmental motion is quite restricted. In a typical DSC scan, for accuracy in reporting, the transitions should be given as onset, peak, or inflection temperatures.

Along with DSC, thermogravimetric analysis (TGA) largely defines the thermal behavior of polymers. TGA records changes in mass that occur when a polymer is heated under specific conditions. Evolved gases may be characterized by a variety of methods (ir, mass spectometry, etc). Weight loss can be determined continuously as the sample is heated at a constant rate or determined at constant temperature vs. time. Typical equipment used is the DuPont 990 thermal analysis unit with a model 951 thermogravimetric analysis unit attached. Approximately 10 mg of sample is used and heated in air or nitrogen gas flow or in vacuum. The rate of weight loss characterizes the thermal stabilities of the polymer as well as suggesting the mode of polymer decomposition, e.g., random chain cleavage or unzipping. Results can also be used to determine activation energies of the specific weight loss process (18).

Probably the simplest laboratory method to determine polymer crystalline melt temperature by loss of birefringence is via a hot stage polarizing microscope. This method gives a rough approximation of the crystalline melt temperature. In determining the melt temperature, the sample is heated on a hot stage microscope, optionally under nitrogen, and the temperature at which the color between crossed optical polarizers disappears is noted. This procedure is applicable to any fiber, film, or other material that exhibits color (birefringence) when placed between crossed optical polarizers and does not decompose at elevated temperatures in the relatively short times required.

There are two sources of birefringence in unoriented polymer samples: birefringence due to strain within the sample, characterized by a brilliant play of colors during the initial heating period and disappears when the temperature approaches the melting point, and birefringence due to sample crystallinity, characterized by the bright yellow or white color of the sample and remains until the crystalline melt temperature is reached. This color disappears rapidly

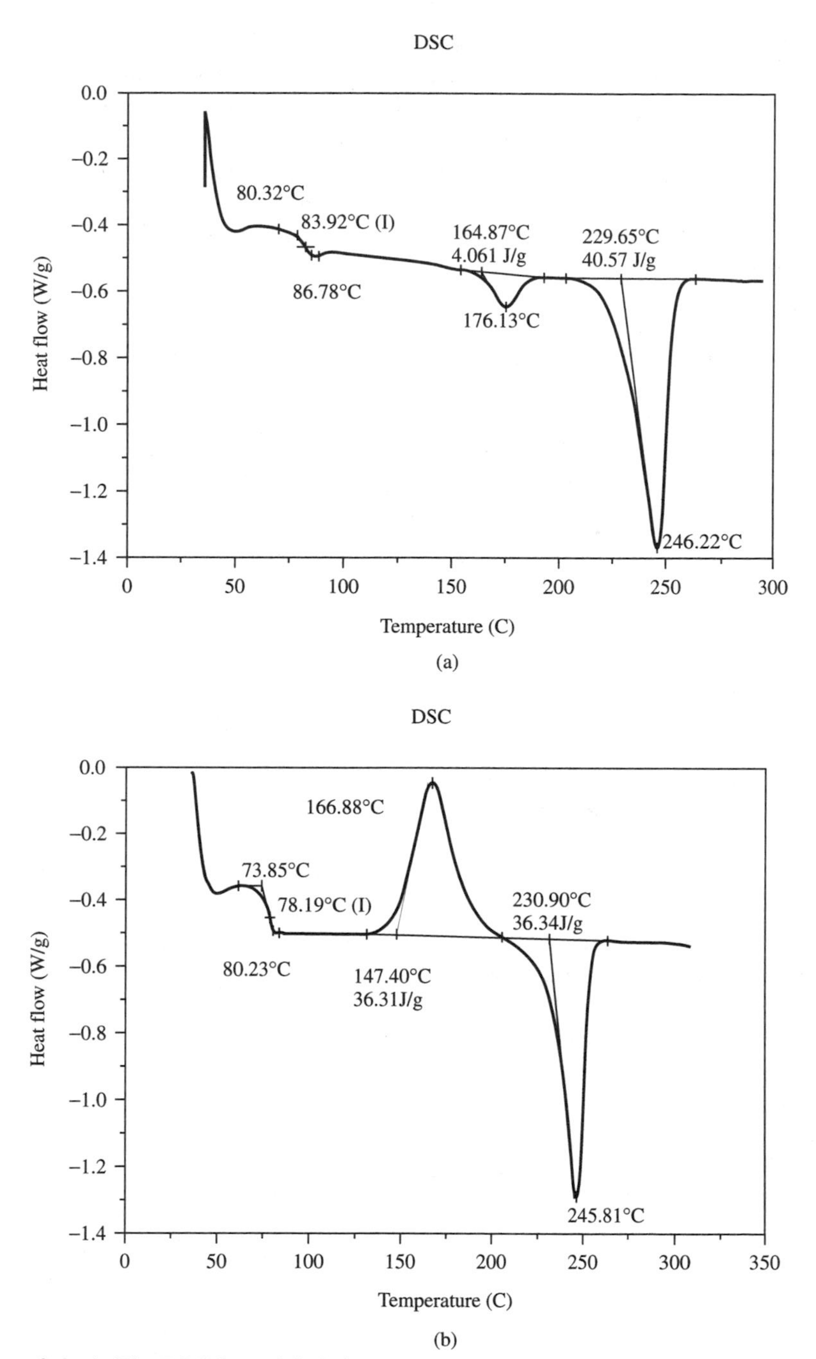

Figure 2.4. A, The DSC for polylethylene terephthalate B, The second cycle for the same chemical after cooling.

over a narrow temperature range and blends with the dark field. Inorganic impurities showing birefringence should be ignored. These are usually high melting substances and can easily be recognized with little practice. For use in this determination, any good microscope equipped with 5× and 10× oculars and objectives, an optical polarizer and analyzer, and an insulated or metal stage is satisfactory. The Unitron Polarizing Microscope, model MPS (United Scientific Co.), which has been modified to seat the Koeffler hot stage accurately, is an example.

2.3.3.B.1. Procedure for Using a Hot Stage Microscope to Determine Melt Temperature

We'll use film as an example for determing T_m with a hot stage microscope. First cut the sample into at least 10 pieces, each about 0.1 × 1 × 1 mm or smaller. Place the pieces in the center of a slide, taking care that the pieces are close to each other but not overlapping. Place a cover glass on top. Place the slide on the stage, and turn on the light. Using the 5× objective and the 10× eyepiece, select a field in which the sample particles are well distributed with little bunching or clumping. Place the heat baffle over the slide and cover the stage with the glass cover.

If the approximate melting point of the sample is known (±10°C.) adjust the heating rate to give a temperature rise of 0.1°C/min at 20°C below the expected value. Allow the sample to anneal for 30 min under these conditions. This is to relieve internal strains in the sample that may cause birefringence. Then adjust the heating rate to give a temperature rise of 1°C/min. Refocus the microscope with the 10× objective and the 10× eyepiece, move the analyzer into position, and note the temperature at which the first polymer particle blends with the dark background; label as T_i. Note the temperature at which the last polymer particle disappears; label as T_f. Disregard any isolated bright spots that persist well above the temperature at which the bulk of the sample particles have lost their color.

When the approximate melting temperature of the polymer sample is not known, refocus the microscope, using the 10× objective and the 10× eyepiece. Move the analyzer into position. Set the heating rate to give a temperature rise of 10°C/min at 250°C. Record the temperature at which the last sample particle disappears and label as T_m. Place the cooling block, which has been precooled in ice water, on the hot stage. Prepare and mount a fresh portion of the sample on the stage as described above.

Anneal the sample for 30 min, using the T_m value just determined as the approximate melt temperature, and then determine the melt temperature as described. As the melted polymer cools, crystallinity will usually reappear and the determination can be repeated. If the sample does not become crystalline again on cooling, treatment with an appropriate solvent or mechanical working may reinduce crystallinity. Orientation in a crystalline polymer does not adversely affect determination of the crystalline melt temperature; in fact, it usually helps, increasing the brightness and making the disappearance of color more apparent.

Calculation. Let T_i be the temperature at which the first sample particle blends with the background and T_f, the temperature at which the last sample particle blends with the background. Then crystalline melt temperature is

$$T_m + (T_i + T_f)/2$$

It should be noticed that amorphous polymers having some orientation of the molecules (as in a stretched film) exhibit birefringence under a polarizing microscope. This will disappear near the glass transition temperature and will not reappear on cooling. It should be noted also that cutting bulk, unoriented polymer will introduce some additional birefringence around the edges due to shear orientation; hence such samples will be more brilliant around the edges when viewed between crossed-polarizers.

2.3.4. Other Thermal Characterization Methods

2.3.4.A. Polymer Melt or Stick Temperature

For many polymers, one of the simplest observations that can be made is the temperature at which a polymer sample simply becomes visually soft or molten. The polymer melt temperature (PMT), or more correctly stick temperature (PST), is defined as that temperature at which a polymer sample becomes molten and leaves a trail when moved with moderate pressure across a hot metal surface. A polymer may become rubbery or soft before becoming molten, particularly if amorphous. When a polymer is of very high molecular weight, it may show an anomalously high PMT because of a high melt viscosity. Some polymers decompose before melting. The temperature at which any such behavior occurs should be noted, as well as the best estimate of the PMT.

The PMT test reveals some practical information about the polymer in terms of fabrication and it can serve as a guide to proper temperatures for molding and melt pressing and as an indication of the thermal stability. For highly crystalline polymers, the PMT may occur rather sharply, over a narrow range, usually a few degrees below the crystalline melt temperature. Amorphous polymers exhibit a PMT over a wider temperature range, frequently with noticeable softening at a lower temperature. The PMT may be determined on a polymer in any form (e.g., powder, film, or a solid plug) manipulated by hand or spatula along a modified temperature gradient Dennis bar (19). The temperature gradient, from cool to hot, is maintained along the bar by differences in bar mass. The heaters are positioned at the thick end of the bar. The temperature gradient along the bar is measured potentiometrically at a number of points. Such a bar may be used to determine PMTs up to 350 to 400°C. The test is somewhat subjective, particularly for amorphous polymers, but has the advantage of speed and simplicity coupled with the observation of potentially useful thermal behavior of the polymer.

2.3.4.B. Heat Deflection Temperature

Heat deflection temperature (HDT) parameter is one of the most commonly cited for the characterization of plastics (20, 21). It refers to the temperature at which a molded bar (5 in. × 0.5 in. × 0.25 – 0.5 in.) supported at both ends, undergoes a deflection of 0.010 in. when exposed to a temperature rise of 2°C/min under a fiber stress of 66 or 264 psi applied across the center of the bar. There are commercially available units for this purpose.

2.3.4.C. Flow Properties

Although crystalline melting and glass temperatures are usually the elements of thermal behavior first encountered in characterizing a polymer, especially for the organic chemist, the flow characteristics of a polymer melt or solution as a function of temperature and applied stress are of great practical importance. The rheologic characterization of polymers is beyond the scope of this book, but at least one method of flow determination should be noted because of its simplicity and the wide application it has had. This is the so-called melt index, which uses a heated barrel to melt the polymer and a variably weighted piston that fits the barrel and extrudes the melt through an orifice at the bottom of the barrel. Essentially any melt-stable polymer can be tested, providing it is not so fluid as to literally run freely from the barrel, or so stiff flowing as to defy extrusion with the application of practical weight (force) levels. The test consists of simply weighing the polymer extruded per time and expressing the results as grams per 10 min. The method is necessarily a one-point determination, because shear stress and rate are not easily varied, whereas in more sophisticated methods (e.g., capillary rheometry at variable shear rates), the shear stress developed over a wider range of shear rates and temperatures can be obtained. The value of the melt index is in the rapid estimation of flow behavior relative to other samples, especially to those of the same polymer type.

2.3.5. Polymer Solubility

The solubility of a polymer is an important part of its characterization, especially for polymers that are not melt stable and are fabricate via solution methods. Molecular weight determination via solution viscosity methods requires that a suitable solvent be found. The determination of chain dimensions, which can be related to polymer microstructure, was discussed earlier. As is to be expected, degree of crystallinity and molecular weight of a polymer affects its solubility. For these reasons, solubility behavior may vary from sample to sample of a given polymer. Cross-linked polymers do not show normal solubility characteristics. At best, such polymers may swell under the influence of certain solvents, but swelling itself is not proof of cross-linking because a polymer on the borderline of solubility may exhibit this behavior. But when a polymer resists solubility in a number of solvents typical for those of its class and is infusible, it is usually considered to be cross-linked, unless there is impelling evidence to the contrary.

In general, aliphatic polyamides are soluble in strong organic acid systems, such as formic or trifluoracetic acids. Aromatic polyamides and aramids are not but are soluble in concentrated sulfuric acid and, to a certain extent, in basic solvents such as dimethylacetamide or *N*-methylpyrrolidinone with saturation levels of anhydrous calcium or lithium chlorides. Polyesters are not usually soluble in mineral acids but are soluble in warm or hot cresols.

Many of the structural aspects of polymer molecules that also affect solubility and melt temperature are those that also affect the same properties in simple organic molecules. For instance, crystallinity, high symmetry, hydrogen bonding, high polarity, chain stiffness, and stereoregularity in the chain contribute to higher melt temperature and reduced solubility than an otherwise similar polymer lacking the feature in question. The following procedure is useful for a rapid, qualitative determination of solubility.

A small amount of polymer, ca. 0.1 g is mixed with 2 to 5 mL of solvent in a test tube and stirred together thoroughly. An indication of solubility is "schlieren" as the mixture is stirred. Solubility is facilitated by using the polymer in a finely divided state. If no sign of solubility is apparent at room temperature, the mixture should be heated to close to the boiling point of the solvent. If the polymer dissolves, the solution should be allowed to cool to see if the polymer remains in solution or precipitates. If the polymer is swollen but not dissolved by the solvent, related solvents should be checked to see if they affect the solution. Screening the polymer in a range of solvents should result in a good candidate solvent. Polymer purification is often effected by dissolving the polymer in a solvent and precipitating it while stirring by pouring the solution slowly into an excess of a nonsolvent that is miscible with the solvent. For careful analytical and characterization work, several precipitations may be necessary.

REFERENCES

1. D. Coffman, G. J. Berchet, W. R. Peterson, and E. W. Spanagel, *J. Poly. Sci.*, **2**, 306 (1947).
2. E. M. Pearce, C. E. Wright, and B. K. Bordoloi, *Laboratory Experiments in Polymer Synthesis and Characterization*, The Pennsylvania State University, University Park, 1979. Z. Grubisic, et al., *J. Poly. Sci., B*, **5**, 753 (1967). M. Kurata et al., J. Brandrup and E. H. Immergut, eds., in *Polymer Handbook*, 2nd ed., Wiley-Interscience, New York, 1975.
3. P. W. Allen, ed., *Techniques of Polymer Characterization*, Academic Press, New York, 1959.
4. H. Mark, *Der Feste Korper*, 1938, 65, (1938). R. Houwink, *J. Prakt. Chem.*, 1957, 15 (1940).
5. H. Staudinger and W. Heuer, *Berichte Chemie*, **63**, 222 (1930).
6. G. Meyerhof, *Advan. Polymer Sci.*, **3**, 59 (1961). M. Kurata and W. H. Stockmeyer, *Advan. Polymer Sci.*, **3**, 196 (1963).
7. J. Brandrup and E. H. Immergut, eds., *Polymer Handbook*, Wiley-Interscience, New York, 1966.
8. ASTM D1601-58T.
9. L. Mandelkern, *Chem. Rev.*, **56**, 903 (1956).

10. P. Fijolka, I. Lenz, and F. Runge, *Makromol. Chem.*, **23**, 60 (1957).
11. J. E. Waltz and G. B. Taylor, *Anal. Chem.*, **19**, 448 (1947).
12. C. L. Ogg, W. L. Porter, and C. O. Willits, *Ind. Eng. Chem., Anal. Educ.*, **17**, 394 (1945).
13. A. Conix, *Makromol. Chem.*, **26**, 226 (1958).
14. I. M. Ward, *Trans. Faraday Soc.*, **53**, 1406 (1957).
15. G. H. Stout and L. H. Jensen, *X-Ray Structural Determination. A Practical Guide*, Macmillan, New York, 1968.
16. R. G. Beaman and F. B. Cramer, *J. Poly. Sci.*, **21**, 223 (1956). R. N. Blomberg, U. S. Pat. 2,965,437 (Dec. 20, 1960) to DuPont.
17. V. P. Privalko, *Polymer*, **19**, 1019 (1978).
18. S. Kim and E. M. Pearce, *Macromol. Chem. Suppl.*, **15**, 187 (1989). T. Ozawa, *Chem. Soc. Jpn. Bull.*, **38**, 1881 (1965). W. Sweeny, *J. Poly. Sci., A*, **30**, 1111 (1992).
19. L. M. Dennis and R. S. Shelton, *J. Am. Chem. Soc.*, **52**, 3128 (1930).
20. R. Houwink, *Plastomers and Elastomers*, Vol. 3, Elsevier, New York, 1948. C. E. Stephenson and A. H. Willbourn, Technical Publication **247**, American Society of Testing Materials, Philadelphia, 1959.
21. ASTM D1525-65T.

3

ADVANCED PROCESSING

3.1. PROCESSING AND BASIC REQUIREMENTS TO MAXIMIZE TENSILE PROPERTIES IN FIBERS

The intent of this chapter is to provide guidance for the processing of polymers, particularly fibers and plastics, to achieve optimum properties. These principles have largely been derived from the study of fibers and are readily applicable to fiber technology. One difficulty is that the control of the variables in plastics processing (i.e., all the various forms of extrusion and molding) is much more difficult than in fiber processing. Fibers are by nature thin, manipulable entities, wherein heat is added or removed rapidly, for example. Plastic entities are thicker, heavier, and have much lower surface to volume ratios than fibers. The latter, therefore, become a better mechanism for establishing the principles and applying them; and for now, a better teaching conveyance.

The processing methods discussed here supplement the simpler routes described in chapter 2 in that they require more elaborate equipment commonly available in commercial semiworks facilities. The bases of these processing methods are described here and in examples in chapter 6. These include how to select processing conditions, such as temperature for melt-processible polymers, solvents for high-melting or melt-unstable polymers, coalescing and quenching conditions for wet-spinning, and drawing and annealing conditions. It seems reasonable to us to address these elements of polymer processing, because most chemists who engage in polymer synthesis will inexorably become involved in the consequences of processing, if not the processing itself, of the polymers they prepare.

Achievement of high tensile strength depends particularly on

- The number of load-bearing chains per unit cross section; which in turn depends on extended-chain orientation and efficient chain packing.
- A low defect level, e.g., minimum voids, chain folds, chain ends, and branching.
- Strong interchain forces, e.g., hydrogen bonding and polar forces vs. van der Vaal forces.
- A high number average molecular weight polymer, because this minimizes the defects from chain ends.

A trade-off usually occurs between high strength, derived from high molecular weight, and ease of processing, derived from lower molecular weight. Crystallinity is not a particularly important factor for tensile strength if the polymer has a high T_g. For example, Technora aramid fiber (Teijin Co.) is not highly crystalline but has tensile strength close to DuPont's aramid fiber Kevlar, which is.

High tensile modulus results from a minimum number of flexible bonds per polymer unit, high chain extension, good chain packing, and crystallinity. High recoverable elongation depends on flexible or rotatable bonds in the polymer chain unit; small crystallite size; and high number average molecular weight, particularly in crystalline polymers. High shear or compressive properties depend on

- Minimum flexible bonds in the polymer unit.
- High interchain forces, e.g. hydrogen bonding or uniform cross-linking.
- Absence of voids or defects.
- High crystallinity.

It should be evident then, that the basic physical and chemical properties of a polymer can sometimes act in concert with, or antagonistically to, the development of desired mechanical properties. In practical terms, if we consider the catastrophic failure of a real fiber and know that orientation, crystal size, and flaws are the limiting factors, then tenacity is principally controlled by flaws and orientation at break. The principal feature controlling elongation is low crystallinity. Thus the requirements for a high-tenacity, high-elongation fiber would be one with high orientation at failure, coupled with low crystallinity or low apparent crystal size. An alternative might be high initial orientation to provide high modulus, combined with low crystallite size, to allow high elongation and orientation at break (1). Technora and, to some extent, Kevlar appear to come close to these latter requirements and have the mechanical properties associated with them.

3.2. MELT PROCESSING OF POLYMERS

Many commercial polymers are melt processed. This leads to lower processing costs and fewer environmental problems (e.g., solvent disposal). Laboratory spinning and processing of melt-stable polymers were described in chapter 2. In this chapter, we will examine more closely how to select the best conditions for processing. First and foremost, we recommend running differential scanning calorimity (DSC) (see Fig. 2.4) and thermogravimetric analysis (TGA) scans in air and nitrogen (first and second cycles) on the polymer to be processed to determine T_g, T_m, T_c, and decomposition temperature. This is especially valuable for polymers, such as poly(lactic acid), that decompose close to the melt temperature. Processing conditions should be started 20 to 30°C above the end of the DSC melt endotherm. This allows the sample to be fully melted. Solid-phase polymerized materials usually contain large and more nearly perfect crystals that often require mechanical mixing and longer hold-up times to melt completely. Incompletely melted crystals produce defects in the extruded sample. Hold-up time in the melt is important, affecting both polymer stability and molecular weight and distribution. Block and other copolymers may randomize on prolonged exposure in the melt, hence minimum melt exposure is desirable.

The crystallization exotherm indicates how crystallinity depends on cooling, rate of cooling, and degree of supercooling. For example, in fiber spinning the crystallization depends greatly on cooling conditions (e.g., if quench is cold gas or liquid), on threadline tension, and on spinning speed. These qualifications also apply to molded and extruded plastic parts. The maximum rate of crystallization from the melt has been reported to be $1.26 \times T_g$ (in K) (2).

To achieve the best final properties in spun fiber, the spinneret should be chosen to allow relaxation of the polymer stream in the capillary and reduce stress buildup in the extruded fiber. The spinneret should be chosen with a tapered inlet and a capillary with length/diameter >2. In the case of drawable fibers, filament birefringence and birefringence spread is important in regard to subsequent drawing. In essence, spun birefringence is a measure of orientation (= draw) produced in the fiber-spinning step and as such reduces subsequent draw potential. It is obvious, then, that the subsequent amount of draw in a fiber bundle is limited by the filament(s) with the highest as-spun birefringence.

For 6-6, nylon spun birefringence (B) is equated to spin draw in the as-spun fiber by the following equation (J. Zimmerman, personal communication)

$$\text{draw} = 1 + 22B + 282B^2$$

Polymer delivery through the aperture should be uniform. Quench-gas flow should also be uniform, and wind-up speed steady. For postdrawable polymers, threadline tension in spinning should be enough to maintain a steady threadline but minimize spun birefringence. For polymers that cannot be post drawn efficiently (e.g., liquid crystalline polyarylates), threadline tension should be high to maximize spin orientation and as-spun tensile properties.

Drawing is controlled by stress level and limited by filament breakage. Drawing results in polymer chain extension and orientation by unfolding lamellar crystals and/or shearing chains past one another. One must recognize that high orientation is not the same as high chain extension; a folded chain in a crystal can be highly oriented but not extended or load bearing. Drawing to maximize tenacity and orientation is best accomplished in steps with increasing temperature zones. Higher temperature reduces the draw stress. However, if too high a temperature is used, especially in the first draw stage, plastic flow can occur and little orientation will result.

With small-scale laboratory equipment, low drawing speeds should be used for uniform heat transfer and control of draw-roll speed fluctuation. This is best accomplished using multistage low-diameter draw rolls (ca. 2 in. in dia), rather than trying to maintain low speed on larger rolls, and by using multistage electrically heated low diameter pins (1 in.) at progressively higher temperatures. In many cases, a final filament/yarn relaxation (1 to 2%) over a hot roll at temperatures 10 to 20°C lower than the final draw temperature will reduce stresses induced in the yarn by drawing. Relaxation is important after the final draw step. Even in fibers such as carbon fiber spun from pitch, a final relaxation step increases final elongation and fiber toughness.

In the case of multifilament yarns, it is important to apply finish to the yarn in spinning to reduce interfilament and yarn-to-roll friction that can result in filament breakage. Simple finishes are water with detergent, silcone emulsions, mineral oils, etc. The choice depends on yarn type. The finish should remain on the surface of the yarn and not penetrate. Obviously one would not use water on a yarn that hydrolyses easily; or use a ketone that plasticizes a polyester; nor should one use highly flammable or toxic liquids. Commercial finishes invariably are proprietary mixtures of lubricants, antioxidants, etc.

The above is meant as a guide. Successful laboratory spinning and processing evolves from know-how and intuition. The same principles also largely apply to plastics extrusion and molding; the range of accessible variables is often less.

The rheologic behavior of molten or dissolved polymer is a quintessential aspect of processing to finished products that deserves more extensive consideration than the brief outline given here. Many excellent reviews and books on the subject are available (3).

Polymers deform and flow under stress. Polymer flow involves different phenomena and is different from viscous flow in ordinary fluids. Polymers possess a yield value, due to entanglements (structure) in the melt or solution, below which no flow occurs. Viscosity varies with shear rate.

$$\text{Shear rate } \ddot{Y} = d\gamma/dt \quad \text{(in reciprocal seconds)}$$

where γ = shear strain. That is, polymer melts and concentrated solutions are non-Newtonian in flow behavior, whereas simple liquids are Newtonian; i.e., the viscosity is independent of shear rate. Knowledge of these factors is important from a practical viewpoint; e.g., fiber spinning uses high shear rates, whereas compression injection molding uses low speeds. The term *viscosity*, relates observed shear rate to imposed shear stress. Newton's viscosity law states:

$$\text{shear stress } \tau = \eta \ddot{Y}$$

where $\ddot{Y}$ = apparent shear rate and η = melt viscosity. Hence $\eta = \tau/\ddot{Y}$ in pascal seconds or poise (1 P = 1 dyne/cm$^2 \cdot$ s). A plot of τ vs. $\ddot{Y}$ for a Newtonian system is linear, i.e., the slope, is constant. In a non-Newtonian system, the slope (viscosity) may decrease (pseudoplastic behavior) or, more rarely, increase (dilatant behavior). When a critical shear stress is required before any flow occurs (Bingham conditions), this indicates that a network "structure" must be fractured before any flow occurs. When viscosity decreases with time at a constant shear rate, the phenomenon is termed *thixotropy*, when the viscosity increases the phenomenon is termed *rheopexy* (4).

An other way to express flow behavior is to use expansion of the power law expression:

$$\tau = A\ddot{Y}^{\beta}$$

Expressed logarithmically

$$\log \tau = \log A + \beta \log \ddot{Y}$$

Thus plotting log τ against log $\ddot{Y}$ should provide a line with a slope of β. Thus $\beta = 1$ for Newtonian flow, <1 for shear thinning, and >1 for shear thickening.

Melt viscosity can be measured by a variety of modes and equipment but the cone and the plate device are preferred for all Newtonian fluids because the rate profile is constant across the instrument (4). If one uses a capillary viscometer, shear stress and rate at the walls should be reported. Viscosity in macromolecular systems significantly depends on temperature, as follows

$$\eta(T) = A_0 \cdot e^{Ea/RT}$$

where $\eta(T)$ = zero shear viscosity at temperature T, Ea = flow activation energy, and A_0 = a constant. Zero shear viscosity means a shear rate low enough so that the behavior is Newtonian. Thus

$$\log \eta(T) = \log A_0 + Ea/RT$$

A plot of log $\eta(T)$ vs. $1/T$ should provide a straight line with slope Ea/R. The slope varies with chain character and does not depend on molecular weight. Stiffer chains have a steeper slope. It should be pointed out that organic polymers have poor thermal conductivity and that the heat of viscous flow should be corrected for. Pressure also affects viscosity, especially at low temperatures

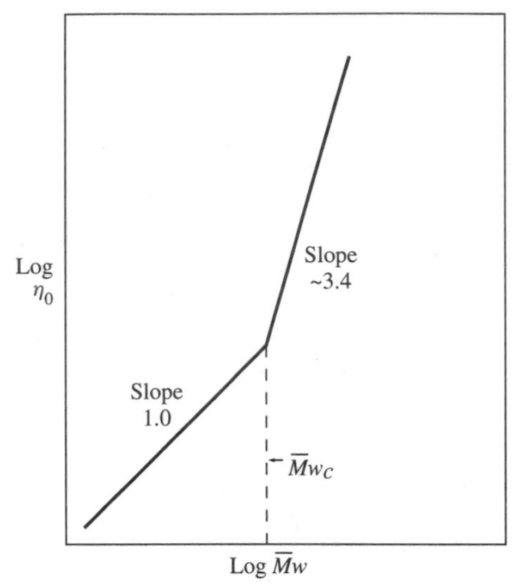

Figure 3.1. Effect of molecular weight on zero shear viscosity.

in the range of T_g by reducing free-volume and increasing viscosity. Effects of molecular weight on zero shear viscosity are pronounced (Fig. 3.1) The inflection point indicates the molecular weight at which entanglements occur; this varies from polymer to polymer and depends on chain stiffness. Polymers are generally useful only above the entanglement point. For any given value of molecular weight, broad distributions show early onsets of shear thinning, but the shear thinning is not as sensitive to shear rate as occurs with narrower distributions.

3.3. SOLUTION PROCESSING OF POLYMERS

Although melt polymer processing is relatively simple, solution processing of high melting or melt-unstable polymers requires a relatively high degree of sophistication. Solution processing is the nearly exclusive domain of making fibers and films. Processing steps and property goals dictate the choice of solvent and polymer concentration. For example, dilute solutions of stiff-chain aromatic polyamides, such as poly(*p*-phenylene terephthalamide), are easily prepared in solvents, such as *N*-methylpyrrolidinone/5% $CaCl_2$ or dimethylsulfoxide (DMSO)/*t*-BuOK/MeOH (20/1/1 by weight). These solutions are isotropic and provide tough, flexible films but do not give the high strength and stiffness properties obtained from lyotropic (liquid crystal) solutions (higher concentrations) of the same polymers in sulfuric acid.

Three solution processing systems are currently used commercially. These are dry-spinning or casting, wet-spinning, and gel-spinning. All of these processes require selection of an appropriate solvent, an optimum polymer concentration for the intended product, a quench bath (except in dry spinning), and conditions for optimum processing.

Dry spinning is a misleading term, because the polymer is dissolved in a solvent and then spun through an orifice into a heated column of nitrogen or air and then wound up or collected as a dry fiber. Hence the term *dry spun*. Currently, in the United States, commercial polymers such as Lycra (DuPont polyurethane/urea elastomer) and Nomex (DuPont poly(*m*-phenyleneisophthalamide) fiber) are dry spun. Polyacrylonitrile, Kevlar (DuPont poly(*p*-phenyleneterephthalamide) are wet spun. Allied-Signal's chain-extended polyethylene is gel spun.

Polymer chains dissolved in solvents do not generally have a rigid geometry and, at concentrations suitable for spinning or other processing, do not unfold completely or become molecularly dispersed. They remain somewhat tangled and curled. The entanglements provide cohesive interaction between the polymer chains, so that they do not behave independently in a flow field. Chains with polar groups also have electronic interactions that further reduce the ability of the dissolved polymer chains to act independently under flow. The floppiness and specific interactions between chains give rise to an elastic component of polymer solutions.

Under flow, the floppy chains and entangled assemblies orient themselves somewhat in the direction of flow, thereby offering less resistance and lower viscosity. The orientation is maintained only as long as the effects of the flow field exist. When the flow field is removed, the units relax to a less oriented state. Flexible-chain polymers relax fast, and stiff-chain polymers much slower. This elastic component must be taken into account when considering the extrusion and kinematics of a polymeric fluid and the processibility of the extruded solution.

Although most polymers dissolve as floppy chains and form microscopically uniform solutions, some important rigid-rod polymers form solutions of liquid crystals. Initially the rod-like molecules dissolve and form normal solutions, but at a critical concentration particular to the polymer, they spontaneously aggregate into highly oriented microscopic liquid crystalline domains. The aggregation is purely shape dependent and is driven by entropic considerations; chain interactions are not required. These effects were predicted by Flory (5).

Anisotropic, or liquid crystalline solutions, are easily oriented even in weak shear fields, and shear thinning occurs at a much lower shear rate than with isotropic solutions of flexible polymers (Fig. 3.2). Relaxation is also slower, because large domains rather than individual chains or small entanglements are involved. It is interesting to note, elastic effects are still pronounced in these ordered fluids due to interconnections between the various ordered domains that become aligned in the stretching field. Differences in viscosity effects are readily apparent by simply stirring both types of solutions in a glass vessel. Isotropic solutions readily climb the stirrer shaft as speed is increased, whereas the effects are much less with anisotropic solutions. Anisotropic solutions also display bright color or fluorescent effects with rapid stirring and shearing.

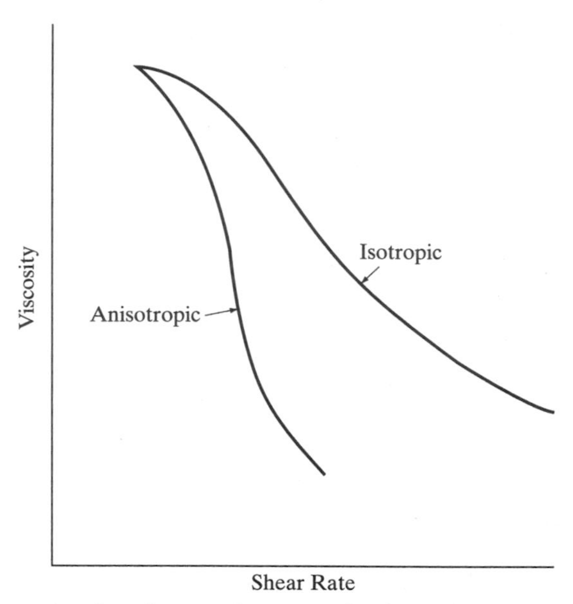

Figure 3.2. Change in viscosity vs. shear rate for isotropic vs. anisotropic polymer solutions.

3.3.1. Conditions for Dry and Conventional Wet Spinning

In this section we will primarily consider solvents for high melting or melt-unstable polymers. Solvents for gel spinning will be considered later. In the ideal situation, the solvent reaction medium will also be the spinning or processing medium. For many aromatic polyamides, good solvent polymerization media are solution 5 to 7% anhydrous calcium or lithium chloride in dimethylacetamide, *N*-methylpyrrolidinone, *N,N'*-dimethylimidazolidinone, and *N,N'*-dimethylpropyleneurea. Relatively stiff chain polymers such as poly(*p*-phenyleneterephthalamide) polymerize well in these systems; but at >5% concentration (concentration at onset of liquid crystal formation), they crystallize and precipitate from solution. Hence such solvents have limited utility as processing solvents. Poly (phenyleneterephthalamide) is currently spun commercially as a 19% solution from 100% sulfuric acid. Hexamethylphosphoramide is also a potent solvent polymerization medium but should be avoided because it is a suspect carcinogen.

For stiff liquid crystal–forming polymers, such as polybenzobisthiazoles and benzobisoxazoles, there are few solvents for the isolated and dried polymer. Methanesulfonic acid and disulfonic acids are probably the best but will dissolve only low polymer concentrations. With these polymers, however, good solutions can be maintained in ≥85% polyphosphoric acid when it is the polymerization medium. Solutions of these high molecular weight polymers are lyotropic above 3 to 5% concentration. The thick rubbery polymer solution can be extruded through fiber spinnerets into water or water and phosphoric acid baths and wound up as fiber. Solvent extraction is completed by water extraction on the wind-up bobbin. For polyesters, useful solvents are cresols, phenol/tetrachloroethane, and pentafluoroisopropanol. For gel spinning (vide infra) of poly(ethylene terephthalate) hot phenyl ether is an effective solvent.

Although the above is directed mainly to solvent systems for the final polymer, selection of an appropriate solvent as the polymerization medium is just as, or even more, important. The solvent must be unreactive to the intermediates and must maintain the formed polymer in solution or at least swell it so that the chain ends have enough mobility to continue polymerization to high molecular weight. This is amply demonstrated in the polymerization of poly(*p*-phenylenediamine terephthalamide) in which the molecular weight increases substantially after the polymer precipitates from solution if it is retained in the reaction medium for several hours more without water quenching.

In fiber spinning from solution, polymer concentration should be maximized (consistent with good flow at the extrusion temperature), to optimize the strength of the thread line, decrease the amount of solvent that has to be removed, and minimize formation of voids from solvent removal. In dry spinning, solvent is removed by extrusion of the solution into a chamber of hot gas, usually nitrogen. Solution concentration is normally in the range of 20 to 50%, and viscosity of the solutions ranges from 500 to 2000 P.

In wet spinning, in which the polymer solution is coagulated in a liquid bath, polymer concentrations range from 3 to 50%, and polymer spinning solutions and bath temperatures from 0° to 150°C (6). The coagulation bath is usually at about the same temperature as the spinning solution but may be higher or lower. The coagulation bath is often defined empirically but is usually composed of solvent and a nonsolvent to effect coagulation. Aqueous coagulation baths often contain salts (e.g., calcium chloride, calcium thiocyanate, and sodium sulfate) to modify rate of coagulation. Wet spinning speeds are relatively slow and limited to about 100 ypm to minimize breakage from the hydrodynamic drag of the coagulating liquid on the moving threadline. Commercial yarns spun by this method include rayon, acrylics, and some spandex.

Wet spinning is the work horse for experimental polymers because of its wide applicability, simplicity of equipment, and ease of operation. Coagulation is most critical and is generally brought about through solvent-nonsolvent diffusional interchange between the spinning (or film casting) solution and the coagulating medium. Solvent leaves as nonsolvent enters. The final equilibrium state is critical and is always governed by the composition and temperature of the coagulation bath.

The coagulation process consists of two sequential transitions; gelation and phase separation or coagulation. The first is a relatively slow but continuous transition of the concentrated polymer solution into an elastic gel accompanied by buildup of fine structure in the nascent fiber (or film). The fiber formed at the initial stage of coagulation has a porous fibrilar network. The next critical step is continuing coagulating and orienting the porous fibrilar network in the direction of stretch. This is usually carried out wet in a warmer medium than the coagulation bath and results in collapse of the voids with concomitant increase in fiber strength. The final stage is drying the fiber under heat and tension. The fibrilar network then compacts owing to capillary attraction between the neighboring fibrils and mechanical strength increases. This is often followed by drawing over hot rolls or a plate to increase orientation and strength.

The important coagulation and spinning variations directly related to fiber formation are spinning dope concentration, coagulation bath temperature, coagulation bath composition, bath pH, jet stretch ratio, stretching and washing bath temperatures, and stretching speed (6). As mentioned, spinning dope concentration should be as high as feasible, compatible with good flowability. Decreasing coagulation bath temperature minimizes voids. Slower coagulation results in thinner skins, better defined membranes, and better structural uniformity. In contrast, high temperatures and increased rates of coagulation produce thick skins and make penetration of the coagulant more difficult. On drawing in the bath, thick skins usually break rather than stretch and result in fiber breakage. Thus bath composition should be high in solvent or swelling agent to keep the spun fiber in the gel state as long as feasible.

In fiber spinning, choice of spinneret capillary is important because it greatly affects the flow characteristics of the polymer stream. As the viscoelastic solution enters the orifice of a spinneret, a shear field is developed at the edges of the hole, and a stretching field is encountered at the entrance to the channel. Stretching at the entrance to the channel comes about because the fluid is accelerating at constant volume. A fluid entering a 3-mil channel, for example, from a 30-mil reservoir will experience a stretch of about 100 times its original length. At practical extrusion rates, the stretch occurs rapidly, and because the spinneret holes are usually short to avoid large pressure drops, the stretched liquid has little opportunity to relieve the stored-in elastic energy until it exits the other end of the channel. The stored-in elastic energy is dissipated as it exits the channel by expansion of the solution diameter. Consequently, the solution suffers a reduction in velocity. This phenomenon is known as jet swell or the bulge and is illustrated in Figure 3.3. It occurs primarily from elastic forces and secondarily from equalizing the cross-sectional velocity gradients induced by shear at the channel wall. The amount of jet swell increases with increasing molecular weight of the polymer because higher molecular weight polymers have more entanglements. Higher polymer concentration leads to greater jet swell because the chains are more entangled. Jet swell also increases with higher flow rates because acceleration of stretch is greater and there is less time to relax in the spinneret channel.

Spinnability of a polymer solution is generally equated to the speed at which a thread line can be wound up relative to the speed at which the solution exits the spinneret channel. The spin : stretch ratio can be expressed as *MV* : *FV*, where *MV* is the linear velocity of the filament at the takeup roll, and *FV* is the velocity of the freely extruding dope at the exit of the spinneret hole (also called free velocity). It is apparent that a high spin : stretch ratio is a measure of the cohesiveness of the threadline and the stability of the fluid

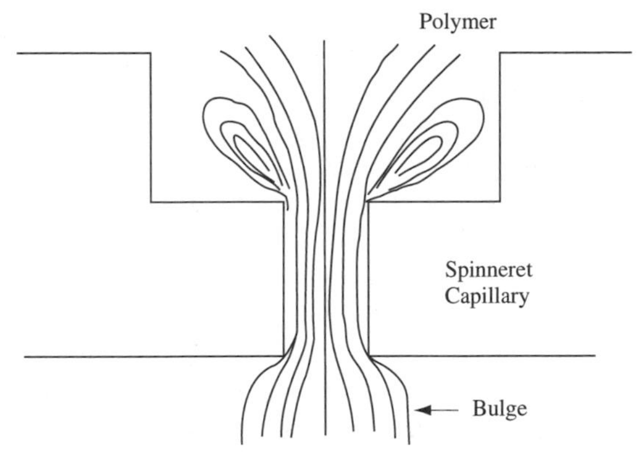

Figure 3.3. Bulge formation on extrusion of polymer solutions.

to elongational acceleration. For isotropic solutions, low free velocities usually yield the most spinnable conditions. It is important to note that a high takeup velocity : free velocity ratio denotes the highest elongational acceleration. Thus the greatest thread line orientation should be obtained under the highest $MV:FV$ conditions. The relationship between orientation and $MV:FV$ ratio for many isotropic systems (e.g., acrylics) shows that orientation increases initially as the ratio increases but then levels off. This occurs as the rapid relaxation of the chains becomes competitive with the orienting forces. For anisotropic systems in which the relaxation is slow, orientation increases more with increasing extrusion : stretch ratio.

Returning to the isotropic systems, as exemplified by acrylics, free velocity and the $MV:FV$ ratio can be influenced by the rate of coagulation. For rapid coagulation in systems where the spinneret is immersed in the coagulating fluid, the free velocity can be surprisingly high. This is because the expanding fluid is rapidly being coagulated as it exits the channel and is restricted from expanding to a value that reflects the stored elastic energy. In such a case, the free velocity is larger than anticipated and elastic stresses are locked within the fiber. Also orientation induced by the spinneret can be retained, particularly in the skin of the fiber resulting in a pronounced gradient in orientation from skin to core. This produces a poorer fiber than desired.

3.3.2. Dry Jet Wet Spinning

Dry jet wet spinning is especially important for spinning anisotropic solutions, such as those of poly(*p*-phenyleneterephthalamide). Because the thread line is extruded into air, hydrodynamic drag is much less than when spun into a liquid, and high spinning speeds are possible. These high speeds result in high elongational flow and generate strong orienting forces in the thread line. This is especially desirable in anisotropic solutions, because orientational relaxation is slow. Orientation on the same order as that achieved from melt spun, solid-state drawing can be achieved as long as coagulation is rapid. Although the swollen fibers from anisotropic dopes have little residual stretchability, higher orientation results when the swollen fiber is stressed and dried under restraining tension. This technique helps maintain orientation of the less crystalline units during densification as solvent and water are removed. Orientation and modulus can also be increased by annealing under tension. Contrary to what occurs in conventional drawing, little compaction of the fiber occurs and structural defects are minimally healed in the process with little or no increase in tenacity.

The size of the air gap in dry jet spinning is usually about 0.25 to 0.5 in., but it is determined empirically by the cohesive strength of the spin dope under spinning conditions and to polymer relaxation time vs. applied stress. Air gap should be increased until fiber breakage occurs. The experimenter should then reduce the gap until acceptable spinning conditions result.

3.3.3. Gel Spinning

One other method of fiber formation is gel spinning. This is currently used commercially to produce high-strength, high-modulus polyethylene fiber. The key requirements of the process are as follows.

- To use high molecular weight polymer to provide a cohesive viscous solution at low polymer concentrations (e.g., 3 to 5%). The low concentration reduces entanglements of the polymer chains in the solution.
- To define a specific solvent that almost instantaneously converts the heated polymer solution to a stretchable gel as the temperature is lowered. The gel immobilizes the dilute disentangled polymer chains and prevents reconcentration and reentanglement of the chains in subsequent processing steps. This process is distinct from the polymer gelation and compaction that occurs in the wet spinning process.
- To extract the solvent from the gelled fiber and compact and extend the polymer chains almost fully by drawing. This provides high tensile modulus and high strength, because many almost fully oriented, compacted chains share the applied loads. Final stretching close to the melt temperature of the polymer heals voids and morphologic defects formed during mass transfer of the solvent.

It is important to understand the rational of gel spinning and its effect on achieving close to theoretical tensile moduli in flexible or semiflexible chain polymers. As mentioned, the requirements for high tensile modulus are high chain extension, good chain packing and crystallinity, and a minimum number of flexible bonds per polymer unit. It should be pointed out that stiff-chain polymers, such as poly(benzobisthiazoles) and poly(*p*-phenylene terephthalamide), meet these requirements and, provided they are spun from anisotropic solutions, will yield high modulus fibers, limited only by imperfect chain alignment and defects.

In conventional melt or solution processing, achieving high chain extension is limited by trapped-chain entanglements. (Consider an analogy of a tangled bunch of yarn; there is no way all the individual elements can be stretched fully until the entanglements are removed, i.e., by separating and minimizing the interaction of the individual lengths of yarn.) For a flexible polymer such as polyethylene, the number of entanglements per chain is estimated to be about 500 at a molecular weight of 1 million (P. Smith, personal communication). For nonrigid molecules, maximum draw ratio λ_{max} should depend on molecular weight and the ratio of extended chain length to the distance between unextended chain ends. However, the draw ratio of melt crystallized polyethylene actually decreases with increasing molecular weight because the number of trapped entanglements increases (6). The easiest way to reduce these trapped entanglements is to decrease the polymer chain concentration in solution and by so doing separate the individual chains (Fig. 3.4) to permit maximum chain extension.

To obtain solution viscosity high enough to allow successful processing to fiber at low polymer concentrations (3 to 5%), requires high molecular weight of essentially linear polymer. Such a polymer has the potential to be drawn highly and to achieve high chain extension (P. Smith personal communication) (Fig. 3.5).

It is important to emphasize in the gel spinning that the polymer chain concentration must not be allowed to increase (e.g., as in dry or conventional wet spinning) while the polymer is mobile, otherwise chain entanglement will reoccur. Retention of low polymer chain concentration while inhibiting mobility results from gelation of the polymer solution while in fiber form. The solvent essentially becomes a nonsolvent and exudes from the gelled fiber, which retains its extruded dimensions. The solvent volume is essentially replaced by voids and the fiber chains are immobilized.

The gelled fiber is then stretched and coalesced to compact the fiber. The solvent is extracted, and the fiber plus void structure is drawn near the melt

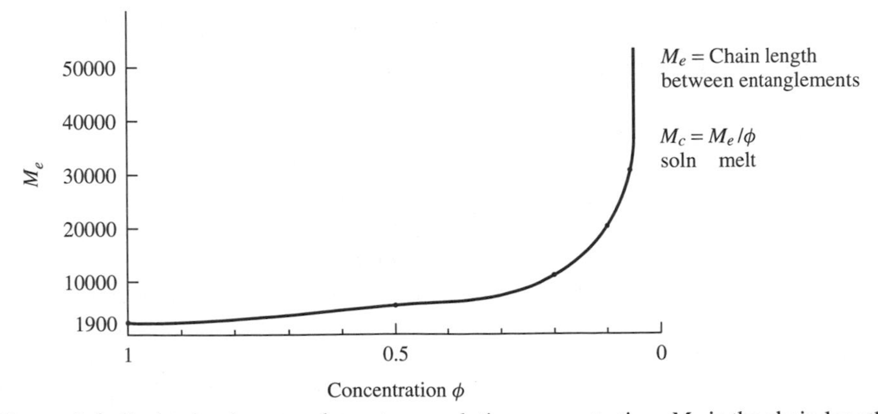

Figure 3.4. Reduction in entanglements vs. solution concentration. M_e is the chain length between entanglements and $M_{c\text{soln}} = M_{c\text{melt}}$.

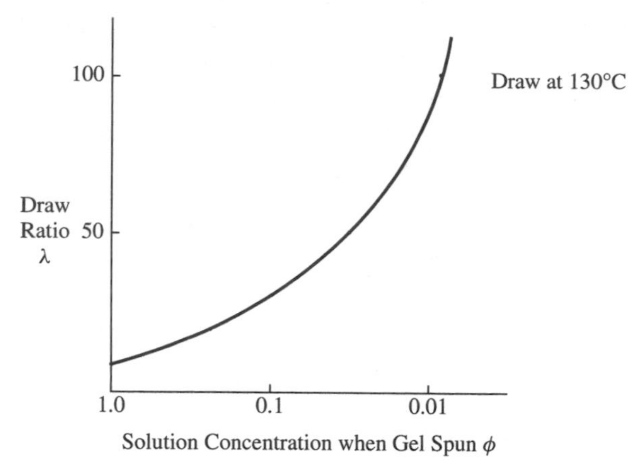

Figure 3.5. Maximum draw ratio for dried gel spun polyethylene fiber.

temperature. Little or no chain entanglement reoccurs in this process, and high chain extension is achieved. The high concentration of packed chains, high orientation, and high crystallinity results in tensile modulus close to theoretical.

The gel-spinning process has been applied most successfully to high molecular weight linear polyethylene, and high tensile properties have been achieved. Gel spinning technology has also been applied to poly(oxymethylene) with good results but little success has been achieved with nylons and polymers such as poly(ethylene terephthalate), possibly because the interchain forces are too strong to permit high chain extension from the gel state.

One further point that may contribute to chain extension in polyethylene is that polyethylene from some solutions crystallizes in a chain-folded lamellar form that unfolds easily under extensional stress. High modulus polyethylene has been prepared by extrusion and drawing sheets of compacted lamellar crystals close to the melt temperature.

3.4. ULTIMATE PROPERTIES OF POLYMERS

Although fibers from gel-spun polyethylene, extruded poly(benzobisthiazole), and solution-spun poly(*p*-phenylene terephthalamide) provide exceptionally high tensile properties, one must ask how close they come to the "ultimate" in properties. The problem is defining *the ultimate*. Can it be realized within the bounds of practical crystallinity highly extended in the absence of flaws at maximum orientation; or is it the tensile properties of a single molecule?

Theoretical tensile modulus, however, can be calculated from shifts in infrared absorption peaks from bond-angle displacement and bond stretching under applied stress and from x-ray lattice distortion under stress (7). The theoretical tensile properties of polymer fiber calculated from bond-angle displacement has been closely approached only in tensile modulus of polymers, such

Table 3.1. Theoretical vs. Demonstrated Tensile Moduli for a Range of Polymer Fibers

Polymer	Theoretical Modulus,[a] GPa		Demonstrated Modulus, GPa
Polyethylene	200–400 (f)	270 (l)	250
Poly(vinyl alcohol)		250 (l)	23–35
6-6 Nylon	190		<16
Poly(ethylene terephthalate)	120, 135		<24
Poly(*p*-phenylene terephthalamide)	180		175
Poly(*m*-phenylene isophthalamide)	115		26
Graphitic carbon	1100		690
Poly(acrylonitrile)	240		29
Polypropylene	28		20

[a] *f*, force constant method; *l* lattice distortion method.

as polyethylene, poly(benzobisthiazoles), poly(benzobisoxazoles), and poly(*p*-phenyleneterephthalamide). Table 3.1 lists theoretical vs. demonstrated tensile moduli for a range of polymers.

It is important to recognize that fiber and yarn tensile properties vary widely in a given sample, depending on testing variables. Measured values of tenacity, elongation, and modulus depend on whether filaments or twisted bundles of filaments are tested, on gauge length, rate of testing, the type of clamps employed, and ambient temperature and humidity. In general, there are two major factors responsible for gauge length and clamp effects; one is structural and the other is a testing feature. The former is the presence of flaws, which statistically are more apt to be tenacity limiting at long gauge lengths; the latter is clamp slippage and clamp deflection, which have a greater affect at short gauge lengths. When both these effects are present, tenacity will be highest at short gauge lengths (more representative of unflawed specimens). Elongation will also be highest at short gauge lengths (due to clamp jaw slippage). On the other hand, modulus will be highest at long gauge lengths, representing minimum slippage and, therefore, will be more nearly correct. It is possible to derive limiting values, free from testing artifacts and flaws, by analysis of gauge length dependence.

It is interesting to note that the limiting "true" modulus derived from tensile test data correlates well with measured sonic modulus at essentially zero elongation (i.e., close to break extension). This implies that the biggest difference between sonic and conventional tensile test modulus is due to slippage inherent in most of the latter measurements (1). The most important structural feature controlling tenacity is the level of orientation at break, next is apparent crystallite size, and the least is severity of flaws. Highest achievable tenacity is controlled by the number of extended or load-bearing chains. Theoretical tensile strength based on chain-bond breakage can be calculated from the Sakurada equation (1).

$$\sigma = (E\gamma/\alpha)^{1/2}$$

where E is Young's modulus, γ is surface energy, and α is the interatomic spacing in the polymer crystal. By experience, σ varies roughly from $E/5$ to $E/15$ or about $E/10$. Thus the ultimate tensile strength should be about 10% of the theoretical tensile modulus.

Unfortunately, in practice, chain length is not infinite or long enough to span the distances between clamps, and failures other than chain breakage occur even when high molecular weight, highly oriented fiber is stressed. Fiber tenacities achieved in practice are about 10% those calculated from C–C bond breakage. Apart from incomplete orientation and defects, failure to achieve theoretical tensile strength in fibers of finite molecular weight can be related to creep or chain slippage.

It has also been proposed that the reason a "perfect" fiber breaks is that it melts when the applied tensile load is great enough to lower the fiber melt temperature to that of the testing temperature (8). Melting may occur microscopically and may lead to local stress concentrations that result in failure. This is similar in theory to the chain-slippage process of failure. Stress-to-failure can be calculated from the following equation:

$$T = [2E\Delta H_u \log_e(T_0/T)]^{1/2}$$

where E = Young's modulus, ΔH_u = heat of fusion per unit volume, T_0 = crystalline melt temperature of the unstressed fiber, and T = melting temperature of the stressed fiber at testing temperature.

REFERENCES

1. J. Ballou, *Am. Chem. Soc. Div. Poly. Chem.* **17**(1), 75 (1976).
2. V. P. Privalko, *Polymer*, **9**, 1019 (1978).
3. J. D. Ferry, *Viscoelastic Properties of Polymers*, 3rd ed., Wiley, New York, 1980. J. J. Akalonis, W. J. MacKnight, and M. C. Shen, *Introduction to Polymer Viscoelasticity*, Wiley, New York, 1972.
4. G. L. Wilkes, *J. Chem. Educ.*, **58**, 880 (1981).
5. J. Flory, *Principles of Polymer Chemistry*, Cornell University Press, Ithaca, New York, 1953.
6. G. Prasad, *Synthetic Fibers*, **14**(1), 6 (1985).
7. K. Tashiro and M. Kobayasi, *Macromolecules*, **24**, 3706 (1991).
8. K. J. Smith, *Polymer Eng. Sc.*, **30**, 437 (1990).

4

POLYAMIDES, UREAS, URETHANES, AND OTHER AMIDE-LINKED POLYMERS

4.1. POLYAMIDES

Polyamides are polymers that contain amide groups as an integral part of the main polymer chain repeat unit. This chapter covers their preparation by melt and solution methods. All these methods are derived from condensation of bifunctional molecules with formation of a new bond and elimination of a small molecule by-product.

$$\mathrm{nAA + nBB \longrightarrow -[-AA{-}BB-]_n- + 2nC}$$

and

$$\mathrm{nAB \longrightarrow -[-A{-}B-]_n- + nC}$$

Polyamides formed by condensation of diamines and dibasic acids are called AA-BB types; those from aminoacid monomers are called AB types. The term *polycondensation* also includes any polymer whose structure indicates that it could have been prepared by a condensation reaction. For example, the addition of a diol to a diisocyanate forms a polyurethane that is structurally a condensation polymer by this definition, although it was prepared, not by condensation, but by an addition reaction without by-product elimination.

$$\mathrm{OCN{-}R{-}NCO + HO{-}R'{-}OH \longrightarrow \left[NH{-}R{-}NH{-}\overset{\overset{\displaystyle O}{\|}}{C}{-}O{-}R'{-}O{-}\overset{\overset{\displaystyle O}{\|}}{C}\right]_n}$$

Before discussing the polymerization processes, it would be well to re-emphasise two of the criteria for successful melt polymerization: close to

100% yield reaction and high purity of intermediates. The criticality of these was discussed in chapter 2. To achieve this kind of efficiency, reactions must be substantially free of chain-terminating side reactions. An exception to this rule is in interfacial polymerization, in which strict stoichiometry is not required, because reactant balance is controlled by their concentration at the interface.

The quantitative aspects of condensation polymerization, e.g., kinetics, molecular weight distribution, and branching, were discussed in chapter 2 and will be mentioned again only briefly with reference to the particular class of polymer being discussed. Polymer preparation and discussion are arranged as far as possible according to the following classes: melt methods, ring opening, interfacial methods, and solution methods.

The most important commercial polymers made by melt polycondensation are 6-6 and 6, nylons and poly(ethylene terephthalate). Nylons have advantages over natural and other synthetic fibers in having relatively low specific gravity, high strength, and good durability. The largest uses for nylons are tires, carpets, hosiery, fabrics, upholstry, and engineering resins.

4.1.1. Nomenclature

A common form of shorthand that serves to identify aliphatic polyamides is to use the numbers that signify the number of carbon atoms in the respective acid and base components. For AA-BB polymers two numbers are used. The first indicates the number of carbon atoms separating the nitrogen atoms in the diamine and the second indicates the number of straight chain carbon atoms in the acid including those of the carboxylic acid groups. For example, 6,6 or 6-6 represents the polyamide from the six carbon hexamethylenediamine and the six carbon adipic acid. In this system the diamine component is always named first. For example, 2-8 represents the polymer from ethylenediamine and suberic acid and 8-2 represents the polymer from octamethylenediamine and oxalic acid.

Ring-containing intermediates are usually coded with single letters or short combinations that represent the ring structure. For example, T and I represent terephthalic and isophthalic acids, respectively. Thus 6-T identifies hexamethylene terephthalamide. For polymer from components such as *p*-bis(aminoethyl)benzene and terephthalic acid the code would be *p*-2P2-T, with P representing the aromatic ring of the diamine component. Likewise the polymer from ethylenediamine and *p*-phenylene diacetic acid would be coded 2-*p*-2B2, with B representing the acid ring component.

The polymer derived from self-condensation of ω-aminocaproic acid or its lactam is coded 6-nylon, because there are six carbon atoms in the main chain. Other amino acids are coded correspondingly.

Polyureas, which can be considered a type of polyamide is named nylon x-1, because the acid component contains only one carbon atom; from hexamethylenediamine and oxalic acid, the designation would be 6-2.

In representing copolymers, the major component is named first, followed by minor components in order of decreasing percentages by weight. Thus a copolymer from hexamethylenediamine, adipic acid, and sebacic acid in which the 6-6 and 6-10 components are present in 95 : 5 weight ratios is coded 6-6/6-10 (95/5), with the two acids randomly distributed in the chain.

4.1.2. Synthetic Methods

Although polyamides have been prepared by many methods, only melt polycondensation, ring opening, and low-temperature solution polymerization have gained commercial importance (1).

4.1.2.A. Direct Amidation by Melt Polymerization

In direct amidation by melt methods a diamine reacts directly with a dicarboxylic acid to form a polymer with elimination of water.

$$H_2N{-}R{-}NH_2 + R'COOH \longrightarrow H_3N^{\oplus}{-}R{-}N^{\oplus}H_3 + {}^{\ominus}OOC{-}R'{-}COO^{\ominus}$$

$$\longrightarrow RNHCOR' + H_2O$$

The reactive groups may be part of a single molecule, (e.g., an amino acid) or they may be on different molecules (e.g., diamines and dicarboxylic acids). In the latter case, a precise balance of reactive groups must be present to ensure that high molecular weight is achieved (see chapter 2). This is not necessary for amino acids because there is a natural balance of amine and carboxylic groups. In the AA-BB system, the precise balance of the acid and amine groups is achieved by isolating and purifying the salt formed by mixing the intermediates in balanced proportions or by controlling the pH of an aqueous common solution of both reactants. For reference, the pH of a 9.5% aqueous solution of 6-6, salt is 7.6 (2).

Less used, but often a useful melt method, is reaction of amines with carboxylic esters, rather than the carboxylic acids. This reaction, which occurs at lower temperatures than the reaction with the carboxylic acid, is particularly useful in the preparation of polyoxamides, because oxalic acid is not particularly thermally stable. The polyoxamide reaction to high molecular weight is usually completed by solid-phase polymerization. Amines also react rapidly with amides with elimination of ammonia or other volatile amines. This method is fast, often competitive with the acid/amine route, and a ready route to polyoxamides. Acyl derivatives of diamines (usually the acetamide) also react with dicarboxylic acids to yield polyamides by acidolysis. This reaction, which is carboxylic acid catalyzed, is important in the preparation of block copolymers.

4.1.2.B. Reaction of Acid Chlorides with Amines

Low-temperature polycondensation of acid chlorides with diamines is a particularly useful method of preparing polyamides from aromatic diamines of relatively low basicity. It is also useful for preparing polymers from amines that are not thermally stable or in reactions in which there is a danger of cross-linking at elevated polymerization temperatures.

Interfacial polymerization was the initial low-temperature route for the preparation of polyamides from diamines and diacid chlorides. Basically, the method consists in reacting a diacid chloride in a water-immiscible solvent with a diamine in water containing a detergent and an inorganic base as hydrogen chloride acceptor. The reaction takes place at the organic side of the interface. Fast stirring maximizes the amount of interface and maximizes the efficiency of the reaction. Overall balance of the reactants is not critical because the balance at the interface is the important feature. The polymerization works well for alphatic (highly basic) amines but not for aromatic amines. The inorganic base (HCl acceptor) can be hydroxide, carbonate, or bicarbonate, depending on the basicity or reactivity of the amine (i.e., weak acid acceptor for weak amines).

The interfacial method is useful for preparation of small polymer samples for evaluation but has limited commercial utility because of broad molecular weight distribution and difficulty in removing salt and detergent contaminants.

A modified "interfacial" polymerization method, which uses water soluble or water miscible solvents (e.g., tetrahydrofuran) as the acid chloride solvent with the detergent eliminated, is useful for preparing polymer from less basic amines (e.g., aromatic amines), and this route has been commercialized in Japan for preparing poly (metaphenylene isophthalamide). The key here is that the selected solvent should dissolve or plasticize the prepared polymer.

The logical extension of interfacial and modified interfacial polymerization was to carry out the polymerization in a 100% organic medium (usually chlorinated hydrocarbons) using organic acid acceptors, such as triethylamine. Of course, it was necessary that the polymerization medium be either a solvent or a swelling agent for the polymer so that chain end mobility and high molecular weight could be achieved. This system is good for making polymers from most diamines but particularly for aromatic amines and other amines of low reactivity, because the acid chloride is not exposed to hydrolysis. Higher molecular weights of crystalline aromatic polyamides, such as poly(*m*-phenyleneisophthalamide), can be achieved if triethylamine hydrochloride (and similar base hydrochlorides) are added to the solvent (3). This apparently results from the potency of the solvent being increased in some cases even beyond that obtained from selected inorganic salts (see below).

Because many aromatic polyamide homopolymers are not particularly soluble in chlorinated hydrocarbons, a vastly improved solution polymerization system was discovered by Sorenson (personal communication) in which the chlorinated hydrocarbon solvent was replaced by weakly basic amide solvents, such as dimethylacetamide (DMAc) and *N*-methylpyrrolidinone (NMP) **without** added acid acceptor. These are good solvents and swelling agents for poly(metaphenyleneisophthalamide) and poly(paraphenylene-terephthalamide), particularly if saturated with anhydrous calcium or lithium chlorides, and also function as acid acceptors. Though weakly basic, such solvents are efficient acid acceptors compared to the reacting diamine (law of mass action) because there are so many more moles of them, especially when the diamine is used up as the polymerization proceeds. If fiber spinning is to be attempted neutralization of the hydrogen chloride complex in solution is recommended with calcium oxide, followed by filtration.

With weak aromatic amines, such as *N*-phenylamines, polymerization can be effected by heating the diamine and the diacid chloride or amino acid hydrochloride acid chloride in an inert solvent at elevated temperatures while nitrogen is passed through the reaction mixture to remove the hydrogen chloride formed (4). This procedure is limited in scope, however, because the solvent reaction media (usually chlorinated aromatic hydrocarbons) are often poor solvents for the polymer and high molecular weights are often difficult to obtain.

4.1.2.C. Ring-Opening Polymerization

Ring-opening polymerization is an effective and commercially important route for preparing aliphatic polyamides from lactams. The ring opening occurs without elimination of another molecule. This is the major commercial route for preparing 6-nylon from caprolactam (5). Water, amino acid, or amine carboxylate is used as an initiator. The water hydrolyzes some of the lactam to amino acid, which is the true initiator. Pure, dry caprolactam does not polymerize, even on prolonged heating at elevated temperatures. Polyamides from lactams are also readily prepared by anionic ring opening polymerization using strong bases, such as sodium hydride as initiator, and an acylating cocatalyst (e.g., acetic anhydride). A most effective, commercially available cocatalyst is *N*-acetylcaprolactam, which can be used in most anionic polymerizations of lactams. Temperature of polymerization depends on the lactam-polyamide equilibrium, and the ceiling temperature and varies with ring strain.

Ring opening polymerization of α-amino acids is readily accomplished by thermal polymerization of *N*-carboanhydrides. High molecular weight polymers are obtained (6). Polyamides have also been prepared by a variety of methods, including addition of an amide to an activated double bond to produce 3-nylons by self-addition of acrylamide (7), by polymerization of the isocyanate group to 1-nylons (8), and by addition of formaldehyde to dinitriles (9).

4.1.3. Factors Influencing Melt Temperatures

Polyamides are high-melting materials whose properties largely result from the high interchain attraction of the hydrogen bonding sites along the chains. The heat of fusion ΔH_u, depends principally on the frequency and strength of the hydrogen bonds between adjacent amide groups. The entropy of fusion ΔS_u, is largely governed by the flexibility of the polymer chain, which reflects the number of conformations a molecule can assume in the liquid state. Obviously, ΔS_u decreases as the polymer chain becomes less flexible, e.g., when methylene groups are replaced by rings.

As a class, polyamides are higher melting than the corresponding polyesters or urethanes (1). The higher melt temperatures of polyamides are due largely to higher heat of fusion, rather than entropy of fusion. In general the distance between amide groups, the lower the melt temperature of the polymer. The qualification must be added that as the number of carbons is increased in either the diamine or the acid unit, the melt temperature decreases in an alternating fashion. The polymers from intermediates with an odd number of carbon atoms have a lower melt temperature than those with an even number. Interestingly, in the AB series, polymers with an odd number of carbon atoms melt at a higher temperature as a class than those with even numbers. The melt temperatures of both AA-BB with an even number of carbon atoms in the main chain and AB polymers with an odd number of carbon atoms in the main chain provide a straight line relationships defined by the equations $T_m(^\circ\mathrm{C}) = 112 + 9.16\,x$ and $112 + 8.3x$, respectively. Amide group concentration x is obtained by dividing the amide group per repeat unit by the number of main chain carbon atoms; the result is multiplied by 100. Thus for 6-6, $x = (2/12) \times 100 = 16.7$. Hence the melt temperature is $112 + 9.16 \times 16.7 = 265^\circ\mathrm{C}$.

Many AA-BB polyamides with an odd number of carbon atoms in the main chain have melt temperatures corresponding to the equation $T_m(^\circ\mathrm{C}) = 112 + 6.85x$, but measured results are often lower than anticipated because the crystallinity of the polymers and often low. The even-numbered AB polyamides do not provide a linear but a semilog relationship corresponding to the equation $T_m(^\circ\mathrm{C}) = 112 + 8.1x^{0.95}$. Interestingly, all the graphs converge at 112°C for zero amide group concentration, which corresponds closely to the melting point of high-pressure polymerized polyethylene (branched). This can be explained by considering a polyamide as an amide copolymer of polyethylene, and as such the convergence to 0% amide would be that of a branched rather than linear polyethylene (Fig. 4.1).

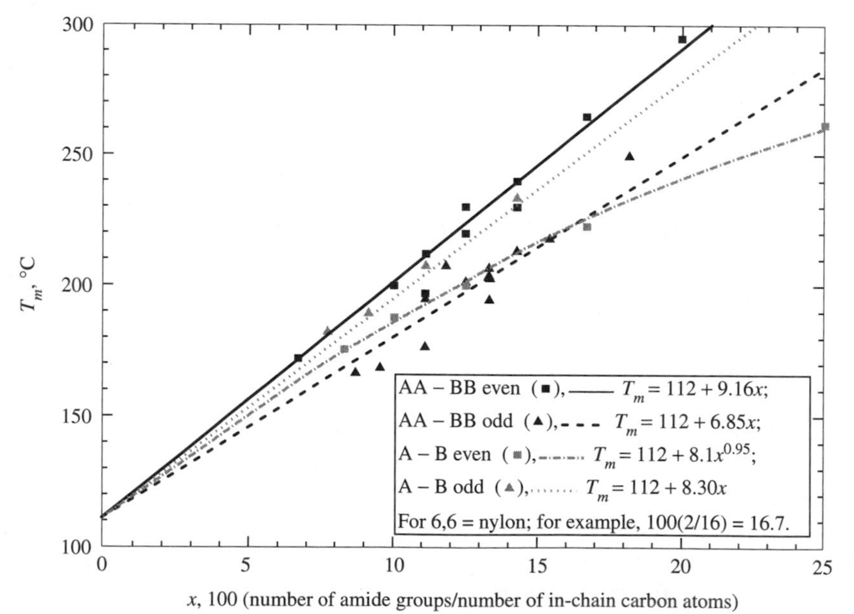

Figure 4.1. T_m vs. amide concentration for even and odd AA–BB and A–B polyamides. AA–BB even (□), ——— $T_m = 112 + 8.95x$; AA–BB odd (Δ), – – – $T_m = 112 + 6.85x$; A–B even (■), —·—· $T_m = 112.7 + 8.25x - 0.092x^2$; A–B odd (▲); ...... $T_m = 112 + 8.30x$. For 6,6-nylon for example, 100(2/16) = 16.7.

Table 4.1. Effect of Substituents on the Melt Temperature of 6,6-Nylon (1)

Nylon	Substituent	T_m,°C
6-6	—	265
6-*N*-Me6	*N*-methyl	145
6-α-Me6	methyl	166
6-α-Et6	ethyl	95
6-α,α'-DiMe6	dimethyl	115
6-β-Me6	methyl	200
6-β-Et6	ethyl	152

Table 4.2. Melt Temperatures of Terephthalamides and Adipamides of Aliphatic Diamines (1)

Nylon	T_m,°C
6-6	265
6-T	371
7-6	220
7-T	341

The melt temperature of polyamides depends not only on the frequency and strength of the hydrogen bonding between amide groups (ΔH_u) but on the chain stiffening (ΔS_u) resulting from tautomerization of the amide group $-\overset{\displaystyle O}{\overset{\|}{C}}-NH- \longleftrightarrow -\underset{\displaystyle OH}{\underset{|}{C}}=N-$. In *N*-substituted polyamides, hydrogen bonding and tautomerism are destroyed and, as a result, melt temperatures are greatly reduced. Introduction of hydrocarbon side chains on the carbon skeleton of polyamides introduces lateral disorder and interferes with the hydrogen bonding; thus the melt temperature decreases with size of the substituent (Table 4.1).

Replacement of an aliphatic main chain segment of a polyamide by a cyclic segment reduces chain flexibility and raises the T_m because of the entropy of melting decreases. The effect on melt temperature is most pronounced with cyclic *p*-aromatic groups, for example, when adipic acid segments are replaced by terephthalic groups (Table 4.2). When the ring is not adjacent to the amide group, the T_m is raised by only 50 to 70°C.

4.1.4. Melt Polymerization of Polyamides

4.1.4.A. Principles of Polyamidation

The principles of polyamidation were discussed in chapter 2 and will be only briefly mentioned here. The number average degree of polymerization is given by $P_n = 1/(1 - p)$, where p is the extent of polymerization. The number average molecular weight is P_n times the weight of the polymer repeat unit. The number average molecular weight $\overline{M}_n$ is given by $\overline{M}_n = 2 \times 10^6/T$, where T is the total number of chain ends. If in the preparation of 6-6 nylon, 99% of the original ends react ($p = 0.99$), the number average degree of polymerization is 100, and the number average molecular weight is, therefore, 11,300 (molecular weight of the repeat unit is 113). Most commercial nylons have a molecular weight at least that high. The total ends T, can be determined by titration, and M_n is determined (assuming there are no nontitratable ends or that their concentration is known).

Molecular weight (MW) determinations frequently employ viscosity measurements. For aliphatic polyamides, such as 6-6, the relative viscosity η_{rel} (or RV), of an 8.4% solution of the polymer in 90% formic acid is often used. Typical commercial values range from 30 to 70. For nylon 6 the relative viscosity of a solution of 1 g polymer in 100 mL 96% sulfuric acid is also used. The values range from 2 to 4.

Inherent viscosity η_{inh} is defined by $\eta_{\text{inh}} = \log_e \eta_{\text{rel}}/c$; where c is in units of grams per 100 mL of solvent. A concentration of 0.5 g/100 mL is assumed when the concentration is not specified in the η_{inh} values.

4.1.4.B. Kinetics

There are contradictory views on the reaction order of the kinetics of polyamidation. Some favor second-order kinetics; but the prevailing view is that for high conversions and low water content carboxyl-catalyzed, third-order kinetics apply. This conclusion is reinforced by many examples of amidation catalysts under these conditions (10).

4.1.5. Experimental Polymer Preparations

Although polyamides can be prepared by a variety of routes, the major synthetic and commercial route is by direct reaction of an amine with a carboxylic acid accompanied by elimination of water. In the case of 6-6, nylon, precise balance of reactive end groups is obtained by isolation and purification of the salt formed by mixing the intermediates. For 6-nylon, caprolactam is hydrolyzed in situ to the amino acid, which then polymerizes with elimination of water. Pure, dry caprolactam does not polymerize when heated alone for prolonged periods. However, amines and carboxylic acids in combination catalyze polymerization, although neither is effective alone. Effective catalysts are aminocaproic acid and 6-6, nylon salt.

Aromatic-aliphatic polyamides and some all-aromatic polyamides can be prepared by melt methods, but the polymers are often branched and are better prepared by solution methods. However, copolymers such as 6-6/6-T that contain 2 to 8% of the latter can probably be made with minimum branching if solid phase polymerization is used early in the polymerization.

The preparation of 6-6 polyamide, is typical of the methods used in melt polymerization. Most aliphatic polyamides can be made by the basic technique described.

1. Preparation of Hexamethylenediamine-Adipic Acid Salt (6-6-Salt) (11)
In a 250-mL Ehrlenmeyer flask, is placed 14.60 g (0.100 mol) adipic acid. The acid is dissolved in 110 mL absolute ethyl alcohol by warming, and then is cooled to room temperature. A solution of 11.83 g (0.012 mol) hexamethylenediamine (boiling point (bp) 90 to 92°C at 14 mm, melting point (mp) 41 to 42°C) in 20 mL absolute ethyl alcohol is added quantitatively to the adipic acid solution. The mixing is accompanied by spontaneous warming. Crystallization soon occurs. After standing overnight, the salt is filtered, washed with cold absolute alcohol, and air dried to constant weight. The yield is 25.5g (97%).

A 2% excess of diamine is used to promote a salt that is rich in diamine, because this is the more volatile component and may be lost during salt drying or during polycondensation. The white crystalline salt melts at 196 to 197°C and has a pH of about 7.6, determined on a 1% solution of salt in water using a pH meter.

A pH tolerance of 0.5 unit is usually acceptable, especially on the high side, because of possible diamine loss, as noted above. Salt imbalance may be corrected by recrystallization or after addition of a small amount of the indicated component. Salts of high and low pH may be mixed to give a balanced composition of the proper pH.

2. Preparation of Poly(hexamethylene Adipamide) (11)

$$\left[H_3\overset{+}{N}-(CH_2)_6-\overset{+}{N}H_3\right]\left[\bar{O}-\overset{O}{\overset{\|}{C}}-(CH_2)_4-\overset{O}{\overset{\|}{C}}-\bar{O}\right] \longrightarrow \left[-NH-(CH_2)_6-NH-\overset{O}{\overset{\|}{C}}-(CH_2)_4-\overset{O}{\overset{\|}{C}}-\right]_n + H_2O$$

A total of 20 g of 6-6 salt is charged into a polymer tube using a polymer tube funnel. A constriction is made in the upper half of the neck of the tube with a glass-blowing torch. The tube is connected to a three-way stopcock, which connects to a vacuum pump and a source of low-pressure nitrogen. The tube is purged of air by alternatingly evacuating to about 0.5 mm and filling with nitrogen. After three or four such cycles, the constriction is sealed shut by means of the torch while the tube is evacuated.

The tube is placed in a steel tube open at the top and immersed in a salt bath heated at 215°C for 1.5 to 2 h. The heated tube is hazardous and should be handled with care. The salt bath is shielded in front, with an access door at the side. The hand and arm used to manipulate the tube should be protected with a Kevlar glove and gauntlet.

The sealed tube step results in the formation of low molecular weight polymer under conditions that prevent the escape of volatile diamine. If a shielded bath is not available, the tube can be heated in a metal tube immersed in a vapor bath or a Woods metal bath. The metal shield should surround the tube completely, although it should be open at the top. The entire apparatus should then be shielded additionally with a transparent safety shield.

After completion of this first heating cycle, the sealed polymer tube is removed from the heating bath and allowed to cool to room temperature. The tube is then opened behind a barrier after first scoring with a file or a glass-cutting tool. In some melt polycondensations, uncondensed gases such as ammonia or carbon dioxide may be evolved during the sealed tube step, and the cold tube may be under considerable pressure.

A neck-bearing a side arm is sealed onto the polymer tube, the tube is clamped in an upright position, and the side arm is connected with pressure tubing to a trap that comprises a 50-mL round-bottom flask fitted with an adapter. The trap is connected via a three-way stopcock, both to a vacuum pump and to a source of nitrogen. The top of the tube is fitted with a short section of rubber tubing. An inlet made up of a 7 to 8 g-mm outer diameter (od) glass capillary tubing drawn to a fine tip is is fitted through the rubber tubing, with the end reaching to the bottom of the polymer tube. The inlet sometimes cannot be brought to the bottom of the tube until the polymer has been remelted. The inlet is connected to a source of low-pressure nitrogen.

The tube is purged of air by alternately evacuating and flushing with nitrogen. It is then heated in a 270°C vapor bath. After 30 to 60 min of heating at atmospheric pressure, the polymer is gradually brought to high vacuum by manipulating the three-way stopcock. The heating is continued under a vacuum of 0.2 to 1.5 mm for about 1 h. The polycondensation is discontinued when visual inspection of the rate of bubble rise indicates that the maximum melt viscosity has been reached.

Nitrogen is introduced through the stopcock, the inlet tube is raised, and the vapor bath is removed. The polymer is cooled under a gentle nitrogen stream that enters through the side arm. The tube is wrapped with a towel during cooling to prevent glass from flying if the polymer tube spontaneously cracks. The tough, white opaque polymer is removed by breaking the tube and separating the adhering glass. The recovery is about 14 g (80%). The inherent viscosity in *m*-cresol at 0.5% concentration is about 1.0 to 1.4. The polymer melts at about 265°C and is soluble in formic acid. Fiber and film may be obtained by melt methods or from formic acid solution.

This method of polycondensation can sometimes be used starting with equivalents of diamine and dicarboxylic acid esters, weighed directly into the polymer tube. In this instance, alcohol is removed rather than water.

A safe and convenient method of carrying out the polyamidation from nylon salt is to use a pipe autoclave, but then visual observation of the polymerization is not possible (12).

Polymers that decompose below their melting temperature can sometimes be made by powder polymerization. The polycondensation is carried out to low molecular weight and the low polymer is ground in a suitable mill. Nylon salts of the type described above, which melt above 225°C, may be powder polymerized directly. The polymer powder should be sieved to a uniform size. The polycondensation is conducted by heating the ground polymer, or high melting salt, under vacuum with a nitrogen bleed to carry away the volatile byproducts. Other types of polymers may be prepared by powder polymerization methods.

The effect of intermolecular hydrogen bonding on polyamide properties is again strikingly demonstrated by the difference in properties between poly(hexamethylene sebacamide) and the same polymer in which 60% of the amide hydrogens have been replaced with alkyl groups. The following polymerization, by which isobutyl is the alkyl substituent on nitrogen, illustrates this. The polymer melt temperature is reduced from 215°C for 6-10 nylon to 145°C for the 60% alkylated 6-10 nylon; and while 6-10 nylon forms fibers with in properties to 6-6 nylon, the N-alkylated polymer forms fibers that are quite elastic.

3. Preparation of Hexamethylenediamine–Sebacic Acid Salt

The subject compound can be made by the directions given in the preparation of 6-6 salt, substituting sebacic for adipic acid and using the same molar quantities of reactants. The pH of a balanced 6-10 salt is about 7.6. The mp of 6-10 salt is 170 to 172°C, and the yield is about 85%. Polymerization is also similar to that for 6-6 nylon, and the mp of 6-10 polymer is about 225°C.

4. Preparation of N,N′-Diisobutylhexamethylenediamine (13)

$$H_2N-(CH_2)_6-NH_2 + O{=}\underset{}{\overset{H}{C}}-\underset{CH_3}{\overset{CH_3}{C}}-H \xrightarrow[-H_2O]{} (CH_3)(CH_3)C{=}N-(CH_2)_6-N{=}C(CH_3)(CH_3)$$

$$\xrightarrow{H_2} (CH_3)_2HC-\underset{H}{N}-(CH_2)_6-\underset{H}{N}-CH(CH_3)_2$$

To 386 g (2.0 mol) of a 60% aqueous hexamethylenediamine solution 288 g (4.0 mol) of distilled isobutyraldehyde is added with stirring. The temperature is kept at 50 to 55°C with an ice bath. The organic dialdimine layer is separated from the water layer and hydrogenated without purification as follows.

Platinum oxide (0.7 g) is reduced in the absence of the hydrogen acceptor by shaking a suspension of the catalyst in 50 ml absolute alcohol with hydrogen for 10 min in a low-pressure Parr catalytic apparatus equipped with a 1-L Pyrex bottle. The dialdimine from above is added to the reduced catalyst in ethanol and the hydrogenation is started at 55 psi and room temperature. When hydrogen is no longer absorbed, the catalyst is removed by filtration, and the N,N'-diisobutylhexamethylenediamine is distilled through a 10-in. helix-packed column at reduced pressure. The fraction boiling at 116 to 117°C/3 mm weighs 365 g (81%) and may be used in the following reaction without further purification.

5. Preparation of N,N′-Diisobutylhexamethylenediamine–Sebacic Acid Salt

To a solution of 48.5 g (0.24 mol) of sebacic acid in 200 mL of absolute ethanol is added 56.5 g (0.246 mol) of N,N'-diisobutylhexamethylenediamine in 100 mL ether. After mixing the solution thoroughly, about 700 mL ether is added. The salt crystallizes on cooling in ice. It is filtered, washed well with ether, and air dried. The pH of a 1% solution of the salt in water is about 6.6. The salt is dissolved in 200 mL absolute ethanol, decolorized with activated charcoal, and filtered. After filtration, 800 mL ether containing 10 g of the diamine is added. After crystallization, the salt is filtered, washed with ether, and dried in a vacuum desiccator. The pH of the salt is now about 7.5. The dry salt melts at 137 to 138°C. The yield is 90 g.

6. Preparation of Partially Isobutylated Poly(hexamethylene sebacate) (14)

$$\left[\begin{matrix} H_2\overset{+}{N}-(CH_2)_6-\overset{+}{N}H_2 \\ | \qquad\qquad\qquad | \\ \text{i-Bu} \qquad\qquad \text{i-Bu} \end{matrix}\right]\left[{}^{-}O-\overset{O}{\overset{\|}{C}}-(CH_2)_8-\overset{O}{\overset{\|}{C}}-O^{-}\right]$$

$$+\left[H_3\overset{+}{N}-(CH_3)_6-\overset{+}{N}H_3\right]\left[{}^{-}O-\overset{O}{\overset{\|}{C}}-(CH_2)_8-\overset{O}{\overset{\|}{C}}-O^{-}\right] \longrightarrow \text{Random copolymer}$$

Using the same technique and equipment described for Prep. 2 for 6-6 nylon, a mixture of 12.9 g dibutyl 6-10 salt (0.03 mol) and 6.36 g 6-10 salt (0.02 mol) is charged to a polymer tube, and the tube is filled with nitrogen and sealed. It is heated in a 202°C vapor bath (*m*-cresol) for 16 h. *N,N′*-Dialkyl diamines react more slowly than the corresponding diprimary amines, and a much longer sealed tube heating stage is required. The polymer tube is cooled and opened cautiously. (Observe the same safety precautions as described for prep. 2). A capillary is introduced to the bottom of the tube, and the tube is flushed with nitrogen. It is then heated in a 218°C vapor bath (naphthalene) for 0.5 h at atmospheric pressure and then transferred to a 275°C vapor bath (methyl naphthyl ether) for 45 min, still at atmospheric pressure and with a slow stream of nitrogen flowing through the capillary. The pressure is reduced to about 1 mm for a period of 6 h. The tube is then cooled under nitrogen, and the polymer is removed after releasing the vacuum with nitrogen and breaking the cooled tube. The polymer has a mp of about 145°C and an inherent viscosity of 0.6 to 0.8 in *m*-cresol. Elastic fibers and films can be melt fabricated. Films can also be cast from 98% formic acid solutions.

4.1.6. Copolymer Preparations

Copolymers can be made by polymerizing together two or more diamines and/or diacids. The polymerization procedure is identical to that for 6-6 nylon. Copolymers can also result from the process of amide interchange, made by melting together two or more polyamides and maintaining in the melt until randomized. The reaction is probably acid (H^+) catalyzed, and the rate is proportional to the first power of the carboxyl group concentration. The mechanism appears to involve reaction of amide with anhydride formed from carboxylic acid end groups. Amidolysis does not occur in the absence of carboxyl. Amide interchange initially forms block copolymers, but with retention in the melt, almost complete randomnization occurs (15).

Generally with random copolymers, properties such as melt temperature pass through a minimum (eutectic) at some composition between those of the homopolymers. Solubility may also show a maximum. Exceptions to these rules are found when the polymers are isomorphous; i.e., fit into the same crystal lattice. Such copolymers show a linear dependence of physical properties on composition. Random copolymers exhibit a single T_g, whereas block copolymers usually exhibit separate glass-transition temperatures equivalent to those of the homopolymers.

In the following preparation of a random copolymer from hexamethylene- and tetramethylenediamines with adipic and sebacic acids, the composition of 0.6 mol of 6-10 and 0.4 mol of 4-6 is used. This gives the minimum melting composition for the 6-10/4-6 composition. It is nearly the maximum in solubility, being slightly exceeded by the 0.5/0.5 molar ratio copolymer.

7. Preparation of Poly(hexamethylene sebacamide-co-tetramethylene adipamide) (0.6/0.4 Molar)

The preparation of 6-10 salt is given in prep. 3. The 4-6 salt is prepared as described for 6-6 salt (prep. 1) using a 2% excess of tetramethylenediamine, which can be purified by distillation in a current of nitrogen and stored under nitrogen. It boils at 158 to 160°C at atmospheric pressure and melts at 27 to 28°C. The 4-6 salt can be obtained in >95% yield. The balanced salt melts at 193 to 194°C. and a 1% aqueous solution has a pH of about 7.1. The polymerization may be conducted as described for the preparation of 6-6 nylon, with the following modifications.

A mixture of 4.685 g (0.02 mol) of 4-6 salt and 9.553 g (0.03 mol) 6-10 salt is charged into a polymer tube that is then evacuated and charged with nitrogen. The tube is sealed and heated at 220°C (methyl salicylate) for 2 h. A side arm is attached, and a vacuum is then gradually applied over a 15-min period. The tube is heated for 1.5 h more at 220°C at a pressure of 1.0 mm or less, while a slow stream of nitrogen is passed through the melt via the capillary tube. The polymer is cooled under nitrogen after raising the capillary tube and is obtained as a tough translucent plug, after breaking the tube. The polymer melts at about 170°C and has an inherent viscosity of 0.7 to 0.8 in *m*-cresol at 0.5% concentration Fibers are slightly crystalline and have a crystalline melt temperature of 179°C. The polymer is soluble in dimethylformamide, 80% ethanol as well as in acidic polyamide solvents (phenols, formic acid, etc.).

Polyamides can be prepared by reaction of a diamine with a diester in a typical aminolysis reaction. Usually phenyl esters or methyl esters are selected because of their greater reactivity. Thiol esters are more reactive than their oxy counterparts. In the following example, a disubstituted phenyl malonate is used; the unsubstituted or monosubstituted malonates do not condense well with diamines.

8. Preparation of Phenyl Di-n-butyl Malonate (16)

$$(n\text{-}C_4H_9)_2\text{—}C(COCl)_2 + 2\,C_6H_5OH \longrightarrow (n\text{-}C_4H_9)_2\text{—}C(C(=O)\text{—}O\text{—}C_6H_5)_2 + 2HCl$$

In a 1-L three-necked flask equipped with a condenser and a dropping funnel, a solution of 123 g (2.46 mol) potassium hydroxide in 500 mL absolute ethanol is heated to reflux, and 100 g (0.38 mol) ethyl di-*n*-butylmalonate is gradually added from the funnel. Hydrolysis takes place almost immediately, and some salt precipitates near the end of the reaction. After complete addition of the ester, the mixture is refluxed for 1 to 2 h. It is then cooled to room temperature and acidified with concentrated hydrochloric acid. The resulting mixture is filtered to remove the potassium chloride, which is washed with ether. The filtrate is extracted with ether, and the ether extractions and washing are combined. The ether solution is extracted with three 200-mL portions of 10% aqueous sodium carbonate. The sodium carbonate solution is treated with activated charcoal, cooled, and acidified with concentrated hydrochloric acid. After cooling to room temperature, the crystallized di-*n*-butylmalonic acid is collected on a filter, washed with cold water, and dried at 50°C. in a vacuum oven. The yield is 78 g (98%); mp, 160°C.

The di-acid (78 g, 0.36 mol) is placed in a 500-mL three-necked flask and 175 g (0.84 mol) phosphorus pentachloride is added portion-wise with occasional cooling of the flask in ice. When the addition is complete and the initial reaction has subsided, the mixture is heated under reflux for 1.5 h. The evolved hydrogen chloride should be trapped or scrubbed. The phosphorus oxychloride is distilled from the reaction at atmospheric pressure (bp 107°C) through a 10-in. Vigreux column, and the residue is fractionated at reduced pressure. The di-*n*-butylmalonyl chloride boils at 160 to 178°C at 85 mm. The yield is 48 g (57%). Then 50 g (0.2 mol) of the acid chloride is heated with 41 g (0.44 mol) phenol at a temperature of 200°C, by means of an oil bath until the evolution of the hydrogen chloride had essentially ceased, which may take 3 h. The reaction is then fractionally distilled at reduced pressure to give the phenyl ester boiling at 191 to 192°C at 2 mm pressure. The phenyl di-*n*-butylmalonate melts at 49°C and can be recrystallized from ethyl alcohol (about 1 g ester in 3 mL alcohol). The yield is 44 g (60%).

9. Preparation of Poly(hexamethylene di-n-butyl malonate) (16)

$$H_2N-(CH_2)_6-NH_2 + (C_4H_9)_2-C(-C(=O)-O-C_6H_5)_2 \longrightarrow$$

$$\left[-HN-(CH_2)_6-NH-\overset{O}{\overset{\|}{C}}-\underset{C_4H_9}{\overset{C_4H_9}{C}}-\overset{O}{\overset{\|}{C}}-\right]_n + 2\ C_6H_5OH$$

A carefully weighed mixture of 2.95 g (0.0254 mol) of hexamethylenediamine and 9.20 g (0.250 mol) of phenyl di-*n*-butyl malonate is placed in a polymer tube, which is then evacuated and filled with nitrogen and finally evacuated as described in the melt preparation of 6-6 nylon. The tube is sealed while evacuated and is then heated for 14 h in a vapor bath at 210°C. The tube is then opened and fitted with a nitrogen capillary bleed reaching to the bottom of the tube and a side arm for distilling. The tube is heated in a 265°C bath for 1 h at atmospheric pressure, for 15 min at 30 mm Hg, and finally for 5 h at 30 mm Hg, all with a slow stream of nitrogen passing through the melt. The tube is allowed to cool to room temperature; it is opened, and the glass is broken away from the polymer plug. Except for mechanical losses in isolation, the yield of polymer is quantitative. The polymer has an inherent viscosity of 0.5 to 0.7 in *m*-cresol at 0.5% concentration and a polymer melt temperature of around 145°C. The polymer can be melt pressed or dry cast to strong films from chloroform/methanol (88/12 weight).

From the standpoint of the diacid involved, the most readily aminolyzed esters are generally those from oxalic acid. For this reason, plus the fact that oxalic acid decomposes on melting, the preparation of polyoxamides usually involves the reaction of an oxalate ester with a diamine in some solvent at room temperature. The reaction in most cases is rapid and exothermic, and a low molecular weight precipitate usually forms.This is then polymerized to high molecular weight in a melt or solid phase polymerization system. Polyoxamides from short chain diamines (<7 carbons) have proved somewhat difficult to prepare in many instances because of a tendency to decompose at the high melt temperatures involved. However, Shorter chain diamines that are laterally substituted have been used successfully to give melt-stable polyoxamides (17; 18) In the example below, the polyoxamide of decamethylenediamine is prepared.

10. Preparation of Decamethylenediamine (19)

$$NC-(CH_2)_8-CN \xrightarrow{(H)} H_2N-(CH_2)_{10}-NH_2$$

Place 250 g sebaconitrile, 250 mL dioxane, and 15 g Raney cobalt catalyst (as an ethanol paste, see below) in a 1-L hydrogen bomb. The bomb is flushed with hydrogen to remove air and pressured to 5000 psi with hydrogen at 120°C. Hydrogen is pressured into the vessel at regular intervals until no more is taken up. The bomb is cooled to room temperature and bled to atmospheric pressure; the solution is filtered free of catalyst. The dioxane is stripped by distillation in a steam bath at water aspiration pressure using a short column. The residue is then fractionally distilled through an 8-in helix-packed column at reduced pressure in a stream of nitrogen to give about 160 g (62%) decamethylenediamine, boiling at 139 to 141°C/12 mm. The mp is 60 to 61°C. If the hydrogenation is carried out in the presence of 180g anhydrous ammonia, yields may be as high as 97%. The diamine should be kept under nitrogen to avoid carbonate formation and be kept free of moisture. The Raney cobalt catalyst used in this preparation can be prepared by suspending 50 g powdered aluminum-cobalt alloy (50/50%) in 300 mL boiling water in a 1-L three-necked flask equipped with a condenser and an efficient stirrer. A solution of 50 g sodium hydroxide in 100 mL water is added slowly with good stirring, and the mixture is boiled for 4 h. After cooling, the supernatant liquid is decanted and the residue is again refluxed with a similar solution of sodium hydroxide. The liquor is again decanted, and the catalyst is washed free from alkali by repeated stirring and decanting with fresh water. The final washing and decanting is done with 95% ethanol. The catalyst is stored under alcohol and should be used as an ethanol paste. It is pyrophoric and should be handled with care.

11. Preparation of Poly(decamethylene oxamide) (20)

$$H_2N(CH_2)_{10}NH_2 + C_4H_9O-\overset{\overset{\large O}{\|}}{C}-\overset{\overset{\large O}{\|}}{C}-OC_4H_9 \longrightarrow \left[-HN(CH_2)_{10}NH-\overset{\overset{\large O}{\|}}{C}-\overset{\overset{\large O}{\|}}{C}-\right]_n + 2C_4H_9OH$$

For use in this preparation, commercial dibutyl oxalate can be freed from acidic impurities by stirring overnight with 10% by weight of dry calcium hydroxide under anhydrous conditions. The mixture is filtered and the filtrate distilled under reduced pressure, bp 84°C/0.85 mm. Decamethylenediamine is prepared and purified as described earlier.

Add all at once 20.22 g (0.10 mol) dibutyl oxalate with stirring under nitrogen to a solution of 17.23 g (0.10 mol) decamethylenediamine in 25 mL toluene (dried over sodium) in a 250-mL three-necked flask equipped with stirrer, adjustable nitrogen inlet tube, and drying tube. The residual oxalate ester is washed in quickly into the flask with another 15 mL of dry toluene. Heat is liberated, and a white solid begins to form in a short time. Stirring is continued until the mass becomes too thick to stir. The solid is a prepolymer with an inherent viscosity of 0.15 to 0.25 (sulfuric acid). Then 2 h after the initial addition, the flask is heated to 270°C in a Woods metal bath with a current of nitrogen continually passed over the reaction mixture. The nitrogen inlet tube is located directly above the reaction mass. The toluene distills as the temperature is raised. The 270°C temperature is maintained for 1 h, then the polymer is cooled under nitrogen. A tough, white plug is obtained. The polymer melts about 240°C and has an inherent viscosity of 0.6 to 0.7 in sulfuric acid at 0.5% concentration. Fibers can be pulled from the melt.

A method of demonstrating the effect of *N*-alkylation in polyamides is the reaction of the polymer with formaldehyde and methanol to give a partially *N*-methoxymethylated product. The resultant decrease in amide hydrogen bonding has a profound effect on the properties of 6-6 nylon. The melt temperature is reduced and the solubility increased. The polymer becomes somewhat rubber-like, depending on degree of substitution. The following directions are for an *N*-methoxymethylated 6-6 having about 36% substitution on nitrogen. A small portion of the substitution is *N*-methylol. The melt temperature is reduced from 265°C to 150°C, and the polymer is then soluble in 80% aqueous ethanol as well as in strong acids.

12. N-Methoxymethylation of Poly(hexamethylene adipamide) (21)

$$\left[-HN(CH_2)_6-NH-\overset{\overset{\large O}{\|}}{C}-(CH_2)_4-\overset{\overset{\large O}{\|}}{C}-\right]_n \xrightarrow[CH_3OH]{CH_2O} \left[-HN(CH_2)_6-\underset{\underset{\underset{\underset{\large CH_3}{|}}{O}}{|}}{\underset{|}{N}}\! -\overset{\overset{\large O}{\|}}{C}-(CH_2)_4-\overset{\overset{\large O}{\|}}{C}-\right]_n$$

(N substituent: $-CH_2-O-CH_3$)

A solution of 60 g 6-6 nylon in 180 g of 90% formic acid is prepared by stirring at 60°C. To this is added a solution of 60 g paraformaldehyde in 60 g methanol, heated to 60°C. The latter solution is prepared by warming a suspension of paraformaldehyde in methanol to 60°C and adding a trace of solid sodium hydroxide, whereupon the solution becomes clear. The rate of addition of the paraformaldehyde solution is slow during the first minute to avoid precipitation of the polymer, then it is increased so the addition is complete in about 3 min. At 10 min after the addition of the paraformaldehyde was begun, another 60 g methanol is added rapidly and the reaction is allowed to proceed for 30 min. (If more than 10 min elapses before adding the second amount of methanol, the degree of substitution increases.) The solution is then poured into 1700 mL water/acetone solution (50/50 vol), and concentrated ammonia is gradually added, causing the *N*-methoxymethylated polymer to precipitate as fine white granules. After filtration, the polymer is washed thoroughly with water and dried over phosphorus pentoxide in a vacuum desiccator. Analysis for methoxyl content by the Zeisel method should give a value of about 7.1%.

Analysis for methylol is carried out by treating a solution of the polymer in 70% ethanol with sodium sulfite and titrating the liberated alkali with acid. Methylol content should be about 1.4%. This corresponds to a total amide substitution of about 36% from methoxymethyl and methylol. The modified polymer melt temperature is about 150°C. It is soluble in 80% ethanol as well as formic acid and *m*-cresol. The polymer chain length is virtually unchanged. A film cast from ethanol is quite elastic, unlike a film from the parent polymer.

Another example of polyamide modification is the reaction of ethylene oxide with 6-6 nylon to give a polyamide in which some of the nitrogens bear poly(ethylene oxide) chains (22). The product, therefore, is a graft copolymer, with many more of the hydrogen bond sites left intact than in the previous example of methoxymethylation. Despite the large modification of 50% by weight of ethylene oxide, the polymer melt temperature is lowered only slightly to about 220°C and solubility is restricted to typical nylon solvents, such as formic acid. The effect of the poly(ethylene oxide) side chain grafts is that of an internal plasticizer, producing a more rubbery, flexible polymer; but one still not possessing distinctly elastic properties. The side chain effect is manifested in a marked lowering of the glass-transition temperature from about 47°C for unmodified 6-6 to one below −40°C for the modified polymer.

In the following reaction, modification should be as high as 50% by weight. It is necessary to start with polymer of low crystallinity, because it is difficult for the ethylene oxide to penetrate crystalline regions. The growth of poly(ethylene oxide) chains instead of formation of *N*-hydroxyethylated polymer points to the preference of ethylene oxide for reaction with hydroxyl over amide hydrogen.

13. Preparation of Poly(ethylene oxide)-Grafted Poly(hexamethylene adipamide) (20)

$$\left[-HN-(CH_2)_6-NH-\overset{O}{\overset{\|}{C}}-(CH_2)_4-\overset{O}{\overset{\|}{C}}-\right]_n + m\,\underset{\diagdown O \diagup}{CH_2-CH_2} \longrightarrow \left[-HN-(CH_2)_6-\underset{(CH_2-CH_2-O-)_mH}{\underset{|}{N}}-\overset{O}{\overset{\|}{C}}-(CH_2)_4-\overset{O}{\overset{\|}{C}}-\right]_n$$

To ensure better reactivity of the polymer, a sample of 6-6 nylon of low crystallinity should be used. Amorphous 6-6 can be prepared by heating it to 280°C in a polymer tube (in a vapor bath) in an atmosphere of nitrogen for a few minutes then quenching the tube in ice water. (**Caution**: use a shield in case the tube shatters.) The polymer should be ground to pass a 30-mesh screen. Then 10 g of the polymer is placed in a stainless-steel bomb and 50 to 60 g ethylene oxide is added to the bomb, which should be chilled to minimize loss of the volatile oxide when it is added. The bomb is sealed and heated in an 80°C thermostated oil bath or jacket behind a barricade for 40 h. Shorter times will result in less take-up of the oxide by the polymer. The bomb is cooled in ice, opened with caution, and allowed to warm to vent the excess ethylene oxide. The polymer remains as a swollen, rubbery mass, which is washed thoroughly with water (in a blender) several times, and then dried over phosphorus pentoxide at 50°C. in a vacuum. The increase in weight may be taken as the amount of ethylene oxide combined as poly(ethylene oxide) side chains. It should be about 50% in this case.

The polymer mp is about 220°C, and the inherent viscosity should be only slightly less than that of the starting polymer, as determined in *m*-cresol for both. Films cast from formic acid are more flexible than ordinary 6-6. The degree of polymerization of the ethylene oxide in the side chains is 5 to 7.

4.1.7. AB Polyamides

The previous polymerization examples were for AA-BB polyamides, but AB polymers are also commercially important. AB polyamides are usually prepared from cyclic lactams or from aminocarboxylic acids. The most important monomer commercially is caprolactam, which is the precursor of 6-nylon. Ends balance is not required when an amino acid undergoes polymerization, because in pure compound reacting-groups are equivalent in number.

14. Polycaprolactam

Caprolactam is readily purified by recrystallization from cyclohexane, followed by drying in a vacuum over P_2O_5 for 2 days or more at room temperature. The mp should be 68 to 69°C. The caprolactam can be polymerized as follows.

$$\text{caprolactam (ring: } H_2C{-}CH_2{-}CH_2{-}C({=}O){-}NH{-}CH_2{-}CH_2) \longrightarrow \left[NHCH_2CH_2CH_2CH_2CH_2\overset{O}{\overset{\|}{C}} \right]_n$$

To a 200-mL resin kettle equipped with an air motor-driven stainless-steel stirrer, a nitrogen inlet, a water cooled condenser, and a stoppered addition port, add 72 g (0.5 mol) caprolactam and 6.56 g (0.05 mol) ω-aminocaproic acid. The system is blanketed with nitrogen, and the temperature is raised to 150°C, using an external heating bath of heavy-duty silicon oil. Stirring is commenced, and the temperature raised to 250 to 260°C (bath temperature) over 30 min. Stirring is continued for 5 h, and a colorless highly viscous melt results, from which fibers can be drawn. The melt is discharged into an aluminum tray where it solidifies. Residues can be scraped from the resin kettle using a 1-in. broad spatula; take care to use heavy gloves to avoid burns. The polyamide melts about 215°C. It contains cyclic oligomers that can be removed by hot aqueous extraction. Inherent viscosity number in *m*-cresol at 0.5% concentration should be 0.4 to 0.6.

An alternative route to the polymerization of caprolactam involves heating the monomer in an autoclave at 250 to 270°C with water and 6-6-nylon salt or aminocaproic acid as initiators. The carboxyl and amine ends produced polymerize. Acetic acid or other end capper may be added to control molecular weight. The conversion reaches 90% with 10% monomer and cyclic oligomers in equilibrium with the polymer. The polymer may be extruded, quenched with water, and cut into chips. Monomer (8 to 9%) and cyclic oligomers (1.5%) are extracted with hot water, and the polymer is dried under vacuum at 60°C before further processing.

An example of an AB polyamide polymerization essentially free from cyclic oligomers is that of 11-nylon from polymerization of 11-aminoundecanoic acid. The monomer is prepared from undecylenic acid, which in turn is a derivative of castor oil. The polymer melts at 185 to 190°C, reflecting a lower degree of hydrogen bonding per unit chain length than that of 6-6 nylon. Solubility is restricted to strong acids and phenols.

15. Preparation of 11-Aminoundecanoic Acid (23)

$$CH_2{=}CH{-}(CH_2)_8{-}CO_2H + HBr \longrightarrow Br{-}(CH_2)_{10}{-}CO_2H$$

$$Br{-}(CH_2)_{10}{-}CO_2H + 2NH_3 \longrightarrow H_2N{-}(CH_2)_{10}{-}CO_2H + NH_4Br$$

A solution of 82 g (0.44 mol) undecylenic acid (mp 24 to 25°C) in 500 mL of olefin-free hexane is stirred and cooled to 0°C. The hexane solution of the acid should be well exposed to air in its preparation, and a trace of benzoyl peroxide should also be added. Hydrogen bromide gas is slowly passed into the reaction flask through an inlet tube that dips below the surface of the hexane. A trap should be provided for unabsorbed hydrogen bromide. When the hydrogen bromide is no longer being absorbed, the mixture is cooled in an ice–salt bath to −20°C, and the solid product is separated by filtration. Some hexane can be evaporated if needed to aid precipitation of the product if little separation occurs at −20°C. The crude 11-bromoacid is twice recrystallized from petroleum ether to give a white crystalline product that melts at 50°C.

A mixture of 100 g (0.38 mol) 11-bromoundecanoic acid, 300 g concentrated ammonium hydroxide, and 200 g ethyl alcohol are stirred together at 30°C for 4 days. The excess ammonia and the alcohol are distilled at reduced pressure on a steam bath, and the residue is taken up in 1500 mL boiling water, which is then cooled in ice. The resulting solid is filtered; washed with water; and if necessary, recrystallized once more from 1-L boiling water. The melting point of the product, after drying in a vacuum oven at 50°C for 5 h and over phosphorus pentoxide in a vacuum desiccator, is 170°C. The yield is 49 g (64%).

16. Preparation of Poly(11-undecanoamide) (23)

$$H_2N{-}(CH_2)_{10}{-}CO_2H \longrightarrow \left[{-}(CH_2)_{10}{-}\overset{\overset{\displaystyle O}{\|}}{C}{-}NH{-}\right]_n + H_2O$$

A total of 25 g purified 11-aminoundecanoic acid is added to a 200-mL three-necked flask equipped with a stainless-steel stirrer and shaft, Claisen head for the distillation of water, and a nitrogen inlet. Nitrogen is passed in to purge the air, and heating is begun by means of a metal or oil bath. The polymerization is heated to 220°C for 10 h while a current of nitrogen is passed through the flask. After raising the stirrer from the molten mass, the reaction is cooled under nitrogen, and the polymer is removed by breaking the flask. The polymerization may be run in a polymer tube using smaller quantities (10 to 15 g) of the amino acid. A nitrogen inlet capillary to the bottom of the tube provides agitation of the melt. The product is obtained in quantitative yield. The polymer melts at 185 to 190°C. The inherent viscosity is 0.6 to 0.7 in *m*-cresol at 0.5% concentraction. Films and fiber can be prepared from the melt.

4.1.8. Polyamides from Diamines and Diisocyanates

The addition of an isocyanate to a carboxylic acid first forms a carboxylic-carbamic anhydride that in most instances, decomposes directly into end products, which are either a mixture of the anhydride of the acid and a urea or a substituted carboxamide and carbon dioxide. Further heating of the anhydride-urea mixture forms the carboxamide and evolves carbon dioxide (24).

$$\mathrm{RNCO + R'COOH \longrightarrow RNH{-}\overset{O}{\overset{\|}{C}}{-}O{-}\overset{O}{\overset{\|}{C}}{-}R' \longrightarrow \tfrac{1}{2}RNHCONHR + \tfrac{1}{2}R'{-}\overset{O}{\overset{\|}{C}}{-}O{-}\overset{O}{\overset{\|}{C}}{-}R'}$$

$$\mathrm{or \longrightarrow R'{-}\overset{O}{\overset{\|}{C}}{-}NHR + CO_2}$$

17. Preparation of Poly(decamethylene sebacamide); 10,10 Polyamide (25)

$$\mathrm{OCN(CH_2)_{10}NCO + HO{-}\overset{O}{\overset{\|}{C}}{-}(CH_2)_8{-}\overset{O}{\overset{\|}{C}}{-}OH \longrightarrow \left[{-}HN{-}(CH_2)_{10}NH{-}\overset{O}{\overset{\|}{C}}{-}(CH_2)_8{-}\overset{O}{\overset{\|}{C}}{-}\right]_n + 2CO_2}$$

In a polymer tube purged of air and flushed with nitrogen and with a capillary inlet tube reaching nearly to the bottom is placed a mixture of 7.5 g sebacic acid and 8.41 g decamethylenediisocyanate (see below). The tube is heated at about 170°C by means of a vapor bath for 1 h at atmospheric pressure while a stream of nitrogen is passed slowly through the melt that forms. A solid eventually forms during this period. The temperature is raised to 222°C. (methyl salicylate) for 3 h. The polymer is then cooled under nitrogen. It has an inherent viscosity of 0.4 in *m*-cresol at 0.5% concentration and is soluble in the usual acidic aliphatic polyamide solvents (formic acid and phenols). The polymer melt temperature is about 185°C. Cold-drawable fiber may be drawn from the melt.

18. Preparation of Decamethylenediisocyanate (26)

$$ClH \cdot H_2N(CH_2)_{10}NH_2 \cdot HCl + COCl_2 \longrightarrow OCN - (CH_2)_{10} - NCO + 4HCl$$

A solution of 86 g (0.5 mol) decamethylenediamine in 1000 mL dry xylene is prepared in a 2-L three-necked flask equipped with a gas inlet tube extending nearly to the bottom of the flask and a condenser with an outlet to a trap or a scrubber for the off-gases of the reaction. Dry hydrogen chloride is passed through the diamine solution until no further precipitation occurs and the solution is saturated. The hydrogen source is replaced with a phosgene cylinder, and the solution is heated to reflux as a slow stream of phosgene is passed through it. (Reaction must be run in a good hood because phosgene is highly toxic.) The off-gases are hydrogen chloride and unreacted phosgene. When almost all the solid has dissolved (about 6 h) the reaction mixture is cooled and filtered. The xylene is distilled at water aspirator pressure through a 10-in. Vigreux column. The residue is then fractionated through the same column at oil pump pressure to give 70 to 75 g (62 to 67% yield) of decamethylenediisocyanate, boiling at 151 to 153°C at 3 mm. The product should be stored under nitrogen in tightly sealed flasks or screw-capped bottles, preferably in a dry box.

4.1.9. Low-Temperature Polymerizations

The preparation of polyamides by low-temperature reaction of diamines and acid chlorides is an important route to high melting polyamides, especially those that are unstable at their melt temperatures. An excellent review on this subject has been written (27).

There are two general methods: interfacial and solution. In contrast to the slow high-temperature polycondensation procedures previously discussed, these reactions proceed at high rates and are capable of giving high yields of polymer at ambient temperatures. The critical requirements for obtaining high molecular weight polymer by low-temperature polymerization are (1) maintaining the reactivity of the acid chloride, (2) selecting a reaction medium that keeps the polymer swollen or in solution, (3) using pure intermediates, (4) maintaining low temperatures (0 to 10°C) during addition of reactants and usually during polymerization, and (5) using a rapid rate of mixing during interfacial polymerization to maximize the interface. Interfacial polymerization involves reacting a diamine or diphenolate in water at the interphase with an acid chloride (or chloroformate) in a nonmiscible organic solvent. The condensation method is similar to the Schotten-Baumann reaction, used in the qualitative identification of organic compounds, and to the Hinsberg reaction of sulfonyl chlorides with primary and secondary amines.

An inorganic base, such as sodium hydroxide, is usually dissolved in the aqueous phase to scavenge the HCl produced by the reaction. With less basic or hindered diamines, weaker acid scavengers such as sodium bicarbonate are often helpful for achieving high molecular weights. The organic solvent need not dissolve the polymer, but should at least maintain it in a swollen state until high molecular weight is achieved. Polymerization takes place at the organic side of the interface and monofunctional intermediates limit molecular weight only when in the organic phase. For example, when one equivalent each of aniline and ammonia in the aqueous phase compete for one equivalent of benzoyl chloride in benzene, only benzanilide and ammonium chloride are formed, even though ammonia is much more basic than aniline and reacts several thousand times faster in homogeneous solution (4).

Interfacial polymerization is highly successful with strongly basic diamines that react fast and provide polymers that are swollen by the organic solvent. Maximum partition of the diamine into the organic phase is desirable for high molecular weights. Similarly, higher molecular weight polymers are obtained from acid chlorides that are least easily hydrolyzed under typical polymerization conditions. With many aliphatic acid chlorides, increasing phase miscibility or partition of the acid chloride into the aqueous phase by use of water-soluble solvents yields lower molecular weight polymer because the rate of hydrolysis of the acid chloride is increased. Less basic diamines and polymers that are not swollen by the organic solvents used for interfacial polymerization usually do not usually yield high molecular weight polymer.

The original workers tackled this problem by two sequential and logical approaches. First, they used an organic phase (e.g., tetrahydrofuran, cyclohexanone, tetramethylene sulfone) that was soluble in or miscible with the aqueous phase but that also swelled or dissolved the polymer and maintained the proximity of the polymer ends for continued polymerization. The expected higher rate of hydrolysis of the acid chloride was counterbalanced by using weak acid acceptors, such as sodium bicarbonate, in the aqueous phase.

A second and better approach was to eliminate the aqueous phase entirely and carry out the polymerization at low temperatures (0 to 10°C) in an organic medium that was a solvent or good swelling agent for the polymer. This required using acid acceptors, such as triethylamine, that were soluble in the organic phase. This method, termed solution polymerization, is highly suitable for diamines of low basicity, such as aromatic diamines, and for acid chlorides that are easily hydrolyzed. Interestingly, Sweeny (3) noted that the trialkyamine hydrochloride formed enhanced the potency of the organic solvent and that higher molecular weight polymers often result when the diamine hydrochloride is used, instead of the free diamine. This generated the tertiary amine hydrochloride in situ (3).

The best approach to solution polymerization, however, was discovered by Sorenson (28). The method involves polymerizing the acid chloride and diamine in a dry, unreactive medium that is both a polymer solvent and an acid acceptor. Typical effective solvents are weak bases such as dimethylacetamide (but not dimethylformamide) and *N*-methylpyrrolidinone. In initial experiments, dry calcium oxide was added to scavenge the halo-acid formed in the reaction. Later, it was found that the calcium chloride formed enhanced the solvent power of the amide solvent just as triethylamine hydrochloride (organic chloride salt) had done for polymerizations in chlorinated hydrocarbons.

Optimum solvents systems now consist of saturated solutions of dry lithium or calcium chlorides in a dry amide solvent. Trialkylamine hydrohalides also enhance the potency of the amide solvents. This method does not work well when using strongly basic aliphatic diamines that are somewhat inactivated by the hydrohalide formed, unless a stronger base is added as acid acceptor.

With the weak bases, solvent and the diamine compete for the hydrogen halide, but solvent hydrohalide is favored because there are so many more moles of solvent (law of mass action) initially and even more as the polymerization proceeds and depletes the diamine concentration. This last polymerization method has been used mostly for preparing all-aromatic polyamides. Examples are given in this chapter and in chapter 6.

4.1.9.A. Low-Temperature Polymerization Preparations

Although interfacial polymerization is rapid, proper solvent system and reaction concentration are essential to ensure sufficient mobility of the growing chain ends to permit high molecular weight to be reached. The great utility of this system is that it is a low-temperature process. Thus polymers that are not stable at their melt temperatures can easily be prepared, and thermally unstable reactant can be used. Little in the way of special equipment is needed, and a household blender provides the correct amount of shear stirring. A variety of polymer types can be made: polyamides, polyphenyl esters, polysulfonamides, and polyurethanes. Substantially different reactivities of the different acid chlorides with the other component may lead to the formation of nonrandom, ordered copolymers (29).

Probably the simplest and most dramatic example of interfacial polymerization is the preparation of 6-6 or 6-10 nylon by the beaker method (30). This process (aptly called the nylon rope trick) consists in carefully pouring an aqueous solution of excess hexamethylenediamine onto a carbon tetrachloride solution of adipoyl or sebacoyl chloride, then steadily pulling away the coherent film of polyamide that forms at the interface. No mixing is required, and the diamine can be used in excess to function as the acid acceptor. With pure intermediates high molecular weights are obtained (inherent viscosity of 1.8 in cresol at 0.5% concentration).

19. Preparation of Poly(decamethylene sebacamide); 6-10 Nylon (31)

$$H_2N-(CH_2)_6-NH_2 + Cl-\overset{\overset{\displaystyle O}{\|}}{C}-(CH_2)_8-\overset{\overset{\displaystyle O}{\|}}{C}-Cl \longrightarrow \left[-HN-(CH_2)_6-NH-\overset{\overset{\displaystyle O}{\|}}{C}-(CH_2)_8-\overset{\overset{\displaystyle O}{\|}}{C}-\right]$$

A solution of 3 mL sebacoyl chloride in a 100-mL distilled tetrachloroethylene is placed in a 200-mL tall form beaker. (The diacid chloride should be distilled material for best results. The bp is 124°C/0.5 mm.; the pot temperature should not exceed 160°C, and the distillation should be as rapid as possible.) A solution of 4.4 g hexamethylenediamine in 50 mL water is carefully poured over the acid solution. (The diamine is handled most conveniently as a standardized stock solution of about 20% in water. The commercially available solid diamine may also be used without further purification.) The polymeric film that forms at the interface of the two solutions is grasped with tweezers and raised from the beaker as a continually forming rope. If a mechanical windup device is placed above the beaker, the polymer may be wound up continuously until one of the reactants is exhausted. The polymer can be washed several times with 50% aqueous ethanol or acetone and dried in a vacuum oven at 60°C.

The inherent viscosity is 0.4 to 1.8 in cresol at 0.5% concentration and the mp is about 215°C. Extremely high melting polyamides can be prepared from short-chain primary allphatic diamines (two to six carbon atoms) and terephthaloyl chloride (32). These polymers require dilute conditions for best preparation. Because of their insolubility, they precipitate rapidly during polymerization and are unswollen by the solvents used. Consequently, there is a tendency for low molecular weight products to form. The polyterephthalamide from ethylenediamine, whose preparation follows, is soluble in none of the usual polyamide solvents (formic acid, cresol) but only in strong acids, such as sulfuric and trifluoroacetic acids.

20. Preparation of Poly(ethylene terephthalamide) (31)

$$H_2NCH_2CH_2NH_2 + Cl-\overset{O}{\overset{\|}{C}}-C_6H_4-\overset{O}{\overset{\|}{C}}-Cl \longrightarrow \left[-HNCH_2CH_2NH-\overset{O}{\overset{\|}{C}}-C_6H_4-\overset{O}{\overset{\|}{C}}-\right]_n + 2HCl$$

For this polymerization, ethylenediamine is purified by drying over potassium hydroxide pellets for 15 h and then fractionally distilling through a 10-in. helix-packed column, taking care to protect the distillate from atmospheric carbon dioxide by means of soda lime–packed drying tubes. Ethylenediamine boils at 117°C/760 mm. Terephthaloyl chloride can be purchased and purified by distilling through a short Vigreux column under vacuum. The bp is 115 to 116°C/3.0 mm. The solid distillate is then recrystallized from hexane (100 g/700 mL). The mp is 81 to 82°C.

A solution of 3.78 g (0.0630 mol) ethylenediamine and 0.126 mol potassium hydroxide from a standardized solution in 4 L water is placed in an 8-L. stainless-steel beaker and stirred by means of an efficient high-speed stirrer. A large spatula is mounted vertically at the edge of the beaker with the blade perpendicular to the wall to act as a baffle for more efficient mixing. Then a solution of 12.79 g (0.0634 mol) terephthaloyl chloride in 1 L methylene chloride is added rapidly to the stirred solution. The polymerization mixture is stirred for 10 min at room temperature. The mixture is then filtered and the polymer is placed in boiling distilled water to remove absorbed methylene chloride. The product is again filtered and washed twice with boiling distilled water. Finally the polymer is dried in a vacuum oven at 80°C. The yield is 9.0 g (75%). The inherent viscosity is about 1.0 in sulfuric acid at 0.5% concentration; this corresponds to a weight average molecular weight of about 18,000. The polymer melts over 400°C; the first DTA transition is at 455°C, which may represent the melt temperature.

The polymer is also soluble in trifluoroacetic acid, and films and fibers can be fabricated from such solutions.

Polyamidation between an acid chloride and a diamine may also be carried out in a single, organic phase in the presence of an organic acid acceptor. It is not necessary that the polymer be soluble in the organic phase; however, it is more convenient if both the polymer and the salt of the acid acceptor remain in solution. The system chloroform-triethylamine is often used.

21. Preparation of Poly(2,5-dimethylpiperazineterephthalamide) Using Triethylamine as an Acid Acceptor (33)

$$ClOC-C_6H_4-COCl + HN(CH(CH_3)CH_2)_2NH \longrightarrow \left[-\overset{O}{\overset{\|}{C}}-C_6H_4-\overset{O}{\overset{\|}{C}}-N(CH(CH_3)CH_2)_2N-\right]_n + HCl$$

Dissolve 2.28 g trans-2,5-dimethylpiperazine and 5.6 mL pure triethylamine in 100 mL of water-washed and sodium sulfate–dried chloroform in a 500-mL Ehrlenmeyer flask. To this mixture, add a solution of 4.06 g terephthaloyl chloride in 80 mL chloroform with swirling. More chloroform (20 mL) is used to wash in the residues of the terephthaloyl chloride. The mixture remains clear, but the temperature rises quickly from 25 to 42°C, and there is an increase in solution viscosity. The solution can be dry cast on glass to produce self-supporting films; however, it must be washed free of salt with water and be solvent free if a tough, flexible product is desired. After 5 min, the solution is coagulated by pouring into hexane with stirring. A fibrous precipitate is obtained, along with crystals of triethylamine hydrochloride. If the precipitation is done slowly with no stirring or only slow stirring, long coarse strings of polymer are obtained, which can readily be torn apart by hand or in a blender. The precipitate is washed well with water and finally with acetone. After drying at 100°C, a 92% yield of polymer is obtained, which has an inherent viscosity of about 3 in *m*-cresol at 0.5% concentration.

Coloration in polymers is usually a sign that degradation has taken place, but sometimes an inherently colored product can be made by inclusion of a chromophoric group in one of the monomers. An example is the azo group present in a fiber-forming polyamide as described below.

22. Preparation of Azodibenzoyl Chloride (34)

$$HO_2C-C_6H_4-N=N-C_6H_4-CO_2H \xrightarrow{SOCl_2} Cl-\overset{O}{\overset{\|}{C}}-C_6H_4-N=N-C_6H_4-\overset{O}{\overset{\|}{C}}-Cl$$

Charge 33 g (0.122 mol) azodibenzoic acid, 150 mL (248 g; 2.08 mol) thionyl chloride, and 0.5 mL triethylamine or dimethylformamide into a 500-mL round-bottom flask equipped with a stirrer and a reflux condenser with a drying tube. Reflux reaction mixture until a dark red solution is obtained, about 4 h. (The off-gas hydrogen chloride must be trapped or scrubbed in a water trap) Excess thionyl chloride is removed using a water pump at a temperature <50°C. Further drying at 1 mm and <50°C gives 37 g (94%) of azodibenzoyl chloride. The acid chloride thus obtained is sufficiently pure for polymerization but can be further purified by recrystallization from dry hexane.

23. Preparation of Poly(azo-4,4′-dibenzoyl-trans-2,5-dimethylpiperazine) (34)

In a household blender jar is placed 4.10 g (0.036 mol) trans-2,5-dimethylpiperazine and 6.31 g (0.06 mol) sodium carbonate and 250 mL distilled water. After brief stirring to dissolve the reactants, 75 mL methylene chloride is added. (Trans-2,5-dimethylpiperazine can be recrystallized from acetone (1 g/mL); mp 117 to 118°C. The mixture is stirred rapidly, then a solution of 9.21 g (0.03 mol) azodibenzoyl chloride in 150 mL methylene chloride (slightly turbid solution) is added to the preformed emulsion in 1 to 2 s, and stirring is continued for 7 to 8 min. Then 300 mL hexane is now added to precipitate the polymer, and stirring is stopped. The polymer is separated by filtration, washed in a blender three times with 200-mL portions of water, and dried at 70°C in a vacuum oven. The yield is 8.6 g (83%) of a bright orange polymer, and the inherent viscosity is about 2.2 in *m*-cresol at 0.5% concentration. It is soluble in 98% formic acid, chloroform, and a 1,1,2-trichloroethane/formic acid (60/40 w/w) azeotrope. A tough, transparent orange film can be cast from a 10% solution in the azeotrope. The film exhibited no color change after 1300 h exposure in a Fade-O-Meter instrument (Atlas Electric Devices Co., Chicago) and is still quite tough.

The effect of *N*-alkyl substitution in polyamides with concurrent loss of hydrogen bonding is demonstrated in preparation 24 with the completely *N*-ethylated polymer (35). The polymer melt temperature is reduced more than 200°C, and the solubility is raised to such a degree that 80% aqueous ethanol becomes a solvent.

24. Preparation of N,N′-diethylethylenediamine (32)

$$ClCH_2CH_2Cl + 2CH_3CH_2NH_2 \xrightarrow{NaOH} CH_3CH_2NHCH_2CH_2NHCH_2CH_3$$

A mixture of 148.5 g (1.5 mol) ethylenedichloride and 450 g (10 mol) ethylamine is heated in a stainless-steel bomb at 100°C for 4 h. The bomb is then allowed to cool, excess ethylamine vented, and the mixture transferred by rinsing with 500 mL water to a 2-L. separator funnel. About 500 g solid potassium hydroxide is added slowly. The diamine separates as an upper oily layer; 20 mL methanol is added to facilitate the separation of the layers. The upper layer is separated and dried over potassium hydroxide pellets overnight. It is fractionated at atmospheric pressure through a precision column to give a forerun boiling at 52 to 138°C, which is discarded, followed by a main fraction, which is the diamine, boiling at 149 to 151°C. The diamine is hygroscopic and must be protected from moisture and carbon dioxide in the air. The yield is 58 g (33%). Also obtained from the distillation is 10 to 20 g *N,N′,N″*-triethylenetriamine; bp 82°C/3 mm.

25. Preparation of Poly(N,N′-diethylethylene terephthalamide) (32)

$$C_2H_5NHCH_2CH_2NHC_2H_5 + Cl-\overset{O}{\overset{\|}{C}}-C_6H_4-\overset{O}{\overset{\|}{C}}-Cl \longrightarrow$$

$$\left[-\underset{}{\overset{C_2H_5}{\overset{|}{N}}}-CH_2CH_2-\underset{C_2H_5}{\underset{|}{N}}-\overset{O}{\overset{\|}{C}}-C_6H_4-\overset{O}{\overset{\|}{C}}-\right]_n + 2HCl$$

A solution of 5.8 g (0.05 mol) of *N,N′*-diethylethylenediamine and 10.6 g (0.1 mol) sodium carbonate in 250 mL water is stirred rapidly in a household blender. A solution of 10.1 g (0.05 mol) terephthaloyl chloride in 80 mL dry, washed chloroform is added all at once; the resultant mixture is stirred for 10 min at high speed. The product, consisting of two clear liquids, is heated in a distilling flask on a steambath to distill and collect the chloroform (toxic) and precipitate the polymer. The polymer/water mixture is filtered, and the solid is washed four times in in a blender with water to give, on drying, 10.1 g (82%) of white polymer with an inherent viscosity of 1.93 in sulfuric acid at 0.5% concentration. The product has a mp of 230°C and is soluble in *m*-cresol, formic acid, chloroform, and 80% aqueous ethanol.

Interfacial polymerization and other low-temperature polymerizations using chlorinated hydrocarbons, although useful for screening a variety of polymer types, have not been successfully adopted for the commercialization of any polymer. For all-aromatic polyamides, such as poly(*m*-phenylene-isophthalamide) and poly(*p*-phenyleneterephthalamide), the method of choice (28) is using an amide solvent, such as NMP or DMAc as both solvent and acid acceptor. The solvent power of these solvents can be boosted significantly by saturating with anhydrous calcium or lithium halides (preferably chlorides). DuPont commercialized Nomex and Kevlar by this route. These polymerizations are described and discussed more fully in chapter 7.

The two polymers referred to above are both highly crystalline species and are somewhat limited in solubility. However, other all-aromatic polyamides of higher solubility can be made in a modified interfacial polymerization using a partially water-soluble solvent such as tetrahydrofuran as an organic phase. The preparation 26 demonstrates such a polymerization.

26. Preparation of Poly(4,4′-diphenylmethane isophthalamide) (31, 36)

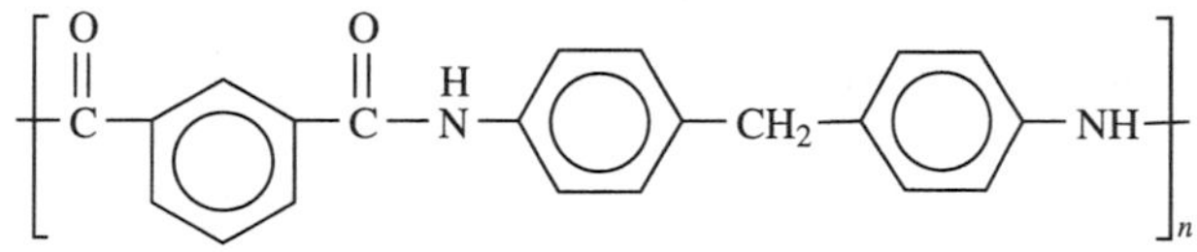

A mixture of 5.95 g (0.03 mol) bis(4-aminophenyl)methane and 6.36 g (0.06 mol) sodium carbonate is dissolved in 150 mL of distilled water and 125 mL tetrahydrofuran. It is then placed in a home blender and stirring is started. A solution of 6.09 g (0.03 mol) isophthaloyl chloride in 50 mL tetrahydrofuran is added from a beaker all at once to the blender with moderate to rapid stirring. Stirring is continued for 10 min; the polymer is then filtered and washed three times in the blender with 200-mL portions of distilled water and dried at 70°C in a vacuum oven. The yield is 9.5 g (97%), and the inherent viscosity in 98% sulfuric acid is 1.86.

27. Poly(m-phenyleneisophthalamide) (31, 36)

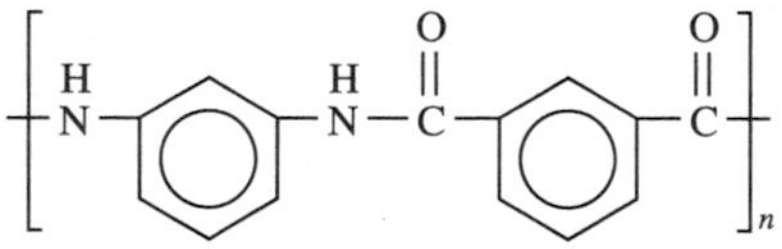

A solution of 4.326 g (0.04 mol) *m*-phenylenediamine and 8.48 g (0.08 mol) sodium carbonate in 120 mL water is placed in a home blender. Vigorous stirring is begun and a solution of 8.12 g (0.04 mol) isophthaloyl chloride in 150 mL 2,4-dimethyltetramethylene sulfone (purified by fractional distillation) is rapidly poured into the blender. The acid chloride solution is rinsed in with an additional 10 mL DMS, and stirring is continued for 5 min. A metastable solution is obtained in the form of an emulsion. The polymer is precipitated by adding 350 mL water and then stirring for another minute. The polymer is filtered and washed three times with water (350 mL) in the blender. After drying at 80 to 90°C under vacuum, the polymer yield is 9.53 g (100%). The inherent viscosity is 2.42 in sulfuric acid at 0.5% concentration.

28. Preparation of Poly(2,2-bis[4-aminophenyl]propaneisophthalamide) (28)

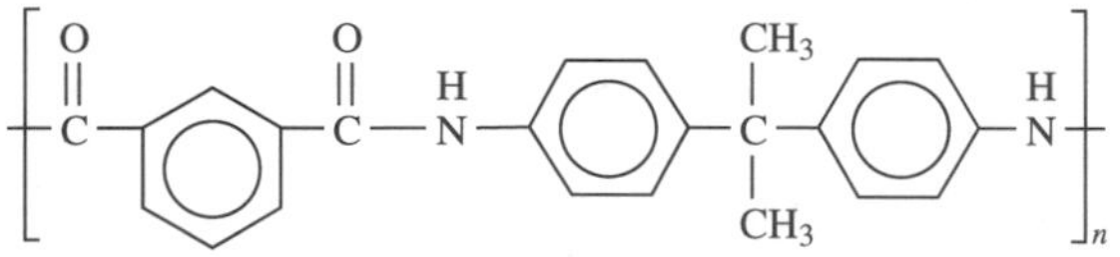

In a flask equipped with a stirrer, nitrogen inlet, and drying tube, 22.63 g (0.1 mol) 2,2 bis(4-aminophenyl) propane, is dissolved in 200 mL anhydrous dimethylacetamide. The solution is cooled to a mush a dry ice/acetone bath. Then 20.3 g (0.10 mol), isophthaloyl chloride dissolved in 18.5 mL toluene is added in one portion. The cooling bath is changed to one of ice/water, and the mixture is stirred for 90 min. During this time, a viscous solution of polymer forms that contained dispersed crystals of dimethylacetamide hydrochloride. The polymer is precipitated by pouring the mixture into rapidly stirred water in the blender. After thorough washing, the polymer is dried at 100°C under vacuum. The yield was 98%, and the inherent viscosity is 1.66 in sulfuric acid at 0.5% concentration.

4.1.9.B. Polysulfonamides by Low Temperature Polymerization

Polysulfonamides can be prepared from aromatic disulfonyl chlorides and aliphatic diamines by interfacial polycondensation (37). Aliphatic disulfonyl chlorides give poorer results. Polysulfonamides with hydrogens on the amide nitrogen are in many cases soluble in strong alkali, just as their monomeric counterparts. With sodium hydroxide as the acid acceptor in the polymerization, branching and cross-linking can result from reaction of sulfonyl chloride with the sulfonamide anion, although conditions can be chosen to minimize these effects.

29. Preparation of m-Benzenedisulfonyl Chloride (37)

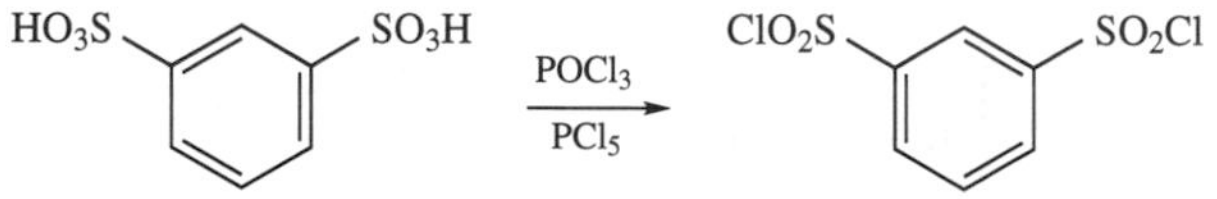

HO_3S–C_6H_4–SO_3H $\xrightarrow[PCl_5]{POCl_3}$ ClO_2S–C_6H_4–SO_2Cl

In a 3-L three-necked flask fitted with condenser, thermometer, and stirrer is placed 1360 g (6.55 mol) ground phosphorus pentachloride and 727 g phosphorus oxychloride. To the stirred mixture is added 770 g (2.94 mol) *m*-benzenedisulfonic acid (90%) over 30 min. The temperature is not allowed to exceed 70°C. The mixture is refluxed for 3 h, then phosphorus oxychloride is distilled first at atmospheric pressure and later under vacuum until approximately 700 mL is collected. The dark liquid reaction mixture is poured with stirring into a 5-L beaker two thirds full of cracked ice. The cold mixture is stirred about 30 min and filtered. The solid is dissolved in 1-L benzene and washed three times with 250-mL portions of 5% sodium bicarbonate solution and once with 250 mL water. After drying over anhydrous calcium sulfate and treating with decolorizing carbon, the filtrate is passed with suction through a 1.5-in-diameter column packed with 15 in activated alumina to remove the last traces of charcoal and to thoroughly dry the solution. The column is washed with 200 mL benzene. To the solution is added 2500 mL olefin-free *n*-hexane, and the oil that separates is cooled to 20°C. and scratched to effect crystallization. The solid is filtered on a large Buchner funnel and washed twice with 500-mL portions of olefin-free *n*-hexane to remove a slight yellow color. After drying in a desiccator over calcium chloride for 3 h under vacuum pump–reduced pressure, the solid weighs 553 g (68%). It melts at 61 to 61.5°C. The filtrate is concentrated to 350 mL and 500 mL hexane is added to give an oil that on seeding gives, after drying, 90 g crystalline material, melting at 61 to 62°C. The total yield is 643 g (80%).

30. Preparation of Poly(hexamethylene-meta-benzenedisulfonamide) (37)

$$H_2N-(CH_2)_6-NH_2 \quad + \quad ClO_2S-C_6H_4-SO_2Cl \longrightarrow$$

$$\left[HN-(CH_2)_6-NH-O_2S-C_6H_4-SO_2 \right]_n + 2HCl$$

A household blender jar is charged with 145 mL distilled water, 20 mL of 10% aqueous Duponol ME surface-active solution, 3.016 g hexamethylenediamine (or enough standardized aqueous diamine to give 0.026 mol), and 5.30 g (0.05 mol) sodium carbonate. To the stirred solution over a period of 20 to 30 s is added 6.88 g (0.025 mol) *m*-benzenedisulfonyl chloride dissolved in 200 mL methylene chloride. The mixture is then stirred 15 min. After adding 100 mL ethanol, the solid is filtered; washed on the funnel with 400 mL water; and then washed in the blender with 200-mL portions of ethanol, acetone, hot water, and acetone again. (**Caution!** If the blender motor has not been modified with a compressed air inlet on the motor housing for flame protection, do not use the flammable solvent washes, instead wash the polymer in a beaker.) The sample is dried at 70 to 75°C in a vacuum oven overnight to give 4.9 g polymer (63%). The inherent viscosity is about 2.0 in sulfuric acid and 1.5 in dimethylformamide at 0.5% concentration. The polymer mp is 160 to 170°C when amorphous and about 200°C when crystalline. The polymer is fairly crystalline as prepared, but melt-pressed films are amorphous. The polymer is also soluble in 6 to 10% sodium hydroxide solutions.

4.2. POLYUREAS

Polyureas are often considered a separate class of polymers; but they are, of course, amides of carbonic acid. Polyureas are generally higher melting than polyamides with a similar amount of separation between functional groups. For example, poly(hexamethylene urea) has a polymer melt temperature of about 295°C, whereas that of poly(hexamethyleneadipamide) is 265°C. The greater extent of hydrogen bonding and higher polarity associated with the urea group are thought to account for the difference.

4.2.1. Synthetic Methods

Polyureas can sometimes be made in melt systems, but low thermal stability limits them to relatively lower melting (<250°C.) polymers. Melt preparation can be accomplished by one these methods:

1. Reaction of a diamine with urea to eliminate ammonia.

$$H_2N{-}R{-}NH_2 + H_2N{-}\overset{\overset{\displaystyle O}{\|}}{C}{-}NH_2 \longrightarrow \left[R{-}NH{-}\overset{\overset{\displaystyle O}{\|}}{C}{-}NH \right]_n + 2NH_3$$

2. Ammonolysis of a bisurethane with elimination of an alcohol.

$$H_2N{-}R{-}NH_2 + EtO\overset{\overset{\displaystyle O}{\|}}{C}{-}NH{-}R{-}NH{-}\overset{\overset{\displaystyle O}{\|}}{C}OEt \longrightarrow \left[R{-}NH{-}\overset{\overset{\displaystyle O}{\|}}{C}{-}NH \right]_n + 2EtOH$$

3. Reaction of aliphatic diamines with carbon oxysulfide (Van der Kerk method (38)).

$$H_2{-}N{-}R{-}NH_2 + S{=}C{=}O \longrightarrow \left[R{-}NH{-}\overset{\overset{\displaystyle O}{\|}}{C}{-}NH \right]_n + H_2S$$

The mechanism of the Van der Kerk polymerization involves the formation first of low polymer during a mild heating cycle (110°C) with the loss of some COS and diamine as well as H_2S. This is followed by a higher temperature cycle to form high polymer.

In general however, high-melting polyureas are made by solution and interfacial methods. If the polymer formed is soluble at room temperature in the reaction medium, no heating may be necessary to complete the reaction, and a viscous polymer solution will result.

Reaction of a disocyanate with a diamine is the most direct and easiest method for achieving high molecular weight polyureas. The best solvents for the high-melting aromatic polyureas are amide solvents such as *N*-methylpyrrolidinone and dimethylacetamide, often containing anhydrous lithium or calcium chlorides. For aliphatic polyureas, the preferred solvent is often a cresol, in which they are quite soluble. The much higher rate of reaction of amines vs. phenols with the isocyanate ensures high molecular weight polymer.

Certain polyureas may be prepared in dimethyl sulfoxide as the solvent using the reaction of a diisocyanate with a monocarboxylic acid. The solvent participates in the reaction, presumably with the mixed anhydride (39) formed from the addition of the acid to the isocyanate group. As has was shown for the monofunctional reactants, the stoichiometry of the reaction is 2 : 1 : 1 for isocyanate : acid : sulfoxide (39). The reaction is considerably different from the 1 : 1 reaction of an acid and an isocyanate, alone or in an inert solvent, that leads ultimately to amide formation (see prep. 24) (24).

Polyureas can also be prepared by interfacial polymerization using phosgene as the diacid chloride. The phosgene must be used in close to stoichiometric amounts, otherwise low molecular weight polymer results. Following the polyamide numeral nomenclature, polyureas are coded n-1 polyamides; where n is the number of carbon atoms in the diamine and 1 is carbonic acid, the one-carbon diacid.

31. Preparation of Bis(γ-aminopropyl) Ether (40)

$$CH_2{=}CHCN + HOCH_2CH_2CN \longrightarrow NCCH_2CH_2OCH_2CH_2CN$$

$$NCCH_2CH_2OCH_2CH_2CN \xrightarrow{(H)} H_2NCH_2CH_2CH_2OCH_2CH_2CH_2NH_2$$

31.A. Bis(β-cyanoethyl) Ether (31)

To a stirred mixture of 177 g (2.5 mol) ethylene cyanhydrin and 6 g of a 20% potassium hydroxide solution in a 1-L three-necked flask equipped with stirrer, dropping funnel, and condenser is added dropwise 132 g (2.5 mol) acrylonitrile over 2.75 h. (**Caution!** Run reaction in a hood.) The reaction is maintained at 40°C during addition. When the addition is complete, the reaction is stirred for a further 18 h at room temperature. It is neutralized with dilute hydrochloric acid and evaporated to dryness under water aspirator pressure (30 mm) on the steam bath. The residue is fractionally distilled to yield about 226 g (80%) bis(β-cyanoethyl) ether, boiling at 159 to 162°C/5 mm.

31.B. Bis(γ-aminopropyl) Ether (31)

To a solution of 86 g (0.69 mol) bis(β-cyanoethyl)ether in 340 mL methanol containing 100 g anhydrous ammonia in a hydrogenation bomb is added 100 g Raney nickel catalyst (41). The dinitrile is hydrogenated at 1500 psi at 100 to 110°C. Hydrogen uptake should be complete in about 30 min. Prolonged heating of the reaction mixture reduces the yield of the desired product by hydrogenolysis of the ether linkage. The solution is filtered free of catalyst, and the methanol and ammonia are removed by distillation on the steam bath at atmospheric pressure. The residue is fractionally distilled through a precision column to give bis(γ-aminopropyl) ether; bp is 72 to 73°C/3 mm. The yield is about 60 g (65%). The possible by product hydrogenolysis γ-aminopropanol boils at 60°C/3 mm, and has the same refractive index ($n^d 1 = 1\ 1.4605$).

32. Preparation of Poly(4-oxaheptamethyleneurea) (42)

$$H_2N(CH_2)_3-O-(CH_2)_3-NH_2 + H_2N-\overset{\overset{\displaystyle O}{\|}}{C}-NH_2 \longrightarrow -\!\!\left[(CH_2)_3-O-(CH_2)_3-NH-\overset{\overset{\displaystyle O}{\|}}{C}-NH\right]_n\!\!- + 2NH_3$$

In a polymer tube with a side arm is placed a mixture of 7.5 g (0.125 mol) urea and 16.5 g (0.125 mol) bis(γ-aminopropyl) ether. Nitrogen is slowly passed through a capillary reaching to the bottom of the tube, and the temperature is raised to 156°C for 1 h, by means of a cyclohexanone vapor bath, during which time ammonia is evolved. The temperature is raised to 231°C in a vapor bath for another 1 h and finally to 255°C for 1 h. During the last 20 min of this part of the heating, an oil pump is cautiously applied; frothing may be serious if the vacuum is applied suddenly. The polymer tube is cooled under nitrogen, the vacuum is released with nitrogen, and the polymer is removed by breaking the tube. The inherent viscosity is about 0.6 in *m*-cresol at 0.5% concentration, and the polymer mp is around 190°C. Strong films may be pressed at or near this temperature.

Polyureas are best prepared in solution by reaction of a diisocyanate and a diamine. By taking advantage of the greater reactivity of isocyanate with amines over phenols and alcohols, the polymerization may be conducted in hydroxylic solvents for the polymer (26). Polydecamethyleneurea can be prepared by a diisocyanate-diamine reaction carried out in *m*-cresol.

33. Preparation of Poly(decamethyleneurea) (26)

$$H_2N-(CH_2)_{10}-NH_2 + OCN-(CH_2)_{10}-NCO \longrightarrow -\!\!\left[(CH_2)_{10}NH-\overset{\overset{\displaystyle O}{\|}}{C}-NH\right]_n\!\!-$$

See preparation 18 for decamethylenediisocyanate. In a 200-ml three-necked flask that has been flushed with nitrogen and equipped with a stirrer, dropping funnel, and condenser (the latter two protected with drying tubes) is placed a solution of 19.0 g (0.11 mol) freshly distilled decamethylenediamine (preparation 10) in 39 ml distilled *m*-cresol. With stirring, 24.8 g (0.11 mol) decamethylenediisocyanate is added over a 10-min period. Much heat is evolved, and a precipitate forms. The dropping funnel is washed with 10 mL *m*-cresol, and the temperature is raised to 218°C for a period of 5 h. The original precipitate dissolves, and the solution becomes viscous. The solution is then permitted to cool and is poured into 1500 mL methanol with vigorous stirring. The polymer, which separates as a white solid, is filtered and washed several times by stirring with ethanol in a household blender. The yield of polymer, after drying at 60°C in a vacuum oven for 15 h, is 38 to 40 g (90 to 95%). The inherent viscosity in *m*-cresol is about 0.3 at 0.5% concentration. Despite this low value, films may be melt pressed and drawable fibers can be melt spun.

Very high melting, soluble polyureas have also been prepared from all-aromatic reactants (43, 44) by a solution method similar to that of preparation 33. If the polymer formed is soluble at room temperature in the reaction medium, no heating may be necessary, and a viscous polymer may result. The following polymerization exemplifies this.

34. Preparation of a Polyurea from Trans-2,5-dimethylpiperazine and Methylene bis(4-phenylisocyanate) (44)

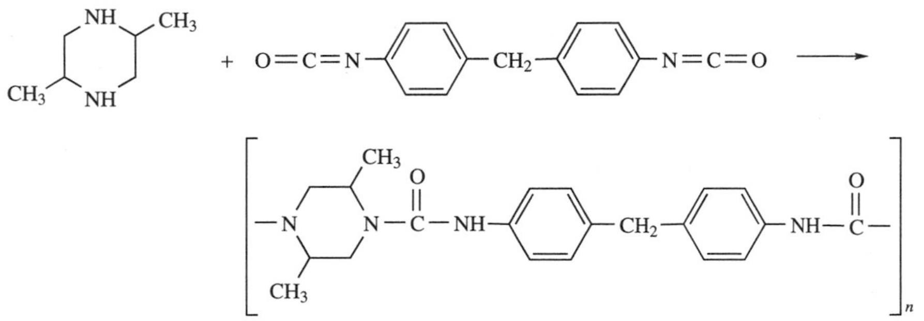

$$\text{trans-2,5-dimethylpiperazine} + O{=}C{=}N{-}C_6H_4{-}CH_2{-}C_6H_4{-}N{=}C{=}O \longrightarrow \left[-N\langle C_4H_6(CH_3)_2 \rangle N{-}\overset{O}{\overset{\|}{C}}{-}NH{-}C_6H_4{-}CH_2{-}C_6H_4{-}NH{-}\overset{O}{\overset{\|}{C}}{-} \right]_n$$

A solution of 3.253 g (0.013 mol) methylenebis(4-phenyl-isocyanate) in 80 mL tetramethylene sulfone/chloroform (70/30 v/v) mixture is placed in a blender jar. A second solution of 1.484 g of trans-2,5-dimethylpiparazine in 80 ml of the same solvent mixture is added rapidly with vigorous stirring. Stirring is continued at high speed for 7 min, and the reaction is stopped by addition of 100 ml of 4 vol % aqueous *n*-butylamine. More water is added to precipitate the polymer, which is then washed once with 50% aqueous acetone and repeatedly with water. The polymer is dried at 80°C under vacuum. The yield is quantitative, and the inherent viscosity is about 3.0 in sulfuric acid at 0.5% concentration. Polyureas can also be prepared by aminolysis of a biurethane, with elimination of an alcohol. Carbonate esters are also reported to form polyureas by reaction with diamines, but the reaction is difficult to control.

35. Preparation of Poly(hexamethylene-decamethyleneurea) Copolymer (45)

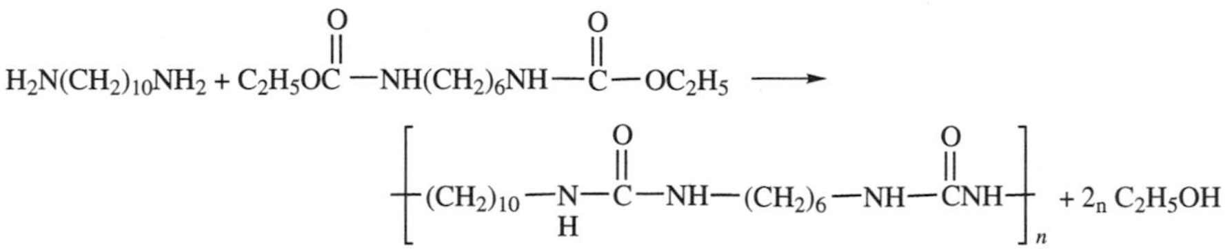

$$H_2N(CH_2)_{10}NH_2 + C_2H_5O\overset{O}{\overset{\|}{C}}{-}NH(CH_2)_6NH{-}\overset{O}{\overset{\|}{C}}{-}OC_2H_5 \longrightarrow$$

$$\left[(CH_2)_{10}{-}\underset{H}{N}{-}\overset{O}{\overset{\|}{C}}{-}NH{-}(CH_2)_6{-}NH{-}\overset{O}{\overset{\|}{C}}NH \right]_n + 2_n\, C_2H_5OH$$

Hexamethylene bis(ethylurethane) is prepared (46) by the simultaneous addition from separate funnels of 130 g (1.2 mol) ethyl chlorocarbonate and 48 g (1.2 mol) sodium hydroxide in 400 mL water to a rapidly stirred solution of 58 g (0.5 mol) hexamethylenediamine in 200 mL ether cooled in an ice bath and maintained at 10°C or less during the addition. The reaction is stirred for 15 min after the addition, and the solid is filtered. It is recrystallized from benzene-petroleum ether. The mp is 84°C.

A mixture of 12.37 g (0.072 mol) decamethylenediamine and 18.70 g (0.072 mol) hexamethylene bis(ethylurethane) is placed in a polymer tube, which is then purged with nitrogen and heated to 202°C (*m*-cresol bath) for 3 h at atmospheric pressure while a slow stream of nitrogen is passed through the melt. The polymer is cooled under nitrogen and is obtained as a tough plug. It has an inherent viscosity of 0.2 to 0.4 in *m*-cresol and a polymer mp of about 170°C. Fibers can be pulled from the melt and film pressed.

High molecular weight polyureas can be prepared from diamines and phosgene provided the addition of the amount of phosgene is carefully monitored. This is illustrated in the next preparation 36.

36. Preparation of Poly(hexamethyleneurea) (47)

$$H_2N(CH_2)_6NH_2 + COCl_2 \longrightarrow \left[(CH_2)_6 - NH\overset{O}{\overset{\|}{C}}NH \right]_n + 2HCl$$

A solution of 5.8 g (0.05 mol) hexamethylenediamine and 4 g (0.10 mol) sodium hydroxide (preferably as an aliquot of a stock solution) in 70 mL water is added with vigorous manual stirring to a solution of 4.95 g (0.05 mol) phosgene in 200 mL dry carbon tetrachloride contained in a wide-mouth Erlenmeyer flask. The polyurea forms rapidly, and heat is evolved. After the reaction has been stirred briskly for 10 min, the carbon tetrachloride is distilled on a steam bath and collected (**Caution! toxic**). The residual polymer is filtered, washed several times with water in a household blender, and air dried overnight. The weight of polymer is 5 g (70%). The inherent viscosity is about 0.90 in *m*-cresol at 0.5% concentration. The mp is about 295°C. Films can be melt pressed.

The solution of phosgene in carbon tetrachloride used in this preparation can be prepared by condensing phosgene into dry carbon tetrachloride in a volumetric flask and adding carbon tetrachloride to complete the volume. The solution is analyzed for grams of phosgene per milliliter of solution by thoroughly shaking an aliquot with excess standard sodium hydroxide and back titrating the excess sodium hydroxide in the aqueous layer using phenolphthalein indicator. The amount of phosgene, for convenience, should be 0.10 to 0.20 g/mL.

4.3. POLYHYDRAZIDES

Just as polyureas are a special branch of polyamides made from the shortest carbonic acid, polyhydrazides are polyamides from the shortest diamine. Because of the high concentration of amide groups they are usually high melting. Melt polymerization of polyhydrazides has not been too successful because, after monoacylation, the residual amine end of the diamine is much less reactive. In addition, many of the polymer products are high melting and intractable.

Melt polymerization of dihydrazides in the presence of dicarboxylic acids yield polymers with unidentified structure (48). Surprisingly, however, the reaction product from the thermal polymerization of adipic acid and hydrazine is reported to be fiber forming (49). The monohydrazide/monoester of sebacic acid has been reported on heating to give a brittle solid that melts at 295 to 300°C (50). High-temperature solution methods of heating stoichiometric amounts of dihydrazides with carboxylic acid esters or acid chlorides in nitrobenzene or xylenols give only low molecular weight polymer (50), perhaps owing to a poor choice of solvent. As was stressed previously, the solvent should dissolve, or at least swell, the polymer formed to attain high molecular weight.

The most successful route to high molecular weight polyhydrazides is the low-temperature solution method of reacting dry dihydrazides with acid chlorides in basic solvents, such as *N*-methylpyrrolidinone, preferably containing dry lithium or calcium chlorides (51). The reaction with diisocyanates to give polysemicarbazides is related to the reaction of dibasic acid derivatives with dihydrazides. High polymer results in solvents such as dimethyl sulfoxide and *N*-methylpyrrolidinone. In many cases the polymer stays in solution and can be cast directly into films or spun into fibers. If instead of a dihydrazide, the diisocyanate is reacted directly with hydrazine, a highly exothermic reaction occurs to form a polyureylene. These are higher melting than the polysemicarbazides because of greater symmetry and higher concentration of amide groups per unit weight. These features also account for the higher solubility of the polysemicarbazides. The corresponding polythiosemicarbazides can be made in high molecular weight in dimethylsulfoxide (DMSO) from isothiocyanates. These polymers have a strong tendency to chelate with various metals such as copper and nickel (52).

Among the other amide related polymers that can be made by reaction of diisocyanates are those with dioximes to give poly (*O*-acyloximes) and with bishydroxylamines to give either *O*- or *N*-acylated hydroxylamines. For example, the reaction of cyclohexanedione dioxime with methylene bis(4-phenyl isocyanate) in DMSO gives viscous solutions of polymer. These solutions could be fabricated into tough fibers and films. The polymers degrade at their melt temperature and disintegrate in boiling water (53).

37. Preparation of Polymer from Methylene Bis(4-phenylisocyanate) and Hydrazine Hydrate: Poly(methane [bis 4-phenyl]ureylene) (54)

$$OCN{-}C_6H_4{-}CH_2{-}C_6H_4{-}NCO + N_2H_4 \cdot H_2O \longrightarrow \left[{-}C_6H_4{-}CH_2{-}C_6H_4{-}NH{-}\overset{O}{\overset{\|}{C}}{-}NH{-}NH{-}\overset{O}{\overset{\|}{C}}{-}NH{-} \right]_n$$

In a 300-mL three-necked flask with stirrer, a mixture of 23.0 g (0.092 mol) methylene bis(4-phenyl-isocyanate) and 100 mL dry dimethylformamide is treated with a solution of 4.60 g (0.092 mol) hydrazine hydrate in 50 mL dimethylformamide with stirring. An immediate exothermic reaction occurs, and the solution becomes viscous. The solution can be cast to a clear tough film by drying in a vacuum oven at 60°C in a stream of nitrogen. The polymer may be precipitated in water, filtered, and washed in a household blender to cut it up. The yield is quantitative. However, the polymer cannot then be redissolved in dimethylformamide or in dimethyl sulfoxide. The polymer decomposes at 300°C without melting. If dianisidinediisocyanate is used in this preparation, the polymer will redissolve after precipitation. The methylene bis(4-phenyl isocyanate) is purified by distillation through a short Vigreux column; bp 142 to 144°C/0.14 mm.

If isophthalic dihydrazide is used in place of hydrazine in preparation 37, the polymer formed is a poly(isophthaloylsemicarbazide). The structure differs from the preceding example by being less symmetrical and having two less urea-type -NH- groups per chemical repeat unit. The increased solubility in dimethylformamide and related solvents and lower polymer melt temperature of the present case reflect this change. The reaction of the diisocyanate with the dihydrazide is initially less vigorous than with hydrazine because of the lower basicity of the hydrazide.

*38. Preparation of Isophthalic Dihydrazide**

$$CH_3O-C(=O)-C_6H_4-C(=O)-OCH_3 + N_2H_4\cdot H_2O \longrightarrow H_2NHN-C(=O)-C_6H_4-C(=O)-NHNH_2$$

In a 4-L Erlenmeyer flask, 194 g (1.0 mol) dimethyl isophthalate, in 500 mL methanol is added to 350 g (7.0 mol) hydrazine hydrate in 2 L methanol. The solution is allowed to stand overnight. The solid that forms is separated by filtration, washed with methanol on the filter, and dried in a vacuum oven at 70°C. The mp is 219 to 220°C, and the yield is about 180 g (93%). The isophthalic dihydrazide may be recrystallized from methanol-water, but this is usually not necessary.

39. Preparation of Polymer from Isophthalic Dihidrazide and Methylene Bis(4-phenyl) Isocyanate (54)

$$OCN-C_6H_4-CH_2-C_6H_4-NCO + H_2NNH-C(=O)-C_6H_4-C(=O)-NHNH_2 \longrightarrow$$

$$\left[-C_6H_4-CH_2-C_6H_4-NH-C(=O)-NHNH-C(=O)-C_6H_4-C(=O)-NHNH-C(=O)-NH-\right]_n$$

* From W. Sweeny, personal communication.

To a solution of 1.94 g (0.01 mol) isophthalic dihydrazide in 50 mL dry dimethyl sulfoxide in a three-necked 100-mL flask equipped with a stirrer and a nitrogen inlet is added at room temperature 2.50 g (0.01 mol) methylene bis(4-phenyl isocyanate). The reaction mixture warms up slightly and becomes viscous quickly. Stirring at room temperature under nitrogen is continued for 2 h. The polymer is isolated by pouring the solution into 300 mL water, filtering, and washing twice with water. The solid is dried in air and has an inherent viscosity in DMSO of about 1.8 at 0.5% concentration. The yield is 3.5 g (78%); the polymer mp is 250°C. The polymer is soluble in cold *N*-methylpyrrolidinone, sulfuric acid, DMSO, and dimethylformamide.

40. Preparation of Methylene Bis(4-phenylisothiocyanate) (52)

$H_2N-C_6H_4-CH_2-C_6H_4-NH_2 + CSCl_2 \longrightarrow SCN-C_6H_4-CH_2-C_6H_4-NCS$

Caution! Thiophosgene is highly toxic and should be handled only in a good hood.

In a three-necked flask cooled with an ice bath is placed 150 g thiophosgene and 1 L ice water. A solution of 87 g 4,4′-diaminodiphenylmethane in 1 L. chloroform is added with stirring over 1 h. The mixture is stirred at 0 to 10°C for an additional 2 h and then at room temperature overnight. The chloroform layer is separated and evaporated to dryness under a stream of nitrogen. The solid residue is dissolved in a mixture consisting of 400 mL benzene and 800 mL cyclohexane at the boiling point. The solution is decolorized, filtered, and allowed to cool. The fine needle-like precipitate is filtered, washed with cold cyclohexane, and recrystallized a second time from benzene-cyclohexane, as described earlier. The yield of pure product (mp 141 to 142°C) is 84 g (66%).

41. Preparation of N,N′-Diaminopiperazine (52, 55)

(1) HN(piperazine)NH + 2H—O—N=O ⟶ ON—N(piperazine)N—NO

(2) ON—N(piperazine)N—NO ⟶ H_2N—N(piperazine)N—NH_2

41.A. N,N′-Dinitrosopiperazine (Carcinogen)

To a solution of 194 g (1 mol) piperazine hexahydrate, 500 mL water, and 250 mL concentrated hydrochloric acid is added at 15 to 25°C, dropwise and with stirring, 150 g (2.1 mol) 97% sodium nitrite in 300 mL water. The mixture is stirred for 2 h and filtered; the solid is washed with water. The yield of *N,N'*-dinitrosopiperazine is 128.5 g (89%); the mp is 162 to 164°C.

41.B. N,N'-Diaminopiperazine Dihydrochloride

To a stirred mixture of 144 g (1 mol) of dinitrosopiperazine, 272 g (4.16 atoms) zinc dust and 1 L water is added dropwise over a period of 2.5 h. 600 mL (10.5 mol) glacial acetic acid, while the temperature is maintained at 20 to 30°C. The mixture is stirred at room temperature overnight and then at 80 to 85°C. for 1 h. On cooling, zinc acetate precipitates. This is filtered and washed with 500 mL cold ethanol, the washings are added to the filtrate. Another 500 mL ethanol is added to the filtrate and 400 g (11 mol) hydrogen chloride is passed in. The mixture is cooled to 0°C, and filtered. Yield, 118 g (62%) *N,N'*-diaminopiperazine dihydrochloride.

41.C. N,N'-Diaminopiperazine

To 1.5 L ethanol is added 453 g potassium hydroxide; the mixture is stirred until the solution is complete. To this is added in portions 378 g (2 mol) diaminopiperazine dihydrochloride. The mixture is heated at reflux for 2 h and filtered to remove insoluble materiaL. On evaporation of the filtrate to a small volume and cooling in ice, a light brown pasty solid precipitates. This is dried on a suction filter (under nitrogen) and recrystallized form 500 mL ethanol/ether (1/1 v/v) to give 140 g (85.5%) of tan crystals. The tan product is sublimed to give 104 g (63.5%) of white *N,N'*-diaminopiperazine; mp 117 to 119°C. Before using in a polymerization, the product should be recrystallized from chlorobenzene.

42. Preparation of a Polythiosemicarbazide (52)

$$H_2N{-}N\langle\text{piperazine}\rangle N{-}NH_2 + SCN{-}C_6H_4{-}CH_2{-}C_6H_4{-}NCS \longrightarrow$$

$$\left[C_6H_4{-}CH_2{-}C_6H_4{-}NH{-}\overset{S}{\overset{\|}{C}}{-}NH{-}N\langle\text{piperazine}\rangle N{-}NH{-}\overset{S}{\overset{\|}{C}}{-}NH \right]$$

In a typical preparation, 56.4 g powdered methylene bis(4-phenyisothiocyanate) is stirred into a solution of 23.2 g diaminopiperazine in 600 mL dimethyl sulfoxide at about 50°C. The mixture rapidly becomes viscous, and heating and stirring are discontinued after 2 h. The next day, the polymer is isolated by precipitating into water and is then chopped up in a home blender, washed thoroughly with water, and dried. The yield is quantitative, and the polymer has an inherent viscosity of over 1.0 in dimethyl sulfoxide. The PMT is about 230°C. The polymerization is quite reproducible and inherent viscosities as high as 1.8 can be obtained with specially purified intermediates, e.g., diaminopiperazine recrystallized from chlorobenzene. The polymer dissolves readily in dimethyl sulfoxide and can be cast to clear, tough films that can be drawn two to three times at about 175°C. The drawn and undrawn samples are amorphous by x-ray examination.

43. Preparation of a Copper Chelate of a Polythiosemicarbazide (52)

$$-NH-C(=S)-NH-N< \; + Cu^{++} \longrightarrow -NH-C(-S-Cu)=N-N<$$

To a solution of 0.4 g copper II chloride in 100 mL dimethylformamide is added a 0.35 g of the polythiosemicarbazide (prep. 42) in 10 mL dimethyl sulfoxide. Then 10 mL triethylamine is added with vigorous stirring. The precipitated polymer is filtered, washed with water, and dried. The polymer is black and contains 13% copper (theory 13.8%).

44. Preparation of Cyclohexanedione Dioxime (53)

$$O=C_6H_8=O + 2NH_2OH \longrightarrow HON=C_6H_8=NOH + 2H_2O$$

A mixture of 39 g (0.35 mol) 1,4-cyclohexanedione, 50 g (0.72 mol) hydroxylamine hydrochloride, 200 mL pyridine, and 200 mL absolute ethanol is refluxed for 4 h. in a 1-L round-bottom flask equipped with a condenser. The resulting solution is poured into a crystallizing dish, and the solvents are allowed to evaporate in a stream of air. Then 400 mL water is added, and the solid is filtered. It is recrystallized from 90% ethanol, giving a white solid that melts at 201 to 202°C.

45. Preparation of Polymer from 1,4-Cyclohexanedioxime and Biphenylene Diisocyanate (53)

$$HON=C_6H_8=NOH + OCN-C_6H_4-C_6H_4-NCO \longrightarrow$$

$$\left[-N=C_6H_8=N-O-\overset{O}{\overset{\|}{C}}-NH-C_6H_4-C_6H_4-NH-C-O-\right]_n$$

A solution of 2.93 g (0.0124 mol) biphenylene diisocyanate (purified by vacuum sublimation and stored in a freezer) in 15 mL dimethylformamide is mixed with 1.76 g (0.0124 mol) cyclohexanedione dioxime in 10 mL dimethylformamide at

80 to 100°C in a 100-mL three-necked flask equipped with stirrer, condenser, and drying tubes. An intermediate reaction is noted, and the solution rapidly becomes viscous. After 30 min, the solution, which has become somewhat cloudy, is poured into water, and the polymer is isolated by filtration. The polymer is then ground up in a blender, washed thoroughly with water, and dried in a high vacuum at 60°C. The yield is 4.35 g (94%). The polymer decomposes without melting above 200°C, dissolves easily in DMSO, and has an inherent viscosity of 0.8 to 1.2 in the same solvent. A 15% solution of the polymer in DMSO yields fibers when extruded into 50% aqueous dimethylformamide.

The dihydrazides of aliphatic and aromatic diacids can be made to undergo reaction in solution with dialdehydes to give high molecular weight polymers (56). It appears necessary that the polymer remain soluble for high molecular weight to be achieved. Polymers made in hexamethylphosphoramide (**Carcinogen!**) from certain combination of reactants were no longer soluble after isolation.

46. Polycondensation of Terephthaldehyde with Adipic and Isophthalic Dihydrazides (56)

$$OHC\text{-}C_6H_4\text{-}CHO + H_2NNHC(=O)\text{-}C_6H_4\text{-}C(=O)NHNH_2 + H_2NNHC(=O){+}CH_2{+}_4C(=O)NHNH_2 \longrightarrow \left[C(=O)\text{-}R\text{-}C(=O)NHN{=}CH\text{-}C_6H_4\text{-}CH{=}N\text{-}NH \right]_x$$

R = $m\text{-}C_6H_4$ and ${+}CH_2{+}_4$

Adipic and isophthalic acid hydrazides are prepared as follows. Stir and reflux 1 mol dimethyl ester with 1.5 L benzene and 500 g hydrazine hydrate for 16 h. The mixture is then cooled. The resulting solid is filtered, recrystallized once from water, and dried at 50°C under vacuum. The yield is about 83%. The mp of the isophthalic dihydrazide is 227°C, and of the adipic, 180°C.

A 250-mL three-necked flask with stirrer and two glass stoppers is charged with 100 mL dimethylsulfoxide (distilled from a Linde 4A molecular seive), 3.48 g adipic acid dihydrazide (0.02 mol), 3.88 g isophthalic acid dihydrazide (0.02 mol) and 5.36 g terephthaldehyde (0.04 mol). The solution is stirred for 48 h at room temperature; the polymer is precipitated in a large amount of methanol, washed repeatedly in a blender with methanol (use compressed air or nitrogen to blanket the motor), and dried in a vacuum at 70°C.

The polymer is soluble in hexamethylphosphoramide (**Carcinogen!**) in which it has an inherent viscosity of about 1.3 (0.25 g/100 mL at 30°C). It may be necessary first to dissolve the polymer with vigorous stirring at 150°C for 30 min.

The PMT is around 300°C, with decomposition. The polymer is essentially amorphous, in contrast to either hydrazide alone; but it is fairly resistant to hydrolysis. Immersion in water for 3 days at 97°C lowers the inherent viscosity to 0.86. Film of the original polymer cast from DMSO has a tensile modulus of 497,000 psi, inducating its stiffness. Its tensile strength and elongation are about 13,000 psi and 22%, respectively.

4.4. POLYURETHANES

The polyurethanes are related in properties to the polyamides because of similar opportunity for interchain hydrogen bonding. Crystallinity is often induced in polyurethanes, or may be present as prepared. The polyurethanes are, however, usually lower melting than polyamides with the same number of atoms in the main chain, but much higher than the corresponding polyesters.

The most practical methods of preparing polyurethanes in the laboratory are by reaction of bischloroformates with diamines and by addition of diols to diisocyanates. Direct melt polycondensation of dicarbamic acids or dicarbonic acids with diols or diamines to form polyurethanes is not practical, because the former are not thermally stable.

Bischloroformates may be prepared from most aliphatic and aromatic diols by reacting them with excess of phosgene at low temperatures. Chloroformates are less reactive toward amines and alcohols than the comparable carboxylic acid chlorides but are more reactive than the sulfonyl chlorides. When they are condensed with diamines, the products are polyurethanes. Examples of polyurethane preparations are given in chapter 5.

REFERENCES

1. W. Sweeny and J. Zimmerman, Polyamides, in *Encyclopedia of Polymer Science and Technology*, vol 10, Wiley-Interscience, New York, 1969.
2. J. Zimmerman, Private Communication.
3. W. Sweeny, U.S. Pat. 3094511 (18 June 1963), U.S. Pat. 3287324 (22 November 1966); and U.S. Pat. 4959453 (25 September 1990), to DuPont, W. Sweeny.
4. U.S. Pat. 3,696,201 (January 3 1967), to DuPont, C. W. Stephens.
5. O. E. Snyder and R. J. Richardson, Polyamide Fibers, to DuPont, in Ref, 1, vol. 10. pages 356-357.
6. D. Coleman and A. C. Farthing, *J. Chem. Soc.*, **1950**, 3213 (1950).
7. U. S. Pat. 2672480 (March 16 1954), to DuPont, A. S. Matlack.
8. V. E. Shashoua, W. Sweeny, and R. F. Tietz, *J. Am. Chem. Soc.*, **82**, 866 (1960).
9. A. Cannepin et al., *J. Poly. Sci.*, **8**, 35 (1952). E. E. Magat et al., *J. Am. Chem. Soc.*, **73**, 1031 (1951).
10. J. Zimmerman, Polyamides, in Ref. 1, vol. 11.
11. D. D. Coffman et al., *J. Poly. Sci.*, **2**, 306 (1947).

12. R. G. Beaman and F. G. Cramer, *J. Poly. Sci.*, **21**, 223 (1956).
13. E. L. Wittbecker et al., *J. Am. Chem. Soc.*, **69**, 579 (1947).
14. E. L. Wittbecker et al., *Ind. Eng. Chem.*, **40**, 875 (1948).
15. L. F. Beste and R. C. Houtz, *J. Poly. Sci.*, **8**, 395 (1952).
16. J. A. Somers, *Man-Made Textiles*, **32**(381), 60 (1956).
17. U.S. Pat. 2704282 (March 15, 1955), to DuPont, G. S. Stamatoff.
18. Brit. Pat. 737939 (October 5, 1955).
19. U.S. Pat. 2166183 (July, 18 1939), to DuPont, F. K. Signaigo.
20. U.S. Pat. 2558031 (July, 26, 1951), to DuPont, S. J. Allen and J. G. N. Drewitt.
21. T. L. Cairns et al., *J. Am. Chem. Soc.*, **71**, 651 (1955).
22. H. C. Haas et al., *J. Poly. Sci.*, **15**, 427 (1955).
23. R. Aelion, *Ann. Chim.*, **3**, 5 (1948) and Brit. Pat. 591027 (Aug. 2, 1947).
24. C. Naegli and A. Tyabji, *Helv. Chim. Acta*, **17**, 931 (1934) and **18**, 142 (1935).
25. Brit. Pat. 543297 (Feb 18, 1942).
26. Brit. Pat. 535139 (March 31, 1941).
27. P. W. Morgan, *Condensation Polymers by Interfacial and Solution Methods*, (no. **10**, Polymer Review Series), Wiley-Interscience, New York. 1965.
28. U.S. Pat. 3063966 (Nov. 13, 1962), to DuPont, S. L. Kwolek, P. W. Morgan, and W. R. Sorenson.
29. D. J. Lyman and S. L. Jung, *J. Poly. Sci.*, **40**, 407 (1959).
30. P. W. Morgan and S. L. Kwolek, *J. Chem. Educ.*, **36**, 182 (1959).
31. H. A. Bruson and T. W. Riener, *J. Am. Chem. Soc.*, **65**, 23 (1943).
32. V. E. Shashoua and W. M. Eareckson, *J. Poly. Sci.*, **40**, 343 (1959).
33. P. W. Morgan and S. L. Kwolek, *J. Poly. Sci., A*, **2**, 185 (1964).
34. U.S. Pat. 2994693 (August 1, 1961), to DuPont, N. Blake and H. W. Hill Jr.
35. W. M. Eareckson, *J. Poly. Sci.*, **40**, 399 (1959). A. Conix, *Ind. Eng. Chem.*, **51**, 147, (1959).
36. U.S. Pat. 3,006,899 (Oct. 31, 1961), to DuPont, H. W. Hill, S. L. Kwolek, and P. W. Morgan.
37. S. A. Sundet et al., *J. Poly. Sci.*, **40**, 389 (1959).
38. G. J. M. Van der Kerke et al., *Rec. Trav. Chim.*, **74**, 1301 (1955).
39. W. R. Sorenson, *J. Org. Chem.*, **24**, 978 (1950).
40. P .F. Wiley, *J. Am. Chem. Soc.*, **68**, 1867 (1946).
41. R. Mozingo, *Org. Synth.*, **21**, 15 (19).
42. Brit. Pat. 530267 (Dec. 9, 1940).
43. M. Katz, U.S. Pat. 2,888,438 (May 26, 1959), to DuPont.
44. S. L. Kwolek, *J. Poly. Sci.*, A, **2**, 5149 (1964).
45. Brit. Pat. 528437 (Oct. 29, 1940).
46. F. B. Cramer and R. G. Beaman, *J. Poly. Sci.*, **21**, 237 (1956).
47. J. F. Klebe, *J. Poly. Sci., B*, **2**, 1079 (1964).
48. J. W. Fisher, *Chem. Ind. (London)*, 244 (1952). U.S Pat. 2,512,633 (June 1950), J. W. Fisher and E. W. Whatley.

49. U.S. Pat. 2,349,979 (May 30, 1944), O. Moldenhauer and H. Bock., Belg. Pat. 443955 (1945).
50. U.S. Pat. 2,615,862 (October 1952), S. B. Mc Farlane Jr. and A. L. Miller.
51. A. H. Frazer and F. T. Wallenberger, *J. Poly. Sci. A*, **2**, 1147 (1964).
52. T. W. Campbell and E. A. Tomic, *J. Poly. Sci.*, **62**, 379 (1962).
53. T. W. Campbell, V. S. Foldi, and R. G. Parrish, *J. Appl. Poly. Sci.*, **2**, 81 (1959).
54. T. W. Campbell, V. S. Foldi, and J. Farago, *J. Appl. Poly. Sci.*, **2**, 155 (1959).
55. A. Schmidt and G. Wichman, *Ber.*, **24**, 3245 (1891).
56. R. H. Michel and W. A. Murphey, *J. Appl. Poly. Sci.*, **7**, 617 (1963).

5

POLYESTERS, POLYCARBONATES, AND POLYURETHANES

5.1. POLYESTERS

This chapter discusses the preparation of polyesters by melt and solution methods. Polyesters from glycols are usually prepared by melt methods and those from bisphenols by melt and solution methods. Akin to nylons, the mechanism of polyesterification is condensation or step growth polymerization. The properties of polyesters vary widely, from all-aliphatic polyesters that are viscous liquids just above room temperature to high-melting products from aromatic acids and bisphenols. Lacking interchain hydrogen bonding, polyesters melt substantially lower than polyamides of like structure. For example, the polyamide 6-6 nylon melts at 265°C, whereas the corresponding polyester poly(hexamethylene adipate) melts at 60°C.

Unlike polyamidation, melt preparation of polyesters from diols and esters or diacids does not require exact reactant balance at the start of polymerization. Indeed, a 20 to 100% molar excess of glycol is normally used. This leads first to low molecular weight hydroxy-ended polymer, which forms high polymer by ester exchange via evolution of glycol. Ester exchange takes place much more easily than amide exchange.

5.1.1. Nomenclature

A convenient nomenclature code for polyesters is the one used here, which is similar to that for polyamides (i.e., enumerating according to the number of carbon atoms in the main chain) except that the letter *G* is added to the enumeration of the hydroxy component. For example 6G-6 is the code for poly(hexamethylene adipate) and 2G-T for the polyester from ethyleneglycol and terephthalic acid. For AB systems the letter *E* is added to the carbon count. For example, the polyester from ω-hydroxycaproic acid is coded, 6-E, and that from *p*-hydroxybenzoic acid is coded 1,4-BE. A similar coding system is used for polyurethanes, but in that system the letter *E* is replaced by a *U* after enumeration of the acid component. Thus the polyurethane from piperazine and ethylene glycol bischloroformate is coded PiP-2U.

5.1.2. Synthetic Methods

Polyesters are usually prepared by high-temperature routes:

1. Direct esterification by reaction of a diol with a dibasic acid.

$$\mathrm{HO{-}CH_2{-}CH_2{-}OH + HOOC{-}R{-}COOH \longrightarrow \left[CH_2{-}CH_2{-}O{-}\overset{\displaystyle O}{\overset{\|}{C}}{-}R{-}\overset{\displaystyle O}{\overset{\|}{C}}O_2 \right]_n + 2_n\,H_2O}$$

2. Transesterification of diols with aliphatic or phenyl esters of dibasic acids.

$$\mathrm{HO{-}CH_2{-}CH_2{-}OH + R'{-}O_2C{-}R{-}CO_2{-}R' \longrightarrow \left[CH_2{-}CH_2{-}O{-}\overset{\displaystyle O}{\overset{\|}{C}}{-}R{-}\overset{\displaystyle O}{\overset{\|}{C}}O_2 \right]_n + 2\,R'OH}$$

3. Ring opening of cyclic esters by ionic or free radical processes.

$$\text{Caprolactone} \longrightarrow \mathrm{\left[(CH_2)_5{-}\overset{\displaystyle O}{\overset{\|}{C}}O \right]_n}$$

4. Double ester exchange between diol diacetates and diesters.

$$\mathrm{R{-}CO_2{-}CH_2{-}CH_2{-}O_2C{-}R + MeO_2C{-}Ph{-}CO_2Me \longrightarrow \left[CH_2{-}CH_2{-}O_2C{-}Ph{-}\overset{\displaystyle O}{\overset{\|}{C}} \right]_n + 2Me{-}O{-}Ac}$$

5. Acidolysis

$$R{-}O_2C{-}Ph{-}CO_2{-}R + HO_2C{-}Ph'{-}CO_2H \longrightarrow \left[Ph{-}O\overset{O}{\overset{\|}{C}}{-}Ph'{-}\overset{O}{\overset{\|}{C}}O \right]_n + 2R{-}CO_2H$$

6. High-temperature reaction of bisphenols with acid chlorides.

$$HO{-}Ph{-}OH + Cl{-}CO{-}Ph'{-}CO{-}Cl \longrightarrow {-}[{-}O{-}Ph{-}O_2C{-}Ph'{-}CO{-}]_n{-} + 2HCl$$

7. Interfacial polymerization of bisphenolates with acid chlorides

$$Na{-}O{-}Ph{-}O{-}Na + Cl{-}CO{-}Ph'{-}CO{-}Cl \longrightarrow \left[Ph{-}O\overset{O}{\overset{\|}{C}}{-}Ph'{-}\overset{O}{\overset{\|}{C}}O \right]_n + 2NaCl$$

5.1.2.A. Direct Esterification by Melt Polymerization

Currently the most important polyester is polyethylene terephthalate, which is widely used in textiles, tire cord, sail cloth, beverage bottles, and moldings. Its importance is derived from its balance of good thermal and physical properties and low cost. Transesterification has been the major commercial route but is now being displaced by direct esterification with readily available pure low-cost terephthalic acid that gives a whiter product and better economics.

The direct esterification synthesis of polyethylene terephthalate consists first in heating a mixture of ethylene glycol and terephthalic acid to a temperature of about 200°C at atmospheric pressure to eliminate water and form diol-ended polyester. This step is self-catalyzed by the terephthalic acid. The second step consists in raising the temperature to about 280°C to effect transesterification, with elimination of glycol, using antimony oxide as catalyst. This is facilitated by carefully controlled agitation and progressive reduction of the pressure to about 1 mm Hg. This route is readily applied to other diol/acid compositions and to hydroxy acids, but usually with downward adjustment of processing temperatures (e.g., in the 150 to 250°C range).

The direct esterification route using terephthalic acid is not easily accomplished in the laboratory because the low solubility of terephthalic acid in the glycol medium retards the rate of reaction. Micronizing increases the surface area of the terephthalic acid and speeds up the rate. In the commercial process, efficient mixing allows the reaction to proceed at reasonable pace.

According to the general theory of step growth polymerization (see chapter 2), of which polyesterification processes are typical, the degree of polymerization (DP) is given by:

$$DP = 1/(1 - p)$$

where p is the extent of the reaction. The number average molecular weight is thus given by:

$$\overline{M}_n = M_0/(1 - p)$$

and

$$\overline{M}_w = M_0(1 + p)/(1 - p)$$

where M_0 is the molecular weight of the repeat unit. Polymerization to high conversions normally requires that good reactant balance is maintained at all stages of the reaction. With hydroxy acids this does not present a problem, provided there are no side reactions, but in the reaction between diols and diacids, the former are often volatile, and some portion may be lost, especially under vacuum, by entrainment with evolved water. Usually about 20% excess glycol is charged to maintain hydroxyl end-group balance by compensating for these physical losses.

A number of studies have been made on the kinetics of direct esterification polymerization (1). These conclude that the rate of reaction of the self-catalyzed process (i.e., carboxylic acid catalyzed) is third order, corresponding to $r = k[\mathrm{COOH}]_2[\mathrm{OH}]$, at least over the later stages of the reaction from $p = 0.8$ to 0.98. However, if the reaction is catalyzed by strong acid (e.g., sulfonic acids), the rate is second order and $r = k'[\mathrm{COOH}][\mathrm{OH}]$. Although direct polyesterification is self-catalyzed by carboxyl groups of the acid components, their concentration is reduced as polymerization proceeds, and other catalysts are often introduced to maintain the reaction rate. Catalysts include toluenesulfonic acid, phosphoric and phosphonic acids, titanium alkoxides, and dialkyl tin oxides (2). Strongly acidic catalysts, however, often promote discoloration and hydrolytic instability if not deactivated or otherwise removed from the final product.

The usual melt synthesis of poly(ethylene terephthalate) does not lead only to high molecular weight linear polymer but also to linear and cyclic oligomers. Poly(ethylene terephthalate) so made contains 1.5 to 1.8% cyclic oligomers, consisting mostly of trimer and lesser amounts of dimer, tetramer, and pentamer (3) and smaller amounts of linear oligomers. A second major by-product is 3-oxapentamethylene terephthalate (oxydiethylene glycol units) formed by the in situ etherification of ethylene glycol or hydroxyethyl end groups. Commonly, polyethylene terephthalate contains about 2.5% of these ether groups.

The conditions of high-temperature polyesterification reactions are sufficiently severe that even with reactants of good thermal stability there is the risk of decomposition and side reactions that may affect the color or stability of the polymeric product. These reactions can arise from heat alone or in conjunction with adventitious oxidation or catalytic impurities. The pattern of thermal decomposition in poly(ethylene terephthalate) has been carefully studied (3). At temperatures around 280°C, which occur in synthesis and fabrication, the primary reaction is scission of the ethylene ester to vinyl ester and carboxyl. The final by-products are carboxyl end groups, aldehydes (which are largely removed in the vacuum cycle), and vinyl esters, all of which produce color in the polymer.

The carboxyl end groups can also affect the long-term stability by catalyzing hydrolysis of the ester linkages.

5.1.2.B. Ester Interchange

The other major commercial polyesterification route, particularly for poly(ethylene terephthalate) is ester interchange. The reaction is expressed stoichiometrically as follows:

$$2\text{HO–R–OH} + \text{R}''\text{O}_2\text{C–R}'\text{–CO}_2\text{R}'' \longrightarrow 2\text{R}''\text{OH} + \text{HO–R–O}_2\text{C–R}'\text{CO}_2\text{–R–OH}$$

$$n\text{HO–R–O}_2\text{CR}'\text{CO}_2\text{–R–OH} \longrightarrow n\text{HO–R–OH} + \text{–(–O–R–O}_2\text{C–R}'\text{–CO–)}_n$$

The simplest method of ester-exchange polymerization is to heat a dicarboxylic ester, a diol, and a polyesterification catalyst with stirring in an inert atmosphere at 150 to 200°C until evolution of the lower boiling alcohol is complete. Raise the temperature and reduce the pressure gradually to 0.5 to 1.0 mm to eliminate glycol and further the polymerization.

In the initial charge, a 10 to 50 mol% excess of the diol is usually added to ensure complete displacement of the alcohol from the carboxylic ester and to provide hydroxyl-ended groups for glycolysis. The final reaction temperature is usually governed by the melt temperature and stability of the polyester product and is normally 10 to 20°C above the melt temperature. This keeps the product molten and allows assessment of viscosity and molecular weight buildup. Final temperatures are usually 230 to 280°C, and total polymerization time is 3 to 4 h.

Originally, poly(ethylene terephthalate) was prepared with a single catalyst (usually lead oxide), but now mixed catalysts are used. One is active for ester exchange and the other for polymerization. Commonly used for ester exchange are oxides, carbonates, or acetates of zinc; manganese; calcium; and magnesium. Antimony trioxide is widely used as the polymerization catalyst. The amount of catalyst used is in the range of 0.05 to 0.1% by weight of the dimethyl terephthalate (4). Normal textile-grade poly(ethylene terephthalate) has an inherent viscosity number of about 0.7. Higher values are difficult to achieve in melt systems because of competitive decomposition reactions but can be achieved by solid-phase polymerization (see chapter 7).

Preparation of polyesters from bisphenols and alkyl esters does not proceed well using direct esterification or ester exchange methods. But high polymer results if phenyl esters are used, especially if potassium or cesium fluoride catalysts are employed (5). The resulting polymer is stable at melt temperatures (usually $T_m < 360°\text{C}$).

$$\text{HO–Ph–R–Ph–OH} + \text{PhO}_2\text{C–R}'\text{–CO}_2\text{Ph} \longrightarrow 2\text{PhOH} + \text{(–O–Ph–R–Ph–O}_2\text{C–R}'\text{–CO–)}_n$$

5.1.2.C. Acidolysis Polymerization

Acidolysis has become an increasingly important method of preparing stiff-chain, liquid crystalline polyesters (discussed in more detail in chapter 8). Acidolysis involves heating an acylated diol, usually a diacetate, with a dicarboxylic acid to effect polymerization via elimination of monobasic acid, usually acetic acid. Apart from reaction temperature, the major factor controlling the rate of reaction is solubility of the dibasic acid in the reaction medium or melt. Again, when using terephthalic acid or other dibasic acids of low solubility, micronizing to increase surface area speeds up the rate of reaction. Ketene formation is sometimes a side reaction (when diacetates of bisphenols are used) that reduces molecular weight because the phenolic groups that result do not polymerize well with carboxylic acids.

5.1.2.D. Polyesters from Diols and Acid Chlorides

Reaction of carboxylic acid chlorides with alcohols is a well-established route to preparing esters. However, if the hydrogen chloride produced is not removed efficiently and if the temperature is not kept low, alkyl halides and dialkyl ethers will be formed as by-products. Reaction of acid chlorides with bisphenols in solution gives a cleaner reaction. Because ether formation is not favored, high temperatures can be used to drive the reaction to high molecular weight, especially if the hydrogen chloride by-product is effectively removed by a flow of inert gas. The controlling factors are *(1)* selection of an inert reaction medium that is a good solvent or swelling agent for the polymer; *(2)* removal of the hydrogen chloride by-product as formed, usually with a stream of nitrogen passed through the reaction mixture; and *(3)* high temperatures (150 to 200°C) to complete the reaction. This method is particularly effective when reacting aromatic bischloroformates with bisphenols or with the normally unreactive *N*,*N*′-diphenyl-*p*-phenylenediamines to produce polycarbonates or polyurethanes.

Interfacial polymerization is an efficient method of preparing polyesters from acid chlorides and bisphenols. Requirements for high polymer are similar to those for polyamides, except that the acid acceptor must be strong enough to form the phenolate. Sodium and potassium hydroxides work well, but the corresponding carbonates do not. Again the major factor controlling molecular weight, apart from purity of the reactants and chemical inertness of the solvent, is the ability of the solvent to dissolve or swell the polymer and keep it in solution until high molecular weight results.

5.1.2.E. Ring-Opening Polymerization

Discussion of ring-opening polymerization will be deferred until chapter 8. Descriptions of ionic and free radical ring-opening polymerizations will be provided.

5.2. POLYCARBONATES

Polycarbonates are the esters of diols and carbonic acid. Only those from biphenols or bisphenols are sufficiently stable and of high enough melt temperature to be of commercial interest. Aromatic polycarbonates are usually made via interfacial polymerization from phosgene or bischloroformates and bisphenolates or from high-temperature reaction of phenyl carbonate and bisphenols.

$$\text{a. Na—O—Ph—R—Ph—O—Na} + \text{Cl—CO—Cl} \longrightarrow \text{—O}\left[\text{Ph—R—Ph—O—}\overset{\overset{\text{O}}{\|}}{\text{C}}\right]_n + 2\text{NaCl}$$

When the acid is ethyleneglycol bischloroformate, ethylene carbonate is eliminated from the melt to give the parent polycarbonate.

$$-[-\text{Ph}-\text{R}-\text{Ph}-\text{O}_2\text{C}-\text{O}-\text{CH}_2-\text{CH}_2-\text{O}-\text{CO}_2-]_n- \longrightarrow$$

$$-[-\text{O}-\text{Ph}-\text{R}-\text{Ph}-\text{O}-\text{CO}-]_n- + \text{ethylene carbonate.}$$

$$\text{b. HO}-\text{Ph}-\text{R}-\text{Ph}-\text{O}-\text{H} + \text{Ph}-\text{O}-\text{CO}-\text{O}-\text{Ph} \longrightarrow$$

$$-[-\text{O}-\text{Ph}-\text{R}-\text{Ph}-\text{O}-\text{CO}-]_n- + \text{nPhOH}$$

Polycarbonates have also been prepared by ring opening of cyclic oligomers of polycarbonates (discussed in chapter 8).

5.3. POLYURETHANES

Polyurethanes are hybrids of polyesters and polyamides and have properties somewhere in between. Interchain hydrogen bonding results in melt temperatures higher than those of closely related polyesters but lower than those of the corresponding polyamides. Polyurethanes are also less thermally stable than polyamides of comparable structure. Melt preparations of polyurethanes that melt above 200°C have not been successful because the polymers are thermally unstable and degrade with elimination of carbon dioxide. In general, however, polyurethanes can be prepared successfully by reaction of diamines with bischloroformates using low-temperature interfacial methods or, in special cases, by high-temperature solution methods and by low-temperature reaction of diols with diisocyanates. Aromatic polyurethanes have been prepared by reaction of biscarbamyl chlorides with diphenols in the melt or by high temperature solution methods. For example, a tough stable polyurethane can be obtained in the melt from reaction of *N*,*N*′-diphenyl-*N*,*N*′-bischlorocarbonylphenylenediamine with diphenols. The AB analog (*N*-phenyl-*N*-chlorocarbonyl-4-hydroxydiphenylamine) on melt polymerization yields a tough polymer of high thermal and hydrolytic stability (6).

Bischloroformates are readily prepared from most aliphatic and aromatic diols by reaction with phosgene at room temperature. Chloroformates are less reactive toward amines than the comparable acid chlorides, but are more reactive than sulfonyl chlorides.

Although alcohols do not react with isocyanates as rapidly as amines, the rate is sufficiently high to permit the formation of high molecular weight polymer from diols and diisocyanates, if proper conditions are used. If melt stability of the resultant polymer permits, the reaction may be carried out without solvent at a sufficiently high temperature to keep the polymer molten. It is common to carry out the reaction in an amide solvent, and high molecular weights are achieved. Elastic fibers based on polyurethanes are discussed in the chapter 10.

5.4. EXPERIMENTAL POLYMER PREPARATIONS

5.4.1. Polyesters

47. Preparation of Poly(ethylene terephthalate) (7)

$$CH_3O-\overset{O}{\overset{\|}{C}}-C_6H_4-\overset{O}{\overset{\|}{C}}-OCH_3 + 2HOCH_2CH_2OH \longrightarrow HOCH_2CH_2O-\overset{O}{\overset{\|}{C}}-C_6H_4-\overset{O}{\overset{\|}{C}}-OCH_2CH_2OH + 2CH_3OH$$

$$HOCH_2CH_2O-\overset{O}{\overset{\|}{C}}-C_6H_4-\overset{O}{\overset{\|}{C}}-OCH_2CH_2OH \longrightarrow \left[-OCH_2CH_2O-\overset{O}{\overset{\|}{C}}-C_6H_4-\overset{O}{\overset{\|}{C}}-\right]_n + HOCH_2CH_2OH$$

This polymerization process is generally applicable to any system in which the monomers and polymers are thermally stable above the polymer melt temperature and the glycol is sufficiently volatile to permit the excess to be completely removed under vacuum. In this preparation there are two ester exchange reactions. The first forms "monomer," from excess glycol and dimethyl terephthalate, with the elimination of methanol. The second eliminates glycol to form polymer. For this preparation, dimethyl terephthalate may be purified by recrystallization from ethanol; the melting point (mp) is 141 to 142°C. The ethylene glycol is purified by dissolving metallic sodium in it (1 g/100 mL) and refluxing in an atmosphere of nitrogen for 1 h, followed by distillation (boiling point (bp) 196 to 197°C).

In a polymer tube bearing a side arm is placed 15.5 g (0.08 mol) dimethyl terephthalate, 11.8 g (0.19 mol) ethylene glycol, 0.025 g calcium acetate dihydrate, and 0.006 g antimony trioxide. The tube is partially immersed in a 197°C vapor bath to melt the mixture, and a capillary tube is introduced that reaches the bottom of the tube. A slow stream of nitrogen is passed through the melt. Methanol is distilled from the mixture over 1 h, after which time the polymer tube is immersed as far as is practical in the vapors of the heating bath. The mixture is heated another 2 h at 197°C. Removal of the last trace of methanol is a requisite for high polymer formation. It may be necessary to heat the side arm during this period to prevent clogging from the distillation of some dimethyl terephthalate.

The polymer tube is now heated by means of a 222°C vapor bath (methyl salicylate) for 20 min and then is transferred to a 283°C vapor bath (dimethyl phthalate). After 10 min, the pressure is reduced to 0.3 mm or less over 15 to 20 min. Safety precautions, especially adequate shielding, should be observed. The polymerization is continued for 3 h; the alteration in the rate of bubble rise from the capillary indicates the change in viscosity. The vacuum line is shut off, and the tube reduced to atmospheric pressure via the nitrogen from the capillary. The polymer tube is wrapped in a towel and is allowed to cool under nitrogen (shattering of the tube may occur as the polymer contracts). The yield of polymer is quantitative if no dimethyl terephthalate was distilled in the early phases of the polymerization. The inherent viscosity in sym-tetrachloroethane/phenol (40/60 w/w) should be 0.6 to 0.7 (0.5% concentration at 30°C). Flexible tough films may be melt pressed and strong cold-drawable fibers pulled from the melt. The polymer melt temperature (PMT) is about 270°C, and the crystalline mp is about 260°C.

This polymerization procedure may be used to prepare a range of polyesters. A variety of catalysts have been used. Tetraisopropyl titanate is one of them and use of it is made in preparation 48 of polyester with a cyclic structure in the glycol and in the acid portion of the repeating unit.

48. Preparation of 1,4-Cyclohexanedicarbinol (8, 9)

Hydrogenate 100 g dimethyl terephthalate in 800 ml, ethanol over 10 g Raney nickel at 200°C and 2000 psi. When hydrogen uptake has ceased, the mixture is filtered, the alcohol distilled at reduced pressure, and the residue fractionally distilled through a 10″. Vigreux column. About 95 g dimethyl hexahydroterephthalate (bp 124°C/5 mm) is obtained. The diester is then hydrogenated over 8 g copper chromite catalyst at 255°C and 4000 psi. When the hydrogenation is completed, the catalyst is separated and the residue distilled, first through a 6-in. Vigreux column, then through a precision distillation column. The bp of the diol is 117 to 120°C at 0.5 mm. It distills as a viscous liquid that partially solidifies on standing. It is a mixture of cis and trans isomers. Chemical reduction of the hexahydro diester has also been carried out with sodium and alcohol (9).

49. Preparation of Poly(1,4-cyclohexanedicarbinyl terephthalate) (10)

$$CH_3O-\overset{O}{\overset{\|}{C}}-C_6H_4-\overset{O}{\overset{\|}{C}}-OCH_3 + 2HOCH_2-C_6H_{10}-CH_2OH \longrightarrow$$

$$2CH_3OH + HOH_2C-C_6H_{10}-CH_2O\overset{O}{\overset{\|}{C}}-C_6H_4-\overset{O}{\overset{\|}{C}}OCH_2-C_6H_{10}-CH_2OH$$

$$\longrightarrow \left[-OH_2C-C_6H_{10}-CH_2O-\overset{O}{\overset{\|}{C}}-C_6H_4-\overset{O}{\overset{\|}{C}}-\right]_n + HOH_2C-C_6H_{10}-CH_2OH$$

A mixture of 25 g (0.148 mol) 1,4-cyclohexanedicarbinol, 13 g (0.067 mol) dimethyl terephthalate, 0.02 g tetraisopropyl titanate, and 0.02 g sodium isopropoxide is charged to a polymer tube. The mixture is heated in a nitrogen stream at 197°C in a vapor bath for 3 h. The tube is then heated at 220°C for 15 min to remove the last of the methanol and begin the polymerization. Heating is continued at 283°C as a vacuum is slowly applied over a 15 min period to bring the pressure to 0.2 mm. The polymerization is complete in about 3 h. The polymer is removed from the tube after cooling. The inherent viscosity is 0.5 to 0.6 (0.5% concentration at 30°C) in tetrachloroethane/phenol (40/60 w/w), and the PMT is 285 to 290°C. It is also soluble in sulfuric acid and in hot *o*-dichlorobenzene. Films can be melt pressed at 285°C, and fibers drawn from the molten polymer.

Polyesters from entirely aliphatic reactants are usually low melting when even of high molecular weight. Although they can be crystalline and fiber forming, they have little use in unmodified form. Direct esterification or ester interchange is commonly used as is variety of catalysts. Preparation 50 condenses dimethyl sebacate with tetramethylene glycol using litharge as catalyst.

50. Preparation of Poly(tetramethylene sebacate) (11)

$$HO(CH_2)_4OH + CH_3O-\overset{\overset{O}{\|}}{C}(CH_2)_8\overset{\overset{O}{\|}}{C}-OCH_3 \longrightarrow \left[-O(CH_2)_4O\overset{\overset{O}{\|}}{C}(CH_8)\overset{\overset{O}{\|}}{C}-\right]_n + 2CH_3OH$$

A mixture of 4.95 g (0.055 mol) tetramethylene glycol, 11.5 g (0.050 mol) dimethyl sebacate, 0.1 g litharge, and 0.1 g di-*t*-butylhydroquinone is placed in a polymer tube with a side arm and a nitrogen capillary inlet reaching to the bottom of the tube. The reaction is heated to about 172°C in a vapor bath for 2 h at atmospheric pressure in a current of nitrogen. The pressure is then slowly reduced over a 4 h to 0.05 mm. The temperature is raised to 215°C for 4 h at the same reduced pressure. The reaction may be heated overnight without harm. The polymer is allowed to cool under nitrogen and is obtained as a white solid (13 g). The inherent viscosity is about 1.0 in chloroform (0.5% concentration at 25°C). The polymer melts at 60 to 65°C. Fibers can be pulled from the melt.

In many difunctional molecules, the tendency to intramolecular cyclization is so great that intermolecular condensation is suppressed. One example of hydroxyacid that has a great tendency to cyclize to a dimer is glycolic acid. However, poly(glycollic ester) can be formed from glycollic acid under favorable conditions (12).

51. Preparation of Poly(glycollic ester) (12)

$$HO-CH_2-CO_2H \longrightarrow \left[-O-CH_2-\overset{\overset{O}{\|}}{C}-\right] + H_2O$$

In a polymer tube equipped with a nitrogen inlet capillary tube and a side arm, 15 g (0.197 mol) hydroxyacetic acid recrystallized from *n*-butyl alcohol (mp 80

$$HOCH_2CO_2H \longrightarrow \text{glycolide (cyclic: } O{-}C({=}O){-}CH_2{-}O{-}C({=}O){-}CH_2) + H_2O$$

$$HOCH_2CO_2H \longrightarrow \left[OCH_2\overset{O}{\overset{\|}{C}} \right]_n + H_2O$$

Schematic 5.1

to 80.5°C) is mixed with 0.015 g triphenylphosphite color stabilizer and 0.001 g antimony trioxide catalyst. The tube is immersed in an ethylene glycol vapor bath at 197°C for 30 to 60 min, during which time the acid melts and water is evolved vigorously. A slow stream of nitrogen is passed through the melted acid by lowering the capillary into the melt. When the reaction has subsided, the pressure is reduced to 1 mm or less as quickly as possible. More water and a small quantity of glycolide (the cyclic dimer) distill during the next 60 to 80 min. It may be necessary to warm the side arm to prevent the glycolide from solidifying. The polymer melt becomes cloudy and begins to crystallize. The polymer tube is quickly shifted to a naphthalene vapor bath (218°C), previously heated to boiling. The polymer melts once more and is heated to 218°C at 1 mm or less for 3 to 4 h. Nitrogen is passed through the melt at a slow rate throughout. During this part of the heating cycle, the polymer may darken to some extent. At the end of this period the melt becomes viscous and may partially solidify. The tube is removed from the bath and is allowed to cool to room temperature under vacuum. The polymer is isolated by breaking the tube after releasing the vacuum. The polymer plug, freed from any glass chips that might adhere, is ground in a mill to pass through a 20-mesh screen. The powder is placed in a 50-mL round-bottom flask equipped with a glass paddle stirrer and a vacuum take off and heated while stirring the solid at a pressure of 1.0 mm or lower in the 218°C vapor bath. The powder polymerization is continued for at least 18 h.

These operations *must* be carried out carefully to obtain high polymer. Coloration can be reduced by using pure monomer and not exceeding 220°C in any part of the polymerization. The polymer may be tan to dark brown in color when removed from the tube. It should weigh 9 to 11 g and have a mp of 230 to 235°C. The inherent viscosity is 0.6 to 1.0 at 0.5% concentration in phenol/trichlorophenol (60/40 w/w). The polymer degrades slowly in this solvent, and the inherent viscosity measurement should be made as soon as feasible. The polymer can be melt pressed into clear, tough films at 240°C. The polymer is degraded severely on prolonged exposure to boiling water.

*52. Preparation of Poly(4-methylenebenzoate)**

$$HO_2C-C_6H_4-CH_2-OH \longrightarrow \left[\overset{O}{\overset{\|}{C}}-C_6H_4-CH_2-O \right]_n$$

The preferred method of polymerization of 4-(ω-hydroxyalkylbenzoic acids) is either direct esterification using sodium acetate as catalyst or ester exchange using manganese acetate catalyst. Polymerization by acidolysis gives only low molecular weight, dark colored products. A total of 20 g (0.191 mol) 4-hydroxymethyl benzoic acid and 1 g (0.012 mol) sodium acetate were mixed together and added to a polymer tube equipped with a nitrogen inlet capillary tube and a vacuum side-arm takeoff. The contents were freed of air by alternately evacuating and purging with nitrogen. Under nitrogen, the tube was immersed in an ethylene glycol vapor bath and heated for 4 h. The tube was then transferred rapidly to a boiling dimethyl phthalate vapor bath (283°C) and after 6 h the pressure the pressure was reduced to 0.7 mm. After an additional 12 h, the melt had become viscous and had a faint yellow color. The tube was removed from the bath and brought to atmospheric pressure by release of vacuum with nitrogen. On cooling, the melt solidified to a white opaque solid with a melt temperature of 175 to 185°C. Tough films can be pressed. The inherent viscosity was 0.6 at 0.5% concentration in tetrachloroethane/phenol (40/60 w/w).

Polyphenyl esters of aliphatic diacids can be prepared by an acidolysis reaction between the free acid and an ester of the bisphenol (13). In the following example, the poly(sebacic ester) of hydroquinone is prepared. The resorcinol polyester can be prepared by the same technique, as can poly(1,4-phenylene succinate). The latter is an "inverted" polyethylene terephthalate; the structures are formally identical, having a *p*-benzene ring separated by an ethylene and two carbonyl groups. They have similar properties.

53. Preparation of Poly(1,4-phenylene sebacate) (13)

$$Ac-O-C_6H_4-OAc + HO_2C-(CH_2)_8-CO_2H \longrightarrow \text{Polyester}$$

$$\downarrow$$

$$\left[O-C_6H_4-O-\overset{O}{\overset{\|}{C}}-(CH_2)_8-\overset{O}{\overset{\|}{C}} \right]_n + 2n\ CH_3CO_2H$$

For this polymerization, *p*-phenylene diacetate can be prepared by dissolving 11 g (0.10 mol) hydroquinone in a solution of 9 g (0.22 mol) sodium hydroxide

* W. Sweeny, personal communication.

in 45 mL water in a 250-mL Erlenmeyer flask. The mixture is cooled in an ice bath and a small quantity of ice is added to the flask. Then 22.4 g (0.22 mol) acetic anhydride is added all at once and the flask shaken vigorously by hand in the ice bath for 7 or 8 min. The white solid is filtered, washed with water, and recrystallized from ethanol. Drying in a vacuum oven at 60°C yielded 17 g product (88%) with a mp of 123 to 124°C.

A mixture of 9.7 g (0.05 mol), *p*-phenylene diacetate 10.10 g (0.05 mol) sebacic acid, and 0.03 g toluene sulfonic acid monohydrate is placed in a polymer tube; the tube is filled with nitrogen by a capillary inlet reaching to the bottom of the polymer tube. The temperature is then raised to 180°C. Acetic acid is distilled as the temperature is slowly raised to 230°C over a period of at least 30 min. The temperature is then raised over 45 min to 280°C while the pressure is slowly reduced in about 10 min to 0.3 mm. The temperature is maintained at 280°C at this pressure for 45 min. Nitrogen is passed slowly through the melt during the heating. The product is a dark, tough solid, with an inherent viscosity in *m*-cresol of 0.5 to 0.6 at 0.5% concentration and a PMT of about 170°C. Films can be pressed at this temperature. Drawable fibers can be pulled from the melt.

The reaction of a diacid chloride and a glycol in an anhydrous melt system provides a method for preparing polyesters that is much faster than either the glycol diacid or the glycol diester condensations. The by-product is hydrogen chloride, which must be removed to prevent alkyl halide formation or etherifiation. The reaction is applicable in most cases to aliphatic glycols having at least a three-atom chain between the hydroxyls and to aromatic acid chlorides, except *o*-phthaloyl chloride.

54. Preparation of Poly(tetramethylene isophthalate) (14)

$$HO(CH_2)_4OH + Cl-\overset{O}{\overset{\|}{C}}-C_6H_4-\overset{O}{\overset{\|}{C}}-Cl \longrightarrow \left[-O(CH_2)_4O-\overset{O}{\overset{\|}{C}}-C_6H_4-\overset{O}{\overset{\|}{C}}-\right]_n + 2HCl$$

A 100-mL three-necked flask equipped with a nitrogen inlet tube extending below the surface of the reaction mixture, a mechanical stirrer, and an exit tube for nitrogen and evolved hydrogen chloride (provision should be made for trapping the HCl) is flushed with nitrogen and charged first with 40.60 g (0.020 mol) isophthaloyl chloride and then 18.02 g (0.20 mol) tetramethylene glycol. The heat of the reaction causes the isophthaloyl chloride to melt. The reaction is stirred vigorously, and nitrogen is passed through the reaction to avoid accumulation of hydrogen chloride, which may bring about formation of tars. On a larger scale, the initial reaction should be controlled by ice cooling to maintain the temperature at 50°C or below. In about 1 h the evolution of hydrogen chloride slows considerably and the mixture begins to solidify. The temperature of the reaction mixture is then raised to 180°C by means of an oil bath and held at that temperature for 1 h. During the last 10 min of the heating cycle, the last of the hydrogen chloride is removed by reducing the pressure to 0.5 to 1.0 mm. The polymer is obtained as a white solid with an inherent viscosity of 0.5 in sym-tetrachloroethane/phenol (40/60 w/w) (5% concentration at 25°C). It is amorphous as formed and has a PMT of 100 to 110°C. It is soluble in 1,2,2-trichloroethane, formic acid, dimethylformamide (DMF), and *m*-cresol. When the polymer is crystallized, the PMT increases to about 140°C, and the polymer is no longer soluble in DMF and formic acid. Films cast from trichloroethane or chloroform are crystalline and quite brittle because of the high degree of crystallinity. Amorphous films can be obtained by pouring the melted polymer onto plates and spreading with a rod or hot casting knife. Crystallization can be accomplished by heating for 3 h at 70°C. Fibers can be pulled from the polymer melt and cold drawn by hand. The drawn sample crystallizes when held under tension. The amorphous fibers and films are somewhat rubbery.

Unlike polymers from aliphatic components, the all-aromatic polyesters are usually high-melting materials. They can conveniently be prepared from the sodium salts of diphenols by interfacial polycondensation (15). Preparation 55 procedure has been used successfully for preparing high molecular weight polymers from a number of bisphenols and appears to be generally applicable for the preparation of polyphenyl esters if the salt of the bisphenol is water soluble. It has been useful for preparing polymers using aromatic acid chlorides, because these are not readily hydrolyzed by the alkaline solution of the bisphenol.

55. Preparation of Poly[2,2,-propane bis(4-phenylisophthalate-co-terephthalate, 50/50)] (15)

A solution of 5.7 g (0.025 mol) diphenylolpropane (bisphenol A) and 2.0 g (0.05 mol) sodium hydroxide in 150 ml water is prepared in a household blender at low-speed stirring to avoid splashing. The bisphenol A (mp 159 to 160°C) was purified by recrystallizing from toluene (80 g/L), rinsing with light petroleum, and drying under vacuum. It is actually better to add the base as a standardized carbonate-free solution, because this gives a more accurate titer of alkali than weighing out pellets. A second solution of 2.54 g (0.0125 mol) isophthaloyl chloride and 2.54 g (0.0125 mol) terephthaloyl chloride in 75 mL chloroform is prepared in a 150-mL Ehrlenmeyer flask. The isophthaloyl chloride can be recrystallized from dry hexane (50 mL hexane for 100 g acid chloride) at 21 to 24°C; mp 42 to 43°C. Ice cooling should not be used. The terephthaloyl chloride can also be recrystallized from hexane (100 g per 700 mL hexane); mp 81 to 82°C. The chloroform must be washed with water to ensure the removal of ethanol, added as stabilizer. The chloroform is then dried over calcium hydride.

To the solution in the blender is then added 15 mL of a 10% aqueus solution of Duponol ME, and the blender is turned to maximum speed. The solution of the acid chlorides is added all at once (the flask rinsed with an additional 10 mL chloroform). The rapidly stirred solution sometimes foams over, and this can be controlled by using a top with a 0.5- to 1-in. center hole and adding the chloroform solution through a powder funnel inserted in the hole. The emulsion is stirred for 5 min, and the blender is stopped. The emulsion is then poured into 1-L acetone to coagulate the polymer and extract the solvents. The polymer is filtered and washed once on the filter with acetone. The granular polymer is transferred back to the blender jar and washed in 500 mL water to remove the salt and dispersing agent. The solid polymer is filtered again and washed on the filter with water. The washing step is repeated twice more, and the polymer is given a final wash with acetone and then dried for 24 h in a vacuum oven at 90°C. The dried polymer weighs 8 to 8.5 g (90 to 97%). The polymer should have a mp of 280°C and an inherent viscosity of 1.8 to 2.2 in sym-tetrachloroethane/phenol (40/60 w/w) at 0.5% concentration. The copolymer is amorphous and soluble in most halogen hydrocarbons, phenols, pyridine, and hot cyclohexanone. Films and fibers can be made from the good solvents cited.

An unusual situation exists in the preparation of polyesters from phenolphthalein by interfacial polymerization (16). The action of alkali on phenolphthalein produces not only colorless phenoxides but also a tautomeric, highly colored quinomethine structure. The colorless to red change is well known in the titration end point of mineral acid solutions with standardized solutions of strong base. Colorless carbinols may also be formed, particularly in the presence of excess alkali. However, as the phenylester is formed via the reaction of phenolphthaleins with diacid chlorides in a two-phase system, the tautomeric equilibrium rapidly and continuously shifts toward the phthalide (closed-ring) structure.

56. Preparation of Polyester from Phenolphthalein and Isophthaloyl Chloride (17)

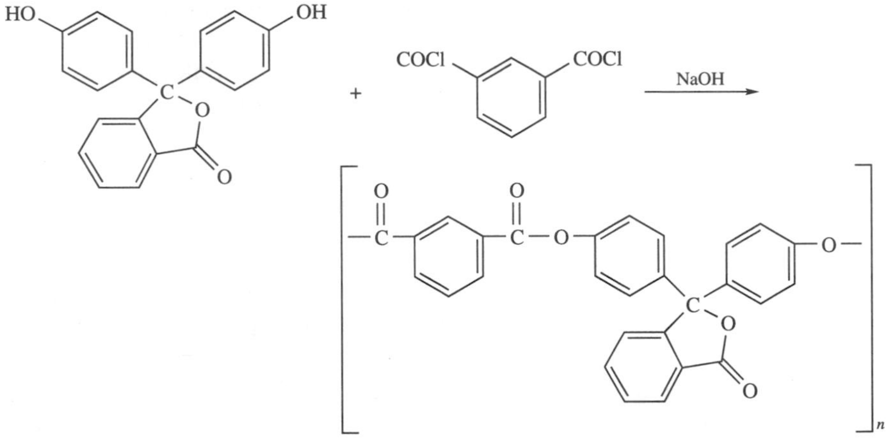

56.A. Interfacial Polycondensation

In a blender jar, 3.18 g (0.01 mol) phenolphthalein and 0.08 g (0.02 mol) sodium hydroxide are dissolved in 100 mL water. The mixture is stirred rapidly, and a solution of 2.03 g (0.01 mol) isophthaloyl chloride in 30 mL 1,2-dichloroethane is added all at once. The bloodred color is quickly reduced to a light pink. Stirring is continued for 5 min. Then hexane 300 mL is added to precipitate the polymer, which is filtered and washed with water; yield 4.22 g inherent viscosity is about 1.0 in sym-tetrachloroethane/phenol (40/60 w/w) at 0.5% concentration.

56.B. Interfacial Polymerization with Phosgene Gas

In a blender jar, 3.98 g (0.0125 mol) phenolphthalein, 1 g (0.025 mol) sodium hydroxide, and 1 g tetraethylammonium chloride are dissolved in 120 mL water. Then 30 mL 1,2-dichloroethane is added. While the mixture is stirred vigorously, phosgene gas is passed in until the color fades. After the first loss of color, the color is returned by adding several drops of a 20% solution of sodium hydroxide, and the phosgene is passed in again. This is repeated two more times. The total time is about 20 min. The polymer is isolated by adding acetone. The precipitate is filtered, washed thoroughly with water, and dried. The yield is 93%. The inherent viscosity is 2 or greater at 0.5% concencentration in sym-tetrachloroethane-phenol (40/60 w/w).

Adipoyl chloride will react at elevated temperatures with hydroquinone in a rigorously anhydrous inert solvent, such as nitrobenzene (**Carcinogen!**) to form high molecular weight polyphenyl ester (17). The evolved hydrogen chloride is removed from the refluxing solvent with the aid of an inert gas sweep, and no acid acceptor is needed. The reaction requires an unusually high degree of purification of the reactants and solvents to produce high molecular weight polymer. The polymer must be soluble in or highly swollen by the hot solvent in which the reaction is run.

57. Preparation of Poly(1,4-phenylene adipate) (17)

$$HO-C_6H_4-OH + Cl-\overset{O}{\overset{\|}{C}}-(CH_2)_4-\overset{O}{\overset{\|}{C}}-Cl \longrightarrow \left[-O-C_6H_4-O-\overset{O}{\overset{\|}{C}}-(CH_2)_4-\overset{O}{\overset{\|}{C}}-\right]_n + 2HCl$$

For this polymerization, hydroquinone is purified by recrystallizing four times from water that has been deoxygenated by boiling and cooling in a stream of nitrogen bubbled through it. The adipoyl chloride used can be commercially available material that has been fractionally distilled twice at reduced pressure in a nitrogen atmosphere using an oil bath not in excess of 150°C as the heat source. The bp at 1 mm is 70 to 72°C; at 10 mm, it is 112 to 115°C. Distillation should be fairly rapid to avoid the decomposition that may result from prolonged heating. The nitrobenzene is purified by washing well with water and drying over calcium chloride. It is then distilled three times from phosphorus pentoxide at atmospheric pressure and then once from the same material at oil pump pressure. It should be stored under nitrogen under anhydrous conditions.

The flask is flamed out and cooled in a current of nitrogen admitted through a gas inlet tube reaching to the bottom of the flask. A mixture of 7.872 g (0.0232 mol) adipoyl chloride, 4.728 g (0.0233 mol) hydroquinone, and 20 mL nitrobenzene is placed in a 100-mL three-necked flask equipped with a condenser protected with a drying tube. The reaction mixture is heated slowly by means of an oil bath to 140 to 147°C. for 2.5 h, then retained at that temperature for an additional 6 h. A slow stream of nitrogen is passed through the reaction mixture during the course of the reaction. Care must be taken to avoid heating above 150°C, because the acid chloride tends to decompose above that temperature. The nitrobenzene is then removed by distillation at oil pump pressure at about 147°C. The solid remaining is dried 2 h further at 147°C and 1 mm or less. The off-white solid obtained has an inherent viscosity of 1.0 to 1.4 in nitrobenzene/phenol (1/1 w/w) at 0.5% concentration and 25°C. The PMT is about 240°C. Fibers can be pulled from the melt.

The acetates of *p*- and *m*-hydroxybenzoic acids can undergo an acidolysis reaction with loss of acetic acid to form high molecular weight copolymers. The meta-isomer can be homopolymerized successfully, but the para-isomer forms an intractable material.

58. Preparation of Poly(m-phenylene carboxylate) (18)

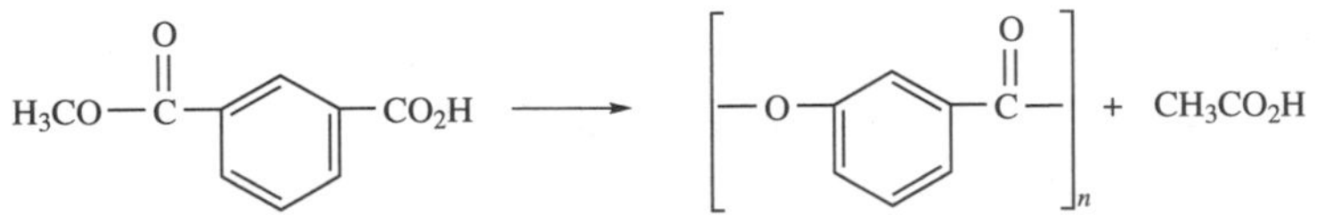

m-Acetoxybenzoic acid is prepared by heating *m*-hydroxybenzoic acid with excess acetic anhydride for 3 h, concentrating the mixture under vacuum, and recrystallizing the solid residue from benzene-ligroin. The mp is 130.5 to 131.5°C. A total of 10 g *m*-acetoxybenzoic acid and a small chip of magnesium (about 0.01% weight of the monomer) are placed in a polymer tube with a side arm and a nitrogen inlet that can be adjusted to reach the bottom of the tube. The tube is flushed with nitrogen and immersed in a 220°C vapor bath. A slow stream of nitrogen is continuously passed through the melt. The pressure is gradually reduced to 60 mm. Acetic acid distills from the tube. After 2 h, the tube is immersed in a 300°C vapor bath, and the pressure is brought to about 0.2 mm. The temperature is maintained until a maximum melt viscosity has been reached, as judged by the rate of bubble rise. The polymer is allowed to cool under nitrogen. The inherent viscosity is about 0.5 in sym-tetrachloroethane/phenol (40/60 w/w) at 0.23% concentration and 30°C. The polymer melts at 185 to 205°C and is soluble in *m*-cresol and *N*,*N*-dimethylaniline. The polymer so obtained can be further powdered polymerized by grinding to fine particle size (0.8 mm or less) and heating to 160 to 170°C under high vacuum. Inherent viscosities of 0.9 result in sym-tetrachloroethane/phenol at 0.23% concentration and 30°C.

High molecular weight poly(phenyl esters) can also be obtained by reaction of diphenols and phenyl esters with elimination of phenol. The reaction is particularly facile when cesium or potassium fluorides are used as catalysts (5).

59. Preparation of Poly[2,2-propane bis(4-phenyl) isophthalate] (5)

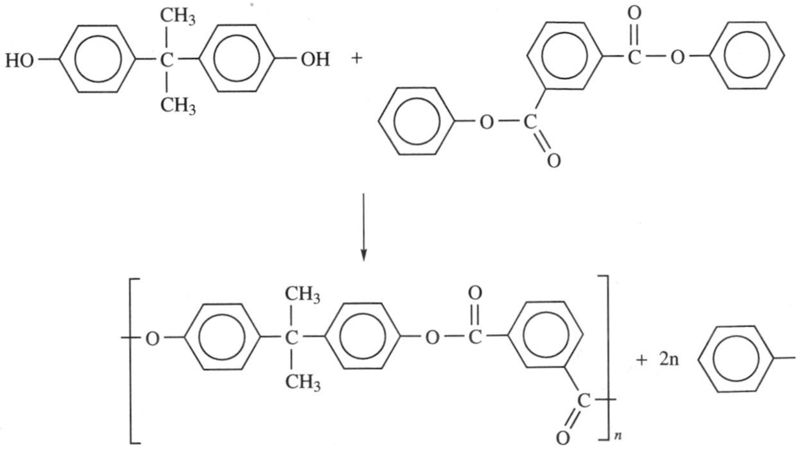

Bisphenol A is sensitive to oxidation and must be pure when used. This is accomplished by vacuum distillation, selecting a center cut, and recrystallizing this from toluene (80 g/L). In addition, bisphenol A should only be heated above about 180°C under vacuum; otherwise partial decomposition occurs.

To 200-mL three-necked flask fitted with a vacuum-tight mechanical stirrer, a gas inlet tube, and a 5-in. Vigreux head were added 22.8 g (0.1 mol) 2,2-propane bis(4-hydroxybenzene) (bisphenol A), 31.8 g (0.1 mol) diphenyl isophthalate, and 0.05 g (0.0003 mol) of dry cesium fluoride. The mixture was freed from air by alternatively evacuating and flushing with nitrogen. About five such operations are enough. The flask is then immersed in a 240°C. Woods metal bath. After about 10 min phenol starts to evolve. A partial vacuum is started to control a steady distillation of the phenol. Over a 30-min period, the vacuum is gradually reduced to about 1 mm pressure, and the temperature raised to 280°C. The temperature is then raised to 320°C over an additional 70 min. The melt is viscous and almost colorless. Nitrogen is then bled into the system and the bath removed. After cooling, the flask is broken, and a tough polymers removed. The polymer melts at about 280°C and gives tough films on melt pressing. The inherent viscosity at 0.5% concentration is 0.47 in pentafluorophenol. This polyesterification procedure is applicable to a range of diphenols and mixtures thereof to provide polymers with a mp in the 250 to 300°C range. Potassium fluoride is slightly less effective as catalyst than cesium fluoride.

5.4.2. Polycarbonates

Phosgene, the simplest diacid chloride, can be used directly in reaction with bisphenols to produce polycarbonates (19). The reaction can be carried out readily in pyridine solution or in aqueous sodium hydroxide. Polycarbonates have also been prepared from bisphenols and dialkyl carbonates in ester-exchange reactions (20).

60. Preparation of Poly(2,2-propanebis[4-phenyl carbonate]) (19)

$$HO-C_6H_4-C(CH_3)_2-C_6H_4-OH + Ph-O-C(=O)-O-Ph \text{ or } Cl-C(=O)-Cl \longrightarrow \left[-O-C_6H_4-C(CH_3)_2-C_6H_4-O-C(=O)-O- \right]_n$$

A 1-L three-necked flask is fitted with an efficient stirrer, a condenser with a calcium chloride drying tube, a thermometer capable of being immersed in the reaction mixture, and a gas inlet tube reaching as nearly to the bottom of the flask as possible for admitting phosgene. The air is displaced from the flask with nitrogen; and the flask is charged with 98 g (0.43 mol) diphenylolpropane (mp 158 to 160 C), recrystallized from 80 g/L toluene, and 700 mL analytical-grade pyridine. When the bisphenol has dissolved, phosgene is bubbled through the solution with stirring at a rate of 1 g/min. The temperature is maintained at 25 to 30°C by means of an ice water bath, applied as necessary. The phosgene flow may be followed by weighing the cylinder periodically or continuously, or a weighed amount of condensed phosgene may be vaporized into the reaction. The theoretical weight of phosgene required is 42.6 g (0.43 mol), but a 10 to 15 wt% excess may be necessary because of loss of unreacted phosgene through the condenser. The exit gases from the reaction should be led to a suitable aqueous alcohol-caustic trap, or to a water scrubber and flushed down a drain with plenty of water. Loss of phosgene can be prevented by moderate stirring to avoid formation of a deep vortex. At about the midpoint of the phosgene consumption, crystals of pyridine hydrochloride begin to form. Toward the end of the reaction, the solution becomes viscous, and the rate of phosgene addition is reduced to a very slow flow. At the end point of the reaction, a yellow to red color may develop, depending of the source and purity of the bisphenol, and the phosgene flow is stopped. The color may be discharged by the addition of a little bisphenol A in pyridine. The polymer is isolated by pouring the mixture into four times its volume of water with vigorous stirring. The polymer is filtered, washed on the filter with water, and suspended in 1 L water at 80°C for 10 min with stirring. It is filtered and washed again, and dried in a vacuum oven at 80°C. The inherent viscosity in sym-tetrachloroethane/phenol (40/60 w/w) (0.5% concentration at 25°C) is 0.6 to 0.8. The PMT is about 240°C, and strong flexible films may be pressed at that temperature. The polymer is generally soluble in chlorinated hydrocarbons.

As an alternative to the above procedure, a methylene chloride–aqueous sodium hydroxide medium may be used as follows. In a 3-L three-necked flask equipped with a stirrer, condenser, and gas inlet tube reaching to the bottom of the flask is placed 137 g (0.6 mol) bisphenol A, 60 g (1.5 mol) sodium hydroxide, 10 g benzyltrimethyl ammonium chloride, 1-L distilled water, and 500 mL methylene chloride. The mixture is stirred rapidly and kept at 20°C with ice cooling while phosgene is passed in at a rate of 2 g/min. Another 40 g sodium hydroxide is added in portions to keep the mixture strongly alkaline. The polymerization is estimated to be complete when a tough skin of polymer forms when a sample of the methylene chloride phase is evaporated. The methylene chloride is totally removed on a steam bath and the coarse polymer washed as described above. The polymer is essentially identical to that obtained from the first method. The weight of polymer is about 140 g (86%). The inherent viscosity is about 0.6, as before.

Polycarbonates can also be prepared in melt condensations with diphenyl carbonates and biphenols. The phenol is distilled as it is displaced. The principal limitation is that the product polymer be melt stable. The reaction can be catalyzed with small amounts of strong base, although too much base causes problems of discoloration, rearrangement, and sometimes cross-linking. Better and safer catalysts are possibly cesium and potassium fluoride (5). The chemistry and physics of polycarbonates have been thoroughly reviewed (20).

61. Preparation of Poly[2,2-propane-bis(4-phenyl)carbonate] by a Melt Process (21)

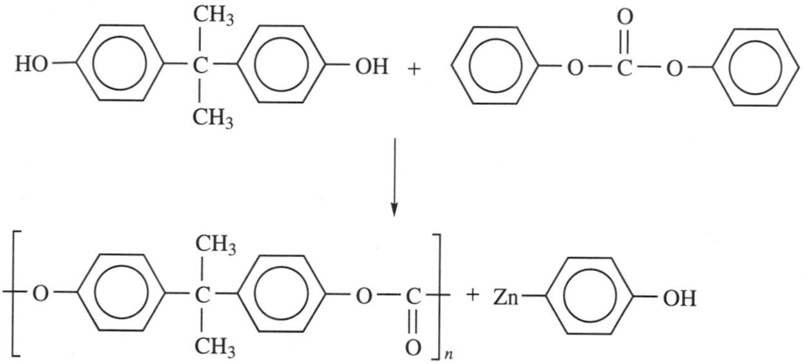

CH$_3$
HO—C—OH + O—C(=O)—O
CH$_3$

↓

CH$_3$
O—C—O—C(=O)
CH$_3$
n
+ Zn—OH

A large polymer tube is flushed with nitrogen and charged with 22.8 g (0.1 mol) bisphenol A (purified by recrystallization from toluene), 23.0 g (0.107 mol), diphenyl carbonate lithium, or about 0.4 mg cesium fluoride. The mixture was melted under nitrogen flow through the customary capillary arrangement at a temperature up to 150°C in a Woods metal bath. The pressure is slowly reduced to 20 mm, and the temperature is then slowly raised to 210°C. It is important to avoid distilling diphenyl carbonate along with the phenol, which is eliminated during this phase. The pressure is cautiously reduced to 0.2 mm, and the temperature is then slowly raised to 250°C over about 1 hr, followed by a rise to 280°C over another 2 h. Polycarbonates, as is typical of ring-containing polyesters, form viscous melts. The tube is removed from the heating bath and cooled under nitrogen. The product is a transparent solid that can be melt pressed or solvent cast (methylene chloride) into tough films. The polymer has an inherent viscosity of about 1.0 (0.5% in methylene chloride).

Pyrolysis of polycarbonates in some instances can yield polyethers (22). Attempts to apply the same reaction to the polycarbonate derived from ethylene bischloroformate and diphenylolpropane did not give a polyether but instead eliminated ethylene carbonate, leaving poly (2,2-bis[4-phenylene]propane carbonate).

62. Pyrolysis of Poly(ethylene-co-2,2-bis[4-phenylene]propane Carbonate) (23)

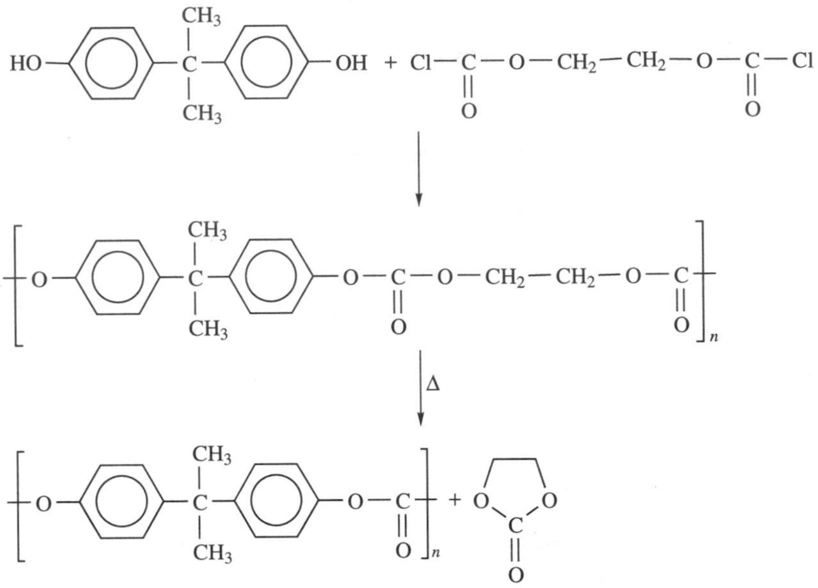

Poly(ethylene-co-2,2-bis[4-phenylene]propane carbonate) is prepared by interfacial polymerization from diphenylolpropane and ethylene bischloroformate (made by reaction of phosgene on ethylene glycol; see prep. 70) using methyl ethyl ketone as the organic phase and aqueous sodium hydroxide as the acid acceptor. The polymer thus obtained is soluble in dimethylformamide and melted at 170°C. The inherent viscosity is 0.9 in tetrachloroethane/phenol (40/60 w/w) at 0.5% concentration and 30°C. The polymer is placed in a distillation flask equipped with a nitrogen inlet capillary tube, a distillation head, a take-off condenser, and a receiver with a vacuum adaptor. The flask is immersed in a Woods metal bath and heated to 280°C under 1 mm pressure. Vigorous effervescence takes place, and a colorless liquid is distilled, which is identified as ethylene carbonate. When the foaming ceases, the flask is cooled under nitrogen, and the polymer is isolated by dissolving in chloroform and reprecipitating in methanol. The precipitate is filtered and dried at 60°C under vacuum. The polymer melts at 240°C, gives tough films, and is identified as poly(2,2-bis[4-phenylene]propane)carbonate). The inherent viscosity is 0.45 in tetrachloroethane/phenol (40/60 w/w) at 0.5% concentration.

5.4.3. Poly(phosphonate Esters)

Polyphosphonate esters can be prepared from phosphonic acid dihalides and glycols. Phenylphosphonyl dichloride, for example, has been condensed with hydroquinone or chlorohydroquinones to give fiber- and film-forming polymers.

63. Preparation of Poly(1,4-phenylene phenyphosphonate) (24)

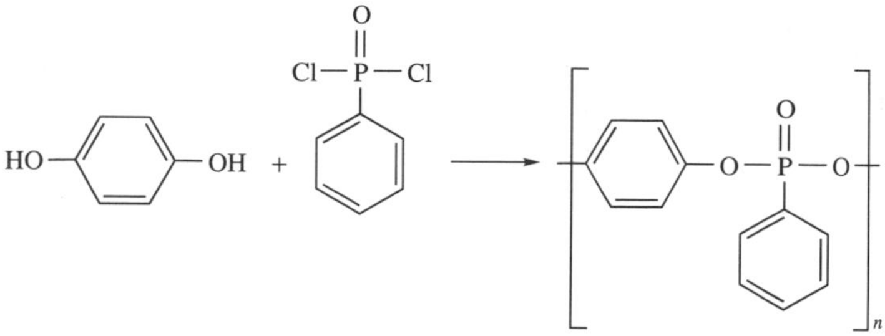

A polymer tube is charged with 10.2 g (0.0525 mol) freshly distilled phenylphosphonyl dichloride (bp 104°C at 4 mm) and 5.50 g (0.050 mol) hydroquinone. The tube is filled with nitrogen and is heated at atmospheric pressure in a 139°C vapor bath for 16 h, during which time a slow steam of nitrogen is passed through the melt by means of a capillary reaching to the bottom of the tube. It is then heated for another 4 h in a 218°C bath. Following this, the tube is heated in a 152°C bath while the pressure is slowly reduced to 1 to 3 mm over a period of 30 min. Heating is continued at this temperature and pressure for 4 h. The tube is then switched to a 242°C bath for 17 h, with a nitrogen flow through the capillary, as in all phases of the polymerization. A final heating stage is carried out at 280 to 290°C for about 4 h at the same reduced pressure. The last heating stage is continued until the viscosity of the melt ceases to increase as judged by the rate of rise of the nitrogen bubbles. The tube is cooled under nitrogen. The polyphosphonate ester has a PMT of around 130°C. Fibers that are cold-drawable may be pulled from the melt. The polymer is soluble in chloroform and ethylene chloride. The inherent viscosity in ethylene chloride at 0.5% concentration is the about 0.4.

64. Preparation of Biphenyldisulfonyl Chloride

To 290 mL phosphorus oxychloride and 2 lb phosphorus trichloride in a 3-L three-necked flask fitted with stirrer and reflux condenser is added 500 g 4,4′-biphenyldisulfonic acid (Eastman technical 90%) over a period of 30 min. The temperature rises to 60°C during this addition. The mixture is stirred and heated at reflux for 3 h and then cooled to room temperature. The slurry is slowly poured over a large amount of crushed ice with stirring. The light tan precipitate is vacuum filtered and sucked dry on the filter. The crude product is extracted four times with 1 L methylene chloride. The extract is washed twice with 250 mL 5% aqueous sodium bicarbonate solution and twice with 250 mL water. After drying over anhydrous sodium sulfate, the solution is concentrated on a steam bath until crystallization starts. Then hexane is added until significant crystallization occurs, and the solution is cooled to 0°C. The white crystalline diacid chloride is filtered; a second crop of crystals can be obtained by concentration and cooling of the filtrate. About 170 g biphenyldisulfonyl chloride (mp 209 to 212°C) is obtained. A further crystallization from methylene chloride and addition of hexane as before raises the melting point of the diacid chloride to 210.5 to 212°C.

65. Preparation of a Polysulfonate from Diphenylolpropane and Biphenyldisulphonyl Chloride (25)

$$ClSO_2-C_6H_4-C_6H_4-SO_2Cl + HO-C_6H_4-C(CH_3)_2-C_6H_4-OH \longrightarrow \left[SO_2-C_6H_4-C_6H_4-SO_2O-C_6H_4-C(CH_3)_2-C_6H_4-O- \right]_n$$

In a 250-mL flask fitted with a stirrer are placed 80 mL 1.495 M sodium hydroxide solution, 13.55 g diphenylolpropane, 1 g tetraethylammonium chloride hydrate, and 1 g detergent such as sodium laurylsulfonate. The mixture is stirred until a clear solution is obtained, then cooled to 10°C. At this point, 21 g biphenyldisulfonyl chloride in 80 mL methylene chloride is added, and vigorous stirring is continued for 1 h. A white dough-like semisolid is separated and boiled with water to remove the methylene chloride. The polysulfonate is obtained in quantitative yield. The inherent viscosity is 1.0 to 1.5 in TCE/phenol (40/60 w/w). The polymer melts at about 210°C.

66. Preparation of 3,5-Dimethyl-4-hydroxybenzenesulfonyl Chloride (26)

$$(H_3C)_2(HO)C_6H_2-SO_3Na + SOCl_2 \longrightarrow (H_3C)_2(HO)C_6H_2-SO_2Cl$$

A solution containing 22.4 g sodium 3,5-dimethyl-4-hydroxybenzenesulfonate in 40 mL dimethylformamide is cooled to 0°C with an ice-salt bath. To this solution 33 g thionyl chloride is added dropwise over 10 min. The cooling bath is removed and the temperature rises to 39°C. After 1 h, the mixture is added to 100 g ice. The white solid formed is filtered and washed several times with ice water. The crude product (19.8 g) is vacuum dried at room temperature and then dissolved in the minimum amount of toluene. An equal volume of hexane is added and the solution cooled to 0°C. The solids formed are filtered and vacuum dried. The white crystals (13.5 g) are further purified by sublimation at 50°C and 0.1 mm pressure. The yield of monomer-grade sulfonyl chloride melting at 134°C is 50 to 60%.

67. Preparation of Poly(3,5-dimethyl-1,4-phenylene Sulfonate) (26)

$$3,5\text{-}(CH_3)_2\text{-}4\text{-}(OH)C_6H_2SO_2Cl \longrightarrow \left[-O-C_6H_2(CH_3)_2-SO_2- \right]_n$$

A solution of 9.2 g 3,5-dimethyl-4-hydroxybenzenesulfonyl chloride in 20 mL dry nitrobenzene (**Carcinogen!**) is prepared in a 100-mL vaccine bottle under nitrogen. To this solution is added over 15 min 6.6 g triethylenediamine in 20 mL nitrobenzene. After standing at room temperature for 15 h, the contents of the bottle are added to ether and the precipitate is washed several times with ether and water. The crude product after vacuum drying weighs 6.7 g (90%). The product is separated into two fractions by extraction with 200 mL acetone. The insoluble material (50%) is high molecular weight polymer with an intrinsic viscosity of 0.5 to 0.9 at 25°C nitrobenzene. A flexible film can be pressed at 260°C and cooled rapidly in ice water. The PMT is 225 to 250°C. The film is amorphous as prepared but can be rendered crystalline by treatment with solvents such as acetone. The acetone-soluble fraction consists of low molecular weight polymer and oligomers.

5.4.4. Polyurethanes

Polyurethanes are related in properties to polyamides because of a similar opportunity for interchain hydrogen bonding. Crystallinity is often induced in polyurethanes or may be present as prepared. They are, however, usually lower melting than the polyamide with the same number of atoms in the chain. The most practical methods of preparing polyurethanes in the lab are reaction of bischloroformates with diamines and addition of a diol to a diisocyanate. Direct melt polycondensation of a diacid with a diol or diamine to form the ester or amide bond, respectively, is not possible in polyurethane preparation because neither the required carbamic acid nor carbonic half-acid ester is a thermally stable compound. Bischloroformates may be prepared from most aliphatic and aromatic diols by reacting them with an excess of phosgene at low temperatures. Chloroformates are less reactive toward amines and alcohols than the comparable carboxylic acid chlorides, but are more reactive than the sulfonyl chlorides.

68. Preparation of Bischloroformates

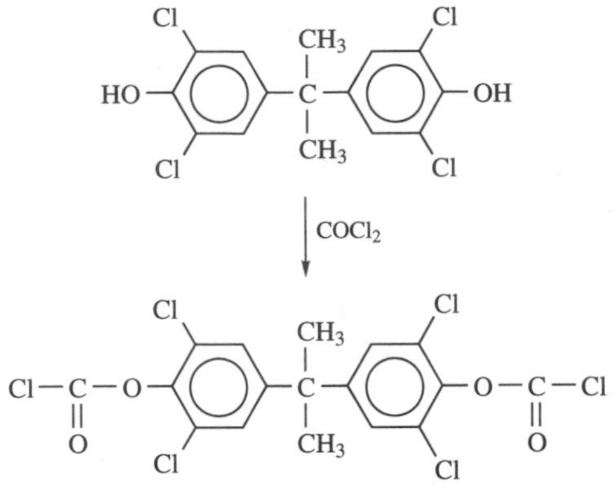

$$\xrightarrow{COCl_2}$$

In an 800-mL round-bottom flask equipped with a stirrer, dropping funnel, and a dry-ice condenser is collected 64 mL (0.9 mol) phosgene. The flask is cooled externally with an ice-salt bath and an additional metal bath is placed beneath the whole setup. The exit from the condenser is connected by tubing to a scrubbing trap through which water flows. Cold tetrahydrofuran (50 mL) is added to the phosgene. A solution of 156 g (0.42 mol) 2,2-bis(4-hydroxy-3,5-dichlorophenyl) propane in a mixture of tetrahydrofuran (100 mL) and dimethylaniline (100 mL) is prepared. This solution is dripped into the stirred phosgene solution over 2 h and then the mixture stirred for another 30 min. The dry ice is removed from the condenser, and the residual phosgene is purged from the system with dry nitrogen and hydrolyzed in the water scrubber. The slurry is then poured into 1-L water and ice. The precipitated bischloroformate is collected by filtration, washed well with cold water, and freed from water by air drying and vacuum drying. The product is recrystallized from hexane and melts at 165 to 165.5°C.

69. Polyurethane from Homopiperazine and the Bischloroformate of Tetrachlorobisphenol-A (27)

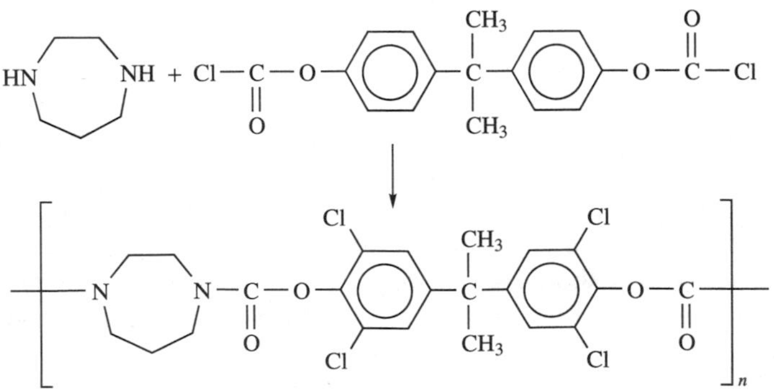

In a blender is placed 2.5 g (0.025 mol) as a 17% aqueous solution, homopiperazine 5.3 g sodium carbonate, and 250 mL water. The blender is then started, and a solution of 12.28 g (0.025 mol) 2,2-bis(4 hydroxy-3,5-dichlorophenyl)propane bischloroformate in 100 mL 1,2-dichloroethane is added as rapidly as possible. The mixture is stirred at high speed for 4 min, and an equal volume of hexane is added to precipitate the polymer. The polymer is washed thoroughly with water by repeated shredding in the blender and filtering. The granular product is dried at 80°C under vacuum. The yield is 12.95 g (97%). The inherent viscosity is 1.75 in *m*-cresol at 0.5% concentration.

The following example is the preparation of a polyurethane from a cyclic secondary diamine, piperazine, and ethylene bischloroformate.

70. Preparation of Ethylene Bischloroformate (28)

$$HOCH_2CH_2OH + COCl_2 \longrightarrow Cl-\overset{\overset{\displaystyle O}{\|}}{C}-O-CH_2CH_2-O-\overset{\overset{\displaystyle O}{\|}}{C}-Cl + 2HCl$$

About 650 mL (about 900 g, 9 mol) phosgene is condensed into an ice-cooled, three-necked flask equipped with a dry-ice condenser, stirrer, and dropping funnel, all suitably protected with drying tubes. To the ice-cooled, stirred liquid is added dropwise 125 g (2 mol) ethylene glycol. The reaction is stirred for 3 to 4 h after the addition is complete. The excess phosgene is allowed to evaporate or is trapped in aqueous alcohol-caustic. The residue is heated at 40 to 50°C at 20 mm for a short time to remove any volatile material. On distillation at 1 to 2 mm through a short path system, 340 g (91%) of ethylene bischloroformate is obtained. It is redistilled through a 10-in. glass helix-packed column to give a product boiling at 71 to 72°C/2.2 mm. If pure ethylene glycol is used, the product obtained after removal of the excess phosgene and other volatiles should be tested in a polymerization before distilling. Polymer-grade material is often obtained without distillation.

71. Preparation of Poly(ethylene N,N′-piperazinedicarboxylate)

$$HN\langle\;\rangle NH + Cl-\overset{\overset{\displaystyle O}{\|}}{C}-OCH_2CH_2O-\overset{\overset{\displaystyle O}{\|}}{C}-Cl \longrightarrow$$

$$\left[-N\langle\;\rangle N-\overset{\overset{\displaystyle O}{\|}}{C}-OCH_2CH_2O-\overset{\overset{\displaystyle O}{\|}}{C}-\right]_n + 2HCl$$

In 100 mL ice-cold water, 4.5 g (0.052 mol) piperazine and 10.6 g (0.10 mol) sodium carbonate are dissolved. It is better to use a standardized stock solution of piperazine and take aliquots than to weigh piperazine hexahydrate because of its variable moisture content. The solution is stirred rapidly in a household blender, and a solution of 6.4 mL (9.35 g, 0.05 mol) ethylene bischloroformate in 30 mL methylene chloride is added all at once. Polymer viscosity will be lower if the acid chloride is added slowly. Using more methylene chloride decreases the viscosity only slightly. The reaction mixture thickens rapidly, and in a few minutes it looks like cottage cheese. At this stage, it is transferred to a 1-L beaker with approximately 500 mL water. The beaker is placed on a hot plate in a hood and the methylene chloride is partially evaporated while stirring occasionally with a stirring rod. The methylene chloride should not be evaporated completely, because then the polymer will set up in tough chunks that are difficult to cut up. As soon as the polymer is solid enough to stick together, it is ready to be chopped. The solid polymer is returned to the blender with some water and chopped into small particles. It is filtered on a Buchner funnel and washed several times with warm water. It is dried in a vacuum oven at 90°C for 24 h. The is yield is 9.5 g (95%). The inherent viscosity in *m*-cresol at 0.5% concentration is between 2.5 and 4.5. The polymer is soluble in formic acid and its mp is 238°C.

Copolyester-urethanes can be made by a variety of techniques. The next example describes the preparation of an alternating ester-urethane by reaction of piperazine with the bischloroformate of bis(hydroxyethyl)terephthalate.

72. Preparation of the Bis(chloroformate) of Bis(hydroxyethyl) Terephthalate (29)

$$HOCH_2CH_2O\overset{O}{\overset{\|}{C}}-C_6H_4-\overset{O}{\overset{\|}{C}}-OCH_2CH_2OH + COCl_2 \longrightarrow$$

$$Cl-\overset{O}{\overset{\|}{C}}OCH_2CH_2O\overset{O}{\overset{\|}{C}}-C_6H_4-\overset{O}{\overset{\|}{C}}-OCH_2CH_2O\overset{O}{\overset{\|}{C}}Cl$$

A round-bottom flask is charged with 97 g dimethyl terephthalate, 372 g ethylene glycol, and 0.3 g calcium acetate monohydrate as a catalyst. From this reaction mixture, methanol is removed by distillation to yield bis(2-hydroxyethyl) terephthalate. Distillation is stopped when the head temperature of the column rises to 150°C. The mixture is cooled to room temperature, poured into 1.5 L distilled water, and filtered. The white solid product is purified by recrystallizing from boiling water. The mp is 107 to 108°C, and the yield is 90 to 95 g. To a mixture of 68.6 g (0.27 mol) bis(2-hydroxyethyl terephthalate) and 700 mL dioxane in a 2-L three necked flask equipped with a stirrer, dry-ice condenser, and a dropping funnel is added a solution of 203 mL phosgene in 300 mL dioxane. There is no apparent reaction, but after stirring 1 h a clear solution results. Stirring is continued for 4 hr while keeping the temperature about 40°C. The dioxane is removed by distillation under reduced pressure. The white crystals that remain are recrystallized from 900 mL cyclohexane and 150 mL benzene. The yield is 91.6 g (90%). The mp is 101 to 104°C, but should rise to 104°C after a second recrystallization.

73. Preparation of an Alternating Copolyester-Urethane (30)

$$Cl-\overset{O}{\overset{\|}{C}}-OCH_2CH_2-O-\overset{O}{\overset{\|}{C}}-C_6H_4-\overset{O}{\overset{\|}{C}}-O-CH_2CH_2-O-\overset{O}{\overset{\|}{C}}-Cl + HN(CH_2CH_2)_2NH$$

$$\downarrow$$

$$\left[-\overset{O}{\overset{\|}{C}}-O-CH_2-CH_2-O-\overset{O}{\overset{\|}{C}}-C_6H_4-\overset{O}{\overset{\|}{C}}-O-CH_2-CH_2-O-\overset{O}{\overset{\|}{C}}-N(CH_2CH_2)_2N-\right]_n$$

A blendor jar is charged with 210 mL distilled water, 15 mL of a 5% solution of a synthetic wetting agent, 50 mL benzene (**Carcinogen!**), 6.36 g sodium carbonate, and 3.1 g piperazine dissolved in 30 mL water. The mixture was stirred rapidly, and a solution of 11.37 g ester bischloroformate in 50 mL benzene is added all at once. Stirring is continued for 20 min, and the formed emulsion is broken by adding the contents of the jar into 500 mL acetone. The product is filtered, washed several times with distilled water, and dried overnight in a vacuum oven at 70°C. Polymer yield is about 10 g. The inherent viscosity is 1.2 to 1.3 in trichloroethane/formic acid (90/10 w/w) at 0.5% concentration. Tough films can be cast from a trichloroethane/formic (90/10 w/w) solution.

Tough, thermally stable aromatic polyurethanes can be prepared by direct reaction by heating selected aromatic chloroformates with *N*-aromatic substituted phenylenediamines or aminophenols. This is illustrated in preparation 74.

74. Preparation of N-Chlorocarbonyl 4-hydroxydiphenylamine (6, 31)

$$C_6H_5-HN-C_6H_4-OH + Cl-\underset{\underset{O}{\|}}{C}-Cl \longrightarrow C_6H_5-\underset{\underset{\overset{|}{C}(=O)Cl}{}}{N}-C_6H_4-OH$$

In a 3-L three-necked flask equipped with a gas inlet tube, reaching to near the bottom of the flask, a dry-ice condenser (equipped with a drying tube and an extension to a flowing water scrubber), and a thermometer is added 100 g redistilled commercially available 4-hydroxydiphenylamine (bp 215°C at 12 mm) and 1200 mL chlorobenzene. The solution is warmed to 90°C, and phosgene is passed at a moderate rate through the solution for 1 h while maintaining the temperature at 90°C. After 1 h a clear solution results. The temperature is raised to 120°C and passage of the phosgene continues for an additional 1.5 h. The solution is then cooled to 10 to 20°C and a colorless solid separates. This is filtered through a sintered glass funnel. The mother liquor is returned to the three-necked flask. The filtered solid is washed with hexane, sucked dry under a dental dam, and then dried in a desiccator under vacuum overnight. The yield is 88 g, mp is 153 to 154°C. The mother liquor is warmed in the original flask to 90°C; another 100 g 4-hydroxydiphenylamine is added and phosgene is again passed at a moderate rate through the solution for 1.5 h at 90°C and for 1.5 hr at 120°C. On cooling to 10°C and working up as before, 108 g of a colorless solid with a mp of 150 to 151°C is obtained.

75. Polyurethane from 4-Hydroxydiphenylamine Chloroformate (31)

$$C_6H_5-N(COCl)-C_6H_4-OH \longrightarrow \left[-C(=O)-N(C_6H_5)-C_6H_4-O- \right]_n$$

In a polymer tube with a nitrogen inlet capillary tube that reached almost to the bottom of the tube is placed 20 g *N*-chlorocarbonyl 4-hydroxydiphenyl amine. The outlet arm of the tube is connected to a vacuum pump through a soda-lime or caustic pellet trap to trap evolved hydrogen chloride. The tube is heated under nitrogen for 15 min at 190°C/25 mm, then 1.25 h at 190°C/1 mm, and finally 3 h at 245°C/1 mm.

Polymerization is rapid and a light yellow, tough glass resulted. This is isolated by cooling the tube under nitrogen breaking the tube, isolating the polymer by dissolving in methylene chloride, precipitating the polymer into excess methanol. The polymer so obtained is tough and melts at 256°C. The inherent viscosity is 0.5 to 0.6 in *m*-cresol at 0.5% concentration. Higher polymer can be obtained by prolonging the heating time at 245°C.

*76. Polyurethane from N,N′-Diphenyl p-Phenylenediamine and Hydroquinone Bischloroformate**

$$HO-C_6H_4-OH + Cl-C(=O)-Cl \longrightarrow Cl-C(=O)-O-C_6H_4-O-C(=O)-Cl$$

$$Cl-C(=O)-O-C_6H_4-O-C(=O)-Cl + H-N(C_6H_5)-C_6H_4-N(C_6H_5)-H$$

$$\downarrow$$

$$\left[-C(=O)-O-C_6H_4-O-C(=O)-N(C_6H_5)-C_6H_4-N(C_6H_5)- \right]_n$$

* C. W. Stephens; personnal communication.

76.A. Intermediates
N,N'-dipheny-p-phenylenediamine obtained from Eastman Chemicals (or Aldrich) is distilled at 100 mm pressure and recrystallized three times from methylene chloride. Hydroquinone bischloroformate is prepared by adding 110 g hydroquinone and 50 g Rohm, and Haas Amberlite IRA-400 (previously dried at 70 to 80°C overnight in a vacuum oven at 100 mm) to a 1-L Hastelloy pressure vessel. The vessel is sealed, after flushing with nitrogen, and then evacuated and cooled to below −40°C with a dry-ice bath. Then 220 g (2.22 mol) phosgene is distilled into the vessel, which is then closed, mounted on a rocker assembly, and heated at 110°C for 8 h, while being rocked to mix the contents. The vessel is cooled, then vented, and the *p*-phenylenebischloroformate is extracted from the resin and any unreacted hydroquinone with three portions of boiling hexane (the extractions are with 2 L, 1 L, and 0.5 L, respectively). The hexane is removed by distillation, leaving a white solid that is dried overnight at room temperature in a vacuum oven. Substantially pure *p*-phenylene bischloroformate (206 g, 88%) with a mp of 98°C is obtained. This method can be used for preparing a variety of aromatic bischloroformates, such as diphenylolpropane and diphenylolmethane bischloroformates

76.B. Polymer
To a three-necked flask, equipped with a gas inlet tube reaching close to the bottom of the flask, a stirrer, and a condenser equipped with a drying tube connected to a water scrubber is added 265 g *N,N'*-diphenyl-*p*-phenylenediamine, 235 g hydroquinone bischloroformate, and 2.5 L dry *o*-dichlorobenzene. The mixture is stirred, nitrogen is passed through at a modest rate, and the solution is raised to reflux over 2.5 h. Heating is continued at reflux for another 3 h until hydrogen chloride evolution ceases. The flask is cooled slightly, and 15 mL phenyl chloroformate is added to end cap the polymer. Reflux is then resumed for a further 3 h. The flask is allowed to cool overnight, and the polymer that had separated is filtered, washed well with acetone, and dried overnight in a vacuum oven at 140°C. The polymer is dried a further 3 h at 200°C at 10 mm. Yield is 411 g (97.5%). The polymer has an inherent viscosity of 0.49 in sulphuric acid at 0.5% concentration and melted at about 265°C. Tough films could be melt pressed at this temperature.

Polycarbamates and polyurethanes are essentially identical and differ in definition of the acid/base components.

77. Preparation of Poly(tetramethylene hexamethylenedicarbamate) in Solution (32)

An 80/20 v/v mixture of chlorobenzene and *o*-dichlorobenzene, both purified by distillation over calcium hydride, is used as the solvent in this reaction. A dry 1-L three-necked flask equipped with a stirrer, nitrogen inlet, and reflux condenser with calcium chloride drying tube is flushed out with nitrogen and is charged with 2.68 g pure tetramethylene glycol (fractionally distilled bp 120°C at 10 mm; mp

19.7°C) in 100 mL solvent. The nitrogen inlet is replaced with a dropping funnel, and the solution is heated to reflux. From the funnel is added 5 g hexamethylene diisocyanate in 50 mL solvent; about half the diisocyanate is added rapidly with vigorous stirring and the remainder over 3 to 4 h. The solution is held at reflux for 1 h after the addition. When the solution is cooled to room temperature, the polymer precipitates and the solvent is decanted. The polymer is dissolved in 50 mL hot dimethylformamide, and 50 mL methanol is added to the still-warm solution. The clear solution is cooled in a refrigerator overnight, and the precipitated polymer is separated by filtration and dried at 0.1 mm pressure in a vacuum desiccator overnight. The inherent viscosity is 0.5 to 0.7 in *m*-cresol at 0.5% concentration.

78. Preparation of Poly(ethylene methylene bis[4-phenylcarbamate]) (33)

$$OCN-C_6H_4-CH_2-C_6H_4-NCO + HOCH_2CH_2OH \longrightarrow \left[-\overset{O}{\overset{\|}{C}}NH-C_6H_4-CH_2-C_6H_4-NH-\overset{O}{\overset{\|}{C}}-OCH_2CH_2O-\right]_n$$

Bis(4-isocyanatophenyl) methane is purified by distillation through a vacuum-jacketed Vigreux column (bp 148 to 150°C/0.12 mm). The ethylene glycol is purified by distillation (bp 79°C/4.4 mm) and should contain <0.05% water. The solvents are purified by distillation: dimethylsulfoxide (bp 66°C/5 mm) and 4-methylpentanone-2 (bp 115°C). In three-necked round bottomed flask equipped with a stirrer and condenser and protected from moisture are placed 40 mL 4-methylpentanone-2 and 25.02 mL methylene bis(4-phenylisocyanate). To this rapidly stirred suspension is added 6.20 g ethylene glycol in 40 mL dimethylsulfoxide. The reaction is heated to 115°C for 1.5 h. The clear, viscous solution is then poured into water to precipitate the polyurethane. The tough, white polymer is chopped up in a home blender, washed with water, then dried in a vacuum oven at 90°C. Inherent viscosity is 1.0 in dimethylformamide. Films may be cast from dimethylformamide or directly from the polymerization solution. The mp is 255°C and the T_g is 90°C.

Although alcohols do not react as fast with isocyanates as do amines, the yield is sufficiently high to permit the formation of high polymer from diol-isocyanate reactions if the proper conditions are used. If the melt stability of the resultant polymer permits, the addition may be carried out without solvent at a sufficiently high temperature to keep the polymer molten. A solvent may be used, providing the polymer is either soluble or is swollen enough to permit reaction of the chain ends to proceed until a high molecular weight is reached.

79. Preparation of Poly(tetramethylene Hexamethylenedicarbamate) by Melt Methods (34)

$$HO-(CH_2)_4-OH + OCN-(CH_2)_6-NCO \longrightarrow \left[-(CH_2)_4-O-\overset{\overset{\displaystyle O}{\|}}{C}-NH(CH_2)_6NH-\overset{\overset{\displaystyle O}{\|}}{C}-O-\right]_n$$

A 250 mL three-necked flask is fitted with a calcium chloride drying tube and an efficient stirring motor with a metal shaft and a paddle. The eventual viscosity of the reaction may break glass-stirring equipment. The flask is charged with 45.0 g (0.05 mol) tetramethylene glycol, and a low-pressure nitrogen flow is passed through the flask above the surface of the liquid through the third neck to displace the air. The tetramethylene glycol purity must be high; it should be subjected to careful fractional distillation (bp 120°C at 10 mm) to give a product that melts at 19.7°C. The nitrogen inlet tube is replaced by a stoppered, pressure-equalizing dropping funnel containing 83.16 g (0.495 mol) hexamethylene diisocyanate. The flask is heated to 50°C in an oil bath, and the diisocyanate is added with rapid stirring over a period of 1 h, during which time the temperature of the bath is raised to 190 to 195°C. The reaction is continued at this temperature until no further increase in viscosity occurs. Good mixing is required to prevent local hot spots and the possible branching or cross-linking that may occur. An insoluble, nonfluid product indicates that the latter has occurred. When the mixture is still molten, the stirrer paddle is raised and the mass allowed to cool. The hard, tough polymer is removed by breaking the flask. The polymer is cut up into pieces and ground through a 20-mesh screen in a mill. It is washed twice with methanol in a home blender and dried at 60°C in a vacuum oven. The yield is quantitative. The PMT is about 180°C, and the inherent viscosity is 0.8 to 1.4 in *m*-cresol (0.5% concentration at 25°C) Othcr solvents are phenol and formamide. Films can be pressed and fiber spun at 190 to 200°C. As an alternative to melt polymerization, the reaction may be carried out in chlorobenzene/*o*-dichlorobenzene solution. The polymer is soluble in the hot mixture but precipitates on cooling.

REFERENCES

1. V. V. Korshak and V. Vinogradov, *Polyesters*, Pergamon Press, Oxford, UK, 1965.
2. A. Fradet and E. Marechal, *Advances in Polymer Science*, vol. 43, Springer-Verlag, New York, 1982.
3. L. H. Bauxbaum, *Angew. Chem. Int. Educ.*, **7**, 182 (1968). D. R. Gaskill et al., *J. Appl. Poly. Sci.*, **39**, 106 (1967).
4. A. L. Cimecioglu et al., *J. Appl. Poly. Sci.*, **32**, 4719 (1986).
5. U. S. Pat. 4,782,131 (1, November 1988), W. Sweeny, to DuPont.
6. U. S. Pat. 3,278,594 (11, October 1966) and U.S. Pat. 3,211,775 (12 October 1965), W. Sweeny and C. W. Stephens, to DuPont.

7. Brit. Pat. 578,079 (June 1946), J. R. Whinfield and J. T. Dickson.
8. G. A. Haggis and L. N. Owens, *J. Chem. Soc.*, 404 (1953).
9. W. A. Lazier and H. R. Arnold, in A. Blatt, ed., *Organic Synthesis*, vol. 2, John Wiley & Sons, New York.
10. U. S. Pat. 2,901,466 (August 25 1959), C. J. Kibler, A. Bell, and J. G. Smith.
11. C. S. Marvel and J. H. Johnson, *J. Am. Chem. Soc.*, **72**, 1674 (1950).
12. U. S. Pat. 2,676,945 (27 April 1954), N. A Higgins, to DuPont.
13. Brit. Pat. 636,429 (Apr. 26, 1950), E. R. Walsgrove and F. Reeder.
14. U. S. Pat. 2,623,034 (December 23, 1952) and U.S. Pat. 2,589,688 (March 18, 1952), P. J. Flory and F. S. Leutner.
15. W. M. Eareckson, *J. Poly. Sci.*, **40**, 399 (1959). A. Conix, *Ind. Eng. Chem.*, **51**, 147 (1959).
16. P. W. Morgan, *Condensation Polymers by Interfacial and Solution Methods*, Wiley-Interscience, New York, 1965. P. W. Morgan, *J. Poly. Sci. A.*, **2**, 437 (1964).
17. K. Yamaguichi et al., *J. Chem. Soc. Jpn., Ind. Chem.*, **58**, 358 (1955) [Chem. Abst. **49**, 14373g, (1955)].
18. R. Gilkey and J. R. Caldwell, *J. Appl. Poly. Sci.*, **2**, 198 (1959).
19. Ger. Pat. 1,046,311 (1958), H. Schnell and L. Bottomburch. H. H. Schnell, *Ind. Eng. Chem.*, **51**, 157 (1959). H. H. Schnell, *Angew. Chem.*, **68**, 633 (1956). Belg. Pat. 546376 (1956) and Belg. Pat. 546377 (1956).
20. H. Schnell, *Chemistry and Physics of Polycarbonates*, Wiley-Interscience, New York, 1964.
21. Ger. Pat. 1,031,512 (1958), H. Schnell and G. Fritz.
22 J. W. Hill, *J. Am. Chem. Soc.*, **57**, 1131 (1935).
23. W. Sweeny, *J. Appl. Poly. Sci. Notes*, **5**, 16 (1961).
24. U. S. Pat. 2,435,252 (February 3, 1948) and U.S. Pat. 2,572,076 (October 23, 1951), A. D. F. Toy.
25. A. Conix and U. Laridon, in Wiesbaden, *Symposium uber Macromolecules* vol. 489, Oct. 1959.
26. W. L. Hall, *J. Org. Chem.*, **31**, 2672 (1966).
27. P. W. Morgan, *J. Appl. Poly. Sci.*, **40**, 1171 (1990).
28. N. Rabjohn, *Am. Chem. Soc.*, **70**, 1181 (1948).
29. E. L. Wittbecker and M. Katz, *J. Poly. Sci.*, **40**, 367 (1959). U. S. Pat. 2,731,446 (January 17, 1956), E. L. Wittbecker, to DuPont.
30. U. S. Pat. 3,036,979 (29 May 1962), E. L. Wittbecker, to DuPont.
31. U. S. Pat. 3,278,594 (11 October, 1966), W. Sweeny, to DuPont.
32. C. S. Marvel and J. H. Johnson, *J. Am. Chem. Soc.*, **72**, 1674 (1950).
33. D. J. Lyman, *J. Poly. Sci.*, **45**, 49 (1960).

6

HIGH TEMPERATURE AND HIGH-PERFORMANCE POLYMERS

This chapter covers a range of polymers with high glass-transition temperatures (T_g), high retention of properties above 200°C, high flame resistance, and/or a high level of tensile properties. The discussion also includes anisotropic melt and and lyotropic liquid crystalline polymers as well as isotropic high-performance polymers. The original impetus for this work was for fibers, films, and plastics to meet military and space needs, including organic and inorganic polymers. Major success to date, however, has been with organic polymers that provide lower densities, greater flexibility, and toughness. In general, a necessary characteristic for tough, thermally stable organic polymers is a high degree of aromaticity.

In chapter 3 we referred to requirements for achieving the "ultimate properties" in polymers. In this chapter we will describe preparations of some of the polymers that come closest to achieving such properties. Although the U.S. Air Force provided much of the driving force for many high-temperature/high-modulus polymers, most of the original discoveries in the high-temperature and liquid crystalline polymer field resulted from basic research on fibers at the DuPont Pioneering Research Laboratory (1–3).

As discussed in chapter 3, the molecular requirements for high strength and high modulus in fibers are *(1)* extended-chain configuration (e.g., all para chain extension) and, in the case of polyamides, all trans amide bonds; *(2)* minimum number of flexible bonds per polymer unit (e.g., condensed ring system); *(3)* high molecular orientation as measured by sonic modulus; *(4)* high molecular weight; and *(5)* freedom from structural defects.

Unless stated otherwise, all inherent viscosities are determined at 30°C. in 0.5% solutions in the solvent stated.

6.1. AROMATIC POLYAMIDES

6.1.1. Polymerization Methods

High molecular weight, all-aromatic crystalline polyamides were first prepared by Sweeny (2) using solution methods in chlorinated hydrocarbons containing dissolved tertiary amine hydrochloride (2) and later in amide solvents containing calcium or lithium chlorides (3). The latter method was discovered by Sorenson alone and has become the standard method for preparing aromatic polyamides. This polymerization chemistry has resulted in a range of aromatic polyamides that are essentially nonflammable and nonmelting. The first commercial aromatic polyamide was poly(*m*-phenylene isophthalamide), introduced by the DuPont Company as Nomex fiber. This has been successful for protective clothing for firemen and racing car drivers and for flight clothing in the military. Preparations for this and other isotropic polyamides are described below.

Melt processible, high melting, high T_g polyamides have been prepared by two routes using 3,4′-bis(aminophenyl) ether as the major amine component. The first route uses diphenyl isophthalate or diphenyl 5-*t*-butyl isophthalate as the major acid component and 5 to 15% of an aliphatic diamine or a diphenyl ester of an aliphatic dicarboxylic acid as the minor component. Polyamidation is catalyzed by fluoride salts (4).

A second novel melt route (5) to high T_g polyamides involves the reaction of aromatic diamines, such as *m*-phenylenediamine and 3,4′-diaminodiphenyl ether, with bis-lactams such as *N*,*N*′-isophthaloyl bis-(2-caprolactam). High molecular weight aromatic copolyamides containing nylon 4, 5, or 6 aliphatic units result from displacement of and some ring opening of the lactam.

A particularly novel route to aromatic polyamides that avoids using acid halides is based on the reaction of aromatic amines with dicarboxylic acids catalyzed by triphenyl phosphite and pyridine (6). Hagashi et al. (6) proposed that the reaction proceeds via the *N*-phosphonium salt of pyridine. The reaction is significantly affected by the tertiary amine used and is catalyzed by tertiary amine hydrochlorides, such as triethylamine hydrochloride (**6.1**). With proper choice of solvent and dehydrating agent this could be an attractive route to the Kevlar-like polyamide from *p*-aminobenzoic acid.

$$P(OPh)_3 + R{-}COOH \xrightarrow{TEA\cdot HCl/Py} \underset{\mathrm{I}}{H{-}\overset{\overset{C_5H_5N^{\oplus}\ \ominus OPh}{|}}{P}(OPh)_2{-}OOC{-}R} \xrightarrow{R'{-}NH_2} R{-}CONH{-}R' + HO{-}P(OPh)_2 + PhOH$$

Schematic 6.1

Aromatic polyamides have also been made in intermediate molecular weight by reaction of aromatic diiodides or bromides, carbon monoxide, and aromatic diamines, such as 4,4′-diaminodiphenyl ether in the presence of a palladium/triphenylphosphine catalyst (7) (**6.2**).

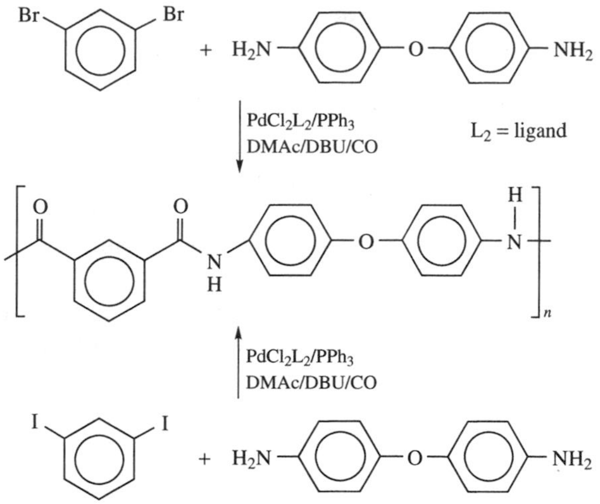

Br Br + H_2N O NH_2

$PdCl_2L_2/PPh_3$ DMAc/DBU/CO L_2 = ligand

O O N H O H N n (1)

$PdCl_2L_2/PPh_3$ DMAc/DBU/CO

I I + H_2N O NH_2

Schematic 6.2

Probably the best known high-performance aromatic polyamide is poly(*p*-phenylene terephthalamide), known as "Kevlar." Initially high molecular weight polymer could not be prepared via low-temperature acid chloride routes either in chlorinated or in amide solvents because the low molecular weight polymer precipitated from solution. Preston and Smith (8) of Monsanto prepared the AB analog poly-*p*-benzamide by low-temperature polymerization of *p*-aminobenzoyl chloride hydrochloride but found the polymer too intractable to spin into fiber. Kwolek et al. (9) followed up on this work and, with great persistence, succeeded in preparing fiber with a tenacity of 7 g/denier (gpd) and an unusually high modulus of 900 gpd. Antal suggested that the polymer was liquid crystalline in solution, based on the polymer solution behavior and schliering. This novel suggestion generated increased activity at DuPont in the area of liquid crystalline polymers, especially for identifying and developing a polymer of potentially lower cost than that from the aminobenzoyl chloride hydrochloride route. Further work in the Pioneering Laboratory by a number of scientists (Bair, Memeger, et al.) found that high molecular weight poly(*p*-phenyleneterephthalamide) could be obtained from terephthaloyl chloride and *p*-phenylenediamine in amide solvents of the "Sorenson" type if the precipitated low molecular weight polymer was retained for at least 4 h in the polymerization medium without quenching with water. Despite being in the solid state, the polymer's chain-ends still had enough mobility to continue polymerizing. The high molecular weight polymer was spun (by Blades) using a 0.25- to 0.5-in. air gap and a 20% solution in 100%

sulfuric acid (Bair), giving outstanding tensile properties. Moral: Do not isolate your polymer until you are sure that polymerization has ceased.

6.1.2. Liquid Crystalline Polymers

As pointed out above, Antal, Kwolek and others recognized that the para-oriented aromatic polyamides were liquid crystalline in solution (lyotropic) above certain critical concentrations (10). Flory (11) had already proposed the requirements for rigid-rod (liquid crystalline) polymers, pointing out that polymer liquid crystallinity is solely a consequence of molecular asymmetry and stiffness and is not due to intermolecular attractions. It is dominated by intermolecular repulsions, and there is a limit to the number of rod-like chains that can be accommodated randomly in solution or a melt. When the limiting concentration is exceeded, a crystalline or liquid crystalline phase separates.

In the liquid crystalline phase, the molecules align to accommodate the higher concentration, and an ordered phase is formed at the expense of the random or isotropic phase. This phase separation is as follows (11):

$$V_p = (8/x)[1 - 2/x] = 8/x - (4/x)^2 = 8/x$$

at high values of x, where V_p = volume fraction of polymer of axial ratio x. The minimum axial ratio for a stable nematic order in a melt of stiff rods is about 6. Theory predicts that anisotropic phases tolerate only negligible concentrations of random-coil molecules, whereas isotropic phases tolerate a fair concentration of rod-like molecules. In rejecting the random coil polymer, the liquid crystalline phase shows the general feature of crystalline materials, namely rejection of foreign material.

Most "rod-like" polymers are not true rods and depart from linearity as x approaches infinity. This is apparent from plots of $\log[\eta]$ vs. $\log M$, where the slope of the curve, i.e, the exponent α in the Mark-Houwink-Sakurada relationship ($[\eta] = KM^{\alpha}$) is close to $1.3 - 1.8$ at low $[\eta]$ but approaches $1 - 0.7$ for random coils at very high $[\eta]$. As flexibility increases, higher solution concentrations are required to achieve the critical V_p concentration for phase separation.

The most favorable conditions for liquid crystallinity are in the melt because of the higher polymer chain concentration. About three aromatic rings with para connecting linkages ($x = 6.4$) should be sufficient to produce liquid crystalline order. With lyotropic polymers, i.e., liquid crystalline in solution, better solvents lead to greater chain extension and facilitate liquid crystallinity at lower concentrations, as predicted by Flory (11).

One measure of rod-like character, or straightness, is persistence length, i.e., the vector sum of the projections of the $(n + i)$th segment (where $i = 0$ to infinity in the nth segment) of a very long chain. The more rod-like the chain, the more the $(n + i)$ bond "remembers" the direction of the nth bond. Another way of looking at the persistence length is to consider it as a measure of how close the isolated chain comes to absolute straightness. Persistence length can be calculated from chain conformation data or derived from polymer solution properties (11, 12).

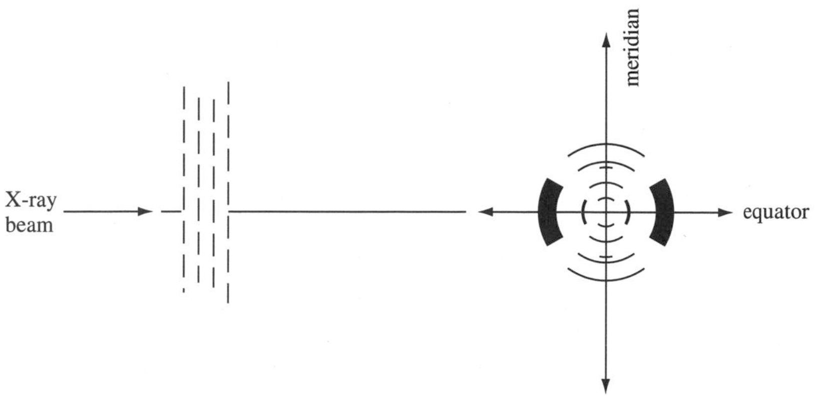

Figure 6.1. Nematic x-ray pattern.

The most common type of polymer liquid crystallinity is called nematic, i.e., the chains pack like straws of various lengths, with lengthwise, but little lateral, order. Those with lengthwise and lateral order are called smectic. An idealized x-ray pattern for a nematic polymer (Fig. 6.1) is characterized by two strong equatorial crescents that correspond to the diffuse outer ring of an unoriented sample. The shorter and sharper the arcs, the higher the degree of orientation in the parallel direction. The meridianal region has a series of short arcs that are attributed to scattering from nonuniformities and lack of register of the chains crosswise. They are considerably weaker than the equatorial arcs and their positions correspond to the repeat distance of the unit lengths.

Para-linked aromatic polyamides represent a class of polymers with rod-like structures that decompose significantly below their melt temperatures but that form lyotropic solutions from which they can be processed. Melt temperatures can be obtained by extrapolation of the melt temperatures of oligomers, as described by Flory (12):

$$\frac{1}{T_m} - \frac{1}{T_m^\circ} = \left(\frac{R}{\Delta H}\right)\left(\frac{2}{X_n}\right)$$

where X_n is the number average degree of polymerization, T_m is the sample melting point in K and T_m° is the absolute melting point. This relationship has been used by Gardner (personal communication) to extrapolate to the theoretical melt temperature of poly-p-benzamide, by plotting the inverse of the T_m of oligomers against the inverse of the number of polymer units and extrapolating to infinite molecular weight (Fig. 6.2). The melt temperatures of poly(p-phenylene terephthalamide) and poly-p-benzamide are estimated to be 727°C and 560°C, respectively.

In aromatic polyamides, chain stiffness results from the aromatic units having coaxial and oppositely extended bonds, enhanced by the partial double bond character of the amide linkage.

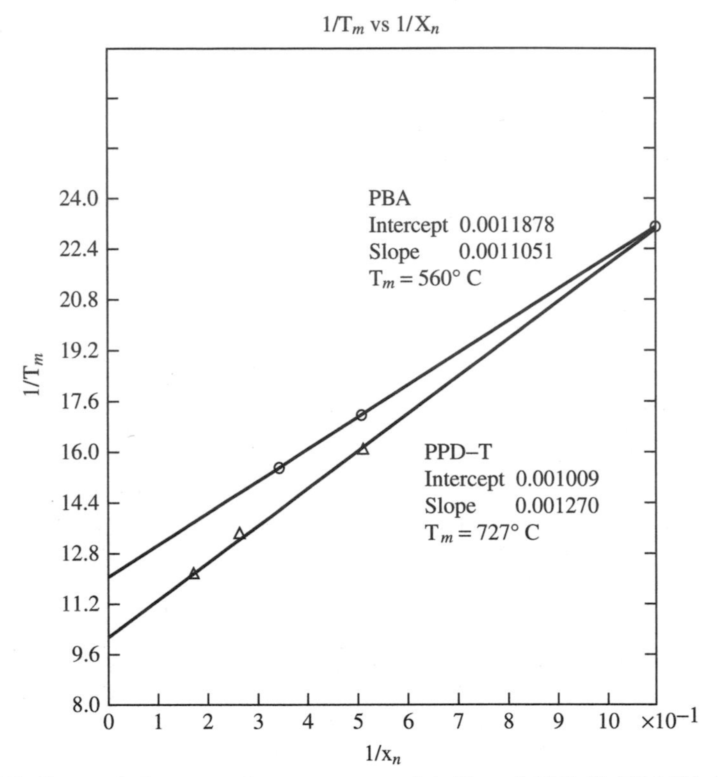

Figure 6.2. Extrapolation to melt temperature of 1,4B and PPD-T. PBA(O) intercept, 0.0011878, slope, 0.0011051, T_m, 560°C. PPD-T (Δ) intercept, 0.001009, slope, 0.001270, T_m, 727°C.

$$-\overset{\overset{\displaystyle O}{\|}}{C}-NH- \longleftrightarrow -(HO)-C{=}N-$$

Rod-like aromatic polyamides are to varying degrees soluble in amide-solvents (e.g., dimethylacetamide, *N*-methylpyrrolidinone, and tetramethylurea, all containing saturated solutions of lithium or calcium chlorides), and in concentrated (101%) sulfuric acid, methanesulfonic acids, and hydrogen fluoride. Solubility in the amide-salt solvents results from association of chloride ions with the amide protons of the polymer and coordination of the solvent amide group with those of the polymer. Solubility in strong acids arises from formation of adducts between acid and amide units and dissolution of the complex with additional acid. Indeed, poly(*p*-phenyleneterephthalamide) forms a complex with 5 mol sulfuric acid. Amazingly the complex does not corrode aluminum.

An unique and powerful basic solvent system composed of dimethylsulfoxide, potassium *t*-butoxide, and methanol (20/1/1 w/w/w) has been discovered for aromatic polyamides (13). This functions by forming and dissolving the amide anion. However, the solutions are not anisotropic, either because the polyanion is not stiff enough or because the chains are pushed too far apart by ionic repulsion. Excellent films can be made from these anionic solutions.

The amide-salt solvents are the best medium for the synthesis of aromatic polyamides and can be used in the direct preparation of liquid crystalline solutions. To obtain high molecular weight polymers in, for example, the *N*-methylpyrrolidone (NMP)/calcium chloride (7.5%) solvent system requires: pure intermediates, dry NMP (distilled over anhydrous calcium chloride or phosphorus pentoxide), calcium chloride dried at 400°C, and flamed-dried equipment to eliminate moisture. Probably most critical is to keep the calcium chloride dry, because it picks up moisture readily. It *must* be kept and handled in a drybox. Achieving high molecular weight polymer also requires continuing polymerization beyond precipitation of the polymer as a crumb or gel. The active ends are usually mobile enough to continue polymerizing in the solid state a few hours longer, resulting in increased molecular weight.

An enhanced polymerization system can be made by adding *N*-methylpyrrolidine as an acid acceptor to the NMP/calcium chloride solvent system. The *N*-methylpyrrolidine hydrochloride that forms enhances the solvent power of the NMP/calcium chloride and, for example, retains poly(*p*-phenylene terephthalamide) in solution for as long as 48 h. Only sufficient (or slightly less) *N*-methylpyrrolidine to take up the hydrogen chloride formed in the polymerization should be used because an excess ties up the acid chloride and results in low polymer being formed. The solution consists of large fibrils (0.05 cm) of polymer, but fibers can be spun and fibrils for paper made by shear precipitation (14).

Very high strength and high modulus fibers have been made from para-oriented aromatic polyamides by "dry-jet" spinning of anisotropic solutions. This allows efficient and high attenuation of the polymer dope before quenching. The high attenuation improves orientation of the polymer chains by extensional flow. Although the liquid crystalline aramide fibers are generally highly crystalline and well oriented and provide a level of tensile modulus approaching the theoretical, the tensile strength, although high, is lower than theoretically calculated ($E/10$). This results from defects within the crystalline lamellae and from partial disorientation of the fiber-core crystals during quenching. Postquench drawing might be expected to improve fiber properties, except that highly crystalline aramid fibers are not readily amenable to drawing because they have high glass-transition and high melt temperatures, and strong interchain hydrogen bonding.

However, as was pointed out in chapter 3, crystallinity is not a prerequisite for high strength in fibers provided the polymer has a reasonably high T_g and the chains are well enough aligned (extended) to provide a high number of load bearing chains. This understanding was capitalized on by Teijin Co. Ltd. in the development of its aramid fiber Technora. The fiber is based on the copolymer of 50/50 mol % of *p*-phenylenediamine and 3,4′-oxydianiline with terephthaloyl chloride. The copolymer is made in and spun from *N*-alkyl amide solvents, such as *N*-methylpyrrolidone, in which it forms isotropic solutions. The polymer is wet-spun from a 10% solution into an aqueous coagulation bath and subsequently drawn 10 times at 490°C to produce high strength, highly oriented fibers (15).

High molecular weight lyotropic poly-*p*-benzamide has been prepared from the aromatic amine hydrochloride acid chloride in amide solvent/salt solutions at ambient temperatures and by a high-temperature route from 4-carboxyphenyl isothiocyanate (16).

Lyotropic polyamides are formed not only from all-aromatic polyamides but also from polymers containing one aromatic component and a short-chain aliphatic component. Thus liquid crystalline polyoxamides, polyfumarates, polyhydrazides, and poly(amide-hydrazides) have been prepared in high molecular weight. As expected, only the para-linked polymers yielded lyotropic polymers, some of which have been spun into high-strength fibers (17). It is interesting to note that 2-T, if made in high molecular weight and properly spun, might rival Kevlar in properties, especially in transverse properties, because of stronger hydrogen bonding. Selection of polymerizing conditions (e.g, tetramethylene sulfone solvent) and spinning solvent [H_2SO_4, methane (di?) sulfonic acid] would be worth exploring.

6.1.3. Preparation of High-Temperature Polyamides

80. Preparation of Bis-1,3-(m-nitrobenzamido)benzene (18, 19)

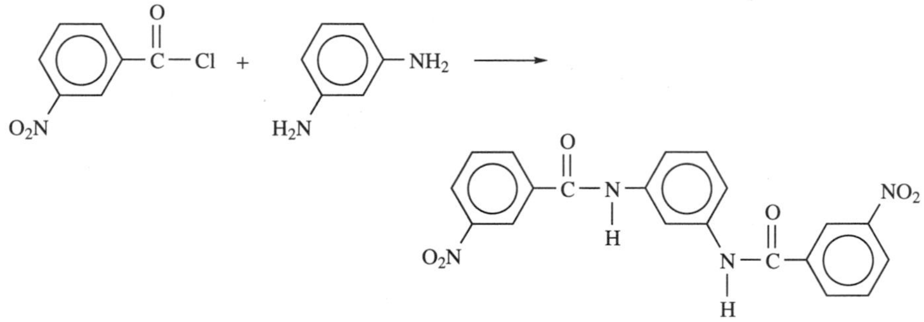

O
C—Cl +
NH₂
O_2N
H_2N
O
C—N
H
NO_2
O
O_2N
N—C
H

A solution of 74.23 g (0.4 mol) *m*-nitrobenzoyl chloride dissolved in 80 mL tetrahydrofuran is poured all at once into a blendor jar containing 21.6 g (0.2 mol) *m*-phenylenediamine and 16 g (0.4 mol) sodium hydroxide in 400 mL ice-cold water. The reaction mixture is stirred rapidly for 5 min, then filtered with suction. The product is washed with dilute hydrochloric acid, dilute sodium carbonate solution, and finally with water. The product is dried for 1 h at 120°C. The dried product is dissolved in dimethylacetamide, decolorized with carbon, and filtered. The product is precipitated from the filtrate with methanol, filtered, and dried. The melting point (mp) is 267–269°C.

81. Preparation of Bis-1,3-(m-aminobenzamido)benzene (18, 19)

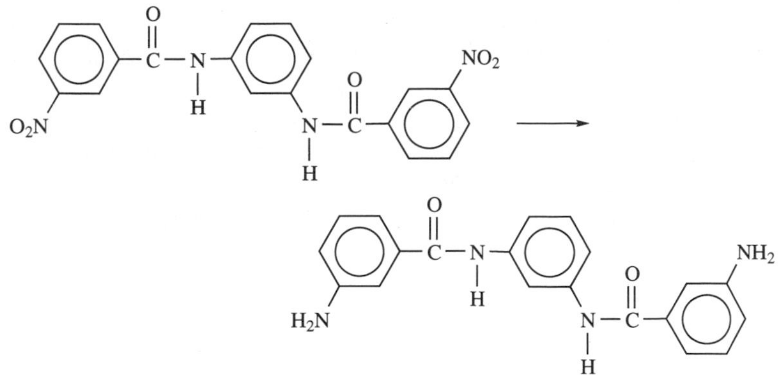

A total of 40.6 g (0.1 mol) bis-1,3-(*m*-nitrobenzamido)benzene is dissolved in dimethylacetamide, and about 1 g 5% palladium on charcoal is added. The solution is placed in a Parr low-pressure reaction apparatus at 60 to 65°C. Hydrogen is introduced until no further pressure drop is observed. The reaction is complete in about 1 h. The catalyst is filtered through a fine-sintered glass funnel. The product is then precipitated from the filtrate as the dihydrochloride by addition of dilute (18%) hydrochloric acid, filtered with suction, and dried. The dihydrochloride is heated in acetone to dissolve any unreacted dinitro compound, refiltered, and dried. The dihydrochloride is added to an aqueous solution of sodium carbonate. The product is filtered, washed with water, and dried. The mp is 212 to 214°C, and the product is sufficiently pure for the next step.

82. Preparation of Ordered Aromatic Copolyamide Poly(isophthalamido-m-phenylamidobenzene) (18, 19)

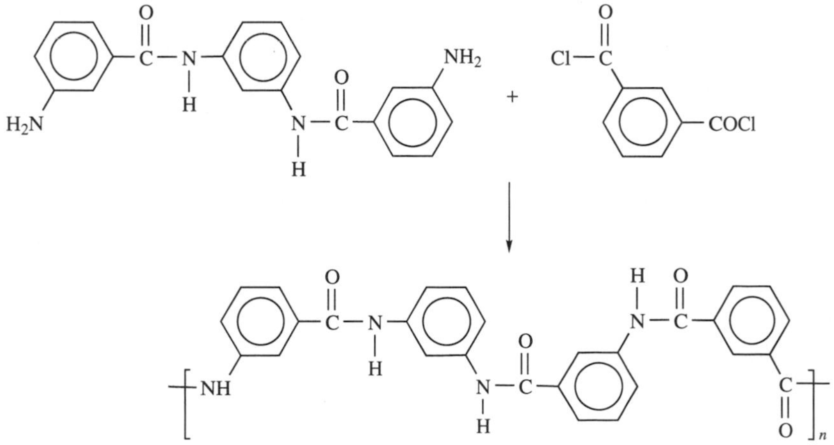

A total of 200 g (0.578 mol) bis-1,3-(*m*-aminobenzamido)benzene is dissolved in 1600 g dimethylacetamide and stirred by an air stirrer in a 3 L resin kettle.

Liquid isophthaloyl chloride is added in portions from a weighing pipet until a clear solution is obtained in a spot test with dimethylaminobenzaldehyde, which gives a yellow color with aromatic amine ends. About 122 g isophthaloyl chloride will be needed (the theoretical amount is 117 g). The solution temperature is maintained between 0 and 25°C during the addition of the isophthaloyl chloride by keeping the solution in a dry-ice bath. Lithium hydroxide is added to the solution until a green color is obtained with bromophenol blue (pH = 3.0 to 4.6). The polymer is precipitated into water. The inherent viscosity in dimethylacetamide (DMAc) containing 4% lithium chloride at 0.5% concentration is about 2.2. Tough films can be cast from solution.

83. Preparation of Polymer from 2,2-Bis(4-aminophenyl)propane and Isophthaloyl Chloride by Solution Method (3)

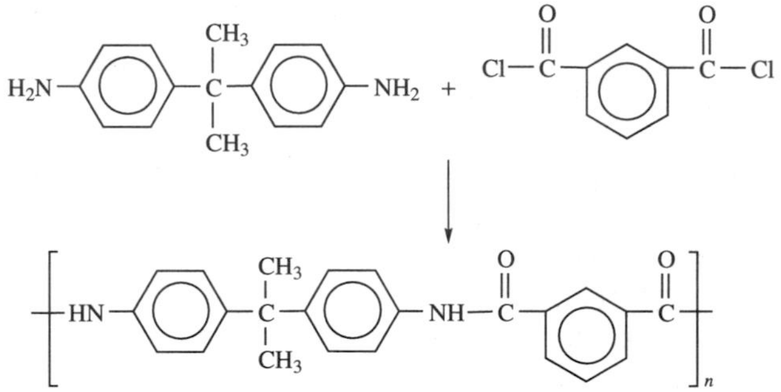

In a flask equipped with stirrer, nitrogen inlet, and drying tube, 23.63 g (0.1 mol) 2, 2-bis(4-aminophenyl)propane, is dissolved in 200 mL anhydrous dimethylacetamide. The solution is cooled to a mush in a dry-ice/acetone bath. Then 20.3 g (0.1 mol) isophthaloyl chloride, dissolved in 18.5 mL toluene is added in one portion, the cooling bath is changed to ice water, and the mixture is stirred for 90 min. During this time, a viscous solution of polymer forms that contains dispersed crystals of dimethylacetamide hydrochloride. The polymer is precipitated by pouring the polymerization mixture into rapidly stirred water in a blender. After thorough washing and drying, a 98% yield of granular polymer is obtained. Inherent viscosity is 1.66 (sulfuric acid) at 0.5% concentration.

84. Preparation of Poly(m-phenyleneisophthalamide) in Chloroform Solution (17)

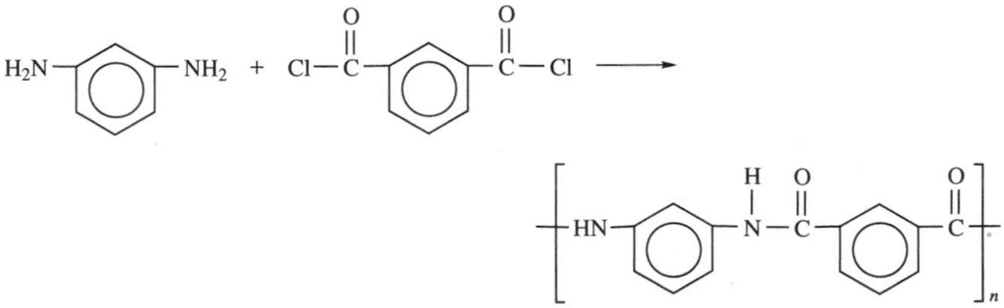

A 200 mL round-bottom flask equipped with a low-speed stirrer and a dropping funnel is charged with 2.163 g. *m*-phenylenediamine, 5.62 mL triethylamine, 5.506 g triethylamine hydrochloride, and 36 mL chloroform. Then 4.06 g isophthaloyl chloride in 14 mL chloroform is added through the dropping funnel over a period of 15 min while the mixture is stirred slowly and maintained at 30°C. An additional 3 mL chloroform is used to rinse the funnel. After 20 min, the reaction mixture is clear and viscous and is poured into a large volume of hexane, yielding a fibrous precipitate. This is collected and washed thoroughly with hot water. After drying, the yield is 99%. The inherent viscosity is 1.9 in DMAc/0.4% lithium chloride at 0.5% concentration.

The next preparations are for polymers that provide fibers with high tensile strength and moduli. The initial polymer preparations are those of poly(*p*-phenyleneterephthalamide), the polymer base for Kevlar.

85. Preparation of Poly(p-phenyleneterephthalamide) (20)

$$H_2N-C_6H_4-NH_2 + Cl-\overset{O}{\overset{\|}{C}}-C_6H_4-\overset{O}{\overset{\|}{C}}-Cl \longrightarrow \left[\overset{H}{\overset{|}{N}}-C_6H_4-\overset{H}{\overset{|}{N}}-\overset{O}{\overset{\|}{C}}-C_6H_4-\overset{O}{\overset{\|}{C}} \right]_n$$

Commercially available 1,4-phenylenediamine and terephthaloyl chloride are purified by vacuum sublimation to essentially white solids with sharp melting points. The terephthaloyl could be further purified, if necessary, by crystallization from hexanc and vacuum drying. Commercially available hexamethylphosphoramide (**Carinogen!** Handle only in hood and with rubber gloves) and *N*-methylpyrrolidinone-2 are purified by vacuum distillation over calcium hydride to clear, colorless liquids with a narrow boiling range.

In an oven-dried, 1-L resin kettle, equipped with an egg-beater stirrer, 11.57 g (0.107 mol) 1,4-phenylenediamine, 280 mL hexamethylphosphoramide, 140 mL and *N*-methylpyrrolidinone-2 are combined anhydrously and stirred for 15 min under dry nitrogen while being cooled in a bath of dry ice and acetone at −15°C. To the resulting cooled solution is added 21.76 g (0.107 mol) powdered terephthaloyl chloride with rapid stirring. The reaction mixture becomes a thick, paste-like gel in about 5 min. Stirring is discontinued after a few minutes, and the thick mixture is allowed to stand overnight under dry nitrogen while gradually warming to room temperature. The mixture is then added with rapid stirring to a blender containing water to wash away the solvent (properly dispose of the hexamethylphosphoramide filtrates), and the polymer is collected by filtration. The washing procedure with water is repeated several times, and finally the polymer is washed with acetone. The polymer is dried in a vacuum oven at 85°C for about 24 h. The dried polymer (24.4 g; 96%) has an inherent viscosity of around 6.90 at 0.5% concentration in sulfuric acid.

86. Poly(chloro-p-phenyleneterephthalamide) (20)

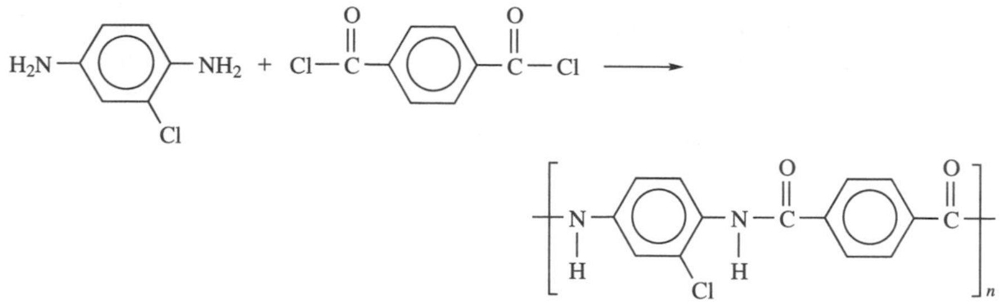

2-Chloro-*p*-phenylenediamine is unstable and is obtained from the commercially available hydrochloride salt. The salt is dissolved in water and the solution is made alkaline to litmus paper with 50% aqueous sodium hydroxide. The mixture is extracted with petroleum ether that is dried over anhydrous sodium sulfate. The solvent is removed under reduced pressure, and the residue is sublimed under vacuum at 80°C to give pale yellow needles (mp 64 to 65°C).

In an oven-dried 1-L resin kettle equipped with an air-driven stirrer, 35 g (0.245 mol) 2-chloro-*p*-phenylenediamine is dissolved in 425 mL hexamethylphosphoramide. The mixture is cooled in an ice-water bath under dry nitrogen, and 49.7 g (0.245 mol) powdered terephthaloyl chloride is added with stirring. Almost immediately, a precipitate forms as a wet paste. The mixture is allowed to stand overnight under dry nitrogen, and the polymer is then isolated by pouring with rapid agitation into a blender containing water. The workup procedure is repeated several times with water and finally with acetone. The polymer is filtered and dried overnight at 78°C in a vacuum oven. The yield is 60 g (90%). The inherent viscosity is 1.13 in sulfuric acid at 0.5% concentration.

Poly(*p*-phenyleneterephthalamide) can also be polymerized to high molecular weight with retention of the polymer in solution for many hours if *N*-methylpyrrolidine is added as an acid acceptor, provided no more than 2 mol *N*-methylpyrrolidine per mole acid chloride is used. This echoes back to the similar behavior of triethylamine in the preparation of poly(*m*-phenyleneisophthalamide) (see prep. 84). The *N*-methylpyrrolidine hydrochloride enhances the solvent power of the NMP solvent/calcium chloride (14).

87. Preparation of Poly(p-phenyleneterephthalamide) in NMP (14)

A resin kettle equipped with a basket stirrer, a nitrogen inlet, and an outlet to which drying tubes are connected is flamed under a stream of nitrogen to remove adsorbed moisture. To the dried kettle are added 15.6 g anhydrous calcium chloride (previously dried at 400°C) and 200 g dry distilled NMP. The mixture is heated to 100°C and stirred until almost all the calcium chloride is dissolved. The solution is cooled with an external ice bath, and 5.4 g (0.05 mol) *p*-phenylenediamine is added and stirred until dissolved. Then 6.2 g (0.073 mol)

NMP is added, and the mixture is stirred for several seconds. Then 10.2 g (0.05 mol) pure recrystallized terephthaloyl chloride is added all at once with good stirring, and the entry funnel is rinsed with an additional 20 mL NMP to ensure total addition. The mixture is stirred rapidly; the cooling bath is removed after 1 min. Stirring is continued at ambient temperature for a further 1.5 h. The reaction mixture changes to a thick fibrous gel composition and remains that way throughout the reaction. A sample of the gel is removed at the end of the reaction time, and after diluting five times with NMP, is examined under a polarizing microscope, showing clumps of fibers and individual fibers about 60 μm in length. A sample of the gel is added to a blender and, after thorough extraction with water is filtered and dried to yield a pulpy material. The inherent viscosity of a 0.5% solution in 98% sulfuric acid at 30°C is 5.03.

The first liquid crystal polyaramide was poly-*p*-benzamide. Preparation of the precursor monomer, *p*-aminobenzoyl chloride hydrochloride, involves a sensitive synthesis. *p*-Aminobenzoic acid is reacted with thionyl chloride to form sulfinylaminobenzoyl chloride, which can be purified by distillation or recrystallization. This is then dissolved in ether and treated with dry hydrogen chloride to yield *p*-aminobenzoyl chloride hydrochloride, which is extremely sensitive to moisture and heat, forming an acid, a dimer, or oligomers. When dissolved in dimethylacetamide or other amide-solvents, the amine group is freed from the hydrochloride and polymerization proceeds (**6.3**).

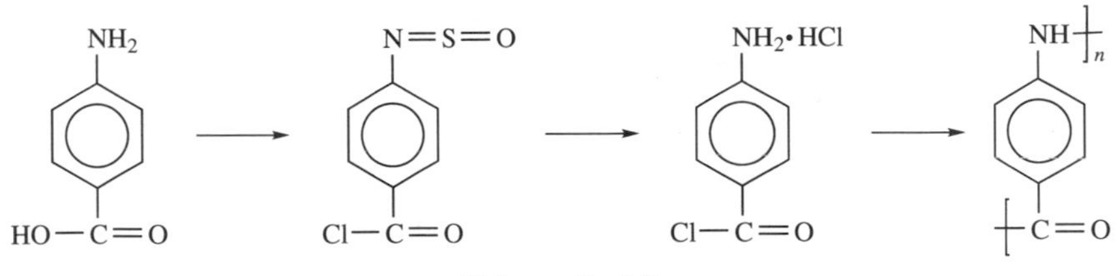

Schematic 6.3

88. Preparation of p-Sulfinamidobenzoyl Chloride (9)

A total of 100 g (0.729 mol) *p*-aminobenzoic acid is refluxed with 450 mL thionyl chloride for 3 h. The excess thionyl chloride is removed under vacuum, and the residue is fractionally distilled. The fraction with boiling point (bp) 107 to 113°C at 1 to 2 mm is collected and redistilled to give the desired product with bp 110 to 111°C at 1.4 mm and mp 31 to 31.5°C.

89. Preparation of p-Aminobenzoyl Chloride Hydrochloride (9)

In a flamed-dried three neck, round-bottom flask equipped with a stirrer, gas inlet tube, and condenser equipped with a drying tube and a lead to a sodium hydroxide absorbing gas system (for the xs hydrogen chloride) is added 1800 mL sodium-dried ether and 100 g *p*-sulfinylaminobenzoyl chloride. The solution is cooled by an external ice bath, and dry hydrogen chloride is passed over the stirred yellow solution for 3 h. A white precipitate forms and after 3 h is filtered in a dry box through a dry sintered-glass funnel. The collected product is washed several times with dry ether, transferred in the dry box into a vacuum desiccator, and dried at reduced pressure. Yield is 93%.

90. Preparation of Poly(p-benzamide) (9)

The amide solvent used, hexamethylphosphoramide (**Carcinogen!**), is purified by distilling over calcium hydride through a spinning band column under vacuum and stored over molecular sieves. Alternatively, *N*,*N*,*N'*,*N'*-tetramethylurea or dimethylacetamide containing 4% dry (at 450°C for 4 h) lithium chloride can be used as a solvent but inherent viscosities are slightly lower. A 500-mL round-bottom, three-neck flask equipped with a stirrer, nitrogen inlet tube, and calcium chloride tube is flamed, purged with nitrogen, sealed, and placed in a drybox. Then 150 mL hexamethylphosphoramide is poured under dry nitrogen in a fume hood into a dry Erlenmeyer flask, which is sealed and cooled in an ice bath. Then 31.6 g (0.165 mol) *p*-aminobenzoyl chloride hydrochloride is weighed in a dry box and transferred to the reaction flask, which is then removed from the dry box, connected to the stirrer motor and dry nitrogen line in the fume hood, and cooled in an ice bath. The hexamethylphosphoramide is poured rapidly into the stirred *p*-aminobenzoyl chloride hydrochloride, and the solution is stirred with cooling for 1 h. The ice bath is removed, and stirring is continued overnight (if gellation occurs stirring is stopped). The reaction mixture is sheared in a blender with excess water and filtered. The collected solid is washed three times with water and once with ethanol. The polymer is dried overnight at 80 to 90°C in a vacuum oven under nitrogen. The yield is quantitative. The inherent viscosity is as high as 2.3 in 98% sulfuric acid at 0.5% concentration.

91. Preparation of Poly-p-benzamide via Pyrolysis of the Adduct of p-Isothiocyanatobenzoic Acid (16)

$$\mathrm{HO\overset{O}{\overset{\|}{C}}-C_6H_4-NH_2 + CS_2} \rightleftharpoons \left[\mathrm{BH^{+}\,{}^{-}O\overset{O}{\overset{\|}{C}}-C_6H_4-NH\overset{S}{\overset{\|}{C}}-S^{-}HB^{+}}\right] \rightleftharpoons$$

VI

$$\left[\mathrm{BH^{+}\,{}^{-}O\overset{O}{\overset{\|}{C}}-C_6H_4-NCS}\right] + [\mathrm{HS^{-}HB^{+}} \rightleftharpoons \mathrm{H2S}\uparrow + \mathrm{B}]$$

VII

$$\Big\Downarrow$$

$$\left[\mathrm{BH^{+}\,{}^{-}O\overset{O}{\overset{\|}{C}}-C_6H_4}\left(\mathrm{NH\overset{S}{\overset{\|}{C}}-O-\overset{O}{\overset{\|}{C}}-C_6H_4}\right)_x\mathrm{-NCS}\right] \xrightarrow{-\mathrm{COS}}$$

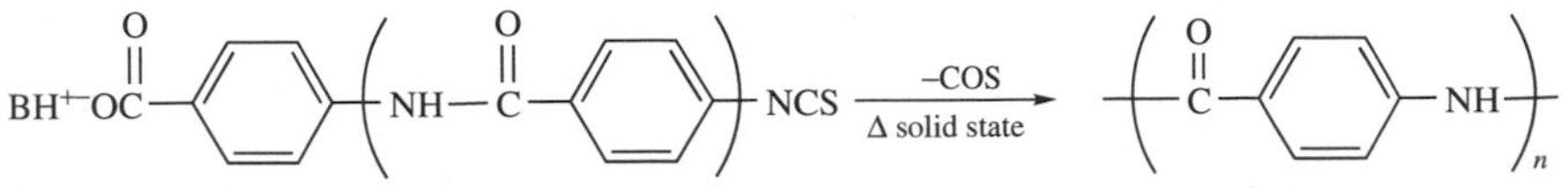

IX

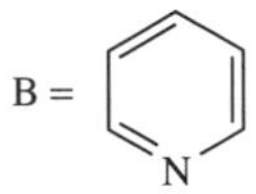

B = pyridine

To a solution of 31.6 g (0.40 mol) pyridine and 39.5 g (0.52 mol) carbon disulfide in 200 mL nitrobenzene contained in a 500-mL three-necked flask with a 2-ft. Allihn condenser, drying tube, mechanical stirrer, and nitrogen purge is added 13.7 g (0.1 mol) *p*-aminobenzoic acid. As the reaction mixture is warmed with a mantle, a clear orange solution is obtained. The reaction is heated rapidly to 110°C, then more slowly to 160°C. Much of the carbon disulfide is lost through evaporation during the heating. After 2 h the reaction becomes cloudy and a yellow precipitate forms. After heating a total of 5 h, the mixture is cooled below 100°C and filtered through a Buchner funnel. The filter cake is washed with toluene and air dried on the filter. Yield is 4 g of a light yellow solid. Infrared (ir) analysis indicates this to be an oligomer of *p*-aminobenzamide.

92. Preparation of Poly(p-benzamide) (16)

To a 25-mm (outer diameter, o.d.) polymer tube is added 1 g of the oligomer obtained from preparation 91. Air was removed from the tube by alternately evacuating and purging with nitrogen. The tube is then heated under nitrogen in an oil bath for 1 h at 180 to 182°C, 0.25 h at 240°C, and 1 h at 256°C, followed by heating 1 h at 400°C. The yield is 0.6 g of a light yellow solid with an ir spectrum identical to the polymer from *p*-aminobenzoyl chloride hydrochloride, i.e., poly-*p*-benzamide. The inherent viscosity is 4.76 in 98% sulfuric acid at 0.5% concentration.

93. Preparation of 50/50 Copolymer from p-Phenylenediamine, 3,4′-Diaminophenyl Ether and Terephthaloyl Chloride (15)

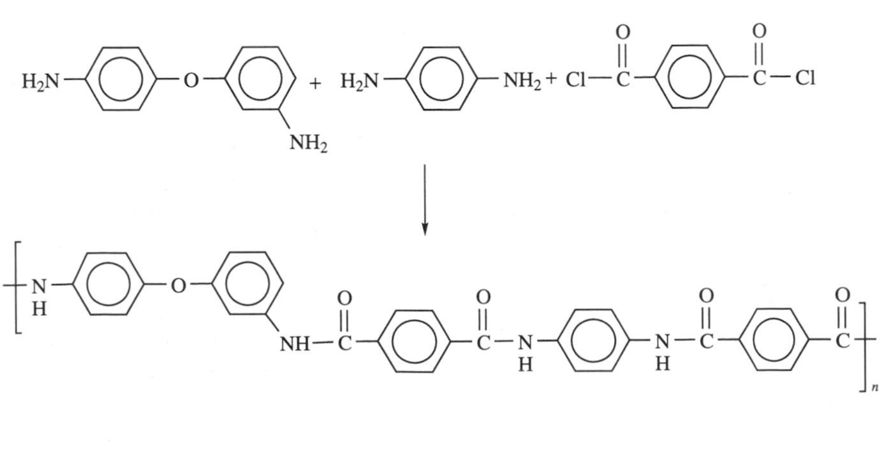

A resin kettle equipped with a basket stirrer, nitrogen inlet and outlet with a calcium chloride drying tube, and thermometer is flamed out with a propane torch while flushing with nitrogen. After cooling under nitrogen, 200 g dry distilled (over calcium hydride) *N*-methylpyrrolidinone, 2.16 g (0.025 mol) pure *p*-phenylenediamine and 5.0 g (0.025 mol) 3,4′-diaminodiphenyl ether are added. The mixture is stirred for about 30 min until the amines dissolved. The kettle is then cooled in an ice-water bath to <10°C, and 10.25 g (0.05 mol) powdered pure terephthaloyl chloride is added all at once with good stirring. Polymerization is continued for 1 h at 10°C, and the temperature is raised to room temperature. Then 3.7 g (0.05 mol) calcium hydroxide is added, and stirring continued for 1 h more at room temperature. The clear polymer solution is filtered and centrifuged, if spinning or film casting is intended. The sample polymer, isolated by precipitating into water and vacuum oven drying, has inherent viscosities (0.5% concentration in NMP or DMAc) in the 2.8 to 3.5 range.

Aromatic polyamides are soluble not only in strong acids but in a novel basic solvent consisting of dimethylsulfoxide (DMSO), potassium or cesium butoxide and an alcohol, preferably methanol. The polyamide anion yields isotropic solutions from which tough film and fiber can be made, though tensile properties are considerably lower than those obtained from anisotropic solutions.

94. General Procedure for Preparing Polyanions from Aromatic Polyamides (13)

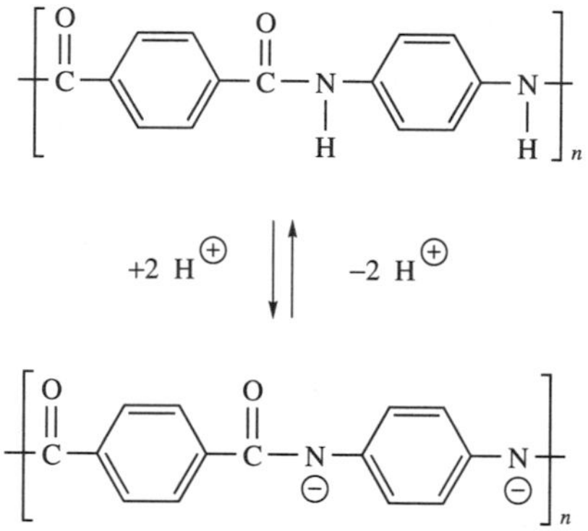

$+2\ H^{\oplus}$ $\quad\rightleftarrows\quad$ $-2\ H^{\oplus}$

Experiments are performed in an inert atmosphere using a nitrogen-filled drybox and by blanketing the reactions with dry nitrogen. DMSO Aldrich Gold Label brand is used with a water content <85 ppm. Potassium and cesium *t*-butoxides are also from Aldrich. Poly(*p*-phenyleneterephthalamide) and poly(*m*-phenyleneisophthalamide) are commercially available Kevlar pulp and Nomex staple fiber from DuPont. Polymer samples are oven dried at 130°C before use.

An argon- or nitrogen-flushed 1-L three-necked flask equipped with a nitrogen inlet, mechanical stirrer, and septum cap on the third neck is charged with 17.1 g (0.15 mol) potassium *t*-butoxide. Then 600 mL DMSO is added by means of a syringe through the septum cap, and 18.2 g poly(*p*-phenyleneterephthalamide), pulp is then added, followed by 19.6 g (24.7 mL, 0.61 mol) methanol. Stirring the slurry for 4 h gives a red homogeneous solution of the polyanion and no detectable unreacted fiber. The concentration of the solution was 2.8 wt%. Films can be made by casting onto, for example, a glass plate. Quenching in an aprotic solvent (e.g., hexane, ether, methylene chloride) or vacuum drying gives films of the polyanion. Quenching in protic solvents (e.g., methanol) gives regenerated polyamide film.

A procedure similar to this with poly(*m*-phenyleneisophthalamide) also gives solutions of the polyanion. In this case, the solution is pale yellow in color. Anion solutions can also be made from poly(benzimidazoles). Aliphatic polyamides usually do not give anion solutions by this procedure.

Aromatic copolyamides can also be made by melt routes. Such polymers, though meltable, have high glass-transition temperature. Preparation 95 is a conventional melt route using phenyl esters. Preparation 96 is a new approach using lactams as leaving groups, some of which ring open during polymerization and are incorporated into the polymer as aliphatic main chain components.

95. Preparation of Aromatic Polyamides by a Melt Route (21)

H_2N—C₆H₄—O—C₆H₄—NH_2 + PhO—C(=O)—C₆H₃(t-Bu)—C(=O)—OPh

Δ | CsF

[—HN—C₆H₄—O—C₆H₄—NH—C(=O)—C₆H₃(t-Bu)—C(=O)—]$_n$

In a round-bottom flask equipped with a distillation head and takeoff and a sidearm fitted with an argon inlet is put 4.0 g (0.02 mol) 3,4′-bis(aminophenyl) ether, 7.5 g (0.02 mol) diphenyl 5-*t*-butyl-isophthalate, and 0.01 g cesium fluoride. The flask and reactants are alternately evacuated and flushed with argon (or nitrogen) to remove any air. Then under a slow steam of argon, the flask is immersed in a Woods metal bath at 260°C. Phenol evolves rapidly, and after 10 min appears completed. The temperature is raised to 280°C, and an oil pump vacuum is slowly applied to assist removal of residual phenol and continue the polymerization. The thick polymer residue is heated at about 0.6 mm pressure for 1.5 h. The bath is then removed, the vacuum released with argon, and the flask immersed in liquid nitrogen to assist removal of the polymer from the glass. On breakage of the flask, the product is a clear, tough, lightbrown polymer. Films can be pressed at 345°C and 10,000 psi. The films are clear and can be hand drawn two to three times on a hot bar at 290°C. The inherent viscosity is 0.52 at 0.5% concentration in sulfuric acid. This technique can be repeated with 0.4 g 3,4′-bis(aminophenyl)ether, 0.4 g 1,12-diaminododecane, 6.3 g, diphenyl isophthalate, and 0.02 g potassium fluoride. A tough, clear polymer that melts at about 285°C results. Prolonged heating at or about 300°C results in reduced fusibility. The inherent viscosity of the polymer at 0.5% concentration measured in DMAc/4% lithium chloride is 1.02. The T_g is 204°C.

The melt polymerization of *N,N*′-isophthaloyl bis(2-pyrrolidone), *N,N*′-isophthaloyl bis(valerolactam), or *N,N*′-isophthaloylbis(caprolactam) with 1,3-phenylenediamine or 3,4′-diaminodiphenylether provides high molecular weight aramid copolymers containing nylon 4, 5, or 6 components, showing that the diamine condensation occurs at both the aroyl and the lactam carbonyls. The insitu generated lactams act as plasticizers during the preparation and processing of the copolymers and can be extracted with methanol.

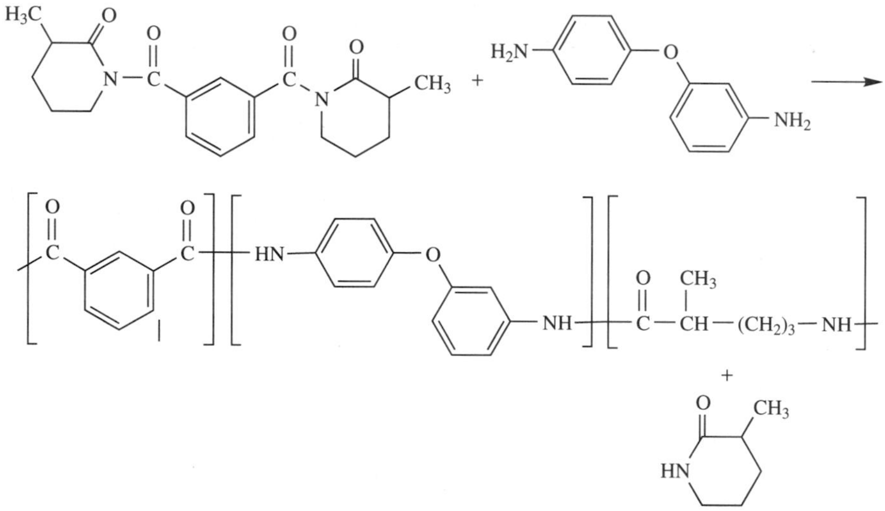

96. Preparation of N,N′-Isophthaloylbislactams (5)

The lactams are prepared by reacting isophthaloyl chloride with the respective lactams in the presence of triethylamine in toluene. The following general procedure is used. To a dry, 3-L, three-neck Morton flask equipped with a mechanical stirrer, reflux condenser, and addition funnel and purged with nitrogen is added 1.5 mol freshly distilled lactam, 152 g (1.5 mol) triethylamine, and 750 mL toluene. The mixture is stirred at 0 to 5°C and a solution of 152 g (0.75 mol) isophthaloyl chloride in 250 mL toluene is added to the addition funnel and added to the reaction mixture over 1 h. The reaction mixture becomes warm, and a white precipitate separates. The mixture is then heated to reflux for 15 min, during which time much of the precipitates dissolves. On cooling to room temperature, the product crystallizes and is collected by filtration. The slurry and the filter cake are rinsed with sufficient toluene to wet the total cake. The product is then added to 1-L deionized water and stirred to remove the tetraethyl ammonium chloride by-product. The product is filtered, and the washing process is repeated three more times. The bislactam products are then washed with dilute aqueous sodium carbonate solution, with water and finally with methanol. The filtered product is then dried in a vacuum oven at 80°C. under a nitrogen sweep. Carboxyl level should be <30 mEq/kg. If necessary for further purification, the bislactams can be recrystallized from acetonitrile/methanol for the bispyrrolidone, methyl ethyl ketone for the bisvalerolactam, and tetrahydrofuran for the biscaprolactam.

97. Polymerization of Isophthaloylbislactams (5)

In a typical polymerization, to a polymer tube with a side arm and a nitrogen bleed is added a mixture of 0.02 mol *N,N′*-isophthaloyl bis(lactam) and 0.02 mol of the aromatic diamine (*m*-phenylenediamine or 3,4′-bis[aminophenyl ether]). The tube is alternately purged with nitrogen or argon and vacuum. The nitrogen inlet is lowered to the bottom of the tube, and the mixture is heated in a Woodsmetal bath at 220 to 280°C, while a slow stream of nitrogen is passed through the reaction. The polymerization is continued for 90 min, then cooled to room temperature to obtain a clear amber-colored polymer. The polymer tube is cooled in liquid nitrogen and the tube broken. The polymer is separated from the glass and extracted three times with 200 mL boiling methanol. A white polymer results and is dried at 80°C in a vacuum oven under a stream of nitrogen. The alphatic component can be determined by NMR (H′). Stirring the mixture has little effect on the polymer molecular weight. Polymer inherent viscosities at 0.5% concentration in the range of 0.7 to 1.0, with M_n ranging from 18,000 to 33,000. Polymers from metaphenylenediamine and isophthalic bis(caprolactone) melted above 380°C and had a 2 to 3% 6-nylon in-chain component. Polymers from 3,4′-bis(aminophenyl)ether and isophthalic bis(caprolactone) melted at 175 to 220°C with a 2 to 7.5% 6-nylon component. Partial removal of the lactams at reduced pressure usually gave higher inherent viscosity numbers and less aliphatic component in the copolymers.

High molecular weight polyamides have been made by coupling *m*-diiodobenzene with aromatic amines in the presence of carbon monoxide and a palladium catalyst.

98. Preparation of Poly(4,4′-bis[phenyl Ether] Isophthalamide from m-Diiodobenzene, 4,4′-Diaminodiphenyl Ether and Carbon Monoxide (6, 7)

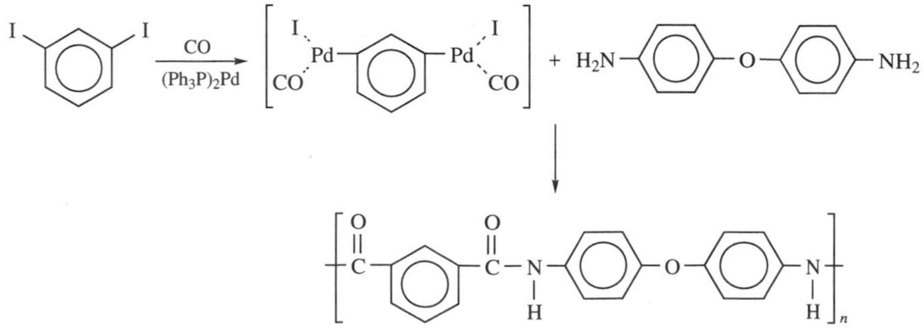

CO, $(Ph_3P)_2Pd$ + H_2N–C₆H₄–O–C₆H₄–NH_2

m-Diiodobenzene (Kodak) and 1,8-diazabicyclo[5.4.0]undec-7-ene (DBU, Aldrich), and 4,4″-diaminodiphenyl ether are purified by vacuum distillation. NMP (Aldrich, anhydrous) is used as received. Triphenylphosphine (Kodak) is recrystallized from hexane and bis(triphenylphosphine)palladium (II) chloride (Aldrich) is used as received.

A clean, dry, 100-mL Fisher-Porter pressure bottle outfitted with a pressure gauge, a pressure release valve, a gas inlet, and a ball valve for sample withdrawal is charged under a helium atmosphere in a dry box with 2.0 g (10 mmol) 4,4″-diaminodiphenyl ether 3.20 g (7 mmol) *m*-diiodobenzene, 40 g NMP, 3.65 g (24 mmol) DBU, 0.312 g (1.2 mmol) triphenylphosphine, and 0.42 g (0.6 mmol) bis(triphenylphosphine)palladium (II) chloride. A magnetic stirbar is added for agitation during the reaction. The reaction vessel and contents are purged three times with carbon monoxide and then pressurized to the desired pressure. The reactor is placed in a stirred oil bath for heating. Samples of the reaction solution are removed at intervals with a syringe to monitor molecular weight growth. After about 2 h the iodo displacement reaction is complete. The polymer solution is filtered, and the polymer is then precipitated into methanol. The resulting polymer is filtered and after thorough washing is dried under vacuum, giving an aromatic polyamide with a number average molecular weight around 24,000 and an inherent viscosity of 0.75 at 0.5% concentration in NMP. The reaction can also be run with dibromobenzene, but this is slower and the molecular weights are slightly lower.

6.2. POLYCARBAMATES

High molecular weight polycarbamates can be made by melt polymerization of *N*,*N*′-aromatic carbamoyl chlorides and biphenols (22). The polymers are tough and hydrolytically stable. Melt temperatures are >255°C and glass-transition temperatures >180°C, depending on structure. Preparation 79 illustrates the polymerization using the AB *p*-hydroxydiphenylamine-*N*-carbonyl chloride monomer.

99. Preparation of Poly (N-Phenyl-p-phenylenecarbamate) (23)

$COCl_2$ → ; Δ, 245°C →

99.A. N-Phenyl p-Hydroxyphenylcarbamoyl Chloride
To 1200 mL warm chlorobenzene in a 3-L three-neck flask equipped with a gas inlet tube that reaches to the bottom of the flask, a stirrer, and a dry-ice condenser equipped with a drying tube leading to a gas scrubber is added 100 g redistilled *p*-anilinophenol (Aldrich). While the solution is maintained at 90°C, phosgene (**Hazard!** Run reaction in a good hood and collect excess phosgene in a caustic trap) is rapidly bubbled through the solution for about 1 h. During the course of the reaction, the initially colorless solution becomes greenish and an oil separates, but the solution again becomes clear and colorless by the end of the hour. The temperature is raised to 120°C, and addition of phosgene is continued for 2 h more. The reaction mixture is then cooled to 10 to 20°C, and a colorless solid precipitates and is removed by filtration (**Caution!** Excess phosgene!) The product is washed with petroleum ether and dried in a desiccator. The *N*-phenyl *p*-hydroxyphenyl carbamoyl chloride melts at 153 to 154°C and should be stored in a sealed jar in a dry box.

99.B. Polymerization
In a 150-mL polymer tube equipped with a nitrogen inlet tube and a vacuum takeoff is added 20 g (0.08 mol) *N*-phenyl *p*-hydroxyphenyl carbamoyl chloride (*N*-chlorocarbonyl *p*-anilinophenol). The tube and its contents are heated to 190°C and maintained at that temperature for 15 min while the pressure is lowered to 25 mm. The pressure is then further reduced to 1 mm while the temperature is maintained at 190°C for 1.5 h and then raised to 245°C and held there for a further 3 h. Polymerization occurs rapidly under these conditions, and the tube is then cooled in liquid nitrogen and the vacuum is released with nitrogen. The tube is broken and the polymer removed from the glass by dissolving in methylene chloride. The polymer is precipitated into methanol, filtered, washed, and dried under vacuum. An off-white polymer resulted that melted at 256°C. Tough films could be pressed at this temperature or cast from solution. The inherent viscosity was 0.6 in *m*-cresol at 0.5% concentration.

6.3. POLYESTERS

Stiff chain polyesters, such as poly-*p*-hydroxybenzoate (PHBA) form thermotropic melts if the melt temperature is lowered sufficiently by copolymerization. PHBA itself can be compression sintered but does not melt without decomposition. Successfully used methods of copolymerization include incorporation of *(1)* bent comonomers (e.g., meta or ortho-substituted benzene structures), *(2)* offset or crankshaft monomers (e.g., 2,6-disubstituted naphthalene groups), *(3)* bulky substituents on the aromatic rings that inhibit chain packing, *(4)* flexible aliphatic spacers between the stiff units, *(5)* easily rotatable links or swivels between the aromatic segments, and *(6)* in the case of biphenylene mesogens, introduction of groups in the 2,2′ positions to inhibit ring planarity. All of these techniques disrupt the linear arrangement of the mesogens and reduce the nematic $\longrightarrow$ isotropic clearing point and melt temperatures.

In contrast to poly(ethyleneterephthalate) (PET), which is essentially Newtonian at low shear rates, thermotropic polyester melts have relatively high zero-shear viscosities but have significant yield stress and show rapid decrease in viscosity at low shear rates. Under normal processing-shear conditions, the flow viscosity of liquid crystalline polyesters can be two orders of magnitude less than that of PET. Because liquid crystal systems have long relaxation times after shearing and maintain a residual degree of orientation, more fluid melts can be formed by heating to a higher temperature and preshearing for a short time, then cooling and extruding. Fiber and molding properties depend markedly on processing conditions. Liquid crystalline melts are ideally suited for injection molding. In larger extrusion moldings, orientation is lower, but still considerable. Die-swell is low in fiber spinning because of the long relaxation times. High extensional flow in fiber spinning results in excellent fibrillar orientation, especially on cooling under stress.

Thermotropic melt polyesters have been processed to fibers of high strength and modulus, similar to those of aramids like Kevlar. However, only polymer (η inh $= 0.8$) of relatively low chain length can be spun successfully, because the melt viscosity is low enough, although still relatively high. The spun polymer gives brittle fiber with high modulus but relatively low tensile and transverse strength. The fiber or polymer requires solid-phase polymerization, close to the melt temperature, to increase molecular weight, crystallinity, orientation and toughness. Solid-phase polymerization can require up to 10 h heating in oxygen-free nitrogen for optimum properties with AA-BB polyester fibers.

A variety of catalysts (primarily basic) has been suggested to accelerate the rate of solid phase polymerization, but balance and type of polymer active ends (see below) appear to be even more important. There are two major routes to liquid crystalline polyesters. The first is acidolysis of a bisphenol acylate (usually the acetate) with a dicarboxylic acid.

$$\text{Ac—O—Ph—R—Ph—O—Ac} + \text{HOOC—R'—COOH} \longrightarrow \left[\text{O—Ph—R—Ph—O—}\overset{\text{O}}{\overset{\|}{\text{C}}}\text{—R'—}\overset{\text{O}}{\overset{\|}{\text{C}}}\text{O}\right]_n + n\,\text{AcOH}$$

The second is phenolysis of a bisphenol with the phenol ester of a dicarboxylic acid.

$$\text{HO—R—OH} + \text{PhO—CO—R'—CO—OPh} \longrightarrow \left[\text{O—R—O—}\overset{\text{O}}{\overset{\|}{\text{C}}}\text{—R'—}\overset{\text{O}}{\overset{\|}{\text{C}}}\right]_n + 2n\,\text{PhOH}$$

The latter method works well for AA-BB poymers but is less available to AB polymers, owing to the difficulty of preparing the phenyl esters of hydroxybenzoic acids. The phenyl-ester exchange is catalyzed by potassium, cesium, or organic fluorides (22).

The acidolysis process has several drawbacks, including insolubility of terephthalic acid (when used) in the melt. Micronized T-acid should be used when possible to speed up the reaction rate. The acetate process also produces a lot of sublimate, especially when vacuum is applied at the later stages of polymerization. Last, the acetate process generates ketene as the temperature is raised. Elimination of ketene results in formation of phenolic-hydroxyl ends that do not polymerize well with carboxylic acid ends.

The importance of type of ends is shown below.

---OAc + HOOC– ⟶ fast –O–CO– polyester

---OH + HOOC– ⟶ slow

---COOH + HOOC– ⟶ fast –CO–O–CO– anhydride

Anhydride groups may be formed in heat-strengthening and become a site of hydrolysis in fiber.

It should be pointed out that total rigidity is not a requirement for thermotropic polyesters. For example, the homopolyester from 3,4′-dihydroxybenzophenone terephthalate can exist in extended or bent configurations. In the melt, a flexible configuration is favored, but in the solid crystalline state and in the liquid crystalline state an extended and stiffer configuration is favored. Melt-spun fibers with high tensile properties ($T/E/M_i = 4.6/4.7/77$ GPa/%/GPa) have been obtained from this polymer (24).

100. Preparation of Poly(3,4′-dihydroxybenzophenone Terephthalate) (24)

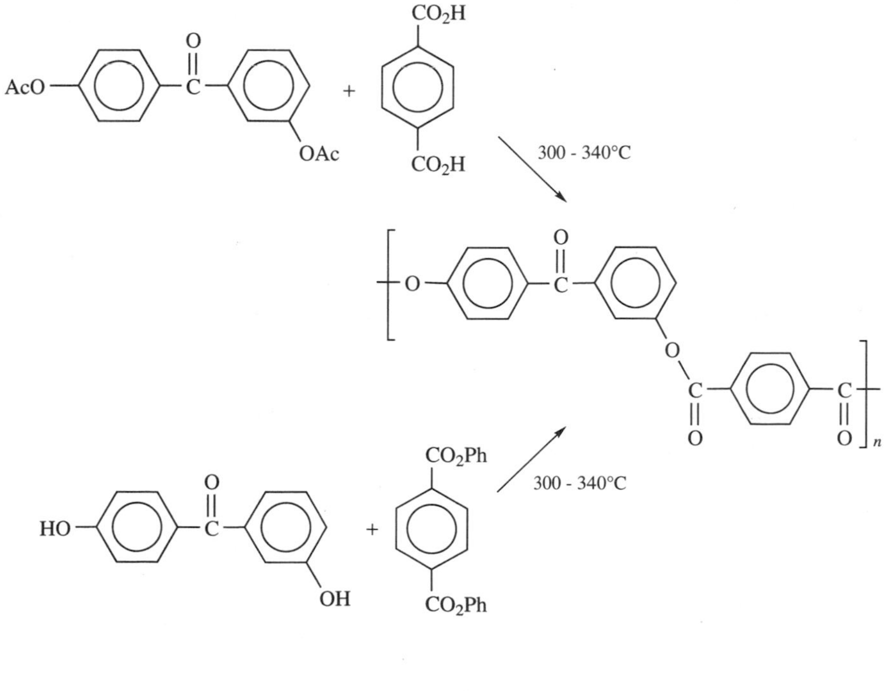

AcO–C₆H₄–C(=O)–C₆H₄–OAc + HO₂C–C₆H₄–CO₂H ⟶ (300 - 340°C)

HO–C₆H₄–C(=O)–C₆H₄–OH + PhO₂C–C₆H₄–CO₂Ph ⟶ (300 - 340°C)

$\left[-O-C_6H_4-C(=O)-C_6H_4-O-C(=O)-C_6H_4-C(=O)- \right]_n$

The aromatic diol is prepared by reacting 110 g (0.8 mol) *m*-hydroxybenzoic acid with 80 g (0.85 mol) phenol in 500 g anhydrous hydrogen fluoride (HF) at 30°C under a pressure of 30 psi boron trifluoride in a Hasteloy autoclave for 4 h. The boron trifluoride and the HF are then vented, and the mixture is poured over 2 kg ice. The product is filtered and washed with water, 5% aqueous sodium bicarbonate solution, and water again before filtering and drying under vacuum at 80°C. The yield is 163 g (95%). The diacetate is prepared from the diol (163 g) by refluxing for 4 h with three times its weight of acetic anhydride and six drops of concentrated sulfuric acid. The reaction mixture is poured into 2 kg ice water, and the solid precipitate is filtered and dried under vacuum at 60°C. The diacetate is recrystallized from methanol to yield 187 g (83%) as colorless crystals; mp 84.5°C.

Pure diol is obtained from the recrystallized diacetate by refluxing overnight in 500 mL ethanol containing 0.8 ml trifluoromethanesulfonic acid. Ethyl acetate is slowly removed by distillation through a 12-in Vigreux column and then ethanol until a final volume of about 250 ml remains in the flask. Then 250 mL water is added to the flask to precipitate the product. This is filtered and recrystallized from boiling ethanol. After filtration, the diol is dried under vacuum and yields white needles, mp 203°C.

The polymerization is carried out by heating 29.8 g (0.1 mol) 3,4-diacetoxybenzophenone, 16.6 g (0.1 mol) finely ground, pure terephthalic acid, and 0.005 g 4-dimethylaminopyridine in a flask equipped with a stirrer, a distillation head with a short Vigreux column, and an inert gas sweep. The flask and contents are voided of oxygen by alternately evacuating and purging with nitrogen or argon. The flask is then blanketed with nitrogen or argon, placed in a Woodsmetal bath at 340°C, and stirred rapidly. Acetic acid evolves and distills through the column. About 10 mL (80%) is collected in the first 15 min of the polymerization. The reaction is heated for 20 min at atmospheric pressure, followed by an additional 25 min at 0.5 mm vacuum. The flask is removed from the bath, and the vacuum is released with nitrogen. The flask is broken, and the polymer removed.

A similar procedure is used in the preparation of the polymer by phenyl ester exchange. In this method, 42.8 g of the diol is reacted with 63.6 g diphenyl terephthalate and 0.02 g dimethylaminopyridine by heating with stirring at 250°C for 30 min, at 275°C for 15 min, at 300°C for 30 min, and at 340°C for 30 min. Phenol is removed as the distillate. The mixture is finally heated at 340°C under 2 mm vacuum for 3 h. The polymer is isolated as described above.

The polyester can also be prepared by low-temperature solution polymerization. In this procedure, 30.45 g of the diol is dissolved in 400 mL dry *N*-methylpyrrolidinone and cooled to 10°C. Then 40 g dry distilled triethylamine is added. The solution is stirred rapidly, and a solution of 30.45 g terephthaloyl chloride in methylene chloride is then added all at once. The reaction temperature rises to 50°C, and a yellow mush is formed. After standing overnight at room temperature, the polymer is dispersed in water, filtered, washed well with hot water, filtered, and finally dried at 100°C, under house vacuum.

The polymers from each of these three methods has inherent viscosities in pentafluorophenol at 0.5% concentration ranging from 0.4 to 2.0. The polymer melt temperatures (PMT) are were 276 to 282°C. The melts are nematic, and the clearing temperatures (nematic $\longrightarrow$ isotropic) are 353 to 360°C. The melt temperature of solid-phase polymeric fiber (annealed 3 h at 190°C, followed by 48 h at 280°C) is 320°C. Fiber can be spun and film pressed at 315 to 330°C. The crystalline, oriented, heat-treated fiber has high tensile properties ($T/E/M_i = 4.6/4.7/77$ GPa/%).

The polyester from hydroquinone and terephthalic acid is too rigid and too high melting to yield useful, melt-processible polyesters. As indicated earlier one route to lower the melt temperature, without significantly affecting the chain rigidity, is to introduce bulky lateral groups that will inhibit chain packing. A useful intermediate for this is phenylhydroquinone.

101. Preparation of Anisotropic Polyester from Phenylhydroquinone (22)

O-Ac, Ph, O-Ac + CO_2H, CO_2H $\xrightarrow[290\text{–}340°C]{\Delta}$ [–O–(Ph)–O–C(=O)–C$_6$H$_4$–C(=O)–]$_n$

Preparation of the precursor phenylquinone involves addition of benzenediazonium nitrate to quinone. This has been described (25). The phenylquinone is converted to crude phenylhydroquinone by hydrogenation over palladium/charcoal in isopropanol at 300 psi and 85°C. Product yields are about 76%. Purification to polymer grade material is accomplished by vacuum distillation through a spinning-band (or other effective) column at 20 mm a 5/1 reflux ratio; the fraction at 223 to 226°C is collected when the foreshot indicates the absence of impurities such as hydroquinone, *o*-phenolphenol, and *m*-phenylphenol. The takeoff and receiver pot should be heated with an ir lamp to keep it molten. The main fraction is redistilled at 10 mm (pot temperature about 250°C) at a 3/1 reflux ratio and the fraction at bp 225 to 226°C is collected.

In a 250-mL three-neck flask equipped with a stirrer, nitrogen inlet port, distillation head, and collection vessel are placed 54 g (0.2 mol) phenylhydroquinone diacetate and 31.6 g (0.19 mol) finely powdered terephthalic acid. The reaction vessel is then thrice evacuated and purged with dry nitrogen. The reaction vessel is then placed in a Woodsmetal bath at 290°C, stirring is begun and the nitrogen flow is maintained. In about 13 min the first acetic acid is collected. After 20 min more, the bath is raised to 300°C. After 27 min, the bath is raised to 310°C, in another 40 min, to 320°C, and in another 35 min, to 340°C. After another 23 min, the nitrogen flow is halted, and vacuum is applied over 8 min to 10 mm Hg. After 2 min, the Woods metal bath is removed, nitrogen is applied to remove the vacuum, and the stirring is stopped. The flask is cooled in liquid nitrogen, the flask is broken, and the polymers isolated by separating it from the glass. The yield is 56 g. The polymer has a mp of about 340°C and an inherent viscosity of 1.03 in pentafluorophenol at 0.5% concentration

102. Preparation of Polymer from Phenylhydroquinone and Diphenyl Chloroterephthalate (22)

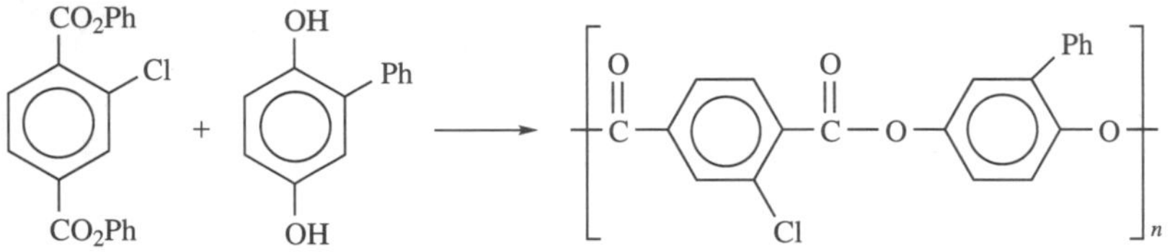

CO_2Ph Cl CO_2Ph + OH Ph OH ⟶ [O=C–(Cl-phenylene)–C(=O)–O–(Ph-phenylene)–O]$_n$

Diphenyl chloroterephthalate is made by reaction of sodium phenolate with chloroterephthaloyl chloride (bp 105°C/1 mm Hg). Preparation of phenylhydroquinone was described earlier. To a 100-mL round-bottom flask equipped with a short stillhead, argon inlet, and 30 mL receiver flask is added 7.05 g (0.02 mol) diphenyl chloroterephthalate, 3.72 g (0.02 mol) phenylhydroquinone, and 0.02 g dry cesium fluoride. The flask and contents are purged with argon and then immersed in a Woodsmetal bath at 260°C. Phenol starts to evolve after 4 min. The temperature is maintained at 260°C for 10 min and then raised to 280°C. After 17 min, vacuum is applied gradually and reduced to 0.3 mm over 12 min Full vacuum is maintained for 1 h. The bath is then removed, and the flask brought to atmospheric pressure with argon. The flask is immersed in liquid nitrogen and broken to remove the polymer. The polymer is light brown and tough. Inherent viscosity is 2.2 at 0.5% concentration in pentafluorophenol.

103. Preparation of Polymer from Phenylhydroquinone, Resorcinol and Diphenyl Terephthalate (22)

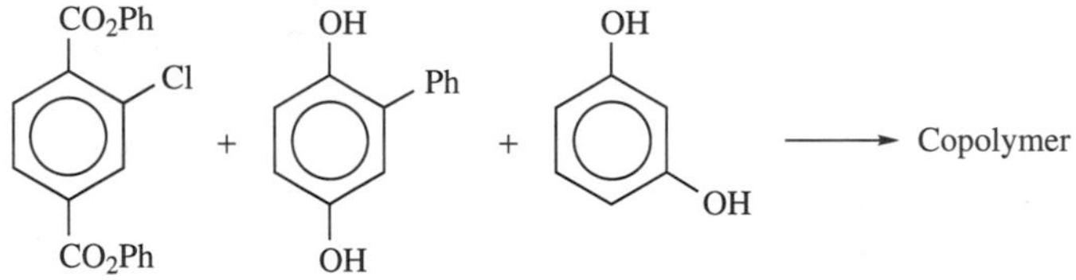

CO_2Ph Cl CO_2Ph + OH Ph OH + OH OH ⟶ Copolymer

Equipment used was the same as that used earlier, but using a 250-mL flask. The ingredients charge is 35 g (0.188 mol) phenylhydroquinone 2.32 g (0.021 mol), resorcinol, 63.6 g (0.2 mol) diphenyl terephthalate, and 0.1 g (0.0017 mol) potassium fluoride. The mixture is purged with nitrogen and heated in a Woods metal bath at 340°C. After about 6 min, phenol evolves, and 30 to 40 mL is collected in 10 to 15 min. After 20 min, vacuum is applied and lowered to 1 mm over 10 min. The reaction is held at 1 mm for 30 min more; the bath is then removed, stirring is stopped, and the flask is cooled under a nitrogen bleed until it reaches atmospheric pressure. The flask is cooled in liquid nitrogen, and broken; the polymer is separated from the glass. The slightly brown polymer is tough and has an inherent viscosity of 2.55 in pentafluorophenol at 0.5% concentration.

6.4. POLYBENZOBISTHIAZOLES AND POLYBENZOBISOXAZOLES

One driving force in polymer chemistry is to prepare polymer products with as close to the ultimate (theoretical) in tensile properties as possible. As has been pointed out, researchers have come close to achieving theoretical tensile modulus, but not tensile strength, because of defects and insufficient numbers of load-bearing chains. A parallel goal has been to prepare polymers with combinations of the highest recorded modulus and highest tensile strength.

Regarding modulus, theoretical calculations (26) predict that poly(benzobisthiazole) should have the highest tensile modulus, followed closely by poly(phenylenepyromellitimide), as shown in Figure 6.3. Neither the poly(benzobisthiazole) nor poly(*p*-phenylenepyromellitimide) has been prepared in high enough molecular weight to determine tensile properties. However, poly(phenylenebenzobisthiazole) and the corresponding bisbenzoxazoles have been made (27) in high molecular weight. Polymerization methods of Wolfe et al. (27) stress that certain combinations of polyphosphoric acid and phosphoric anhydride are required to obtain high molecular weight poly(phenylenebenzobisthiazoles) or bisbenzoxazoles. However, high molecular weight poly(phenylenebenzobisthiazole) can be made from 1,4-diaminobenzene-2,5-dithiol, and terephthalonitrile using commercially available polyphosphoric acid as dehydrant without added phosphoric anhydride (28).

The poly(phenylenebenzobisthiazoles) and bisoxazoles are soluble as made in polyphosphoric acid (PPA) and give lyotropic solutions with shear opalescence at low (3%) solution concentrations. The polymers should be processed directly

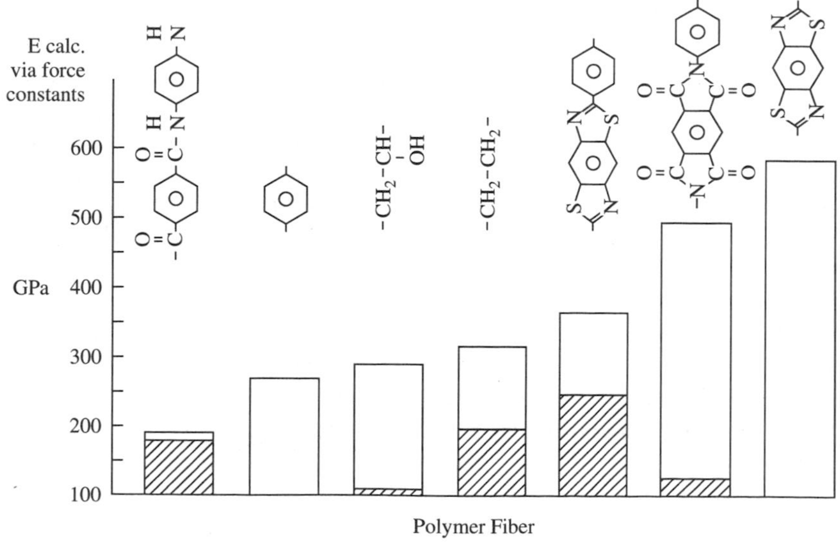

Figure 6.3. Theoretical (open bars) vs. archieved (hatched bars) tensile modul. *E* calculated via force constants.

from the PPA solutions without isolation. The solutions have extremely high solution viscosities even under shearing conditions. After aqueous isolation, the polymers are no longer soluble in PPA and can be dissolved only in strong acids, such as methanesulfonic or chlorosulfonic acids, but not in sulfuric acid. Solutions of the poly(phenylenebenzobisthiazoles) and bisoxazoles in polyphosphoric or methanesulfonic acids have been dry-jet spun into water to produce highly oriented as-spun fibers (27, 28). After heat treatment, by annealing under tension through a tubular oven at 525 to 650°C, fibers with tensile strength and modulus superior to those of known aramids result. Poly(benzoxazoles) from aliphatic acids are isotropic and many are melt-processible. Such polymers can be melt prepared from the diaminodiols and dibasic acids (29).

104. Preparation of Poly(benzobisthiazoles) (27)

HS, $NH_2.HCl$, $HCl.H_2N$, SH + CO_2H, CO_2H

Polyphosphoric acid/P_2O_5

S, N, N, S, n

2,5-Diamino-1,4-benzenedithiol dihydrochloride is made from *p*-phenylenebis-(thiourea). To a 3-L flask are added 170 g (1.57 mol) *p*-phenylenediamine, 1.5 L deaerated water, 300 mL concentrated hydrochloric acid, and 15 g activated charcoal. The mixture is warmed to 50°C and transferred with filtration to another 3-L flask. Then 485 g (6.38 mol) ammonium thiocyanate is added, and the mixture is stirred at 90 to 100°C for 24 h. A yellow, granular product begins to precipitate after about 2 h. The mixture is allowed to cool and the product is filtered. The solid is washed with 1-L hot water and then dried at 100°C under vacuum. Yield is 340 g (95.6%).

To a stirred suspension of 340 g (1.5 mol) of the above product in 1.6 L dry chloroform is added a solution of 550 g (3.47 mol) bromine in 225 mL chloroform so that the pot temperature does not rise above 50°C. The orange slurry is stirred at room temperature overnight and then heated at reflux for 24 h. The mixture is allowed to cool under a slow stream of argon. The orange product is filtered and washed with 675 mL chloroform, dried in air, then stirred in 225 g aqueous sodium bisulfite in 1.75 L water. The yellow solid is filtered, washed with 565 mL concentrated ammonium hydroxide and then with 1.50 L water. The crude product is recrystallized twice from 15.75 L glacial acetic acid. After drying to constant weight under reduced pressure at 85°C, 180 g

(54%) 2,6-diaminobenzo[1,2-d : 4,5-d′]bisthiazole is obtained as feathery needles, mp >350°C.

The diaminobenzobisthiazole is converted to 2,5-diamino-1,4-benzenedithiol dihydrochloride as follows. To a 2-L flask equipped with a stopcock at the bottom and containing 750 mL deaerated water, is added 685 g (10.5 mol) 85.9% potassium hydroxide. The solution is allowed to cool slightly and then 170 g (0.764 mol) 2,6-diaminobenzo[1,2-d : 4,5-d′]bisthiazole is added while the flask is swept with a stream of argon. The mixture is heated to reflux under argon for 5 h. The yellow solution is allowed to cool overnight with stirring. The solution is then cooled to 15°C, and the resulting yellow needles, while maintaining under argon, are transferred to an air-free box under argon and filtered. To avoid air contamination, the crystals are sucked dry in the air-free box under a heavy dental dam. Still in the air-free box under argon, the potassium salt is dissolved in 375 mL deaerated water and filtered directly into an evacuated, stirred flask containing 1.25 L deaerated water and 1.25 L concentrated hydrochloric acid. Fine colorless crystals of the dihydrochloride separate rapidly. These are filtered in the air-free box under argon, washed with, 650 mL methanol and dried to constant weight at room temperature under reduced pressure with an argon bleed. The yield is 159 g (85%). The mp is 200–210°C with decomposition.

Polymerization is carried out in a 2.5-L resin kettle to which are added 130 g (0.467 mol) of the dihydrochloride (above) and 1 kg 115% commercially available polyphosphoric acid (previously deaerated by heating to 100°C with stirring overnight under vacuum with an argon bleed and then cooling to room temperature under argon). The mixture is stirred slowly at room temperature under argon for 24 h and then at 60 to 70°C for 36 h. The resulting solution is clear with no evidence of bubbles. Then 88.15 g (0.531 mol) Finely ground and sieved pure terephthalic acid is and incorporated into the solution by rapid stirring at 110°C under argon. An additional charge of 1.38 kg deaerated polyphosphoric acid is then added. The mixture is heated at 110 to 165°C for 5 h. A slight vacuum is applied to assist in removal of the hydrogen chloride, and foaming is controlled by intermittent breaking of the vacuum with argon. After the initial period, heating is continued at 165°C for 12 h, 180°C for 12 h and 195°C for 12 h. The mixture becomes stir opalescent after about 6 h. The solution is green with a yellow-green opalescence when sheared. The hot solution (**Gloves!**) is removed from the resin kettle with aid of a broad blade spatula onto heavy-duty Teflon-coated aluminum foil (Polymer on the Teflon side!), sealed, and stored in a dry box until used for film or fiber extrusion. A portion of the polymer is isolated by shearing in water in a home blender until all the polyphosphoric acid is removed, then drying at 80 to 100°C under vacuum. Inherent viscosities in methanesulfonic acid at 0.1% concentration ranged from 10 to 30, depending on the purity of the starting material and in elimination of oxygen during polymerization.

High molecular weight poly(phenylenebenzobisthiazoles) can be made from terephthalonitrile and 2,5-diaminobenzene-1,4-dithiol dihydrochloride using commercially available polyphosphoric acid (85% phosphoric anhydride) from Fluka without the need to boost with additional phosphoric anhydride (28).

105. Preparation of Poly(benzobisthiazole) from Terephthalonitrile and 2,5-Diaminobenzene-1,4-dithiol Dihydrochloride (28)

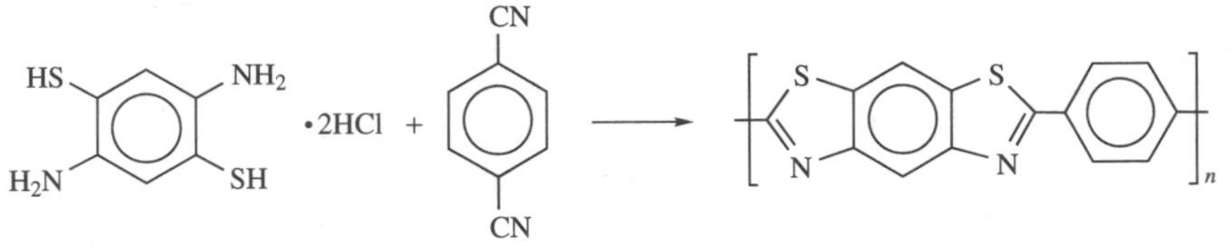

A 200-mL resin kettle equipped with a stainless-steel basket stirrer, argon inlet, and a short condenser fitted with a drying tube is flamed free of moisture with a propane torch while a stream of argon is passed through. Under a slow stream of argon is added 114.5 g polyphosphoric acid (87% from Fluka) and heated overnight at 100°C under argon to deaerate. Then 6.4 g (0.05 mol) Pure terephthalonitrile is added over 5 min with stirring. The temperature is raised to 110°C and kept there for 30 min, then 12.2 g (0.05 mol) diaminobenzenedithiol dihydrochloride is added over 1.5 h. Then 0.05 g 4-dimethyl-aminopyridine is added as catalyst; the temperature and heating are continued for 1.5 h at 120°C, 1 h at 140°C, 18 h at 160°C and 48 h at 180°C. A further 0.05 g dimethylaminopyridine is added and heating is continued at 180°C, for a further 21 h. A purple viscous gum results, which is removed from the kettle, while still hot, with a broad blade spatula. The polymer is stored as before in a drybox in heavyduty Teflon-coated aluminum foil. A sample of the polymer is shredded in water in a home blender until free of polyphosphoric acid. After drying, the sample has an inherent viscosity of 29.5 at 0.1% concentration in methanesulfonic acid.

Poly(phenylenebenzobisthiazole) copolymers that give lyotropic solutions as-made in polyphosphoric acid are easily prepared from terephthaloyl and halo-terephthaloyl chlorides. The halides are thermally labile in the polymers and provide a route to cross-linking when heated above 350°C. Regarding the halo-homopolymers, only the chloro and fluoro polymers are lyotropic (30)

106. Random Halo-Copolymers of Polybenzobisthiazoles (28)

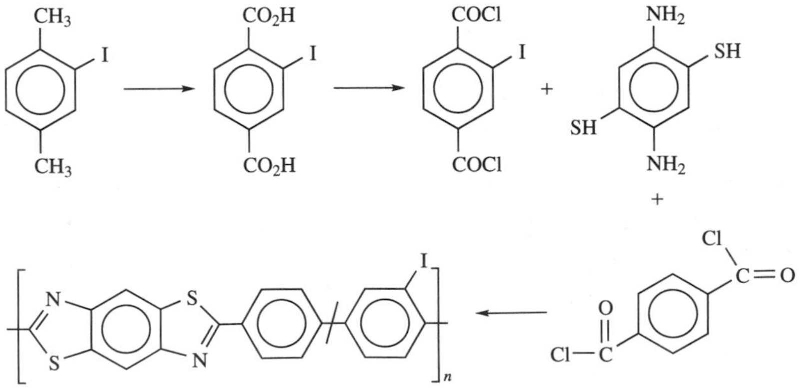

Chloroterephthaloyl chloride (bp 105°C/1 mm Hg) is made by reaction of the acid is thionyl chloride. The acid was made by oxidation of chloro-*p*-xylene with aqueous sodium dichromate in the presence of tetrabutylammonium hydrogen sulfate in an autoclave with vigorous agitation at 250°C. Bromoterephthaloyl chloride (bp 98 to 104°C/ 0.3 to 0.4 mm Hg) is made by reaction of the bromo acid (Lancaster Synthesis) with thionyl chloride.

The acid precursor to iodoterephthaloyl chloride is made by basic (NaOH) permanganate oxidation of 2-iodo-*p*-toluic acid (30). The iodo-acid melts at 299°C. The acid chloride is made by refluxing with excess thionyl chloride (DMF catalyst). Removal of the excess thionyl chloride leaves a red oil, which is distilled through a 24-in spinning band column to give a slightly pink fraction (bp 114 to 115°C/0.6 mm). The fraction is distilled twice more (bp 110 to 112°C/0.4 mm) but is still pink in the receiver even though the distillate appears slightly yellow. It seems that traces of iodine are being eliminated and contaminating the distillate. On heating, the main distilled fraction (pink) in a flask for 3 h under 0.3 mm vacuum at 60°C, the iodine is removed by sublimation, leaving a bright yellow pot liquor. Purity was 97% by gas chromatography (GC).

The following general procedure is used for preparing copolymers of the various halo-terephthalic acids and terephthalic acid. The apparatus consists of a 200-mL resin-kettle (3-in dia) equipped with a basket stirrer, argon inlet, air condenser outlet, drying tubes, and a stoppered-port for chemicals addition. The equipment is flamed out with a propane torch to remove adsorbed moisture. Then 114.6 g polyphosphoric acid (Fluka, 85% phosphoric anhydride) is added and deaerated by heating at 100 to 120°C (oilbath temperature) overnight with slow stirring under a stream of argon. The temperature is raised to 120°C, and 0.03 g 4-dimethylaminopyridine is added as catalyst, followed by the addition of 5.06 g (0.0249 mol) terephthaloyl chloride over 15 min. This is followed by the addition of 8.11 g (0.0246 mol) iodo-terephthaloyl chloride over 15 min. The temperature is raised to 110°C for 30 min, and 11.89 g (0.0485 mol) 2,5-diaminobenzene-1,4-benzenedithiol dihydrochloride is added over 2.5 h. A further 0.1 g of the dimethylaminopyridine is added, and heating is continued for 1 h at 120°C, 1.5 h at 140°C, 16.5 h at 160°C, and 27.5 h at 180°C. A further 0.05 g dimethylaminopyridine is added and heating is continued for 21 h at 180°C, and 7.5 h at 190°C. A thick isotropic polymer solution results. A sample of the polymer is isolated by shearing in water in a home blender till free of PPA, followed by drying under vacuum. The inherent viscosity at 0.1% concentration in methanesulfonic acid is 22.6. The major portion of the polymer is stored as before in heavy-duty Teflon-coated aluminum foil in a dry box until ready for fiber spinning or film casting.

A procedure similar to this gives an inherent viscosity of 30 dL/g for the 50/50 bromo-terephthalic/terephthalic acid copolymer.

Polybenzthiazoles and polybenzoxazoles can be prepared by melt polymerization, provided the intermediates are properly chosen. Melt processible polymers can be made from aliphatic/aromatic intermediates.

107. Melt Polymerization of Polybenzoxazole from 3,3-Dihydroxybenzidine and Mixed Dibasic Acids (29)

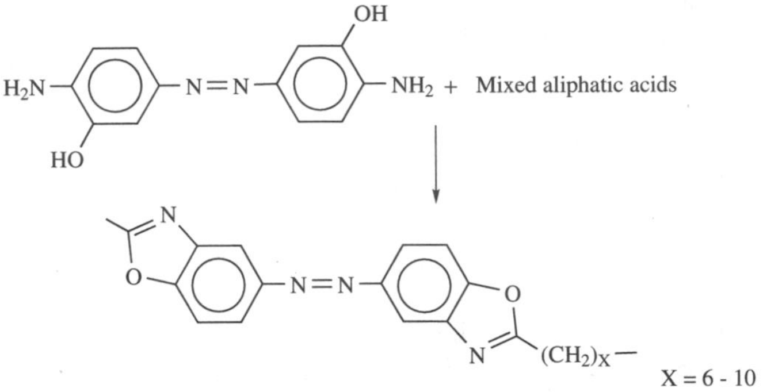

OH
H_2N — N=N — NH_2 + Mixed aliphatic acids
HO

N
O — N=N — O
N $(CH_2)_X$—
X = 6 - 10

A thick-walled polymer tube fitted with a nitrogen capillary bleed and a vacuum take-off is charged 10 g 3,3′-dihydroxybenzidine (**Carcinogen!**) and 9.39 g mixed dibasic acids 81% suberic acid, 17% azelaic acid and 1% sebacic acid. The tube is flushed with nitrogen and immersed in a *m*-cresol vapor bath for 1 h, while a slow stream of nitrogen is passed through the melt. Most of the water evolves and is collected (**handle with caution!**). The tube is then in a methyl naphthyl ether bath (275°C) and heated for a further 3 to 4 h. A viscous polymer should result. Vacuum may be applied (behind a thick-walled shield) if the polymer is not too viscous. The polymer tube should be removed, cooled, and broken to isolate the polymer. The PMT is about 255°C, and tough films can be pressed at 270 to 280°C and quenching. On heating, the quenched film to about 150°C, a tough crystalline film results. The inherent viscosity at 0.5% concentration in *m*-cresol is 1 to 1.2 dL/g.

6.5. POLYBENZIMIDAZOLES

Surprisingly little reference has been made to poly(benzobisimidazole), possibly because the tetra-aminobenzene intermediate is oxidatively unstable. However the AB monomer, from 3,4-diaminobenzoic acid, is easily made and can be polymerized to high polymer in polyphosphoric acid alone or in methanesulfonic acid/phosphoric anhydride (W. Sweeny, personal communication). Solutions are isotropic and tensile properties are similar to those obtained from poly(metaphenylene isophthalamide). Moisture regain is high (9%), and fire-, hydrolytic-, and thermal-stabilities are very good.

Air Force–sponsored research programs on thermally stable fibers in the 1960s resulted in Celanese Co. preparing and commercializing the poly(benzimidazole) from 3,3′-diaminobenzidine and isophthalic acid as "Celazole." The synthesis involves condensation of diphenyl isophthalate with the tetraamine to give a prepolymer, followed by solid-phase polymerization to high polymer. This is then spun from dimethylacetamide (or other *N*-alkylated amide solvents) close to the boiling point to yield fiber with relatively low tensile but excellent thermal properties. Fiber stability is enhanced by sulfonation to produce a zwitterion type of structure with very low shrinkage at high temperatures (31). Thermally stable poly(benzobenzimidazole) ladder polymers have also been made by reaction of tetraamines and dianhydride.

*108. Preparation of Polybenzimidazole from 3,4-Diaminobenzoic Acid in Methanesulfonic Acid**

CO_2H NH_2 NH_2 —PPA or CH_3SO_3H→ H N N n

A 250-mL three-necked round bottom flask equipped with a Teflon stirrer, air-condenser, nitrogen inlet, and drying tubes is flamed under nitrogen to remove adsorbed moisture. A 9% solution of phosphoric anhydride in 63 g methanesulfonic acid is added, followed by 3.04 g 3,4-diaminobenzoic acid. The flask is immersed and stirred in an oilbath at 120°C for 30 min, raised to 160°C, and held at this temperature for 8.5 h. After about 1 h, the solution becomes purple and viscous. At the end of the heating period, the solution is cooled to room temperature and is very viscous. The polymer is isolated by precipating into water in a blender. The washing is repeated several times to free the polymer from methanesulfonic acid. The isolated polymer is then washed several times with acetone and dried at 60°C under house vacuum. A fibrous, olive-colored polymer results. Inherent viscosity at 0.5% concentration in 98% sulfuric acid is 2.63.

109. Preparation of Polybenzimidazole from 3,4-Diaminobenzoic Acid in Polyphosphoric Acid (32)

A three-necked flask equipped with stirrer, argon inlet, air condenser, and drying tubes is flamed under argon to remove adsorbed moisture. To it is then added 433 g 87% polyphosphoric acid (Fluka) and 54 g phosphoric anhydride, the mixture is heated for 1.5 h at 115°C with slow stirring under argon to remove dissolved air. Then 20 g phosphorous acid (optional) is added, and stirring is continued until it dissolves (5 to 10 min). This is followed by the addition of 32.4 g 3,4-diaminobenzoic acid. The temperature is raised to 150°C, and stirring continued for 20 h under argon. A viscous solution results. The solution is centrifuged to remove bubbles and is used directly for the wet-spinning of fibers. A sample of the polymer is isolated by washing with water in a house blender until free of PPA. The inherent viscosity is 2.3 at 0.1% concentration in methanesulfonic acid.

6.6. POLYIMIDES

Aromatic polyimides have attracted attention for many years because of their high thermal stability and high resistance to oxidative degradation. Fiber properties and

* W. Sweeny, personal communication.

thermal stability are generally superior to those of polybenzimidazoles. The high melting temperatures of the polyimides appears due to the polar attractions between the electron-deficient anhydride rings and the electron-rich diamine-bearing rings.

Commercial polymers were initially made by reaction of a diamine with pyromellitic dianhydride in amide solvents to form polyamic acids, which were then converted thermally, or chemically with acetic anhydride, to the polyimide. The driving force for intramolecular ring closure to the pyromellitimide ring system is great and minimizes cross-linking, but imidization is often incomplete due to steric constrictions. Behavior of the amic acids in solution is complex and depends highly on the solvent used. Amide interchange and molecular weight equilibration often occur. It is possible that addition of phosphorus pentoxide to the amide solution during polymerization could accelerate the formation of imide. Hydrolytic stability of pyromellitimides to strong bases greatly depends on the structure of the diamine. The most stable polymers are from electron-rich diamines such as 4,4′-diaminodiphenyl ether and 4,4′-diaminodiphenyl sulfide. Less stable polymers are derived from diamines with electron withdrawing groups, such as 4,4′-diaminodiphenyl sulfone.

As has been discussed earlier, stiff-chain polyimides potentially can provide fiber with high tensile properties, particularly modulus. Melt-processible polyimides can be made directly in the melt from a diester of pyromellitic acid and long chain aliphatic diamines. The stiff, inflexible pyromellitimide ring system gives rise to high melting polymers, despite the lack of hydrogen bonding. The stiffness of the polymer chain is reflected in the high T_g of the polyimides. For example, the polyimide from nonamethylenediamine and pyromellitic anhydride has a T_g about 100°C, whereas that from 6-6 nylon is about 50°C (32).

Other melt-processible polyimides have been made by scientists at the Langley (U.S. Air Force) Research Center from 4,4′-bis(3,3′-diaminophenoxy) biphenyl and pyromellitic dianhydride. These are high melting >300°C, and are of interest as fibers and matrix resins for compounds.

110. Preparation of Poly(nonamethylenepyromellitimide) (32)

$$\mathrm{MeO{-}C(=O)},\ \mathrm{HO{-}C(=O)},\ \mathrm{CO_2H},\ \mathrm{CO_2CH_3}\ \text{(on benzene ring)} + \mathrm{H_2N{-}(CH_2)_9{-}NH_2} \longrightarrow \left[\mathrm{N(C{=}O)_2C_6H_2(C{=}O)_2N{-}(CH_2)_9}\right]_n$$

In a 100-mL three-necked round-bottom flask protected by a drying tube is placed 6.08 g (0.0279 mol) pyromellitic dianhydride and 30 mL absolute methanol. (Pyromellitic dianhydride may be recrystallized from acetic anhydride.) The mixture is swirled by hand and gently warmed on the steam bath until a clear solution results. Then 4.414 g (0.0279 mol) nonamethylenediamine is added quantitatively, and the methanol distilled on the steam bath. The residual salt, which should be dry enough for easy handling, is transferred by means of a long-stemmed funnel to a polymer tube with a side arm. The tube is fitted with a capillary inlet reaching almost to the bottom. The tube is purged by alternatively evacuating and filling with nitrogen. At atmospheric pressure and in a slow stream of nitrogen, the temperature is raised to 139°C by immersion in a *m*-xylene bath for 2 h, during which time water and methanol are driven off. The polymer tube is then transferred to a 325°C vapor bath for another 2 h. The tube is cooled after first raising the capillary above the level of the melt. The polymer is removed by breaking the tube. The PMT is about 325°C, and the inherent viscosity is 0.8 to 1.2 in *m*-cresol at 0.5% concentration. Tough films can be pressed at 340°C.

The nonamethylenediamine used above can be prepared as follows (33). In a 500-mL three-necked flask equipped with gas inlet tube and straight distillation head with condenser is placed a mixture of 250 g azelaic acid and 10 g polyphosphoric acid. The temperature is raised to 120°C by means of a Woodsmetal bath to melt the acid. Anhydrous ammonia is passed through the reaction at a fairly rapid rate by means of a gas inlet tube that reaches nearly to the bottom of the flask. The temperature is raised to 290 to 300°C for a period of 8 h while the ammonia is added. Aqueous ammonia distills from the reaction. The residue in the flask is fractionated at reduced pressure to give about 150 g (75%) azaleonitrile (bp 145 to 150°C/2 mm), which solidifies on standing. Reduction of the dinitrile to the diamine is carried out in an autoclave in dioxane over Raney cobalt catalyst at 5000 psi and 120°C. Nonamethylenediamine boils at 80 to 82°C/3 mm.

Polyimides from an aromatic diamine and a bisanhydride are thermally quite stable. The first step in the synthesis is the formation of a soluble amide acid. This can be cast to a film and imide formation brought about by heating to elevated temperatures.

111. Preparation of Poly(pyromellitimide) of 4,4′-Diaminodiphenyl Ether (32, 34)

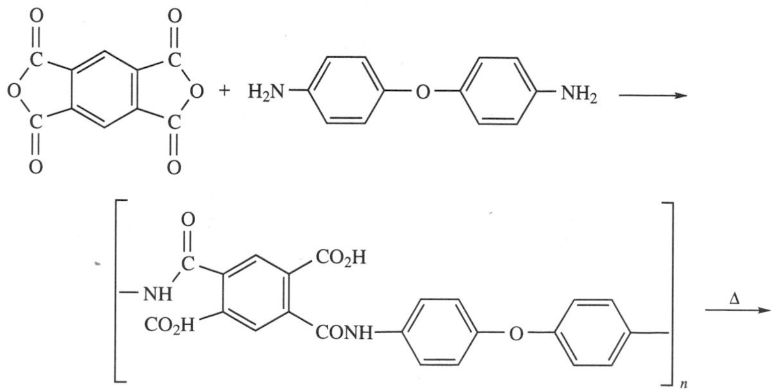

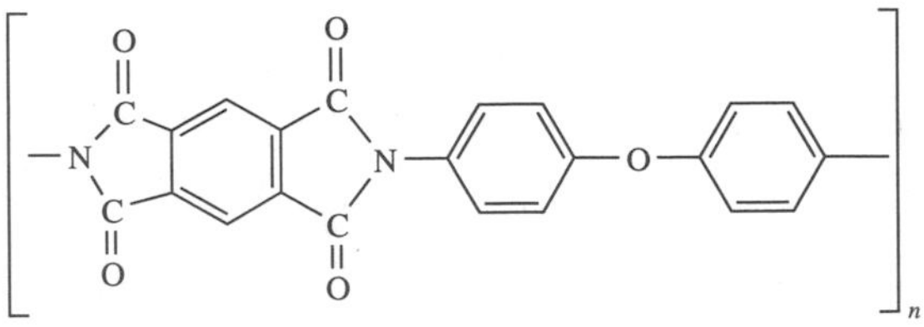

Materials and equipment must be anhydrous. Pyromellitic dianhydride is preferably dried at 180°C for 24 h in vacuo with a nitrogen bleed. 4,4′-Diaminodiphenyl ether may be dried under the same conditions, at 100°C, but is usually sufficiently anhydrous. Anhydrous pyridine is prepared by redistilling from maleic anhydride and passage through a column of 5-Å angstrom molecular sieves; its water content should not exceed 100 ppm. Then 66.7 g (0.33 mol) 4,4′-diaminodiphenyl ether is dissolved in 665 mL pyridine in a 2-L resin flask, with a slow current of dry nitrogen passing over the liquid. The solution is cooled to 10 to 15°C. Then 69.8 g (0.32 mol) pyromellitic dianhydride is added to the solution in four equal portions. Time is allowed after each addition for complete reaction, indicated by the disappearance of all solid particles. External cooling is applied to maintain the temperature $<$40°C. The polymerization is completed by careful, portionwise addition of small amounts of a 10% solution of the dianhydride in pyridine near the equivalent point. The viscosity increases sharply so that care must be exercised to avoid formation of a high viscosity gel. This procedure typically yields a 17% solution of polyamic acid with an inherent viscosity of 1.0 (0.5 g/100 mol in pyridine or dimethylacetamide) and a solution viscosity of 2,500 P. Attractive solvents include dimethylacetamide, *N*-methylpyrrolidinone, and pyridine; fibers can be processed.

6.7. POLYOXADIAZOLES

Aromatic poly-1,3,4-oxadiazoles and thiadiazoles give fiber and films with outstanding thermal and hydrolyticstabilities. However, these products do burn in a direct flame. The all-aromatic polymers can be readily prepared by thermal conversion from tractable polyhydrazide precursor films and fiber.

112. Preparation of a Poly(oxadiazole) from an Alternating Copolyhydrazide of Terephthalic and Isophthalic Acids (35, 36)

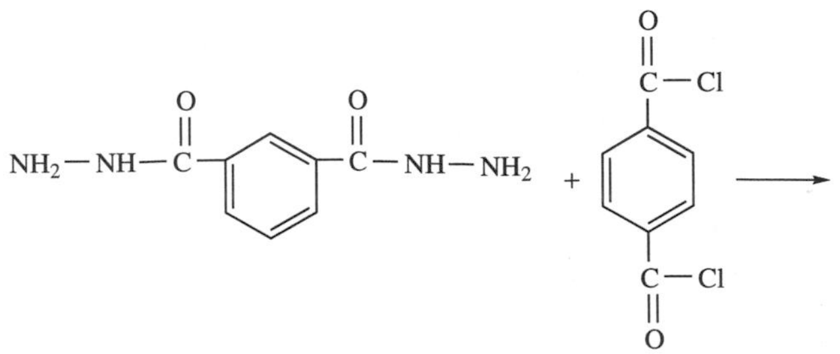

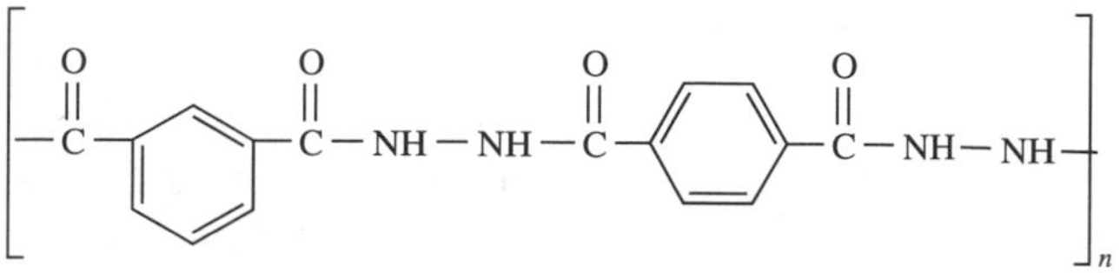

$$\left[-\overset{O}{\overset{\|}{C}}-C_6H_4-\overset{O}{\overset{\|}{C}}-NH-NH-\overset{O}{\overset{\|}{C}}-C_6H_4-\overset{O}{\overset{\|}{C}}-NH-NH- \right]_n$$

The alternating polyhydrazide copolymer is made as follows. It is important in this procedure to avoid all traces of moisture. Viscosity can be regulated by moisture content of the solvent. In a nitrogen-filled dry box, 250 mL hexamethylphosphoramide (**carcinogen!**) with <80 ppm water and 19.10 g (0.0984 mol) isophthalic dihydrazide (dried 48 h at 120°C and 1 mm) are placed in a 500-mL resin kettle fitted with a nitrogen bleed, drying tube, air-driven shear stirrer, and stoppered openings for addition of dry solids. The sealed system is transferred to a hood for connection to a nitrogen line and attachment of the stirrer to an air motor. The hydrazide is dissolved by stirring at a water bath temperature of 50°C. When solution is complete, the hot water is replaced with an ice bath. After cooling to ice-bath temperature, the first portion of 10 g terephthaloyl chloride is rapidly added with stirring. After cooling with stirring for 30 min, a second portion of 10 g of terephthaloyl chloride is added (total 20 g, 0.094 mol). The solution rapidly becomes thick and excessively difficult to stir. After 1 h cooling with stirring, the ice bath is removed and slow stirring is continued for another 1 h The viscous polymer solution is poured into distilled water, with rapid stirring in a home blender. The polymer is collected on a sintered glass funnel, washed twice with water in the blender, and then with methanol. After drying overnight in a vacuum oven under nitrogen, the polymer has an inherent viscosity of 1.8 at 0.5% concentration in hexamethylphosphoramide. The PMT is 390°C.

The polyoxadiazole is prepared from this dihydrazide by heating in a vacuum oven under nitrogen for 24 h at 275°C, followed by 24 h at 320°C. It is important to maintain an oxygen-free system. The product is a golden yellow material, with an inherent viscosity of 1.0 in concentrated sulfuric acid. The polymer does not melt or degrade below 450°C.

113. Preparation of Poly(N-phenyltriazole) (37)

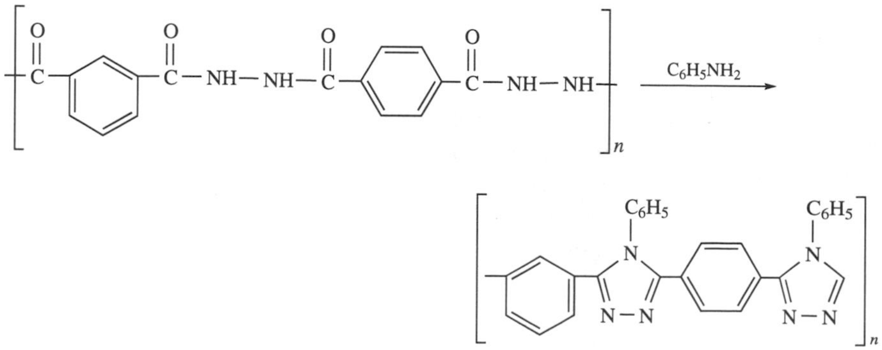

$$\left[-\overset{O}{\overset{\|}{C}}-C_6H_4-\overset{O}{\overset{\|}{C}}-NH-NH-\overset{O}{\overset{\|}{C}}-C_6H_4-\overset{O}{\overset{\|}{C}}-NH-NH- \right]_n \xrightarrow{C_6H_5NH_2}$$

$$\left[-C_6H_4-\text{(triazole, N-}C_6H_5\text{)}-C_6H_4-\text{(triazole, N-}C_6H_5\text{)}- \right]_n$$

A 3-kg portion of polyphosphoric acid is heated to 150°C and stirred under dry nitrogen for 2 h to deaerate. Then 751 g (8.1 mol) aniline is added dropwise with stirring at such a rate that the temperature of the reaction mixture does not exceed 190°C. The temperature of the resulting solution is adjusted to 175°C, and 97.2 g of η about 3.0 poly(*m,p*-phenylenehydrazide) is then added. The mixture is stirred and heated at 174 to 176°C for 140 h, during which time the polyhydrazide dissolves to form a homogeneous solution. The clear, light orange, fairly viscous solution is poured hot into deionized water in two operating 1-gal. blenders. Sodium hydroxide pellets are added cautiously to each resulting hot slurry until the agglomeration point is reached. The mixture is filtered, and the combined residue is reslurried in a blender, first in 5% aqueous sodium hydroxide solution and then twice in deionized water. The residue is then extracted continuously with hot ethanol in Soxhlet extractors until the effluent liquid is colorless. The combined residue is then dried in a vacuum oven at 80 to 120°C.

There should be no formic acid–insoluble prepolymer present in this product. This can be checked by dissolving 0.2 g of the product in 20 mL boiling 98 to 100% formic acid in a centrifuge tube. The solution is cooled and centrifuged. The supernatant solution is decanted, and 10 mL fresh formic acid is added. The mixture is heated to the boil and then filtered through a tared coarse-grade fritted glass filter. The transfer to the filter is aided by additional formic acid wash. The residue is dried on the filter and weighed. The percentage formic acid insoluble product may then be calculated.

The poly(*m,p*-phenylene-4-phenyl-1,2,4-triazole) obtained from this reaction should have a number average molecular weight of about 28,000. If lower inherent viscosity polyhydrazide is used, the triazole will have correspondingly lower molecular weight.

Hydrazine may be condensed at high temperatures and under pressure with a dicarboxylic acid, ester, diamine, or a dihydrazide to produce high polymer through formation of the 4-amino-1,2,4-triazole ring (38). An advantage is gained by using the dihydrazide of the acid because a vacuum cycle is not needed in the polymerization. The polyaminotriazoles from aliphatic acids ($C > 6$) have been melt spun into fibers with good tensile properties and high dyeability with acid dyes resulting from the pendant amine group.

6.8. POLYQUINOXALINES

Polyquinoxalines have been synthesised from tetraamines and bis (1,2-dicarbonyl) compounds to give high-melting, thermally stable polymers that are highly resistant to acids, bases, and oxidizing agents (39). Polymerization is usually carried out in two stages. The first consists of solution polymerization in hexamethylphosphoramide (**Carcinogen**) under nitrogen followed by isolation of the prepolymer and futher polymerization by heating at 350 to 400°C under reduced pressure. The polymers are highly colored, show only trace crystallinity, and are soluble only in concentrated sulfuric acid.

114. Preparation of Precursors for Polyquinoxalines

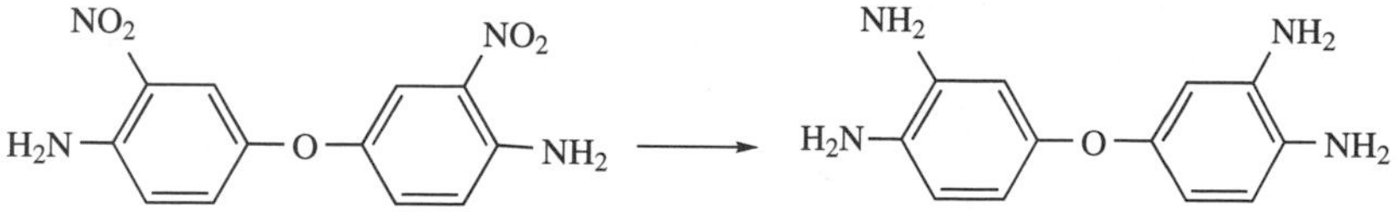

114.A. 4,4′-Diacetamidodiphenyl Ether

To a solution of 140 g (0.70 mol) *p*-oxydianiline in 500 mL glacial acetic acid is added dropwise 173 g (1.6 mol) acetic anhydride at such a rate to maintain a temperature of 50 to 60°C. The temperature is maintained at 90 to 100°C for an additional 1 h; then the solution is allowed to stand overnight. The precipitate that forms is collected and dried in air, then under reduced pressure for 24 h to give 147 g 4,4′-diacetamidodiphenyl ether. The filtrate from the above reaction is poured into 1 kg ice, and the precipitate is collected by filtration and dried under reduced pressure to give an additional 12 g acylated product. The combined yield is about 80%. The product melts at 227 to 228°C.

114.B. 3,3′-Dinitro-4,4′-diacetamidodiphenyl Ether

To 700 mL cold acetic anhydride is slowly added 95 mL colorless 70% nitric acid at such a rate to keep the temperature $<$10°C (**Extreme caution!**). The acetyl nitrate formed is cooled to 0°C and 75 g (0.265 mol) 4,4′-diacetamidodiphenyl ether is added in small portions while the temperature is maintained between 10 and 15°C. The yellow mixture is stirred for 30 min at room temperature and then poured slowly into 3-L of a 1 : 1 mixture of ice and water. The yellow precipitate is filtered to give a quantitative yield of 3,3′-dinitro-4,4′-diacetamidodiphenyl ether.(mp 210 to 213°C).

114.C. 3,3′-Dinitro-4,4′-aminodiphenyl Ether

To a solution of 199 g (0.53 mol) 3,3′-dinitro-4,4′-diacetamidodiphenyl ether in 1.3 L methanol is added a solution of 84 g potassium hydroxide in 300 mL methanol, dropwise with stirring. After 1 h, an additional 56 g potassium hydroxide in methanol is added, and the mixture is stirred for 3 h more. The mixture is poured into 5 L water, and an orange solid precipitates and is collected by filtration. The precipitate (mp 174 to 177°C) is recrystallized from 95% ethanol to give about 130 g 3,3′-dinitro-4,4′-diaminodiphenyl ether.

114.D. 3,3′-4,4′-Tetraaminodiphenyl Ether

To a warmed, vigorously stirred solution of 240 g stannous chloride dihydrate in 500 mL concentrated hydrochloric acid is added 46.4 g (0.116 mol) 3,3′dinitro-4,4′-diaminodiphenyl ether at such a rate as to maintain the temperature at 50 to

60°C. The mixture is heated at 65 to 70°C for 3 h, and then cooled to −10°C to yield a pink solid. The tetrahydrochloride is collected by filtration and dissolved in 300 mL hot water. To the solution is added 300 mL concentrated hydrochloric acid. Cooling produces white needles of the tetrahydrochloride salt, which are collected and pressed dry under a stream of nitrogen. The salt is then dissolved in water and added dropwise to a stirred solution of 60 g sodium hydroxide in 300 mL deoxygenated water, which is cooled in an ice bath. The gray white precipitate (45 g) is collected by filtration under nitrogen, washed with cold water, and dried under reduced pressure to yield 31 g (84%) of product, (mp 150 to 151°C). The tetraamine is purified by sublimation before each polymerization.

*114.E. Preparation of 4,4′-Diglyoxalyldiphenyl Ether Dihydrate (39)**

In a 5-L three-necked flask equipped with a reflux condenser, a sealed mercury stirrer, and a dropping funnel, 249 g (1.875 mol) anhydrous aluminum chloride, is suspended in 2-L tetrachloroethane **(Caution, suspected carcinogen)**. Then 147 g (1.875 mol) acetyl chloride, and 100 g (0.625 mol) diphenyl ether are dissolved in 100 mL tetrachloroethane and added to the rapidly stirred suspension over a period of 20 min. The temperature is raised to 60°C on a steam bath and kept at 60°C for 4 h. The reaction mixture is then cooled to room temperature and poured carefully into 800 mL. cold water. The water layer is extracted with four 100-mL portions of ether; the ether extract is added to the tetrachloroethane layer. The solvent is removed under reduced pressure, and on standing a yellow solid crystallizes (mp 96 to 98°C). The yellow solid is recrystallized from 95% ethanol, using charcoal to remove the color. The yield of the purified white 4,4′-diacetyldiphenyl ether is about 126 g of (63%) and has a mp of 98 to 99°C.

To a solution containing 29.7 g (0.268 mol) selenium dioxide dissolved in 150 mL dioxane and 4 mL water containing 3 drops of hydrochloric acid is added 34 g (0.134 mol) of 4,4′-diacetyldiphenyl ether. On addition of the diacetyl adduct, the solution turns red; and after refluxing for 6 h, it turns black, indicating precipitation of selenium. Filtration of the hot reaction mixture gives a red solution. The dioxane mixture is heated to reflux with 4 g charcoal. After filtration, a yellow liquid is obtained, and addition of water, a light tan solid precipitates. The compound is recrystallized from a 50/50 dioxane/water mixture to give about 20 g of a white solid (mp 120 to 121°C).

The synthesis of the polyquinoxaline from the diglyoxal and the tetraamine is carried out in two stages. The first consists of polymerization under nitrogen in hexamethylphosphoramide (**Cacinogen!**). The hexamethylphosphoramide (HMPA) is freshly distilled under nitrogen before transfer to the reaction vessel. The monomers must also be protected from the air and transferred quickly from a nitrogen atmosphere storage to the nitrogen swept reaction vessel. The workup of the polymers is to first carry out a precipitation with methanol and then an extraction with benzene by means of a Soxhlet extractor to remove any occluded HMPA.

* Thanks to Professor Stille, Dr Williamson & Dr Arnold for details not appearing in the reference.

115. Preparation of Poly(2,2′-[4,4′-oxydiphenylene]-6,6′-oxydiquinoxaline) (39)

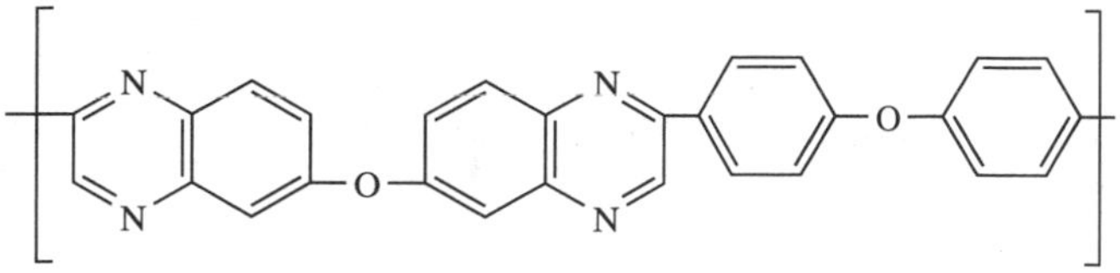

To a solution containing 5.682 g (0.01785 mol) 4,4′-diglyoxaldiphenyl ether dihydrate dissolved in 100 mL HMPA is added 4.111 g (0.01785 mol) 3,3′,4,4′-tetraaminodiphenyl ether dissolved in 70 mL HMPA. The weighing of the monomers and the monomers addition are carried out under a nitrogen atmosphere in a dry box. After stirring in the dry box for 30 min at 30°C, the flask is heated to 100°C for 2.5 h, then overnight at 160°C. The polymer is precipitated by the addition of 700 mL absolute methanol. The yellow polymer is washed with methanol and dried under reduced pressure over phosphorus pentoxide. The inherent viscosity is about 0.7 in hexamethylphosphoramide. (**Carcinogen!**). The polymer is then treated to a second high-temperature heating cycle to complete the cyclization. This is accomplished by heating the polymer in a 50-mL rotating flask at 375°C under 0.1 mm pressure for 2 h Two 8-mm-dia. steel ball bearings are added to the flask to facilitate mixing.

6.9. POLYAZOMETHINES

The literature on poly(azomethines) is extensive and is a segment of polymer chemistry in which many researchers have tried to make high molecular weight polymer but have failed because of low solubility or infusibility of the products. Successful synthesis of high molecular weight, tractable polymer resulted from astute selection of polymer compositions (40). Melt processible polymers with rod-like structures can be made from unsymmetrically substituted aromatic diamines and aromatic dialdehydes. The polymer from 2-methyl *p*-phenylenediamine and terephthalaldehyde has been studied extensively and has been melt-spun to give fiber tenacity/tensile modulus values of 38/1012 gpd (4.6/124 GPa) after heat treatment. The aldehyde/amine reaction is rapid and endcapping is necessary to inhibit further molecular weight increase during melt processing. Polymerization is readily initiated in water-free solvents, such as dimethylacetamide, hexamethylphosphoramide, *N*-methylpyrrolidinone, ethanol, or benzene. No catalyst is necessary, but removal of water expedites polymerization. High molecular weights are attained by conducting further polymerization at elevated temperatures in the solid or molten state, optionally at reduced pressure.

A range of polymers with melting or softening temperatures suitable for melt processing can be achieved by appropriate ring substitution, copolymerization, and/or introduction of limited chain flexibility. For preparations in the molten state, the amine exchange reaction of diamines with bis(azomethines) is a convenient and well-controlled process that yields polymer of high molecular weight.

116. Preparation of Poly(azomethines) from Aromatic Dialdehydes and Aromatic Diamines (40)

116.A. Intermediates

Terephthaldehyde, (Aldrich Chemical Co.) is purified by sublimation at reduced pressure. The melting point is 115°C, and the product is about 95% pure. 2-Chloroterephthaldehyde is prepared from 2-chloro-*p*-xylene according to the procedure of Naylor (41). The aldehyde is recrystallized from hexane and melted at 79 to 80.5°C. 2-Methyl-1,4-phenylenediamine, the diamine, is prepared from the dihydrochloride by treating the salt with concentrated aqueous ammonium hydroxide, separating the diamine as an oil, and distilling the oil from granular zinc (5 to 10 mol%) through an efficient column. The water-white diamine is collected in 50-mL flasks under nitrogen for storage in a drybox. The flasks are shielded from light by enclosing them in aluminum foil. The flasks are opened only for use in a nitrogen atmosphere. The pure diamine melts at 64°C and has a bp of 274°C/760 torr. Diaminodiphenylethane is purchased from Eastman Chemical Co. and sublimed at 140°C/0.25 torr before to use. The mp is 138 to 140°C.

116.B. Polymerization Methods

116.B.1. SOLUTION POLYMERIZATION IN AMIDE SOLVENTS

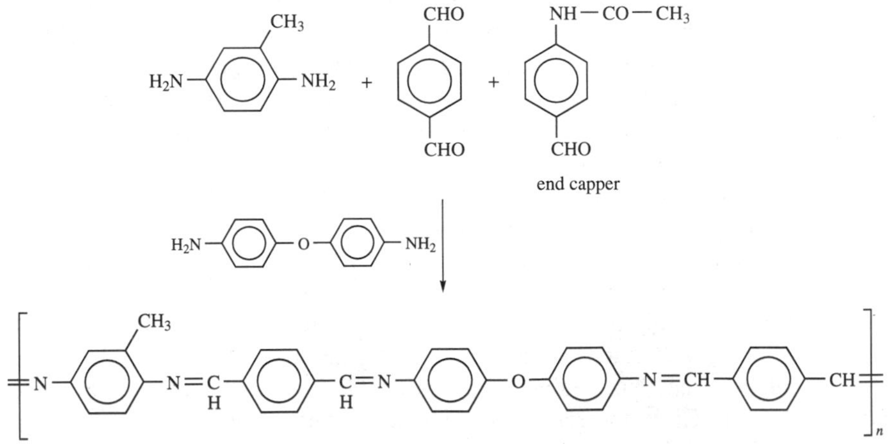

CH_3 H_2N NH_2 + CHO CHO + NH—CO—CH_3 CHO

end capper

H_2N O NH_2

CH_3 N N=C H C=N H O N=CH CH n

A total of 4.40 g (0.036 mol) 2-methyl-1,4-phenylenediamine, 0.80 g (0.004 mol) bis(aminophenyl) ether, and 0.04 g-acetamidobenzaldehyde are dissolved in an anhydrous mixture of 25 mL hexamethylphosphoramide 25 mL and

N-methylpyrrolidinone containing 2 g anhydrous lithium chloride. To this solution is added 5.36 g (0.04 mol) terephthalaldehyde with stirring. After 16 h at room temperature, the reaction mixture is too viscous to stir. The reaction mixture is added to water in a home blender with stirring. The product is collected by filtration, washed well with water and then ethanol, and dried at 80°C, under vacuum. The yield is 100%, and the PMT is 306°C. The inherent viscosity is 1.1 at 0.5% concentration in 98% sulfuric acid.

116.B.2. MELT POLYMERIZATION

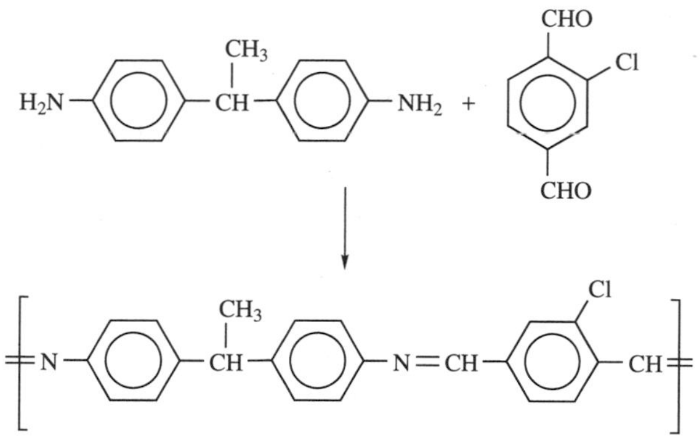

CHO
CH_3
Cl
H_2N CH NH_2 +
CHO

Cl
CH_3
N CH N=CH CH

A total of 4.24 g (0.02 mol) 4,4′-diaminodiphenylethane and 3.36 g (0.02 mol)2-chloroterephthaldehyde are placed in a 250-mL three-necked round-bottom flask equipped with stirrer, nitrogen inlet, and distilling head. The temperature is raised to 275°C. over 30 min while the reaction mixture is stirred and by-product water is collected. The polymer melt is cooled to room temperature. The polymer is removed from the flask, ground in a laboratory mill, washed with acetone, and dried in a vacuum oven at 80°C. The orange-colored polymer has an inherent viscosity of 0.94 in sulfuric acid at 0.5% concentration and a PMT of 260°C.

REFERENCES

1. U.S. Pat. 3,094,511 18, June (1963), H. W. Hill, S. L. Kwolek, and W. Sweeny, to DuPont.
2. U.S. Pat. 3,287,324 (22, November 1966), W. Sweeny, to DuPont.
3. U.S. Pat 3,063,966 (Nov. 13, 1962), W. R. Sorenson et al. to DuPont.
4. U.S. Pat 4,980,446 (December 25 1990), W. Sweeny, to DuPont.
5. G. Singh et al., *Macromolecules*, **25**, 6095 (1992).
6. F. Higashi et al., *J. Poly. Sci. A*, **18**, 1711 (1980); **21(A)**, 3338 (1983).
7. S. R. Turner et al., *Macromolecules*, **25**, 4819 (1992).
8. Brit. Pat. 1,008,854 (Nov. 3, 1965) and Belg. Pat. 637,260 (1963), J. Preston and R. W. Smith, to Monsanto.
9. S. L. Kwolek et al., *Macromolecules*, **10**, 1390 (1977).

10. S. L. Kwolek et al., Liquid Crystalline Polymers. In *Encyclopedia of Polymer Science and Engineering*, Vol. 19, 1987, John Wiley & Sons Inc., New York.
11. P. J. Flory and M. Gordon, eds. *Liquid Crystal Polymers. I*, [Advances in Polymer Science, Vol. 59], Springer-Verlag, New York.
12. P. J. Flory, *Principles of Polymer Chemistry*, Cornell University Press, Ithaca, New York, 1953.
13. R. R. Burch, Y. Kim, and W. Sweeny, *Macromolecules*, **23**, 1065 (1990).
14. U.S. Pat. 4,959,453 25, September (to DuPont) (1990), W. Sweeny.
15. S. Ozawa, *Polymer J.*, **19**, 119 (1987).
16. W. Memeger, *Macromolecules*, **9**, 195 (1976).
17. P. W. Morgan, *Macromolecules*, **10**, 1381 (1977).
18. J. Preston, *J. Poly. Sci. B*, **2**, 1171 (1964).
19. J. Preston et al., Paper presented at the High Temperature Fiber Symposium, Phoenix, Anz., Jan. 1966.
20. T. J. Bair et al., *Macromolecules*, **10**, 1396 (1977).
21. U.S. Pat. 5,077,377 (31 December 1991), W. Sweeny, to DuPont.
22. U.S. Pat. 4,782,131 (1 November 1988) W. Sweeny, to DuPont.
23. U.S. Pat. 3,278,594 (11 October 1966), W. Sweeny, to DuPont.
24. R. S. Irwin et al., *Macromolecules*, **22**, 1065 (1989).
25. R. J. W. Reynolds and J. A. Van Allan, *Org. Synth. Coll.*, **4**, 15 (1963).
26. K. Tashiro and M. Kobiyashi, *Macromolecules*, **24**, 3706 (1991).
27. J. Wolfe et al., *Macromolecules*, **14**, 915 (1981). U.S. Pat. 4,806,688 (21 February 1989), M. N. Inbasekaran and R. M. Strom, to Dow Chemical Co.
28. W. Sweeny, *J. Poly. Sci. A*, **30**, 1111 (1992).
29. U.S. Pat. 2,904,537 (Sept. 15, 1959), K. C. Brinker et al. to DuPont.
30. Abbes, *Berichte Chemie*, **26**, 2951 (1983).
31. Polybenzimidazoles, in *Encyclopedia. of Polymer Science and Engineering*, vol. 11, A. Buckley, D. E. Stnetz, and G. A. Serad, (Celancse Co.). J. Wiley & Sons New York 1987.
32. W. M. Edwards, U.S. 3,179,614, 3179634 (April 20, 1965), to DuPont.
33. E. L. Wittbecker and P. W. Morgan, *J. Poly. Sci.*, **40**, 289 (1960).
34. C. E. Sroog et al., *J. Poly. Sci. A*, **3**, 1373 (1965).
35. A. H. Frazer and F. T. Wallenberger, *J. Poly. Sci. A*, **2** 1147 (1964).
36. A. H. Frazer, F. T. Wallenberger, *J. Poly. Sci. A.*, **2**, 1157 (1964).
37. J. R. Holsten and M. R. Lillyquist, *J. Poly. Sci. A.*, **2**, 3904 (1965).
38. J. W. Fisher, *Chem. Ind.*, **71**, 244 (1952)
39. J. K. Stille et al., *J. Poly. Sci, A*, **3**, 1013 (1965).
40. P. W. Morgan et al., *Macromolecules*, **20**, 729 (1987).
41. J. R. Naylor, *J. Chem. Soc.*, **1952**, 4085 (1952).

7

SOLID-PHASE POLYMERIZATION

The term *solid state* or *solid-phase polymerization* applies largely to prepolymers that contain end groups able to react with each other. This usually means AABB or AB polymers. Two different types of polymerization processes are incorporated in the solid state concept. The first involves a process to increase the molecular weight of a polymer by heating in the solid state above the glass-transition temperature T_g but below the melting or decomposition temperature. The second involves polymerizing a crystalline monomer in the solid state directly to crystalline polymer. Success of the latter type of polymerization depends on lattice spacing and overall packing alignment of the monomers in the unit cell.

In addressing the first type, the key to achieving high molecular weight polymer is first to understand the principles involved. In melt polymerization, or in solution polymerization when relatively low molecular weight polymer precipitates, the rate of polymerization slows as the local concentration of active ends diminishes. Reconcentration is inhibited in the melt by the high-melt viscosity. If the polymer is isolated and crystallized, "defects" such as polymer ends are excluded from the crystalline phase and are concentrated in the "amorphous," or less crystalline, regions. This higher concentration of active ends, under the right conditions (e.g., heating in the solid state without melting), can again polymerize at a rate at least equal to that initially achieved in the melt. As polymerization continues in the solid state, 'new' polymer is incorporated into the crystalline phase, concentrating the remaining ends again into the amorphous phase and so the process continues.

Simplistically, in this process it appears necessary to crystallize the polymer and then to heat it above the T_g but below the melt temperature T_m. However, successful laboratory practice requires more than this. To achieve high molecular weight polymer it is essential to have a good balance of active ends in the prepolymer (which should be checked analytically before proceeding) and a finely powdered prepolymer sieved to a uniform size. This allows uniform heat transfer within particles, minimizes sticking, and enhances the polymerization of polymers such as polyethylene terephthalate, for which polymerization and propensity to stick are affected by diffusion of a significant amount of byproduct (e.g., glycol).

The solid-phase polymerization temperature should be ramped slowly from above the crystallization temperature to below the start of the T_m. Polymerization rate is enhanced if an inert gas such as nitrogen or argon is passed through the prepolymer powder or chips and/or if the pressure is reduced. A key step recommended before embarking on any solid-phase polymerization is to run a differential scanning calorimetric assessment (first and second cycles) of the prepolymer to determine at what temperature the polymer crystallizes, to detect any unusual exotherms or endotherms that might suggest sticking or fusing, and to determine starting and peak temperatures of the melt endotherm.

Solid-phase polymerization should be started 10 to 20°C below any endotherm that indicates fusing or sticking (e.g., caused by melting of a low molecular weight polymer fraction) and slowly ramped to the tangent temperature of the first downward slope of the first main melt endotherm. Solid-phase polymerization increases crystallite size and perfection and raises melt temperatures and enthalpies higher than those of the starting polymer, often making remelting difficult and perhaps causing problems and defects in spun fibers if complete melting is not achieved. Thorough mixing (e.g., in a screw melter) is recommended to avoid this problem.

Figures 7.1 and 7.2 show first- and second-cycle differential scanning calorimitry (DSC) scans of commercial poly(ethylene terephthalate) (PET) flake. The first scan shows a T_g of 83.9°C (onset 80.3°C); a phase-transition endotherm at 176.1°C (164.9°C onset), which is possibly melting of imperfect crystallites or of low molecular weight polymer; and a T_m of 246.2°C (onset 229.7°C). The second cycle is a rerun of the same polymer after rapid cooling. This shows a T_g of 78.2°C (onset 73.9°C), a crystallization exotherm at 166.9°C (onset 147.4°C), and a T_m at 245.8°C (onset 230.9°C). The endotherm at 176°C is absent. The lower ΔH of the T_m of the second cycle reflects a lower order of crystallinity after rapid recooling.

Based on the DSC scans, to solid-phase polymerize this sample of PET efficiently, one would first increase the crystallinity of the finely powdered polymer by heating at about 170°C under a flow of nitrogen, followed by raising the temperature to about 180°C, with gas or mechanical agitation, to minimize the lower melting fraction and prevent sticking. Heating would then be ramped slowly to 230 to 235°C and maintained there for several hours, until the desired molecular weight was achieved. A broader molecular weight distribution usually results but equilibrates to the more normal 2/1 distribution when the polymer is remelted.

Commercial solid phase polymerization, particularly of polyester for tire cord and bottles, uses polymer flake or chips of fairly uniform size, and stirs or rotates the chips while heating, often under reduced pressure. It is important that the

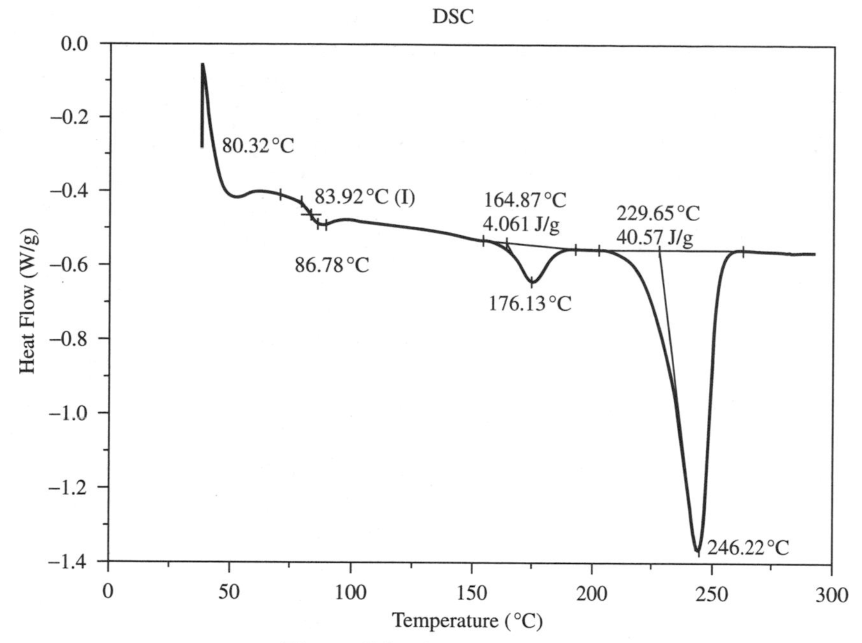

Figure 7.1. First-cycle DSC.

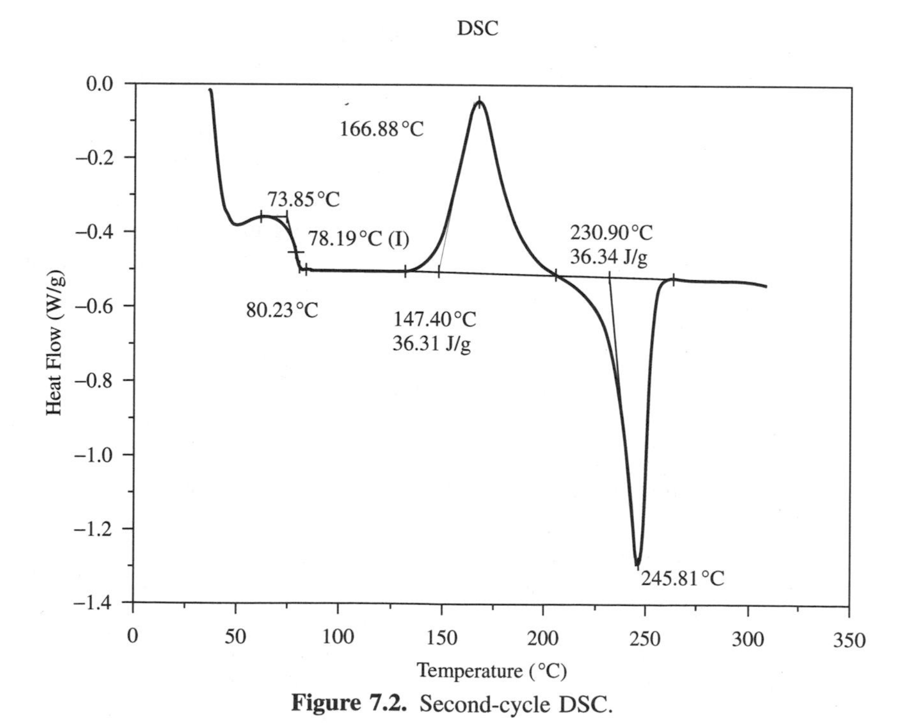

Figure 7.2. Second-cycle DSC.

material remains particulate and does not coalesce. The molecular weights of polyesters processed this way are considerably higher than those from all-melt processes, and molecular weight distributions deviate from the most probable values obtained from melt systems ($\overline{M}_w/\overline{M}_n > 2$).

The following describes a simple laboratory solid-phase polymerization of PET. Clear bottle-grade PET flake (inherent viscosity about 1.2 dL/g at 0.5% concentraction in trichloroethane/phenol solvent; $\overline{M}_w 1 \times 10^5$) is micronized to <250 μm particle size. The polymer is spread out about 0.25-in thick in an aluminum pan and placed in an oven and blanketed with a stream of dry nitrogen. The sample is first heated at 100°C to remove adsorbed moisture; then the temperature raised to 235°C over 2 h and maintained at that temperature for 36 h. After cooling under nitrogen, the sample is stored in a polyethylene jar under nitrogen to prevent re-adsorption of moisture. A sample of the polymer should have an inherent viscosity of about 2.5 dL/g ($\overline{M}_w 4 \times 10^5$).

Instead of micronizing the PET polymer, fine samples can be obtained by crystallizing the polymer from solution in nitrobenzene (1). Crystal aggregates of 2 to 10 μm are obtained. Polymerization is then carried out at 253°C for 24 h under vacuum. The intrinsic viscosities of the sample before and after the polymerization are reported to be 0.67 and 2.41 dL/g. The initial crystallinity of samples treated this way is high because of the solvent crystallization, and higher temperatures can be used in the solid-phase polymerization step. Most condensation polymers can be solid-phase polymerized to higher molecular weight. For example, the intrinsic viscosity of 6-6 nylon prepolymer can be raised from about 0.25 dL/g (0.5% concentration in *m*-cresol) to 0.9 by heating at 259°C in the solid state under nitrogen for 3 h (2). In other studies, a 6-6 nylon prepolymer of 2,500 molecular weight was polymerized at 216°C for 4 h to a molecular weight of 15,000 (2).

Polymerization in the solid state is also an important route for increasing the molecular weight of polymers that decompose close to their melt temperatures. In fact, it is one of few routes to high molecular weight polyoxamides. For example, the polyoxamide prepared from 2-methylhexamethylenediamine and dibutyl oxalate (with 0.1% SbF_3 catalyst added) was polymerized in the solid state at 245°C for 2 h, increasing intrinsic viscosity from 0.45 to 1.5 dL/g (2). This polyoxamide is unstable in the melt and only low molecular weight polymer is obtained by solution methods. Studies of the kinetics of solid-phase polymerization of nylons 6, 6-6, 6-10, and 6-12 indicate that the rate is controlled to a certain extent by the diffusion of reactive end groups, particularly carboxyl (2). This suggests that the activation energies of diffusion and amidation are about the same.

High-temperature solid-phase polymerization also can be carried out on nonfusible polymers, provided the ends are mobile and give the desired polymer linkage. For example, poly(*p*-phenyleneterephthalamide) prepolymer with amine and carboxyl ends does not increase significantly in molecular weight when subjected to high temperature in the solid phase because carboxyl ends are lost by decarboxylation. However, prepolymer containing phenyl ester group ends (–COOPh instead of –COOH) solid-phase polymerize readily to higher molecular weight (3).

The second broad type of solid-phase polymerization involves transforming a solid-state monomer to polymer without apparently going through a liquid state. Success of the polymerization depends on monomer lattice spacings and overall packing and positioning of the monomer in the unit cell. If the polymerization proceeds, the propagation direction is controlled by the crystal structure and the symmetry of the monomer. If the geometric arrangement is unfavorable, the polymerization may proceed in arbitrary directions giving amorphous products. A reaction that leads to a product that is crystallographically related to that of the monomer is called topotactic and the process is topotaxy (4). For crystal-to-crystal transformation by solid-state polymerization, the following rules apply: the distance between reacting unsaturated bonds of neighboring molecules should not exceed 0.4 nm and the mutual orientation of the unsaturated bonds should be close to parallel and preferably coplanar (5).

Some solid-phase polymerizations depend on the diffusion of the monomer and give amorphous products. This type includes high-energy irradiation of crystalline vinyl or acrylic monomers at low temperatures. Localized melting, however, may occur. Other solid-state polymerizations include inclusion- or clathrate-polymerization in which monomers are included and polymerized in the channel-like cavities of specific host crystals. Polymerization is usually initiated by high-energy irradiation and the mechanism is a free radical (6). Typical clathrate hosts are urea and thiourea. Typical monomers are butadiene, acrylonitrile, and vinyl chloride. Polymers are stereoregular and regioselective and often melt higher than the same polymers prepared by conventional routes.

Probably the most interesting type of solid-state polymerizations are those that proceed without appreciable translational diffusion of the monomers. These are termed *diffusionless*, or *topochemical polymerizations*. In diffusionless polymerization the monomers, by virtue of their spatial proximity and mutual

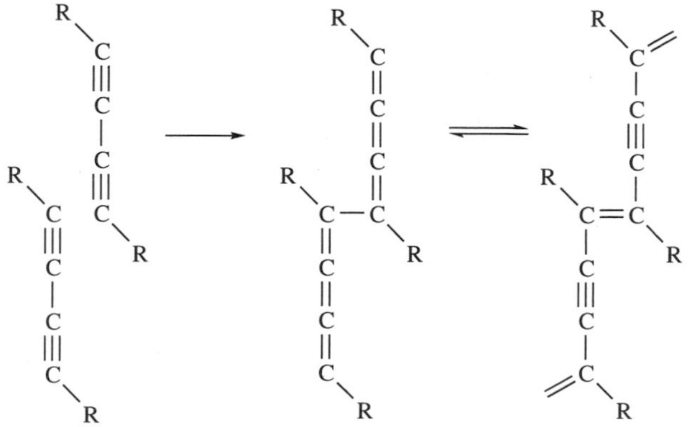

Figure 7.3. Topochemical polymerization of diacetylene.

orientation in the crystal state, react with each other and produce fully extended polymer chains along well-defined crystallographic directions. Typical monomers are specific diacetylenes and diolefines.

In the crystal, the polymerization reaction occurs by simple rotational and torsional movements of the active groups. A typical polymerization of the diolefinic group occurs via a [2 + 2] photocycloaddition of the double bonds of the nearest neighbor molecules to form cyclobutane linkages. The crystal symmetry of the monomer is largely retained (7). Polymerization of selected diacetylenes provides the best examples of topochemical polymerization (8). Polymerization of the diacetylene monomer crystals leads to fully ordered and fully extended polyconjugated polymer chains (Fig. 7.3).

We do not describe any preparations involving topochemical polymerization of the diacetylenes, not only because of difficulties in monomer preparations but primarily because they entail use of high-energy radiation initiators (e.g., ^{60}Co radiation).

REFERENCES

1. M. Ito and K Takahashi, *J. Appl. Poly. Sci.*, **40**, 1257 (1990).
2. J. Zimmerman, Polyamides, in *Encyclopedia of Polymer Science and Engineering*, vol. 11, John Wiley & Sons, New York, 1988. C. D. Papaspyrides, *Chem. Abs.*, **117**, 8 (1992).
3. J. Fitzgerald et al., *Macromolecules*, **24**, 3295 (1991).
4. Solid-State Polymerization in *Encyclopedia of Polymer Science and Engineering*, vol. 15, John Wiley & Sons, New York, 1989.
5. F. M. Hirshfeld and G. M. J. Schmidt, *J. Poly. Sci.*, **48**, 37 (1960). M. D. Cohen and G. M. J. Schmidt, *J. Chem. Soc.*, **23**, 271 (1969).
6. K. Takamoto and M. Miata, *J. Macromol. Sci. Rev. Macromol. Chem.*, **C18**, 83 (1980). M. Farina et al., *J. Am. Chem. Soc.*, **89**, 507 (1967).
7. H. Nakanishi et al., *Proc. R. Soc. London*, **A369**, 307 (1980). M. Hasegawa and Y. Suzuki, *J. Poly. Sci. Poly. Lett. Educ.*, **5**, 813 (1969).
8. G. Wegner, in W. E. Hatfield, ed., *Molecular Metals*, Plenum Press, New York, 1979. G. Wentz et al., *Macromolecules*, **17**, 837 (1984). Diacetylene Polymers, in *Encyclopedia of polymer science and engineering*, vol. 4, John Wiley & Sons, New York, 1988. H. J. Cantow, ed., *Advances in Polymer Science*, vol. 63, Springer-Verlag, Berlin.

8

RING-OPENING POLYMERIZATION

A variety of different ring-opening polymerizations and mechanisms have been described in the literature (1). The common feature is relief of ring strain to produce linear polymer. Most examples of this type of polymerization are those of a heterocyclic ring opened with anionic or cationic catalysts. However, there are also many examples of free-radical ring opening polymerization, e.g., of cyclic ketene acetals to produce polyesters (2). A wide variety of ring compounds have also been opened to produce polymers with all-carbon backbones. Examples of these include cyclopentene, norbornene, and 1,3-bridged cyclobutanes (3). In this chapter the various types of ring-opening polymerizations will be discussed in separate sections according to monomer and product. Ring opening has been a popular and chemically fascinating route to polymers, as the examples given in this chapter make evident.

8.1. CYCLIC AMIDES

Ring opening and polymerizability of cyclic amides vary greatly with ring size and substituents. As noted, ease of polymerization varies with ring strain. This relationship between polymerizability and configurational strain has been studied extensively by Hall et al. (4) for polycyclic and bridged rings as well as for simpler systems.

Enthalpies of polymerization for several lactams are shown are as follows (5):

Monomer	Ring Carbons	ΔH, Kcal/mol
Caprolactam	6	−3.2
Enantholactam	7	−5.2
Capryllactam	8	−9.6
Laurolactam	12	−1.4

Polymerization of lactams is a thermally controlled equilibrium process between polymer and cyclic monomer; and in some cases, there are ceiling temperatures above which lactam only is favored. For comparison, caprolactam, a seven-membered ring, polymerizes readily at high temperatures with anionic initiators. On the other-hand butyrolactam, a five-membered ring, and valerolactam, a six-membered ring, polymerize only at relatively low temperatures and reconvert to monomer in the presence of the polymerizing catalyst at higher temperatures, e.g., 60 to 80°C. It should be pointed out that in the high temperature (water- or base-catalysed) polymerization of caprolactam, the molten polymer is in equilibrium with about 10% of the cyclic monomer.

Alkyl or aryl substituents on the ring carbons and especially on the nitrogen of lactams inhibit polymerization and shift the equilibrium toward monomer (4, 6). Lactams with heteroatoms beta to the carbonyl group are usually thermally unstable. There also appears to be no relationship between rate of hydrolysis and polymerizability of lactams (7).

Currently, the most important ring-opening polymerization is that of caprolactam to produce 6-nylon. Commercially, caprolactam is polymerized in a high-temperature aqueous system using nylon 6-6 salt or aminohexanoic acid as catalysts. (See chapter 4). In this chapter we will focus on the anionic polymerization of caprolactam and cyclic amides. With alkali metals or hydrides, the polymerization is characterized by an induction period that decreases with increasing temperature and is about 5 min at 265°C. The induction period is followed by extremely rapid polymerization and an increase in temperature. At this point, the monomer concentration falls to an equilibrium value and viscosity is at a maximum. If the polymer is kept at the elevated temperature, the viscosity falls rapidly as a result of redistribution of molecular weight from a broad to the most probable equilibrium distribution ($M_w/M_n = 2$). This results from interchange of the reactive ends with chain amide groups (transamidation). Understanding of the mechanism of anionic polymerization of lactams was provided by the discovery that *N*-acyl or *N*-aroyl lactams and acylating agents strongly accelerated the polymerization of lactams (8). The mechanism can be interpreted as a nucleophilic attack of the lactam anion on the ring carbonyl group of the imide (i.e., the ring carbon of the acylated lactam). This is the most electrophilic site in the polymerizing system. A new *N*-acyl lactam site is created and chain growth occurs via continued attack on these recreated centers. Acyl lactams themselves do not polymerize.

Use of bifunctional activators (e.g., from caprolactam acylated with diacid chlorides) yield narrower molecular weight distributions.

It is obvious from the mechanism of anionic polymerization (**8.1**) that the amide anions can be formed not only from the monomeric caprolactam but also from amide sites in the polymer chain. Thus branched polymer and cyclic oligomers form. It is obvious, too, that *N*-substituted lactams are not polymerized via the basic catalyst route. The equilibrium polymer/monomer/cyclic oligomer content is about the same in both base-catalyzed and hydrolytic-polymerization processes run at high temperatures. However, when anionic polymers are prepared

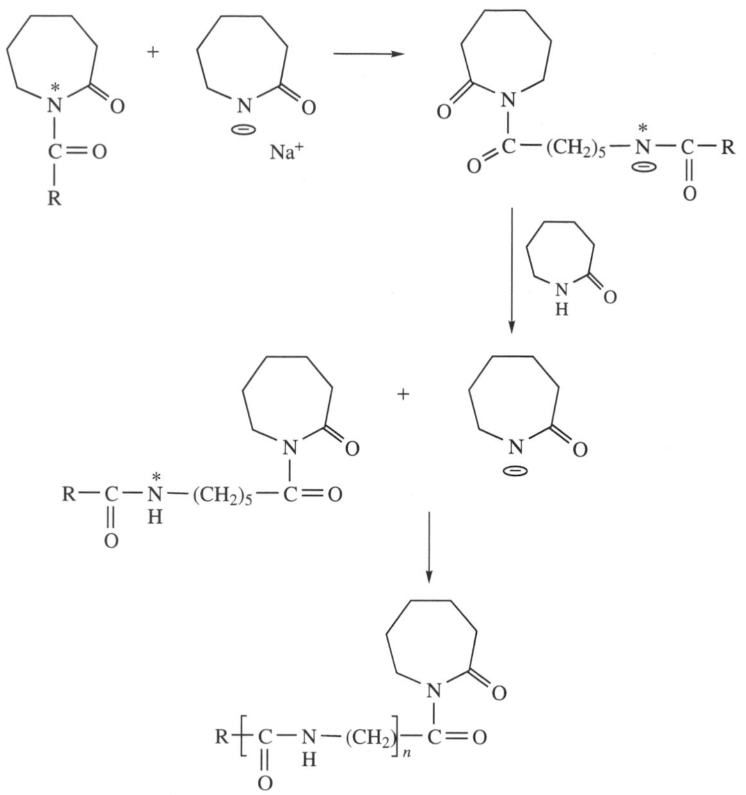

Schematic 8.1

below their melt temperature T_m, the monomer content is much lower than expected from extrapolation of the equilibrium monomer content/temperature curve obtained at higher temperatures (9) (Fig. 8.1). This suggests that only the amorphous polymer is in equilibrium with monomer and that the crystalline part is not.

Low monomer content is valuable from a commercial viewpoint and solid moldings have been made in situ via the anionic process, even in large shapes. The overall contraction resulting from polymerization and crystallization is balanced by the thermal expansion during the adiabatic molding process. Internal strains are minimized and are lower than those generated by the process of remelting polymer and molding into a part.

Linear high molecular weight polyamides, 6, 7, and 6-6, have been prepared via the rapid ring-opening polymerizations of their respective lactams, catalyzed by super nonionic poly(aminophosphazene) and prophosphatrane bases (10, 11). These polyphosphazenes appear to mimic catalysis by conventional anionic catalysts but with the advantage of tolerating some degree of moisture. Of interest is that the catalyst can be recovered by methanol extraction and after isolation can be reused for further polymerizations. A typical polyphosphazene (P_4-*t*-Bu)

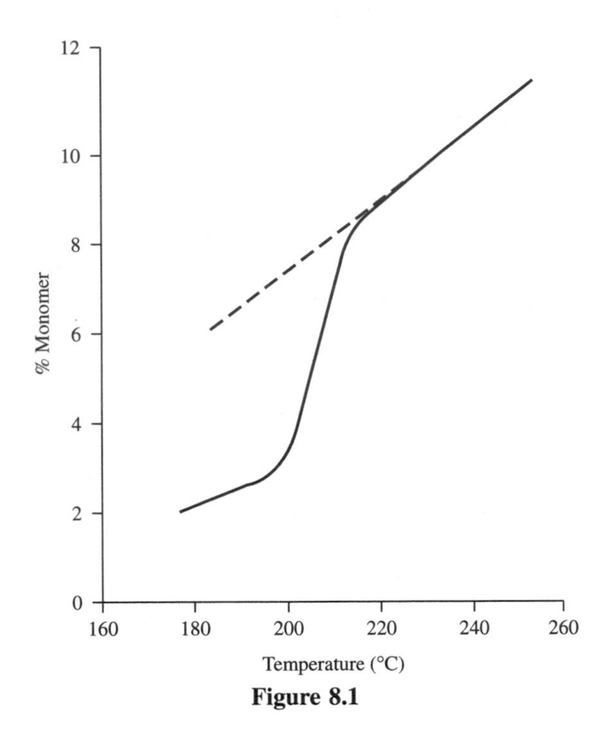

Figure 8.1

from Aldrich is formulated as follows:

$$[(Me_2N)_3P{=}N]_3\text{-}P{=}N\text{-}t\text{-}Bu$$

117. Fast Polymerization of ε-Caprolactam to 6-Nylon using Anionic Catalyst and N-Acetylcaprolactam As Cocatalyst (12–14)

$H_2C—CH_2$, H_2C, CH_2, H_2C, $C{=}O$, NH (ring) $\longrightarrow$ $\left[—NHCH_2CH_2CH_2CH_2CH_2\overset{O}{\overset{\|}{C}}—\right]_n$ $\longrightarrow$

In a glass polymer tube is placed 25 g caprolactam, which is then melted on a steam bath under nitrogen. To the molten caprolactam is added 0.6 g sodium hydride, which is allowed to dissolve to give the *N*-sodio derivative. The mixture may be maintained at 139°C in a boiling xylene vapor bath for several hours, but no polymer is produced. To the molten caprolactam containing the sodio derivative is now added 0.33 g *N*-acetyl caprolactam at 139°C. The tube is shaken to mix the contents that solidify rapidly. After 30 min, the tube is cooled and broken, and the polymer is ground in a Wiley mill. The polymer is then extracted with hot water and dried to yield high molecular weight 6-nylon in yields of 80%. The inherent viscosity is about 1.0 in *m*-cresol at 0.5% concentration.

118. Polymerization of Caprolactam Using Water as Catalyst (15, 16)

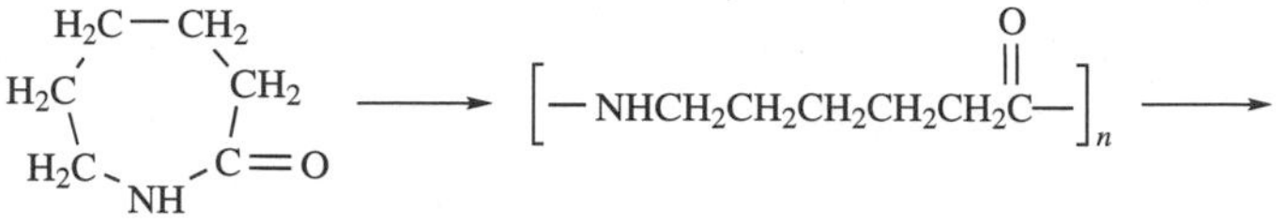

$$\text{caprolactam (cyclic: } H_2C{-}CH_2{-}CH_2{-}C{=}O{-}NH{-}H_2C{-}H_2C) \longrightarrow \left[-NHCH_2CH_2CH_2CH_2CH_2\overset{O}{\overset{\|}{C}}-\right]_n \longrightarrow$$

The water catalyzed polymerization of caprolactam does not involve total hydrolysis of the caprolactam to the amino acid followed by a typical polycondensation reaction. The polymer grows by reaction of generated amino acid with the cyclic monomer. The polymerization of caprolactam is carried out in two stages. In a polymer tube is placed 56 g of purified caprolactam. Then water 1 mL is added and the tube is sealed under nitrogen. The polymer tube is heated to 250°C (**Shield!**) and maintained at this temperature for 6 h. The tube is then cooled, opened cautiously, and reheated under a stream of nitrogen to 250 to 255°C. The polymerization is continued under nitrogen for another 2 h. The polymer is now a viscous melt that cools to a tough plug.

119. Polymerization of Caprolactam Using ε-Aminocaproic Acid as Initiator

Aminocaproic acid or 6-6-nylon salt (see chapter 4) is an effective initiator for the polymerization of caprolactam instead of water (17). In a polymer tube under nitrogen are placed 56 g pure caprolactam and 5 g ε-aminocaproic acid. The polymer tube is heated on a steam bath to melt the caprolactam, and the polymer tube is shaken to mix the ingredients. Still under nitrogen, the tube is transferred to a Woods metal bath at 150°C, and the temperature raised to 250°C over 1 h and held at that temperature for a further 4 h. The bath is lowered, and the contents are an almost colorless, viscous melt. The tube is cooled behind a shield and the polymer removed by fracturing the tube. The polymer is ground in a Wiley mill and extracted with hot water to remove cyclic oligomers. The polymer is filtered and dried under vacuum. Tough fibers can be drawn from the melt at about 250°C.

The caprolactam used in the experiments above is purified before use by crystallizing twice from cyclohexane. It is then stored in a vacuum desiccator at room temperature over phosphoric anhydride for 48 h preferably at pressures below 0.1 mm. After this treatment, the water content should be below 0.15% as determined by Karl Fischer titration.

120. Low-Temperature Polymerization of γ-Butyrolactam to 4-Nylon (14, 18)

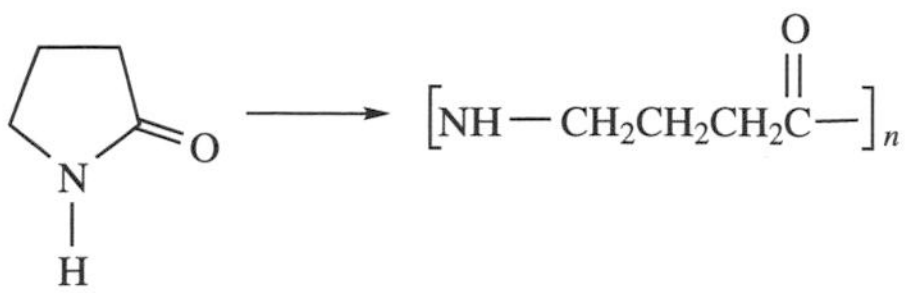

$$\gamma\text{-butyrolactam (N–H)} \longrightarrow \left[NH-CH_2CH_2CH_2\overset{O}{\overset{\|}{C}}-\right]_n$$

In a 1-L three-necked round-bottom flask under dry nitrogen is placed 500 g freshly distilled γ-butyrolactam. It is essential to protect the freshly distilled lactam from moisture because this interferes with the polymerization. The flask is heated on a mantle to 80°C and 5 g metallic potassium (**Danger!**) weighed under dry hexane is added over 1 h. About 100 mL monomer is distilled from the flask under vacuum to remove by-products of the reaction of the potassium with the pyrrolidone. The flask, which now contains about 400 g lactam and several grams of catalytic *N*-potassium butyrolactam, is allowed to cool to room temperature, still under nitrogen and protected from moisture. Polymerization begins when the contents of the flask reach about 50°C. The flask is allowed to stand for 12 h, then the polymer cake is removed and broken up. It is frequently necessary to use a hacksaw to cut the chunk into smaller pieces that can be ground in a Wiley mill. The ground polymer is washed thoroughly with water and then dried under vacuum at 70°C. The yield is about 300 g, and the polymer has an inherent viscosity of 1.2 to 1.8 at 0.5% concentration in *m*-cresol. The polymer is soluble in formic acid and tough fiber and film can be fabricated from these solutions.

The polymerizability of butyrolactam is largely a function of its purity. Commonly an activator such as *N*-acetylpyrrolidone is used to remove any induction period and ensure smooth polymerization. The activator is used roughly at one fourth the molar concentration of the potassium salt for absolutely pure monomer.

121. Preparation of 4-Nylon using an Activator (16)

$$\text{(2-pyrrolidone)} \xrightarrow[AC_2O]{NaH} \left[NH-CH_2-CH_2-CH_2-\overset{\overset{O}{\|}}{C}- \right]_n$$

Since *N*-acetylpyrrolidone is not readily available, it is prepared in situ here. A total of 0.10 g acetic anhydride is added to a solution of 0.13 g sodium hydride dispersion (54% in mineral oil) in 6.5 g pure, dry pyrrolidone contained in a dry polymer tube and blanketed with dry nitrogen. An exothermic reaction ensues, leading to the rapid formation of a hard white plug of polymer. This is broken into small pieces with a knife and hammer and extracted with water and acetone to give 5.7 g (88%) poly-γ-butyramide. The inherent viscosity in *m*-cresol is 0.88 at 0.5% concentration.

The reaction should not be run this way on a much larger scale because the exothermic nature will lead to excessive temperature rises that will cause the polymerization to get out of control.

Although most of the emphasis on polymerization of lactams has been with caprolactam because of its commercial importance, considerable effort has been directed to polymerization of 4,4-disubstituted β-propiolactams because they are said to provide fibers with “silk-like” properties (19).

β-Propiolactams are conveniently synthesized by condensing a variety of olefins with chlorosulfonyl isocyanate, followed by acid hydrolysis of the chlorosulfonyl group. The synthesis works well with the following types of olefins: $RR'C{=}CH_2$, $RR'C{=}CHR''$, $RR'C{=}CR''R'''$, and styrene. The nitrogen always adds to the most substituted carbon. The α, α-disubstituted lactams cannot be synthesized by the same route but are prepared by cyclization of 1,1-dialkyl-2-aminopropionic acid ester with the Grignard methyl magnesium iodide (20).

β-lactams polymerize well in high temperature and anionic processes. The latter generally give higher molecular weight polymer and proceed well in solutions of dimethylsulfoxide or hexamethylphosphoramide using potassium *t*-butoxide as initiator and *N*-acetylcaprolactam as activator. Fibers and films have been prepared from only a few of the β-lactam polymers because they generally decompose close to their melt temperatures and are soluble only in solvents that are unsuitable for solution fabrication (19). It should be stressed again that acyl-lactam activators appear essential for the anionic catalyzed polymerization of α-piperidone (5-nylon) and pyrrolidone (4-nylon). Both polymers revert to cyclic monomer in the melt, especially if traces of catalyst remain. Basic catalyst and impurities must be removed before any attempt at melt fabrication. However, 4-nylon has been made and melt spun under carbon dioxide atmosphere (21). In addition to a carbon dioxide atmosphere, high polymer has been catalyzed using quaternary ammonium salts or potassium pyrrolidonate catalyst (20, 21).

122. Preparation of Ethyl 1,1-Diethyl-3-aminopropionate (20)

$$(C_2H_5)_2C(CN)(CO_2C_2H_5) \xrightarrow{H_2} (C_2H_5)_2C(CH_2NH_2)(CO_2C_2H_5)$$

A solution of 150 g ethyl 1,1-diethyl cyanoacetate in 300 mL absolute ethanol is hydrogenated at 80°C and 600 psi over 100 g Raney nickel. After about 3 h, hydrogenation should be complete. The catalyst is filtered (**Caution! Fire!**), and the alcohol is removed in a vacuum. The residual oil is treated with water, and HCl added until the solution is acid. The solution is extracted with ether, and the aqueous phase is separated, treated with charcoal, and filtered. The aqueous filtrate is made alkaline with saturated sodium carbonate, and the aminoester extracted with ether. The ether solution is washed with water, dried, and distilled. The aminoester is obtained as an oil (bp 85 to 95°C/8 mm). Yield is 115 to 120 g. The distillate is pure enough for the next step.

123. Preparation of α,α-Diethyl-β-Propiolactam (20)

$$NH_2CH_2{-}C(C_2H_5)_2{-}CO_2C_2H_5 + CH_3MgI \longrightarrow \text{3,3-}(C_2H_5)_2\text{-azetidin-2-one (NH, C=O)}$$

In a 2-L three-necked round-bottom flask equipped with stirrer, efficient reflux condenser, thermometer, nitrogen inlet, and dropping funnel is placed 0.5 mol methyl magnesium bromide (commercial product or synthesized in situ) in about

300 mL dry ether. The solution is cooled to 0 to 5°C, and 25 g ethyl 1,1-diethyl-2-aminopropionate in 200 mL anhydrous ether is added with stirring over 40 min. The 0 to 5°C temperature should be maintained during addition. The mixture is then refluxed, cooled to 0°C and treated dropwise with 100 mL 5% aqueous ammonium chloride. It is warmed to room temperature and stirred for about 15 min; the ether solution is then separated. The aqueous phase is extracted with five 50-mL portions of ether. The combined ether extracts are washed with 10% sodium thiosulfate, dilute hydrochloric acid, and water. The ether solution is dried over anhydrous sodium sulfate and filtered, and the ether is evaporated. The residual oil is distilled to yield about 18 g (90%+) of product (bp 91 to 96°C/0.8 mm).

124. Polymerization of α,α-Diethyl-β-propiolactam (19, 22)

$$H_3CH_2C-\underset{\text{(ring: C(}CH_2CH_3\text{)–C(=O)–NH–}CH_2\text{)}}{} \longrightarrow \left[-NHCH_2-\overset{C_2H_5}{\underset{C_2H_5}{C}}-\overset{O}{\overset{\|}{C}}- \right]_n$$

A solution of 25 g lactam is dissolved in 150 mL pure dimethyl sulfoxide in a 250-mL three-necked flask fitted with a nitrogen inlet, stirrer, and thermometer. The flask and all ingredients must be dry. To ensure completely anhydrous conditions, some of the solvent may be distilled off in vacuum; however, some monomer may be lost. To the solution of the monomer is added 1 g potassium pyrrolidone and 3 mg oxalyl dipyrrolidone (*N*-acetyl caprolactam may also be satisfactory) at room temperature. Polymerization is exothermic, and the temperature is maintained at 20 to 25°C with a cooling bath. After 2 to 3 h, the gel-like product is mixed with 300 mL acetic acid. The product is filtered, washed with water, and dried. The polymer is a white powder, with inherent viscosity of 1 to 2 measured in sulfuric acid at 0.5% concentration. The polymer melts at about 200°C and can be pressed into tough films.

β,β-Disubstituted propiolactams have been examined extensively, and the dimethyl polymer has been reported to have properties similar to silk (4). The disubstitution adjacent to the nitrogen makes such polymer particularly resistant to degradation from light. The lactam monomers are prepared by condensation of appropriate olefins (e.g., isobutene) with chlorosulfonyl isocyanate, which in turn is prepared from cyanogen chloride and sulfur dioxide (19). The nitrogen always adds to the most substituted carbon atom.

As is usual in anionic polymerization, the highest molecular weight is obtained at low temperatures (0 to 30°C) because chain transfer is limited. Effective polymerization conditions for β-lactams are monomer concentration, 20 to 25%; solvent, dimethylsulfoxide, or hexamethylphosphoramide; temperature 0 to 30°C; strongly basic initiators such as potassium pyrrolidone or potassium *t*-butoxide; and activator such as *N*-acetyl-caprolactam or bifunctional activators to yield narrower molecular weight distributions.

125. Polmerization of Caprolactam Using Poly(aminophosphazene) P_4-t-Bu Catalyst

125.A. Intermediates

The phosphene base was obtained from Fluka as a 1-M solution in hexane. Caprolactam from Aldrich was vacuum distilled three times (bp 137°C/10 mm Hg) before use.

125.B. Polymerization

A thermally dried 15-mL round-bottom flask with a 2-cm inner diameter (1d) neck is transferred to a dry box, and 2.5 mL (0.25 mol%) of the 1-M P_4-*t*-Bu in hexane is added. The hexane is evaporated in the dry box with a nitrogen bleed and slight vacuum. Then 2.5 g (20 mmol) caprolactam is added to the flask in the dry box and the flask equipped with a dried three-necked adapter, a glass stirrer shaft with a paddle stirrer blade, and a drying tube. The equipment is transferred to a hood and connected to a nitrogen line and a mechanical stirrer motor and controller. The flask is then lowered into a preheated Woods metal bath at 270°C and slow stirring started. After about 5 min the clear water white melt thickens and wrapps around the stirrer. After 20 min, the Woods metal bath is lowered, and the flask is allowed to cool to room temperature. The polymer product shrinks from the flask wall and adheres to the stirrer. The flask is broken, and the polymer removed from the stirrer using a sharp-bladed knife (**Gloves!**). The polymer is ground in a Wiley micromill to pass a 20-mesh screen. The polymer is then extracted overnight with 100 mL methanol. After drying in a vacuum oven at 100°C under nitrogen, the polymer yield is 2.12 g (85%). η_{inh} in *m*-cresol is 1.85 at 0.5% concentration T_m 218°C.

8.2. SULTAMS

The ring opening of propiosultam (23, 24) has been carried out using a base catalyst and an *N*-sulfonyl derivative of the monomer in a manner, and presumably by a mechanism, similar to that postulated for lactams. Little is known about the mechanism of polymerization of sultams in contrast to lactams. The polypropiosultam analog of 3-nylon, melts at 257 to 260°C with decomposition.

126. Preparation of Propiosultam (23, 24)

$$CH_3\overset{\overset{\displaystyle O}{\|}}{C}{-}SH + CH_2{=}CHCH_2Cl \longrightarrow \underset{\text{I}}{CH_3\overset{\overset{\displaystyle O}{\|}}{C}{-}S{-}CH_2CH_2CH_2Cl}$$

$$\text{I} + Cl_2 + H_2O \longrightarrow \underset{\text{II}}{ClCH_2CH_2CH_2SO_2Cl}$$

3-Chloropropyl thiolacetate is prepared from the peroxide-catalyzed addition of thiol acetic acid to allyl chloride. A mixture of 76.2 g thiolacetic acid, 76.5 g allyl chloride, and 0.5 g benzoyl peroxide are heated in a 500-mL three-necked flask with stirring under nitrogen at 60°C for 6 h. Vacuum distillation gives 120 g chloropropyl thiol-acetate (79%) as a colorless oil (bp 83 to 84°C/10 mm).

To 596 g (3.9 mol) 3-chloropropyl thiolacetate in a 5-L three-necked flask equipped with gas inlet tube, stirrer, and air condenser equipped with a lead to a sodium hydroxide gas trap is added 1-L water and 500 g crushed ice. The flask is cooled in an ice bath and the contents stirred rapidly while chlorine is passed below the surface at a rapid rate. The temperature is maintained at 5 to 10°C by addition of more ice. The mixture changes from colorless to yellow to pink and then to orange, which persists for about 1 h, finally fading to a colorless turbidity. When a persistent color of chlorine appears in the mixture, the reaction is stopped, the aqueous supernatancy decanted, washed twice with ether, and discarded. The residual heavy oil in the flask and the ether extracts are combined, washed with water, and dried over anhydrous sodium sulfate. After filtering and distilling the ether, the residual yellow oil is distilled to give about 590 g (87%) of chloropropylsulfonyl chloride (bp 107°C/6 mm).

A solution of 177.5 g (1.0 mol) sulfonyl chloride in 500 mL ether is added to a stirred mixture of 138 mL concentrated aqueous ammonia and 1-L ether in a 3-L flask cooled in an ice bath. The rate of addition is such as to maintain the temperature at 5°C. The mixture is stirred for a further 30 min after the addition is complete. The ether solution is dried as above, filtered, and enough low-boiling petroleum ether added to give a faint turbidity. On cooling in a freezer, about 56 g white needles is obtained, mp 64°C. This is the chloropropylsulfonamide. The organic filtrate and aqueous layer are evaporated, and the solids combined and extracted three times with 500 mL hot benzene (**Carcinogen!**) Cooling gives 48 g more of the sulfonamide.

To a solution of 115 g (0.73 mol) sulfonamide in 1-L absolute ethanol (freshly distilled from KOH) is added 40.9 g (0.73 mol) potassium hydroxide in 200 mL absolute ethanol. The solution becomes turbid and is refluxed for 45 min, when a white precipitate forms and the solution becomes neutral. A solution of 2 g KOH in ethanol is added and refluxing is continued for 30 min, at which point the basic mixture is cooled and filtered, and the salt washed with two 25-mL portions of ethanol. The filtrate and washings are neutralized with concentrated HCl and again filtered, and the alcohol is removed at reduced pressure to give 89.5 g of a light yellow oil, with some KCl still present. Distillation of the oil gives about 3.5 g of a vile-smelling liquid, 74 g of an odorless yellow liquid (bp 156 to 157°C/2 mm), and a black residue. The main product solidifies, (mp 23°C). The main product is treated with decolorizing charcoal and redistilled to give 65 g propanesultam (bp 157°C/2 mm). The product can be purified further by low-temperature crystallization from ether-alcohol.

N-Phenylsulfonylpropanesultam is prepared by adding 5.0 g sultam and 5.2 mL phenylsulfonyl chloride to 16 mL pyridine, allowing to stand 72 h, and pouring into 250 mL ice water. The tan precipitate is recrystallized from ethanol, decolorized with charcoal in an acetone solution, and recrystallized from ethanol to give 1.5 g white plates (mp 170°C).

127. Polymerization of Propiosultam (23)

$$\text{propanesultam} \xrightarrow[\text{PhO-SO-N(propanesultam)}]{\text{NaH}} -\!\!\!\![\!-(CH_2)_3\!-SO_2NH-]_n$$

In a small, dry reaction vessel continually flushed with nitrogen and equipped with a nitrogen inlet and outlet is placed 0.0103 g *N*-phenylsulfonylpropanesultam and 1.1 g purified propansultam. At room temperature, the mixture is liquid, and dry nitrogen can conveniently be bubbled through it to agitate and dissolve the cocatalyst sulfonyl sultam. The about 3 mg sodium hydride is added under nitrogen, the vessel is shaken until all the hydride has reacted, and the mixture placed in an oil bath at 130°C for 18 h under nitrogen. After cooling, the hard, slightly yellow cake is ground in water in a mortar, and the solid is collected by centrifuging. The solid is then washed twice with water, then acetone, and then dried at 60°C in vacuum. About 0.7 g polysulfonamide is obtained (mp 257 to 260°C), with an intrinsic viscosity of 0.7 at 0.5% concentration in concentrated sulfuric acid. Cold-drawable fibers can be drawn from the polymer melt. The acidity of the sulfonamide hydrogens permits easy alkylation of the nitrogen. *N*-Substitution results in lower melting temperature and increased solubility.

128. N-Methylation of Poly(propylene-1,3-sulfonamide) (25)

$$-[(CH_2)_3SO_2NH]_n- \xrightarrow[(CH_3O)_2SO_2]{\text{NaOH}} -[(CH_2)_3SO_2NH]-[(CH_2)_3SO_2N(CH_3)]-$$

To a solution of 11.7 g sodium hydroxide in 230 mL water in a home blender is added 3 g poly(propylene-1,3-sulfonamide). The polymer dissolves on stirring, and then 1.88 g dimethyl sulfate (**Toxic!**) is added when the temperature of the stirred solution is 35°C. The mixture is stirred vigorously, and the temperature rises to about 52°C. When the exotherm subsides, 20 mL aqueous concentrated ammonia is added to react with the residual alkylating agent, then concentrated hydrochloric acid is added in sufficient amount to precipitate the polymer and cause the slurry to remain acidic after standing. The polymer is separated by centrifuging and washed extensively with water, ethanol, and ether and finally dried in vacuum at 55°C. A clear film can be pressed at about 190°C, and the inherent viscosity is about the same as the starting polymer. The polymer is about 60% *N*-methylated and the polymer melt temperature (PMT) is between 150 and 185°C, depending on the uniformity of the alkylation reaction throughout the molecular system.

8.3. *N*-CARBOANHYDRIDES OF α-AMINO ACIDS

Many naturally occurring fibers are polypeptides, i.e., polyamides based on α-amino acids. The synthesis of high molecular weight polypeptides with a predetermined sequence of amino acids is extremely laborious and not suited to simple polymer preparation. However, amino acids can be converted to cyclic derivatives, *N*-carboanhydrides, that can be polymerized by ring opening and loss of carbon dioxide to give poly(2-nylons).

N-Carboanhydrides (NCAs) can be polymerized with a variety of catalysts, such as water, amines, and strong bases (e.g., sodium methoxide). Polymerization occurs readily with or without *N*-substitution provided reasonably strong bases such as alkoxides are used. Ring opening and propagation occur with loss of carbon dioxide. The complex character of NCA polymerization and the diversity of mechanistic views advanced have been treated at length (26).

129. Preparation of 4-Benzyloxazolid-2,5-dione (27–29)

$$C_6H_5CH_2{-}CH(NH_2){-}C({=}O)OH \longrightarrow C_6H_5{-}CH_2{-}HC{-}C({=}O){-}O{-}C({=}O){-}HN \text{ (ring)}$$

Caution! Phosgene is very dangerous and must be used carefully in a well-ventilated hood. All off-gases should be adequately scrubbed.

A total of 20 g DL-phenylalanine is dissolved in 400 mL pure dioxane and treated with a slow stream of phosgene for 2 h at 40°C. Excess phosgene and solvent are removed with a stream of dry air or nitrogen, and the residue is heated at about 40°C in a vacuum. The residual solid is recrystallized from ethyl acetate–petroleum ether. The yield is about 60% of material that melts that at 127°C. It is moisture sensitive and should be kept in a sealed container under nitrogen.

130. Preparation of 4,4-Dimethyloxazolid-2,5-dione (27–29)

$$CH_3{-}C(CH_3)(NH_2){-}CO_2H \longrightarrow CH_3{-}C(CH_3){-}C({=}O){-}O{-}C({=}O){-}NH \text{ (ring)}$$

A total of 15 g isobutyric acid in 400 mL pure dioxane is treated with a slow stream of phosgene at 50°C for 9 h. The dioxane and excess phosgene are removed with a stream of dry air or nitrogen. The oily residue solidifies when warmed at 40°C in a vacuum. The solid is dissolved in the minimum amount of hot chloroform, which should be purified just before use to remove water and ethanol stabilizer. (The ethanol is removed by washing the chloroform with water and drying over anhydrous sodium sulfate.) The chloroform solution is filtered and treated with three volumes of petroleum ether. The product crystallizes and is filtered and dried under vacuum and is kept out of contact with moist air. The yield is about 80% and the product melts at 95 to 97°C.

131. Copolymerization of 4,4-Dimethyloxazolid-2,5-dione and 4-Benzyl-DL-oxazolid-2,5-dione (27, 29)

```
      CH3   O                  H         O
      |    //                   \       //
CH3 — C — C        +   C6H5CH2 — C — C        ————>  Random copolymer
      /     \                    /     \
    NH       O                 NH       O
       \   /                      \   /
         C                          C
         ||                         ||
         O                          O
```

A solution of 2.6 g (0.02 mol) 4,4-dimethyloxazolide-2,5-dione and 3.8 g (0.02 mol) of 4-benzyl-DL-oxazolide-2,5-dione in 70 mL benzene is treated with a solution of 0.5 mL water in 1 mL purified dioxane. The solution viscosity increases perceptibly over 6 days. The solution may be poured onto a glass plate and evaporated to a thin polymeric film. Similarly, a 1 : 2 copolymer of these reactants may be prepared. Thus in 70 mL benzene is placed 1.3 g (0.01 mol) 4,4-dimethyl-oxazolide-2,5-dione, 3.8 g (0.02 mol) of 4-benzyl-DL-oxazolide-2,5-dione, and 0.25 mL dioxane containing 1% aniline. Polymerization proceeds over a period of about 10 days. At the end of this period, the solution is clear and may be cast to clear, tough film that can be stretched and oriented by drawing.

When possible, most workers in the polypeptide field employ phosgene in the synthesis of NCAs. However, it is possible to prepare this ring by a route that avoids the use of phosgene and this may be advantageous for some workers.

132. Preparation of γ-Benzyl-L-glutamate (30)

$$HO_2C(CH_2)_2CH(NH_2){\cdot}CO_2H + C_6H_5CH_2OH \longrightarrow C_6H_5CH_2O_2C(CH_2)_2CH(NH)_2{\cdot}CO_2H$$

To a mixture of 70 g L-glutamic acid and 250 mL benzyl alcohol is added 70 mL concentrated hydrochloric acid; the mixture is warmed until the solution just boils and becomes homogeneous. (Overheating is undesirable because excessive amounts of diester are formed.) After standing for 2 h, the crystalline slurry is added to 4.5 L acetone. The crystals of ester hydrochloride that separate completely in another 2 h are collected, washed with acetone, and dissolved in 500 mL water. The solution is cooled in ice and neutralized by the careful addition of solid sodium bicarbonate, using vigorous stirring, until pH 8 is obtained. After filtration, the crude amino acid γ-ester is washed with ice water and crystallized from hot water. The pure glutamate is collected, washed with ice water and acetone, and dried in a vacuum oven at 60°C. Yield is about 30 g (25%) of pure ester.

133. Preparation of γ-N-Benzyloxycarbonyl-γ-Benzyl-L-glutamate

$$C_6H_5CH_2O_2C(CH_2)_2CH(NH_2)\cdot CO_2H + C_6H_5CH_2OCOCl \longrightarrow C_6H_5CH_2O_2C(CH_2)_2CH(NH\cdot CO\cdot OCH_2C_6H_5)\cdot CO_2H$$

To 2 L water at 65°C in a 5-L beaker are added 20 g sodium bicarbonate and 23.7 g γ-benzyl-L-glutamate; the mixture is stirred until dissolved. To the stirred solution is quickly added 20 mL fresh benzyl chloroformate (**Fume Hood!**), and the stirring is continued for a further 3 h. By-products are extracted with 500 mL ether, and the aqueous layer is filtered if necessary and acidified to congo red paper with hydrochloric acid. The oil formed is extracted with 500 mL ether and dried over anhydrous sodium sulfate, filtered, and evaporated in vacuo. The solid or viscous oily residue remaining is dissolved in carbon tetrachloride (**Carcinogen!**) and the solution is concentrated in vacuo to about 300 mL. On standing overnight, *N*-benzyloxycarbonyl-γ-benzyl-L-glutamate crystallizes. The yield is about 20 g (55%)

134. Preparation of 1,4-(Butyl-γ-benzyloxycarbonyl)-oxazolid-2,5-dione-γ-benzyl-L-glutamate N-Carboanhydride

A mixture of 20 g *N*-benzyloxycarbonyl-γ-benzyl-L-glutamate and 20 mL freshly distilled thionyl chloride is warmed in a flask, equipped with an air condenser and a calcium chloride drying tube, to 60 to 70°C until a distinct yellow-orange color forms. The color changes from white through yellow to reddish brown; too prolonged a reaction results in a product that is impossible to purify, and too short a contact time leads to poor yields. After cooling, the reaction mixture is solidified under 60 to 80°C petroleum ether, filtered, and washed with cyclohexane. It is repeatedly recrystallized from hot ethyl acetate (1 mL/g dry product) by cooling in ice until colorless and chloride free. The chloride impurities, which are ionizable, can be detected by decomposing about 0.1-g samples of monomer with boiling dilute nitric acid and adding silver nitrate. It is essential that this test be negative if high molecular weight polymers are to be obtained at the next stage. Typical yield from two recrystalizations is about 7 g (50%). Note that the monomer is moisture sensitive and that it is advisable to convert monomer to polymer at the earliest opportunity.

135. Preparation of Poly(γ-benzyl-L-glutamate)

$$C_6H_5CH_2O_2C(CH_2)_2HC\text{—}C(=O)\text{—}O\text{—}C(=O)\text{—}NH \text{ (ring)} \xrightarrow{n\ Bu_3N} \left[\text{—}NH\cdot CH((CH_2)_2\cdot CO_2CH_2C_6H_5)\cdot CO\text{—}\right]_n + CO_2$$

For the polymerization, pure dry ethyl acetate, methylene chloride, and tri-*n*-butylamine are required. Ethyl acetate is dried over potassium carbonate for

24 h and distilled. Methylene chloride of satisfactory purity is obtained as the residue after distilling one third by volume of the general-purpose reagent. This removes water, hydrogen chloride, and acid chlorides. Tri-n-butylamine that has been allowed to stand over barium oxide for 24 h is distilled at reduced pressure (86°C/15 mm), the middle fraction being collected. A standard solution of this initiator solution is prepared from 0.6 mL amine in methylene chloride.

To polymerize the monomer, 5 g L-4-(butyl-γ-benzyloxycarbonyl)oxazolid-2,5-dione is dissolved by warming with 5 mL ethyl acetate; to this solution is rapidly added 10 mL methylene chloride and 4 mL initiator solution, avoiding as far as possible the separation of solid monomer on the walls of the vessel. The reaction mixture is then refluxed on a water bath for 30 min, after which the polymerization is completed at ambient temperature overnight. The resulting polymer is precipitated by running the reaction mixture into 500 mL methanol. The polymer is collected, washed with methanol, and dried in vacuo at 40°C. If this is the desired end product of the synthesis, a more convenient physical form can be obtained by dissolving the polymer in 25 mL chloroform and casting on a clean glass sheet. After solvent evaporation, the polymer film can be floated off with a little water and the film dried in vacuo. A typical yield is 4 g (95%).

136. Preparation of Poly(α-L-glutamic acid)

$$\left[-\text{NH}\cdot\underset{}{\overset{(\text{CH}_2)_2\text{CO}_2\text{CH}_2\text{C}_6\text{H}_5}{\overset{|}{\text{CH}}}}\cdot\text{CO}- \right]_n \xrightarrow{\text{HBr}} \left[-\text{NH}\cdot\overset{(\text{CH}_2)_2\text{CO}_2\text{H}}{\overset{|}{\text{CH}}}\cdot\text{CO}- \right]_n + n\text{C}_6\text{H}_5\text{CH}_2\text{Br}$$

A 1 to 2% solution of poly-γ-benzyl-L-glutamate in distilled, sodium-dried benzene (**Carcinogen!**) is made and through this is passed a stream of pure dry hydrogen bromide (HBr) until poly-α-L-glutamic acid starts to precipitate. The solution is left standing for 8 h (from the start of the HBr injection). The polymer solution is then added to excess petroleum ether; the polymer is precipitated, filtered, and washed well with petroleum ether. The product can be purified by dissolving in 5% aqueous sodium bicarbonate, filtering, and precipitating by pouring into 2N hydrochloric acid. The product is filtered, washed well with water, and dried in-vacuo. Yields are quantitative. It is important to recognize that that hydrolytic cleavage of the peptide will reduce the degree of polymerization significantly, hence moisture *must* be excluded while the polymer is in the presence of the hydrogen bromide.

8.4. CYCLIC ESTERS

This section includes a variety of ring-opening procedures yielding polyesters, polyurethanes, and polycarbonates. They include ring opening of cyclic lactones, anhydrosulfites, iminocarbonates, ketene acetals, and cyclic carbonates. Preferred catalysts vary for opening the different ring structures, and mechanisms cover ionic to free radical. Most of the lactones can be polymerized cationically and anionically. Depending on the type of initiator, polymerization is induced by alkyl or acyl cleavage. Ring opening depends not only on ring size but also on number, size, and position of any substituents.

It is interesting to note that although cyclic carboanhydrides of α-hydroxy acids, unlike the *N*-carboanhydrides described above, do not polymerize to high molecular weight, it is possible to prepare a polyester by the thermal polymerization of an anhydrosulfite. Thus α-hydroxyisobutyric acid gives with thionyl chloride the α-hydroxyisobutyric anhydrosulfite, which polymerizes when heated, with elimination of sulfur dioxide.

137. Preparation of α-Hydroxyisobutyric Acid Anhydrosulfite (31, 32)

In a three-necked flask equipped with a stirrer, reflux condenser, and a dropping funnel is placed 1000 g thionyl chloride. The thionyl chloride is cooled in an ice-salt bath to about 0°C, and 312 g α-hydroxyisobutyric acid is added. The reaction system is attached to a water pump, and a pressure of 100 to 200 mm is maintained in the flask with stirring to remove evolved hydrogen chloride. After 18 h at 0°C and 100 to 200 mm pressure, the flask is allowed to warm to room temperature at this pressure. Excess thionyl chloride is then distilled rapidly through a short still-head, then the anhydrosulfite is obtained (bp 41 to 48°C/8 mm). The yield of crude product is about 309 g (69%). The anhydrosulfite is purified by distilling through an efficient fractionating column and collecting the fraction (bp 53 to 55°C/16 mm).

138. Prepolymerization of the above Anhydrosulfite

To remove chance initiators from the anhydrosulfite, the material is maintained at reflux for 145 h under reduced pressure so that the reflux temperature is about 55°C. Under these conditions, about 10% of the anhydrosulfite will polymerize to low molecular material. This procedure effectively removes chance impurities that might give erratic results during polymerization. Redistillation of the unpolymerized material purified in this manner gives a product with a refractive index of 1.4298.

139. Polymerization of α-Hydroxyisobutyric Acid Anhydrosulfite (31)

$$(CH_3)_2C(-O-)-C(=O)-O-S(=O)-O\ \text{(ring)} \xrightarrow{-SO_2} \left[-O-C(CH_3)_2-C(=O)- \right]_n$$

A 300-mL round-bottom flask is dried by baking in an oven at 110 to 150°C for about 3 h. To the flask, cooled under nitrogen, is added 150 mL benzene (**Carcinogen!**). To ensure dryness of the flask and contents, about 50 mL of the benzene is distilled from the flask. The flask is then cooled under nitrogen in an ice bath until the benzene has frozen and 50 g the pure anhydrosulfite is added. The reaction mixture is refluxed for 52 h under nitrogen to give a cloudy, colorless gel. This is filtered to give a solid polymer of molecular weight >100,000. It melts at 240°C and has and inherent viscosity of 1.5 at 0.5% concentration in a solvent consisting of 58.8 parts phenol and 41.2 parts 2,4,6-trichlorophenol. The polymer can be pressed to a clear, colorless film that can be stretched and oriented. Fibers can also be melt spun.

Chlorobenzene may also be used as the reaction medium and the polymer remains in solution at the boiling point. Thus 49.2 g anhydrosulfite is distilled into frozen dry chlorobenzene. The reaction mixture is maintained at reflux and after 7.5 h the polymer solution may be cast to a film or poured into alcohol and precipitated as a white solid with an inherent viscosity of 1.5 at 0.5% concentration in the solvent mixture described above. It is essential in either procedure to have absolutely dry equipment and reagents otherwise the polymerization will not proceed to high molecular weight.

In general, unsubstituted polyesters derived from lactones are low melting and tend to revert to monomer. Certain lactones, such as propiolactones, however, deserve special mention. Propiolactone, a four-membered ring, is prepared commercially by the condensation of ketene and formaldehyde. It polymerizes in the presence of a variety of catalysts and is so liable to spontaneous exothermic polymerization in an uninhibited condition that it constitutes a laboratory hazard. The polymer, poly(β-hydroxy propionate) melts below 100°C and is difficult to obtain in high molecular weight (33). Although the polymer from propiolactone is low melting, gem disubstitution on the α-carbon raises the melt temperature of the polymer markedly. Thus α, α-bis-(chloromethyl)-β-propiolactone, prepared from silver trichloropivalate, polymerizes in the presence of alkali catalyst to a fiber-forming polymer melting at over 300°C (34).

α,α-Dimethyl and α,α-diethyl-β-propiolactones are easily prepared and polymerized. The polymers have been melt spun to fibers that have attractive properties. The key drawback, however, is elimination of small amounts of cyclic monomer when heated at or close to the melting temperature. The monomer is considered to be a suspect carcinogen.

140. Preparation of α, α-Bis(chloromethyl)-β-propiolactone (34)

$$(ClCH_2)_3C\cdot CO_2Ag \longrightarrow ClCH_2-\underset{CH_2-O}{\overset{CH_2Cl}{\overset{|}{C}}}-\underset{|}{C}=O$$

In a 1-L round-bottom flask equipped with an efficient wide-bore condenser and set up in an efficient hood is placed 47.9 g (0.25 mol) trichloropentaerythritol. Then 100 mL concentrated nitric acid is added, and the mixture is warmed cautiously, preferably with an IR lamp. The chlorohydrin dissolves, then two layers appear and evidence for initiation of a reaction is noted. The flask is rapidly lowered into a cold water bath to moderate the violent reaction that quickly develops. After the reaction moderates and evolution of nitrogen oxides have nearly ceased, the flask is warmed cautiously until no more brown fumes evolve; then the clear, colorless solution is poured into water to give a quantitative yield of crude β,β',β''- trichloropivalic acid (mp 108 to 110°C). Recrystallization from petroleum-ether gives a product that melts at 113°C (35).

This is a violent reaction that should not be run on a larger scale. The operator should be protected by shields and gauntlets at all times.

Anhydrous, finely powdered silver salt of β,β',β''-trichloropivalic acid is heated cautiously in an oil-jacketed distillation unit in a slow stream of nitrogen at a pressure of 0.2 to 0.3 mm. A liquid begins to distill slowly when the jacket temperature reaches 105°C and more rapidly at 110°C. The liquid soon begins to crystallize in the receiver. The jacket temperature is maintained between 110 and 115°C until the distillation slackens and then is slowly raised to 150°C. Little distillation occurs above a jacket temperature of 125°C. The solid distillate is collected and consists of pure α,α-bis(chloromethyl)-β-propiolactone (mp 35 to 36°C). This is unchanged after recrystallization from a mixture of n-hexane and benzene.

141. Polymerization of α,α-Bis(chloromethyl)-β-propiolactone (34)

$$\begin{array}{c} CH_2Cl \\ | \\ CH_2-C-CH_2Cl \\ |\quad\;\; | \\ O-C=O \end{array} \longrightarrow \left[-OCH_2-\overset{\displaystyle CH_2Cl}{\underset{\displaystyle CH_2Cl}{\overset{|}{\underset{|}{C}}}}-\overset{\displaystyle O}{\overset{\|}{C}}- \right]_n$$

In a baked dry flask under nitrogen 150 g α,α-bis-(chloromethyl)-β-propiolactone is heated to 40°C until it melts, and then 0.1 g finely powdered dry potassium hydroxide is added with stirring. Heating and stirring are continued at 40°C for about 15 min, then raised to 50°C. The mixture soon becomes turbid owing to polymer separation and is completely solid in about 2 h. Heating, without stirring, is continued for a further 4 h at 50°C. The temperature is then raised to 100°C in a slow stream of nitrogen at 0.1 mm to remove any volatile material. The product is a tough, white polymer, softening at about 300°C and gives a viscous melt from which fibers may be spun or drawn.

Synthesis and polymerization of a range of α,α-disubstituted β-lactones have been described (33, 36). Blume (37) developed simpler and more elegant methods for the monomer syntheses and polymerization, which will be described later.

142. Preparation and Polymerization of α,α-Diethyl-β-propiolactone (36)

$$NH_2-CH_2-\overset{\displaystyle C_2H_5}{\underset{\displaystyle C_2H_5}{\overset{|}{\underset{|}{C}}}}-CO_2H \xrightarrow{HONO} \text{(α,α-diethyl-β-propiolactone: } C_2H_5,\ C_2H_5 \text{ on four-membered ring with } C{=}O \text{ and ring } O)$$

A total of 80 g ethyl α,α-diethyl-β-aminopropionate is dissolved in 400 mL 6-N hydrochloric acid and refluxed for 22 h. After cooling, the solution is neutralized with 50% sodium hydroxide to pH 7 (about 100 mL alkali). To the solution is added 200 mL glacial acetic acid, and the mixture is cooled to 0°C. A solution of 70 g sodium nitrite in 250 mL water is added over 30 min, while keeping the temperature at 0 to 5°C. The mixture is stirred another 30 min and an oil separates. This is extracted into ether or benzene, dried over sodium sulfate, and distilled. The product boils at 67 to 69°C/0.6 mm. The yield is 35 to 40 g. If a test sample does not polymerize, the product should be redistilled.

A total of 20 mg triethylenediamine is dissolved in 10 g of the lactone contained in a dry polymer tube that is sealed. The tube is placed in an oil bath. At 50°C, polymerization is complete in a few minutes; at 100°C, only seconds are required. A white solid is obtained, melting at about 235°C. It is highly crystalline and has an inherent viscosity of 1 to 2 in trifluoroacetic acid at 0.5% concentration. Clear films that crystallize rapidly may be pressed.

143. Preparation of Pivalolactone (37)

$$HO-CH_2-C(CH_3)_2-CO_2H \longrightarrow \text{pivalolactone (3,3-dimethyl-2-oxetanone)}$$

All equipment should be rinsed in dilute hydrochloric acid and oven dried before use to inhibit premature polymerization of the pivalolactone product. In a 250-mL round-bottom flask with a stoppered opening attached to a 15-in spinning band column is placed a Teflon-coated magnetic bar and 29.53 g (0.25 mol) hydroxypivalic acid and 32.225 g (0.1125 mol) methyl orthoisophthalate. Stirring is started, and the flask is heated in an oil bath at 120 to 125°C at a pressure of 50 mm. The methanol by-product is collected at a high reflux ratio over a period of 4 to 5 h. The pressure is then lowered to 12 mm and the fraction that boils at 39 to 42°C is collected. This is pivalolactone. About 17 g is collected. The oil bath temperature is slowly raised to 215°C, and a small amount of dimethyl isophthalate (bp 145 to 148°C/12 mm) is collected. Blume (37) submitted a preparation of pivalolactone from pivalic acid and trimethyl orthobenzoate.

144. Polymerization of Pivalolactone (37)

$$\text{pivalolactone} \longrightarrow \left[-\underset{\substack{\| \\ O}}{C}-C(CH_3)_2-CH_2-O-\right]_n$$

The pivalolactone is freshly distilled from acid-washed and oven-dried equipment through a short Vigreux column. The polymerization equipment is likewise washed and dried. To a dry 300-mL three-necked flask equipped with a mechanical stirrer, a reflux condenser, and a nitrogen capillary inlet is added 10 g pivalolactone, 100 mL dry hexane, and 0.1 mL 1-M tetrabutylammonium hydroxide. The slow stream of nitrogen is continued, and the solution heated to reflux. A fine white suspension appears after a few minutes and increases as the polymerization continues. After about 2 h at reflux, the reaction is cooled and the polymer filtered through a medium-sintered glass funnel. The polymer is washed several times with hexane, air dried, and then dried overnight in a vacuum oven at 60°C. The yield is about 80%. The inherent viscosity in trifluoroacetic acid at 25°C and 0.5% concentration is about 1.6. The crystalline mp is 235°C.

Propiothiolactones can likewise be prepared and polymerized, in some cases giving rubbery polymers (38).

145. Preparation of 2,2-Diethyl-3-propiothiolactone (38)

$$Br-CH_2-C(C_2H_5)_2-CO_2H \longrightarrow Br-CH_2-C(C_2H_5)_2-COCl \xrightarrow{Na_2S} (H_5C_2)(C_2H_5)C\langle CH_2-S-C{=}O\rangle \longrightarrow \left[-C(=O)-C(C_2H_5)_2-CH_2-S- \right]_n$$

In a 3-L flask equipped with stirrer, condenser, thermometer, and HCl absorption system are added 800 g (3.8 mol) 2-ethyl-2-bromomethylbutyric acid and 1250 g thionyl chloride. The contents are refluxed gently for 3.2 h, during which time the temperature rises from 58 to 97°C. The excess thionyl chloride is stripped off through a Claisen head by increasing the temperature of the flask to 150°C. The residual liquid is distilled at 15 mm Hg. The desired product, 2-ethyl-2-bromomethylbutyryl chloride, boils at 93 to 94°C/15 mm Hg and has a refractive index of 1.4829 at 27°C. The yield is about 834 g (96%).

In a 3-L flask equipped with stirrer, condenser with drying tube, thermometer, and addition funnel is placed 160 g thoroughly dried sodium sulfide with an estimated sodium content of 1.86 mol, and 1100 mL dimethylacetamide (DMAc). A solution of 400 g (1.76 mol) 2-ethyl-2-bromomethylbutyryl chloride in 400 mL DMAc is added over a period of 1 h, while maintaining the reaction mixture at 17 to 23°C with external cooling. The reaction mixture is stirrred at room temperature for 1.5 h, is then poured with stirring into 3 kg crushed ice, and is stirred for a further 1 h. The mixture is extracted three times with 800-L ether, and the combined extracts are washed twice with 800-L portions of water, twice with 800 mL 5% sodium bicarbonate, and twice more with 800 mL water. The ether solution is then dried over anhydrous sodium sulfate. The solution is filtered, the ether removed by distillation on a steam bath, and the residue distilled through a Claisen head at 70 to 76°C at 10 mm. The yield of crude product is 119 g (47%).

This is distilled through a spinning band column (60 cm) to give pure product (bp 78°C/10 mm Hg).

146. Polymerization of 2,2-Diethyl-3-propiothiolactone (38)

A quantity of 0.04 mL 1-normal tetrabutylammonium hexyldimethylacetate (commercially available as Versatic acid salt from Shell Development Co.) in benzene (**Carcinogen!**) is placed in a dry 25-mL flask; the flask is heated gently to remove the benzene. The flask is charged with 3.5 g (0.024 mol) 2,2-diethyl-3-propiothiolactone, after which the flask is swirled to mix the ingredients and is then heated on a steam bath for 19 h. The resulting product is a rubbery polymer. It is dissolved in 75 mL of tetrahydrofuran and precipitated by addition of 200 mL water. After standing overnight in the solvent mixture, the rubbery mixture becomes hard. It is pulverized twice by agitation with water in a high-speed mixer, filtered and dried at 1 mm pressure. The yield is 2.4 g (68%), and the inherent viscosity in toluene is 0.85 at 0.5% concentration. Elastic fibers can be pulled from the melt at about 180°C.

Using the same preparative procedure as above, 360 g (1.68 mol) 2-bromomethyl-2-methyl-butyryl chloride and 145 g thoroughly dried sodium sulfide with an estimated sodium sulfide content of 1.69 mol are reacted in 1300 mL DMAc. The product, 2-methyl-2-ethyl-3-propiothiolactone, is obtained in a 72% yield (bp 62°C/12 mm) and a refractive index of 1.4850 at 27°C. This thiolactone is polymerized in the same manner as for the diethylthiolactone and is isolated as a foamy, white rubbery solid having an inherent viscosity of 1.02 in toluene at 0.5% concentration. A sample could be melt pressed into a clear elastic film with elongation as high as 790% with an initial modulus of 0.015 /denier. Fiber could also be melt spun at about 200°C at 30 f/min, giving a filament with a 460% elongation and a modulus of 0.023 gpd.

A dimeric lactone, glycolide, can be polymerized to a polyester of α-hydroxyacetic acid in the presence of antimony fluoride. The polyester does not tend to revert to glycolide under normal conditions.

147. Preparation of Glycolide (39)

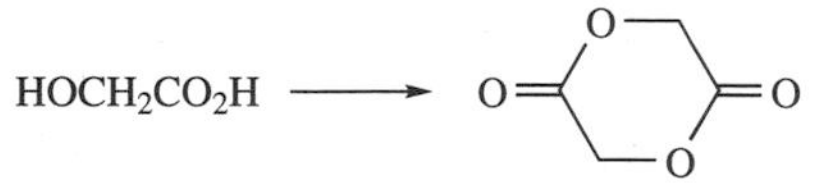

$$HOCH_2CO_2H \longrightarrow$$ O=⟨ring: O, O⟩=O

In a round-bottom flask, 400 g hydroxyacetic acid is heated at atmospheric pressure until the temperature of the liquid is 175 to 185°C. The temperature is maintained in this range for 2 h, or slightly longer, until water ceases to distill. The pressure is then reduced over 30 min to about 150 mm, and the temperature is maintained for an additional 2 h. The residue is poured into an enamel pan where it solidifies to a white, brittle solid. The solid is low molecular weight poly(hydroxyacetic ester), which is depolymerized to glycolide. A three-necked reaction vessel is equipped with a stirrer and a neck suitable for introduction of powdered low molecular polymer and also equipped with a takeoff for distilling the glycolide as it is prepared. The equipment is swept with a steady stream of nitrogen gas, and the receiver for the glycolide is cooled in an ice bath. After the ice-cooled receiver, three ice-cooled traps are placed in series to catch any glycolide that is carried beyond the receiver by the nitrogen stream. Then 100 g of the powdered low molecular weight polyester is thoroughly mixed with 1 g antimony trioxide and placed in a supply vessel connected to the inlet neck of the three-necked flask. The polymer is introduced from the supply vessel into the reaction vessel that is maintained at 270 to 285°C at a pressure of 12 to 15 mm. The solid is added at the rate of about 20 g/h to give a 93% yield of the glycolide. The glycolide is recrystallized from about 2 vol. ethyl acetate with charcoal.

148. Polymerization of Glycolide to Poly(hydroxyacetic Acid) (39)

$$\text{glycolide} \longrightarrow \left[-CH_2-\overset{\overset{\displaystyle O}{\|}}{C}-O- \right]_n$$

A mixture of 60 g pure dry glycolide is placed in a reaction vessel under nitrogen in the presence of 0.03% by weight of antimony trifluoride. The vessel is heated to 195°C by an oil bath and the contents stirred for 1 h. The viscosity of the melt increases rapidly during this period and is too viscous to stir at the end of the hour. Heating is continued for a further 1 h; then the temperature is raised to 230°C for a further 30 min. The resulting polymer has high molecular weight and can be fabricated into drawable fibers and film.

148.A. Poly(lactic Acid)

There is considerable driving force to identify polymers derived from biochemical rather than from petroleum resources. A leading candidate is poly(lactic acid), derived from corn sugar fermentation. The polymer is biodegradable and has potential uses in packaging, medical, and other commercial uses. Lactic acid is optically active, and can be separated into D, L, racemic, and meso forms. The polymers are usually prepared from the lactide (cyclic dimer) from separated L-acid. Recently there has been much activity to combine the polymerization of the racemic lactide with enhanced separation of the D or L forms using stereo-directing aluminum catalysts (40). Spassky et al. (40), indeed, effected a kinetic resolution and obtained polymer with 88% enantiomeric enrichment in D-units at a 19% conversion. The polymerization appeared to have "living" features with narrow molecular weight distribution ($M_w/M_n = 1.05$ to 1.3) (Fig. 8.2; see also chapter 9 for isotactic polymers).

Following up on this work, and using a similar catalyst (replacing O-Me with O-Pr), Coates and Ovitt (41) polymerized the meso-lactide to highly syndiotactic poly(lactic acid). Baker et al. (42), using the same catalyst as Ovitt and Coates, effected the polymerization of racemic lactide to an equivalent mixture of isotactic

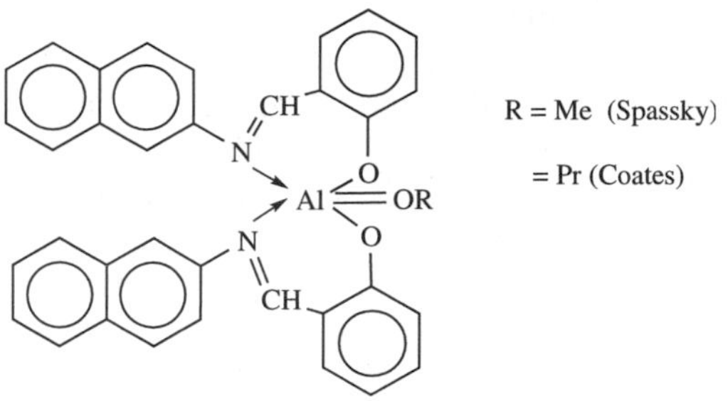

R = Me (Spassky)
= Pr (Coates)

Figure 8.2

D and L-poly(lactic acid). Further details of this interesting work can be obtained from the publications cited. Almost concurrently with these publications, two major companies announced a joint venture to commercially produce polylactide useful in fibers and plastics. Polymerization of the lactides with the aluminum catalyst appears to be straightforward and consists in heating the lactide and catalyst (75/1; mol/mol ratio) in toluene at 70°C.

148.B. Macrocyclic Polycarbonates

Advances in ring-opening polymerizations have been made with macrocyclic polycarbonates (43). In these cases, ring strain is low and the driving force is entropic. The preferred catalyst for these ring-opening polymerizations is tetrabutylammonium tetraphenyl borate. The macrocyclic monomers are synthesized by a pseudo-high-dilution triethylamine-catalyzed hydrolysis/condensation reaction of bisphenol A bis(chloroformate) in methylene chloride solvent using sodium hydroxide as acid acceptor (**8.2**)

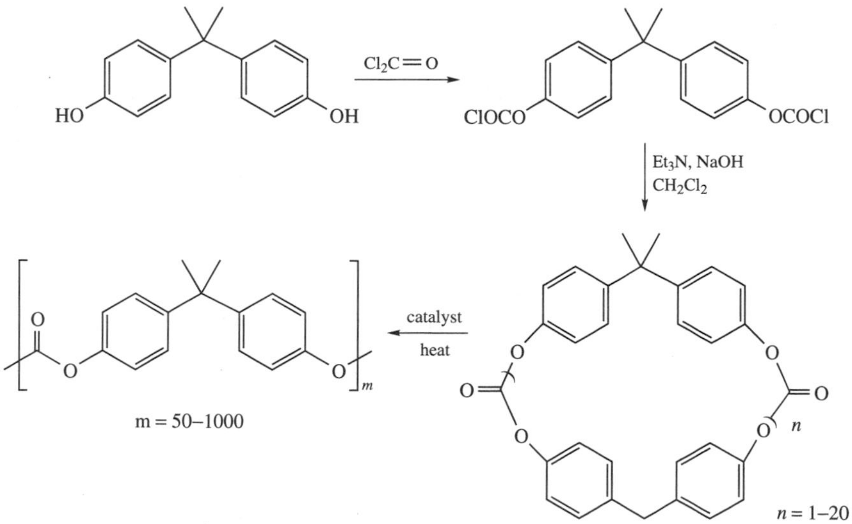

m = 50–1000

n = 1–20

Schematic 8.2

149. Preparation of Macrocylic Aromatic Carbonates (43)

Reagent-grade solvents and chemicals are used without further purification. Dow "Parabis" is used as a source of bisphenol A. Purified, crystalline bisphenol A bischloroformate is prepared by reaction of bisphenol A in methylene chloride with phosgene at 0°C, using diethylaniline as acid acceptor (44).

A 1-L. Morton flask equipped with a mechanical stirrer and condenser is charged with 200 mL methylene chloride, 7.0 mL water, 3 mL 9.75 M NaOH (29 mmol), and 2.4 mL triethylamine (17.25 mmol). The mixture is heated to reflux and stirred vigorously. A solution of bisphenol A bis(chloroformate) (200 mL 1.0 mol in methylene chloride) is added subsurface over the tip of the impeller at 6.7 mL/min by use of a peristaltic pump. Concurrently, 59 mL 9.75 M NaOH (575 mmol) is delivered over 25 min by use of a dropping funnel, and 2.4 mL triethylamine is added over 28 min via a syringe pump. Within 10 min after complete bis(chloroformate) addition, the phases are separated, and the organic layer washed with 1.0 M HCl and then with water three times. After drying over sodium sulfate, the organic layer is concentrated in vacuo to give almost a quantitative yield of product containing 85% cyclics by high-performance liquid chromatograply (HPLC) analysis. A typical distribution of cyclic oligomers is 5% dimer, 18% trimer, 16% tetramer, 12% pentamer, 9% hexamer, and 25% higher cyclics. Washing the crude cyclics/polymer products with 5 vol acetone provides a solution of pure cyclics in about 75% yield. The higher molecular weight polymer and macrocyclic carbonates with more than about 15 repeat units are insoluble in acetone. The mixed macrocyclic oligomers melt at 200 to 210°C.

150. Polymerization of Oligomer to Linear Polycarbonate (43)

In a dried 50-mL flask under nitrogen, is placed 10 g mixed cyclic oligomers. Then 0.02 g lithium phenoxide is added and the flask heated to 250°C (Woods metal bath) for 30 min. On cooling, a tough linear polymer results.

8.5. CYCLIC POLYURETHANES

Polyurethanes have been prepared by ring opening of cyclic iminocarbonates, prepared by condensation of disodium glycoxide with phenyl imidophosgene (45).

151. Preparation of Disodium Ethyleneglycoxide

A 500-mL three-necked flask is equipped with magnetic stirrer, condenser, nitrogen tube inlet, and solid addition flask (125-mL Erlenmeyer flask attached by means of large-diameter Gooch tubing). The flask is charged with 22.3 g (0.36 mol) fractionated ethylene glycol and 300 mL dry tetrahydrofuran. From the addition flask is added slowly 34.6 g dark gray 50% sodium hydride suspended in mineral oil (0.72 mol pure NaH). The reaction mixture is stirred under nitrogen overnight before filtration in a dry box. The light gray product is dried in a vacuum desiccator. The yield is 36 g (95%).

152. Preparation of Phenylimidophosgene (45)

$$C_6H_5-NCS + Cl_2 \longrightarrow C_6H_5-N{=}CCl_2$$

I

A 1-L three-necked flask equipped with stopper, condenser, and fritted gas-inlet tube is charged with 300 g (2.2 mol) phenyl isothiocyanate and 275 mL carbon tetrachloride (**Carcinogen!**). Chlorine is bubbled into the reaction mixture at a slow rate. The solution turns a deep cherry and a great deal of heat evolves and ice bath cooling is required. Chlorine is added until 310 g (4.4 mol) has been absorbed. Any chlorine escaping is trapped in aqueous caustic. The carbon tetrachloride and sulfur chloride are distilled at atmospheric pressure at 80°C. The product is then fractionated under water aspirator pressure. The fraction boiling at 98 to 107°C/26 mm is collected. This material is fractionated through an 18-cm Vigreux column, and the fraction boiling at 104°C/26 mm is collected. The yield of pale yellow liquid is 316 g (82%) and is stored under nitrogen in a dry box.

153. Preparation of Ethylene N-Phenyliminocarbonate (46)

$$\text{I} + \text{II} \longrightarrow C_6H_5-N{=}C\langle O-CH_2-CH_2-O\rangle$$

III

A 2-L three-necked flask is equipped with magnetic stirrer, condenser, dropping funnel, and stopper and is protected from moisture. The flask is charged with 80 g (0.75 mol) of the sodium ethyleneglyoxide II suspended in 600 mL benzene (**Carcinogen!**). From the dropping funnel is addded a solution of 132 g (0.75 mol) of the iminophosgene in 600 mL benzene over a 10 to 15-min period. The heat of the reaction is sufficient to reflux the benzene. After the addition is complete, a heating mantle is used to reflux the reaction mixture for a further 2 h. Prolonged addition time or reflux leads to polymeric side products. After cooling, the sodium chloride is extracted by washing twice with 1-L portions of water. The benzene is removed in a rotary evaporator after drying over anhydrous magnesium sulfate. The light yellow solid residue (yield 83 g) melts at 72 to 75°C. Recrystallizing twice from ether gives about 75 g colorless iminocarbonate (mp 76 to 77°C). This is stored under nitrogen in a dry box.

154. Polymerization of N-phenyliminocarbonate to Poly(ethylene-N-phenyl Urethane) (47)

$$\text{III} \longrightarrow (-N(C_6H_5)-CO-O-CH_2-CH_2-)_n$$

IV

Total 76 g ethylene *N*-phenyliminocarbonate is dried at 40°C in vacuo (nitrogen bleed) and placed in a previously dried 100-mL flask protected from moisture by a drying tube. The monomer is melted by heating the flask in an oil bath at 90°C. To the melt is added from a microliter syringe about 10 μL gaseous phosphorus pentafluoride (about 0.01% by weight of monomer). Polymerization proceeds smoothly to give a white solid mass after 16 h. An additional 26 h at 90°C is allowed for complete polymerization. The flask is broken, and the extremely hard polymer is chopped into small pieces. The polymer is dissolved in 1500 mL hot chloroform with the aid of a shearing stirrer. The solution is pressure filtered through felt to remove insoluble material and then poured into 3 L acetone to precipitate the polymer. The polymer is filtered and dried in a vacuum (nitrogen bleed) at room temperature for 60 h, then at 100°C for 2 h. The yield of white polymer is 38 g. (50%). The mp is 190°C, and the inherent viscosity is 2.36 in chloroform and 2.18 at 0.5% concentration in trifluoroacetic acid. Clear film can be pressed at 180°C, and fibers spun at 200°C. These can be drawn up to 200% over a hot pin. The thermal decomposition temperature of the polymer can be raised from 240 to 276°C (differential scanning calorimitry; DSC) by end capping with acetic anhydride. This is done by refluxing with acetic anhydride in pyridine solvent. The end-capped polymer has a mp of 184°C.

8.6. CYCLIC ETHERS

Practically all oxiranes (epoxides) and oxetanes (trimethylene oxides) will polymerize.

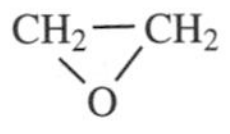

CH_2-CH_2 (bridged by O)

Oxirane

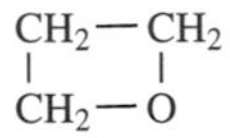

CH_2-CH_2 / CH_2-O

Oxetane

H_2C-CH_2 / H_2C O CH_2

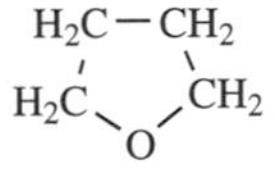

Tetrahydrofuran

Polymers ranging from low molecular weight syrups to tough, high molecular weight solids can be obtained, depending on conditions and type of catalyst. Normally, ionic catalysts, either cationic or anionic, are the only effective types. Tetrahydrofuran (THF) will polymerize readily but only with cationic catalysts. Antimony pentachloride was the preferred catalyst to polymerize THF but has been displaced by phosphorus pentafluoride, which gives high molecular weight polymer (48).

All of the unsubstituted polyethers, except the polyacetal derived from formaldehyde, melt in the range of 35 to 65°C. As will be seen later, substitution on the chain may tend to raise the melt temperature. Alkylated derivatives of THF have resisted polymerization. On the other hand, a bicyclic epoxide (1,4-epoxycyclohexane), which may be considered as a disubstituted THF, does polymerize, probably to relieve the strain in the bridge structure. The resulting polymer has a high melt temperature (>400°C) (49, 50).

Epoxides will polymerize with either cationic or anionic initiators to give linear polymers as high as 1 million in molecular weight. Lower molecular weight polymers (e.g., 1000 to 3000) are normally hydroxyl terminated at both chain ends and serve as difunctional molecules for further reaction to higher molecular weight, e.g., in preparing Spandex or urethane foams. Polyethylene oxide of several million molecular weight can be prepared using pure strontium carbonate (free from trace amounts of nitrates, chlorate, bisulfate, and other ions). Trace amounts of water (e.g., 0.5%) are desirable for effective polymerization (51–53).

155. Polymerization of Ethylene Oxide (52)

$$Sr(OH)_2 + CO_2 \longrightarrow SrCO_3 + H_2O$$

$$\underset{\text{(ethylene oxide, } CH_2-CH_2 \text{ bridged by O)}}{CH_2-CH_2} \xrightarrow{SrCO_3} [-OCH_2CH_2-]_n$$

155.A. Preparation of Active Strontium Carbonate Catalyst

A 22% aqueous solution of pure strontium hydroxide is prepared at 90°C in distilled water. A stream of carbon dioxide gas is introduced under the surface of the solution and allowed to proceed until precipitation of the carbonate is complete. The solution is filtered and the filtered solid is washed with distilled water. The product is dried to a water content of not less than 0.5 to 1.0%.

155.B. Polymerization

In a polymer tube is placed 50 mL redistilled ethylene oxide containing <50 ppm aldehyde. To the tube is then added about 0.2 g strontium carbonate prepared as above. The tube is sealed and heated to 50°C and maintained at this temperature. After an induction period of about 90 min, polymerization, begins. During the polymerization, the tube should be rocked or rotated to enhance mixing but not by hand. At the end of the induction period, polymerization may become very rapid, sometimes causing the tube to shatter hence (**Caution!**). Once the polymerization has started it should be complete in about 2 h.

Using a commercial strontium carbonate from Baker (CP grade) and drying to 0.5 to 1.0% water and using concentrations of 1.5 g per 50 mL ethylene oxide gives a less active catalyst with a longer induction period (20 h) and a less violent polymerization.

The poly(ethylene oxide) is soluble in chloroform, acetonitrile, and other such solvents. Viscous solutions can be prepared in these solvents. The polymer is also miscible with water at room temperature and extremely viscous solutions can be prepared. Propylene oxide may also be polymerized with basic or acidic catalysts. The polymer has regularly spaced asymmetric carbon atoms and hence should be capable of existing in random or stereoregular configuration. Using a complex iron catalyst, a research group from Dow Chemical Co. has shown that polymerization of DL-propylene oxide will give a solid, crystalline high molecular weight product (51–53).

156. Polymerization of Propylene Oxide (54–56)

$$\underset{\text{H}}{\overset{\text{CH}_3}{}}\!\!>\!\text{C}\!-\!\text{C}\!<\!\!\underset{\text{H}}{\overset{\text{H}}{}} \;(\text{O bridge}) \xrightarrow{Fe^{+3}} \left[-\text{O}-\underset{|}{\overset{\text{CH}_3}{\text{CH}}}-\text{CH}_2-\right]_n \longrightarrow$$

156.A. Preparation of a Ferric Chloride Complex Catalyst
A complex catalyst is prepared in a polymer tube by dissolving 1.0 g anhydrous ferric chloride in 5 mL diethyl ether and gradually adding 1.0 g liquid propylene oxide with agitation and cooling below 60°C. When the condensation of the ferric chloride and the propylene oxide is complete, the product is warmed in a vacuum to remove volatile matter leaving a semisolid brown residue.

156.B. Polymerization
To the catalyst residue prepared above is added 100 g propylene oxide. The tube is cooled under nitrogen and sealed. The mixture is heated at 80°C with agitation for 88 h, at which time polymerization is complete. In this way 94 g of a brown rubbery material results. This is dissolved in hot acetone and sufficient concentrated HCl is added to convert the iron complex present into soluble ferric chloride. The solution is chilled to −20°C, and solid polymer crystallizes from solution and is separated by filtration. The polymer is precipitated twice in acetone in the same manner giving about 25 g pure white polymer. The polymer melts at about 70°C and has a molecular weight in the 100,000 to 150,000 range. The polymer is soluble in a range of solvents (acetone, benzene, THF) and can be melt pressed into film that can be cold drawn. Epoxides with several symmetrical substituents exhibit chain stiffening, have high melt temperatures, are highly crystalline, and are insoluble in most solvents.

157. Preparation of 1,1,2,2-Tetramethylethylene Oxide (57)

$$(CH_3)_2C=C(CH_3)_2 \xrightarrow[Na_2CO_3]{CH_3CO_3H} (CH_3)_2C\overset{O}{-}C(CH_3)_2$$

In a 2-L three-necked flask, equipped with a stirrer, a condenser, and a dropping funnel and cooled with an ice bath is placed 300 g anhydrous sodium carbonate, 300 mL methylene chloride, and 168 g freshly distilled tetramethylethylene. The mixture is stirred, and the temperature is held at 5 to 10°C. To the mixture is added 372 g 40% peracetic acid at such a rate to keep the temperature below 10°C. The addition takes about 6 h. The mixture is stirred vigorously throughout. Stirring is continued for a further 1 h, and 500 mL water is then added. The organic phase is extracted with two 200-mL portions of methylene chloride. The methylene chloride is washed with ferrous sulfate solution until it no longer gives a positive peroxide test. The product, tetramethylethylene oxide boils at 90 to 91°C at atmospheric pressure. The yield is about 70%.

158. Polymerization of 1,1,2,2-Tetramethylethylene Oxide (57)

$$(CH_3)_2\overset{O}{\overbrace{C-C}}(CH_3)_2 \xrightarrow{BF_3} \left[-O-\underset{CH_3}{\overset{CH_3}{|}}\!\!C-\underset{CH_3}{\overset{CH_3}{|}}\!\!C- \right]_n$$

In a 300-mL round-bottom flask equipped with drying tube and a gas inlet and cooled in dry ice is placed 5 mL tetramethylethylene oxide and 100 mL dry methyl chloride. To this solution is added 1 mL boron trifluoride etherate. The mixture is kept at dry ice temperature for 24 h. The solid is filtered, washed with methylene chloride, alcohol, and then dried. The polymer yield is quantitative. The polymer is a hard white solid that does not melt below 300°C. It is unaffected by boiling with solvent such as xylene, THF, dimethylformamde (DMF), and dioxan.

A large number of polyethers derived from phenyl and naphthyl glycidyl ethers have been studied (58, 59). Dibutyl zinc-water is the catalyst used and is superior to dibutyl zinc alone. Other complexes such as aluminum alkyls and alkoxides are also quite effective (60, 61).

159. Polymerization of Phenyl Glycidyl Ether (58, 59)

$$C_6H_5-O-CH_2\overset{O}{\overbrace{CH-CH_2}} \longrightarrow \left(\underset{\underset{O-C_6H_5}{|}}{\underset{CH_2}{|}}\!\!\!CH-CH_2-O \right)_n$$

Phenyl glycidyl ether (from Shell Chemical) is fractionally distilled through an efficient column to give a fraction that boils at 141°C/30 mm. It should show no high or low peaks in gas chromatography. A toluene solvent is dried by refluxing over sodium for 24 h, distilled through a short Vigreux column, and stored under nitrogen. The polymerization is carried out by adding the components by means of dry, nitrogen-flushed syringes to a nitrogen-flushed glass ampoule with constricted neck, topped with a four-way tube in which the lateral tube acts as a nitrogen inlet and exit to exclude moisture and air while the components are added through a serum cap in the top vertical tube. The bottom vertical tube is attached to the reaction ampoule, which is shaped like a wider version of the polymer tubes used for melt polymerizations. To the ampoule is charged with 10 g phenyl glycidyl ether, 15 g toluene, 0.0151 mL water (via microliter syringe), and 1.0 mL of a toluene solution of dibutyl zinc containing 0.15 g/mL of the catalyst. The latter two components are both added in amounts of 0.000838 mol. The neck of the ampoule is sealed off under nitrogen, and the container is rotated in an air oven or in an oil bath for 24 h. The polymer is removed from the tube, chopped in a blender with 200 mL toluene to give a finely divided, swollen product, and precipitated into 3 L ethanol. The product is filtered, washed repeatedly in a blender with ethanol, and finally dried at 60°C in a vacuum oven for 16 h. The polymer is crystalline by x-ray and melts at 240°C. The reduced viscosity in *p*-chlorophenol containing 25 α-pinene at 47°C is about 10 (0.2 g/100 mL). It is necessary to heat the solution to 14°C for 30 min to effect solution.

Polymerization of unsubstituted oxetane is not straightforward because the parent compound is prone to rearrange in acid media to allyl alcohol. Successful polymerization to high molecular weight linear polymer was first reported by Rose (62), but since the synthesis is quite tedious it will not be described here. Polyethers from oxetane and its gem disubstituted derivatives are crystalline and reasonably high melting if the gem substituents are bulky or halomethyl. This results from chain stiffening (62).

160. Preparation of Trichloropentaerythritol (63)

$$(HOCH_2)_4C \longrightarrow (ClCH_2)_3{-}C{-}CH_2OAc \longrightarrow (ClCH_2)_3{-}CH_2OH$$

Hydrogen chloride gas is bubbled into a mixture of 600 g acetic acid and 100 g water at 0°C until a total of 176 g (4.9 mol) is adsorbed. The mixture is charged into a 1-L. Hastelloy bomb, together with 200 g (1.5 mol) pentaerythritol. The bomb is sealed and heated to 160°C for 8 h. Due to the corrosive nature of this mixture, it is best to back the stainless-steel rupture disk of the bomb with a thin sheet of Teflon film resin and then with platinum foil to prevent the disk from corroding. After 8 h, the bomb is cooled to room temperature, and the reaction mixture is diluted with water. The trichloropentaerythritol acetate is isolated by extracting with methylene chloride. The solvent is removed and the residual oil is refluxed overnight with 500 mL methanol and 50 mL concentrated HCl. The next day, the mixture of methyl acetate and methanol is distilled slowly. The residual trichloropentaerythritol acetate crystallizes and is then filtered, washed with water, and dried. The crude product weighs about 275 g and melts at 60 to 63°C. This product can be used directly for the preparation of bis(chloromethyl)oxetane.

161. Preparation of 3,3-Bis(chloromethyl)oxetane (63)

$$(ClCH_2)_3{-}CH_2OH \longrightarrow ClCH_2{-}\overset{CH_2Cl}{\underset{}{C}}\text{(oxetane ring, O)}$$

A mixture of 275 g trichloropentaerythritol, 500 mL methanol, 60 mL water, and 80 g potassium hydroxide is refluxed for 18 h. An equal volume of water is added, and the heavy oil is separated with two 100-mL portions of ethyl ether. The ether extract is dried over anhydrous calcium chloride and distilled through an efficient column to give about 125 g pure cyclic ether (bp 101°C/27 mm or 62°C/4 mm). Immediately before polymerization, this product should be fractionated through an efficient column to ensure purity.

162. Polymerization of 3,3-Bis(chloromethyl)oxetane (63)

$$(ClCH_2)_2C(CH_2)_2O \longrightarrow \left[-O-CH_2-C(CH_2Cl)_2-CH_2- \right]_n$$

Into a 500-mL three-necked flask equipped with a stirrer, a dry-ice condenser, and a gas inlet tube is condensed 100 mL anhydrous methyl chloride. Then 25 g freshly distilled bischloromethyloxetane is added all at once. The dry-ice bath is removed and the mixture is allowed to reflux (−25°C). Into the mixture is introduced a trace of phosphorus pentafluoride gas. After a short induction period, polymerization takes place rapidly. Ordinarily, the polymerization is uneventful and the solid polymer precipitates in the first few minutes. **Occasionally, the polymerization is quite violent** and the contents of the reaction flask may be ejected through the top of the condenser. However, nothing more serious occurs on this scale of reaction. The methyl chloride is allowed to evaporate, and the solid polymer is isolated and washed several times with methanol. The polymer is obtained as a spongy, white solid with an inherent viscosity of about 1.0 in hexamethylphosphoramide (HMPA) (**Carcinogen**!). The crystalline polymer melts at 177°C and can be pressed into clear, tough, drawable films at 175 to 200°C. The polymer is readily soluble in HMPA and has limited solubility in DMF or hot cyclohexanone. The bis(chloromethyl)oxetane can be used to prepare the difluoro- and the diiodo-counterparts.

163. Preparation of Bis(fluoromethyl)oxetane (64)

$$(ClCH_2)_2C(CH_2)_2O \xrightarrow{KF} (FCH_2)_2C(CH_2)_2O$$

In a three-necked 1-L flask, equipped with stirrer and a reflux condenser is placed a mixture of 156 g anhydrous powdered potassium fluoride and 156 g bis-(chloromethyl)oxetane. To it is added 340 mL anhydrous ethylene glycol, and the mixture is heated with a metal heating bath to 160°C with vigorous and efficient stirring. The mixture is allowed to reflux and is slowly distilled. The distillate separates into two layers, the lower consisting of bis(fluoromethyl)oxetane. The product is obtained in a crude yield of 70%. The crude distillate is mixed with twice its volume of water, and the mixture is again distilled. The organic product is entrained with the water vapor and is separated and dried over magnesium sulfate. It is again distilled in an anhydrous condition to give about a 40% yield of pure product (bp 56°C/29 mm).

164. Preparation of Bis(iodomethyl)oxetane (63)

$$ClCH_2\text{-}C(CH_2Cl)(\text{oxetane}) \xrightarrow{NaI} ICH_2\text{-}C(CH_2I)(\text{oxetane})$$

A mixture of 15.5 g 3,3-bis(chlorimethyl)oxetane, 150 mL methyl ethyl ketone, and 35 g dry sodium iodide is refluxed for 24 h. The solution is then cooled, filtered, and the solvent partially removed by evaporation. The residue solidifies on standing and is recrystallized from cyclohexane. The yield is 30 g (89%) of coarse, colorless, very dense crystals with a mp of 50°C.

The fluoromethyl and iodomethyl oxetanes can be polymerized in a similar manner to the bischloromethyloxetane using phosphorus pentafluoride as initiator. The phosphorus pentafluoride can be prepared on a laboratory scale by the thermal decomposition of an aryldiazonium hexafluorophosphate. Between 3 and 12 g salt is heated at 150 to 160°C in a distillation setup, and the phosphorus pentafluoride is swept into the polymerization vessel with a stream of dry nitrogen. Between 1 and 2 h are required to decompose all the phosphate salt under these conditions Thietanes can be polymerized in a similar manner. Ring opening of 3,3-dimethylthietane with boron trifluoride etherate has been described (65).

165. Synthesis of 5,5-Dimethyl-1,3-dioxane-2-one and 3,3-Dimethylthietane (66, 67)

$$HOCH_2\text{—}C(CH_3)_2\text{—}CH_2OH + (C_2H_5O)_2CO \longrightarrow (CH_3)_2C(CH_2O)_2C{=}O + C_2H_5OH$$

$$(CH_3)_2C(CH_2O)_2C{=}O + KSCN \longrightarrow (CH_3)_2C(CH_2)_2S + CO_2 + KOCN$$

A mixture of 104 g (1 mol) 2,2-dimethyl-1,3-propanediol, 130 g (1.1 mol) diethylcarbonate, and 0.5 g sodium dissolved in 8 mL absolute ethanol is heated with stirring in a 500-mL, round-bottom, three-necked flask provided with a Vigreux column and distillation head. Heating is carried out to maintain the distillate at a boiling point below 80°C (pot temperature 90 to 150°C) and is continued for about 3 h, when nearly the theoretical amount of alcohol is obtained. The mixture is allowed to cool to room temperature and is taken up in an equal volume of benzene (**Carcinogen!**). The benzene solution is washed three times with water and dried over calcium chloride. After removing the benzene, the residue is distilled at reduced pressure (bp 116 to 118°C/1 mm) to give 95.7 g crude product. This is dissolved in 95 mL benzene at 50°C and precipitated with petroleum ether. The resulting white crystals are filtered and air dried, followed by 2 h in a 60°C vacuum oven to yield 74 g product (mp 109 to 110°C).

The 3,3-dimethylthietane is prepared by heating the cyclic carbonate (above) with an equivalent amount of potassium thiocyanate to 170 to 185°C and removing the cyclic sulfides by distillation as formed. Dimethylthietane boils at 116°C at atmospheric pressure. The yield is 41%.

166. Preparation of Poly(2,2-dimethyltrimethylene Sulfide) and Conversion to the Sulfone (65)

$$\text{3,3-dimethylthietane} \longrightarrow \left[-S-CH_2-\underset{CH_3}{\overset{CH_3}{\underset{|}{\overset{|}{C}}}}-CH_2- \right]_n \xrightarrow{H_2O_2} \left[-SO_2-CH_2-\underset{CH_3}{\overset{CH_3}{\underset{|}{\overset{|}{C}}}}-CH_2- \right]_n$$

About 2.7 mL 3,3-dimethylthietane is placed in a test tube stoppered with a serum bottle cap and cooled to 0°C in an ice bath. Then 0.15 mL boron trifluoride-etherate is added through the stopper with a small syringe. The solution is warmed to room temperature and set aside. After 16 h, a clear, viscous syrup results, which is left standing for 2 days and gives a immobile gel. This is the polysulfide. The gel is dissolved with gentle warming in 20 mL trifluoroacetic acid. The solution is cooled in ice and 10 mL 30% hydrogen peroxide is added in three portions with stirring. An exothermic reaction occurs, the color changes to yellow, to bright red, and finally back to colorless. The polymer separates as a soft gel. This dissolves on warming, and the resulting clear solution is heated on the steam bath for 2 h. The polymer is precipitated into water, washed twice with water in a home blender, rinsed with alcohol on a funnel, and dried in a vacuum oven at 60°C overnight. The yield of fine, white powdery polymer is 1.4 g. The mp is 260°C (with decomposition). The inherent viscosity in *m*-cresol is 0.57 at 0.5% concentration. A brittle film can be pressed at 152°C. Crystallinity by x-ray is high.

Compared to oxetanes and epoxides, polymerization of tetrahydrofuran proceeds rather slowly. Ordinarily, the products are viscous oils but very high molecular weight solids are obtained using phosphorus pentafluoride as initiator (48).

167. Polymerization of Tetrahydrofuran (48)

$$\text{tetrahydrofuran} \xrightarrow{PF_5} [-OCH_2CH_2CH_2CH_2-]_n$$

The tetrahydrofuran is purified by refluxing over solid sodium hydroxide, distilling under nitrogen, then refluxing over lithium aluminum hydride and distilling immediately before use. To about 350 g purified tetrahydrofuran in a suitable size vessel under nitrogen, is added 1 g of a solid phosphorus pentafluoride-tetrahydrofuran coordination complex. The latter is prepared by saturating THF with phosphorus pentafluoride at 0°C and subliming the resulting solid at 70°C and 0.02 mm pressure. The mixture of THF and initiator is maintained at 30°C for about 6 h to effect polymerization. The resulting solid, colorless polymer is heated with water to destroy initiator residue and then is dissolved in more THF. The polymer is recovered by pouring the THF solution into water with violent agitation, preferably in a high-speed mixer such as a home blender, adapted with the motor blanketed in nitrogen to minimize chance of fire. A white polymer in yield of about 59% is obtained after airdrying. The polymer has an inherent viscosity of 3.6 in benzene, which corresponds to a molecular weight of about 329,000. The polymer can be molded at 100 to 230°C to a clear, tough film that crystallizes on standing. Above the crystalline melting point of 45°C, the polymer films take on a rubbery appearance and feel but maintain their toughness. Poly(tetramethylene oxide) of high molecular weight is crystalline, melting at about 45°C. Again, the tremendous effect on properties of stiffening a polymer chain is demonstrated by the polymerization of 1,4-epoxycyclohexane. The polymer is highly crystalline, but melts above 400°C (49, 50).

168. Preparation of 1,4-Epoxycyclohexane (68)

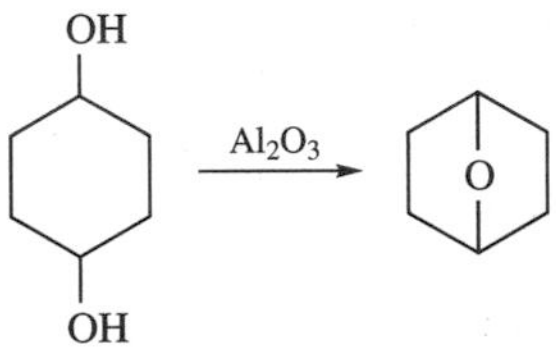

OH

Al_2O_3

O

OH

A mixture of cis-and trans-cyclohexanediols obtained by the hydrogenation of hydroquinone or purchased should be distilled before use to obtain a dry product. The diol mixture boils at about 146°C/18 mm. To 432 g distilled diol is added an equal weight of activated alumina. The intimately mixed solids are placed in an ordinary distilling apparatus, and the solids heated slowly for about 6 h at such a temperature that a distillate is obtained slowly. The product, about 200 g, is isolated, dried over anhydrous potassium carbonate, and distilled, preferably through a precision distilling column. The pure 1,4-epoxycyclohexane boils at 119°C. The yield is about 41%.

169. Polymerization of 1,4-Epoxycyclohexane (54)

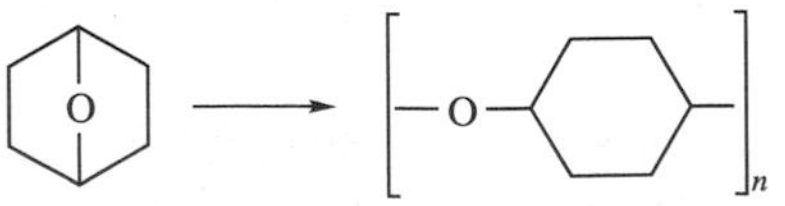

O

O

n

We have selected two catalyst systems effective in the polymerization of 1,4-epoxycyclohexane to high molecular weight. With the first, a mixture of 1 part of the epoxycyclohexane and 2 parts of nitrobenzene is cooled to −30°C. To the mixture is added phosphorus pentafluoride gas in a quantity of 1 to 5 mol%. The mixture is maintained at −30°C for roughly 100 h, when a solid mass is formed. This is removed and broken up with acetone, filtered, washed repeatedly with water and acetone, and dried. The yield is 50%. The inherent viscosity is about 1 in tetrachloroethane/phenol (66/100 w/w) at 0.5% concentration.

An alternative method uses a catalyst combination of ferric chloride-thionyl chloride. Thus to 50 g epoxycyclohexane, maintained at 0°C, is added 0.015 g anhydrous ferric chloride as a 10% solution in dry ether, followed by 0.062 g thionyl chloride, also added as a 10% solution in anhydrous ether. The mixture is stirred and maintained in a stoppered flask at 0°C for 18 h. After this period, the polymer is mixed with alcohol, and the solid filtered. The polymer is washed repeatedly with alcohol and water and dried. The yield is 37% and the inherent viscosity is about 0.6.

Both the above polymers are highly crystalline by x-ray, and the pattern does not disappear below 400°C. The polymer is insoluble in most solvents, but is soluble in tetrachloroethane/phenol (66/100 w/w). Films can be cast from solutions in this system, but the dry films are usually brittle because of the high crystallinity, even though the inherent viscosity is high.

8.7. POLYMERIZATION OF CYCLIC HYDROCARBONS

Certain carbocycles, may be bulk polymerized over a molybdenum catalyst, giving an all-cis polyene, e.g., cyclopentene can be made to form poly-*cis* (1,5-pentene). The generalized name for this kind of reaction is *olefin metathesis*, whether to form polymers or to synthesize small molecules from noncyclic olefins. Olefin metathesis (51) describes the interchange of carbon atoms between pairs of double bonds as shown in **8.3**. The reaction is metal catalyzed in which there is fragmentation at one double bond and recombination of the fragments with those of another double bond to form different molecules.

$$\mathrm{RRC{=}CRR'} \rightleftharpoons \mathrm{RRC{=}CRR} + \underset{\text{cis and trans forms}}{\mathrm{RR'C{=}CRR'}}$$

Schematic 8.3

The reaction extends to a large number of acyclic and cyclic olefins and to unsaturated organic compounds. The catalysts cover a range of complexes and oxides of metals in groups IV to VIII in the periodic table. Side reactions can be avoided by proper choice of solvents (e.g. chlorobenzene), basic additives to eliminate cationic side reactions, choice of catalyst/cocatalyst and running the reaction at as low a temperature a feasible. Typical metathesis reaction are shown in Figure 8.3. The subject is too broad to cover meaningfully in this

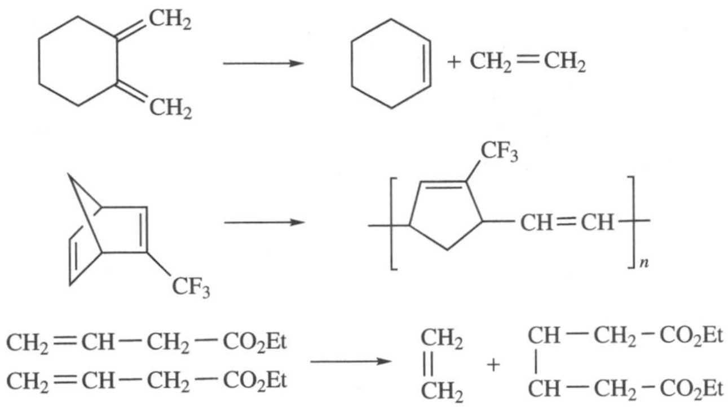

$$CH_2=CH-CH_2-CO_2Et + CH_2=CH-CH_2-CO_2Et \longrightarrow CH_2=CH_2 + EtO_2C-CH_2-CH=CH-CH_2-CO_2Et$$

Figure 8.3

book, and readers are directed to a broad reviews given by Drägujan et al. and by Ivin (51).

170. Ring Opening Polymerization of Cyclopentene via Metathesis (69)

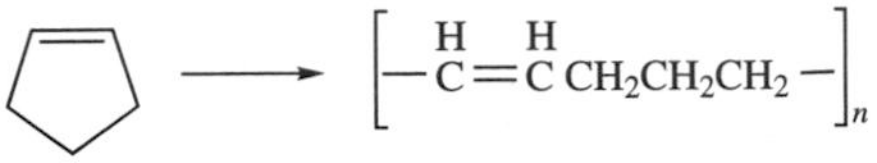

$$\longrightarrow \left[-\overset{H}{C}=\overset{H}{C}\,CH_2CH_2CH_2-\right]_n$$

A 250-mL three-necked flask is equipped with a stirrer and a nitrogen inlet; the air is replaced with dry nitrogen. The reaction vessel is charged with 20 g pure dry pentene (the commercial product has to be refluxed under nitrogen, rectified over sodium wire, and stored under nitrogen) and then cooled to −40°C with a dry ice–acetone bath. The catalyst is formed in the cooled monomer by adding under stirring 0.160 g (0.59 mmol) pure molybdenum pentachloride (powdered and weighed under dry nitrogen) followed by 1.18 mmol triethylaluminum.

Immediately after the addition of the triethylaluminum, the reaction mixture assumes a darkbrown color and polymerization starts. The polymerization is continued for 4 h at −40°C with stirring; by this time, the reaction mixture appears like a brown lump of soft rubber. The reaction is terminated by adding 100 mL methanol and stirring vigorously for 30 min. The mother liquor is discarded, and the polymer is dissolved by stirring with 200 mL benzene. The addition of a small quantity of antioxidant, such as phenyl-β-naphthylamine, is recommended to prevent cross-linking. When the polymer is dissolved (the solution is usually free of gel but should be filtered to remove catalyst residues and any other insolubles). The polymer is reprecipitated by pouring the solution into excess methanol and stirring vigorously. The polymer is dried under reduced pressure at room temperature.

To obtain higher purity and to remove lower molecular weight fractions, the polymer can be redissolved in benzene and precipitated into acetone containing an antioxidant. The yield of dry polymer is about 9 g (45%), and the intrinsic viscosity is in the range of 1.8 to 2.0 in toluene at 30°C. The polymer is amorphous at room temperature, even after stretching. Infrared examination shows more than 99% of the double bonds are of the cis type, the remainder are trans or vinyl end groups.

The *cis*-1,5-poly(pentene-1)—(it could be called alternating poly(trimethylene-co-*cis*-vinylene)—thus obtained is a translucent solid with rubber-like properties. It is soluble in many hydrocarbons at room temperature but insoluble in alcohols and ketones, even at elevated temperatures. The polymer cross-links in air at room temperature, much as polymers from conjugated dienes that give significant 1,4 linkage, and should be stored under nitrogen at low (<0°C) temperatures, especially in the presence of a phenolic antioxidant additive, such as by precipitating into methanol (abve) containing bisphenol A.

8.8. SILICONES

A class of polymers of considerable commercial importance is based on linear, cyclic, or cross-linked arrangement of alternating silicon and oxygen atoms, where the silcon is substituted with organic radicals or hydrogen. These are called organopolysiloxanes, or simply silicone polymers. The usual procedure to prepare silicone polymers is to hydrolyze, either singly or in the appropriate combinations, compounds of the type R_3SiCl, R_2SiCl_2, $RSiCl_3$, and $SiCl_4$ depending on the kind of product desired. The intermediates in the reaction are believed to be the corresponding silanols, e.g., R_2 $Si(OH)_2$, which condense rapidly with the formation of the -[Si-O-Si-]n linkage. In addition to linear polymer, cyclic forms, in which n ranges from 3 to 9, are often encountered. These can be converted to high molecular weight linear polymer with alkaline catalysts. Although many of the silicone "polymers" are often of low molecular weight, their relationship to higher molecular weight silicones warrants inclusion here. The linear silicones, $(CH_3)_3Si[OSi(CH_3)_2]_n$-$OSi(CH_3)_2$, where n is fairly small, form the basis of the well known silcone oils. The cyclic silicones, formed in hydrolysis reactions of the silane dihalides, especially $[(CH_3)_2\ SiO]_n$, where $n = 3$ to 4, are convertible to high molecular weight linear silcone elastomers. Various curing techniques are available for converting linear and cyclic materials to crosslinked elastomers and resins.

171. Preparation of Cyclic Polysiloxanes (70)

$$CH_3-\underset{\underset{Cl}{|}}{\overset{\overset{Cl}{|}}{Si}}-CH_3 \xrightarrow{H_2O} \left[\underset{\underset{CH_3}{|}}{\overset{\overset{CH_3}{|}}{Si}}-O\right]_{3\text{-}9}$$

171.A

From a dropping funnel protected with a drying tube, 200 mL dimethyldichlorosilane is added slowly to 600 mL of vigorously stirred water maintained at 15 to 20°C. When the addition is finished, the oily organic layer is taken up in 150 mL ethyl ether, separated from the water phase, and dried over magnesium sulfate. The ether layer is filtered, and the ether removed by evaporation. The oily residue contains cyclic products plus some high molecular weight material, probably linear and cyclic. The material from the ether layer, about 100 mL, is fractionated in a precision column to isolate the individual components of the mixture. Trimer (bp 134°C) and tetramer (bp 175°C) can be distilled conveniently at atmospheric pressure. Pentamer (bp 101°C/20 mm) and hexamer (bp 128°C/20 mm). Trimer to hexamer is half the total product.

171.B

The higher molecular weight residue in the still pot is a viscous oil that is pyrolyzed to trimer and tetramer by heating in a slow stream of nitrogen by means of a Woods metal bath to 350°C with a Claisen head and a condenser set up for distillation. Up to 350°C, only trace quantities of distillate appear; from 350 to 400°C, the liquid in the flask begins to boil, with the distillate temperature being 135 to 210°C. Continued heating at 400°C results in almost the entire contents of the flask being distilled. The distillate, about 400 mL, forms a mixture of crystals and liquid and consists about 44% of the cyclic trimer and 24% of the tetramer, with the remainder being pentamer and above. The mixture can be fractionated into its components.

These reactions demonstrate the tendency of the (R_2-Si-O) unit to form cyclic structures under the condition given. The reverse of this latter process, namely formation of low molecular weight linear polymer from low molecular cyclic siloxanes, can be accomplished by an equilibrium reaction in the presence of sulfuric acid and is demonstrated in preparation 172.

172. Preparation of Linear Polysiloxanes (70)

$$[(CH_3)_2SiO]_4 \xrightarrow{H_2SO_4} [(CH_3)_2SiO]_n$$

Place 20 mL octamethylcyclotetrasiloxane in a stoppered flask or bottle with 3.7 mL concentrated sulfuric acid and 10 mL ethyl ether; shake at room temperature for 24 h. The mixture becomes viscous. Then 20 mL ether and 10 mL water are added, and the mixture is shaken for 1 h. The lower aqueous layer is drawn off, and the ether solution is washed three times with 10-mL portions of water and dried over anhydrous potassium carbonate. The ether layer is distilled from the solution through a Claisen head, and the temperature of the distillation flask is raised by means of a metal bath to 310°C, during which time a small quantity of distillate forms. The residue in the flask is a clear, viscous oil that is soluble in many hydrocarbon and ether solvents. If purified tetramer is used as the starting material, a cryoscopic molecular weight determination in cyclohexane indicates a value of about 2740, or 37 dimethylsiloxane units.

By carrying out this reaction on the cyclic tetramer in the presence of a definite amount of the hexamethyldisiloxane, one can prepare linear polymers of the structure $(CH_3)_3\ SiO[(CH_3)_2\ SiO\}_n\ Si(CH_3)_3$, where n is determined by the amount of chain-terminating $(CH_3)_3SiOSiO(CH_3)_3$ used. It is these linear polysiloxanes that form the basis of methyl silcone oils. A variety of products is possible with a wide range of viscosities, depending on the value of n. They are distinguished by their small change in viscosity over a wide range of temperatures, quite unlike petroleum oils.

The other linear polysiloxanes mentioned are terminated with hydroxyl groups, which may condense further on heating and alter the molecular weight and viscosity as a result. The advantage of the $(CH_3)_3$ SiO terminated polysiloxanes over these is the stability to heat conferred by the trimethylsiloxy group.

173. Preparation of Linear Polysiloxanes Terminated with Trimethylsiloxy Groups (70)

$$[(CH_3)_2SiO]_4 + [(CH_3)_3Si]_2O \xrightarrow{H_2SO_4} (CH_3)_3SiO[(CH_3)_2SiO]_nSi(CH_3)_3$$

A total of 20 mL of the cyclic tetramer from previous preparations is mixed with 0.4 mL hexamethyldisiloxane and shaken for 24 h with 0.8 mL concentrated sulfuric acid. After this time, 5 mL water is added, and shaking is continued another 1 h. The mixture is centrifuged, and the two layers are separated. The viscous upper layer of silicone oil has a viscosity of about 130 cs at 40°C.

Linear silicone polymers, which may have molecular weights of 1 million or more, are conveniently prepared from base-catalyzed ring-opening polymerization of the cyclic trimer or tetramer. The gummy products are soluble in aromatic hydrocarbons and may be compounded, molded, and cured, much like natural rubber.

174. Preparation of Polydimethylsiloxane (71)

$$\text{cyclo-}[(CH_3)_2Si{-}O]_4 \xrightarrow{KOH} \left[-\overset{CH_3}{\underset{CH_3}{\overset{|}{\underset{|}{Si}}}}-O- \right]_n$$

A catalyst solution is prepared by dissolving 1.0 g potassium hydroxide (dried at 70°C in vacuum overnight) in 400 mL dry isopropanol in a flask protected with a drying tube and equipped for distillation. Isopropanol is distilled until the volume of the solution is reduced to 250 mL Then 0.1 mL of this solution is transferred by a pipet to a 100-mL three-necked flask purged with nitrogen and fitted with a good mechanical stirrer and a nitrogen inlet. The isopropanol is removed in a stream of nitrogen with moderate heating. Then 29.6 g octamethylcyclotetrasiloxane is added; the flask is equipped with a drying tube and heated by an oil bath to 165°C with stirring for 1 to 5 h, without the nitrogen, during which time the reaction mass will become a highly viscous gum. The polymer is transferred to an evaporating dish and heated in an oven at 150°C for 3 h. The polymer is soluble in benzene and toluene and has an inherent viscosity of about 0.7 in toluene. Catalyst residues may be removed by washing a solution of the polymer with very dilute hydrochloric acid.

Silicone rubbers are noted for their flexibility over a wide range of temperatures (−90 to 300°C) and their resistance to moisture, air, and weathering. The polymers so obtained must be properly compounded with fillers and curing agents to form cross-linked or vulcanized products. Using a rubber mill, a vulcanized rubber may be obtained by blending 20 g polydimethylsiloxane, 12 g diatomaceous earth, 17.5 g titanium dioxide, 0.3 g benzoyl peroxide, and 2.0 g 2,5-di-*t*-butylquinone. The mix may then be molded or shaped and cured by heating at 250°C in an oven for 24 h.

Probably one of the most unequivocal examples of the much sought after ladder structure in polymers is the high molecular weight, double-chained, *cis*-syndiotactic polyphenylsilesquioxane, in which cyclotetrasiloxane chains are fused cis-anti-cis (68, 72, 73). This linear, soluble polymer is the therodynamically favored product from the base-catalyzed equilibrium of products from the hydrolysis of phenyltrichlorosilane. Linear, double-stranded polymer is unexpected from a trifunctional monomer according to the classic view of polycondensation as the random reaction of equally reactive functional groups. The validity of the assumption of equal reactivity is doubtful in the sesquisiloxane case; the intramolecular cyclization tendency is also evidently great and contributes to linearity of the product here.

175. Preparation of a "Ladder" Poly(phenylsilesquioxane) (68, 72, 73)

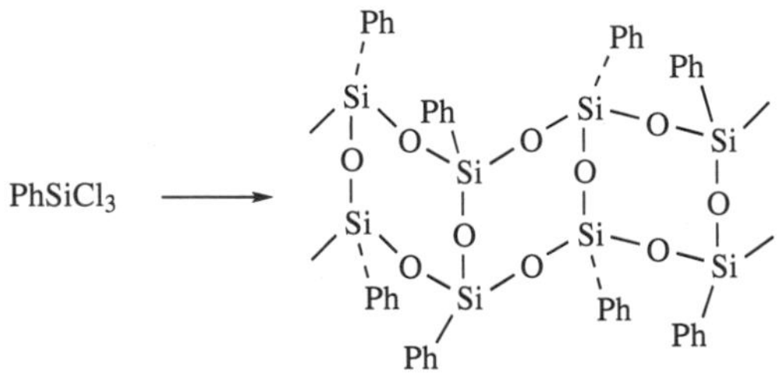

$PhSiCl_3 \longrightarrow$

A total of 300 g phenyltrichlorosilane, diluted with an equal volume of toluene, is run into water with stirring. After removal of the acid layer, the toluene solution of the hydrolysate is distilled to remove residual water and acid, along with 130 g toluene. Next, 0.14 g powdered potassium hydroxide is added, and refluxing and trapping of the evolved water are continued for 9 h. The solution is cooled and filtered to remove the crystalline octamer; and the filtrate is added to 1500 mL methanol to precipitate the 155 g phenylsilesquioxane prepolymer. This is low molecular weight cage-terminated "ladder" polymer with an intrinsic viscosity of 0.12 dL/g in benzene ($M_n = 14{,}000$).

Preparation of higher polymer requires equilibration at higher concentrations. Thus a mixture of 0.5 g of the above prepolymer (which still contains unneutralized base), 0.5 mL benzene, and 0.13 g Dowtherm A is heated for 1 h at 250°C in a loosely stoppered 16 × 150-mL test tube. The resulting tough, frothy mass is dissolved in benzene containing a drop of acetic acid and precipitated into excess methanol to give 0.33 g polymer with an intrinsic viscosity in benzene of 4 dL/g, ($M_w = 4$ million). The polymer is soluble in benzene, THF, and methylene chloride, and orientable films can be solvent cast. The polymer is of poor crystallinity and is not meltable.

8.9. CYCLIC PHOSPHORUS COMPOUNDS

Phosphonitrilic chloride is obtained by the reaction of phosphorus pentachloride and ammonium chloride. It does not exist in the monomeric state but is isolated in the form of ring compounds, predominently trimer and tetramer. The purified cyclic materials can be polymerized quite readily to linear polymer with the structure $\text{-[-N} = \text{P(Cl}_2\text{)–]}_n$, which is very similar in properties to ordinary rubber. It has been called inorganic rubber. Pieces of film can be stretched and will give a typical fiber diagram (74). The polymer is hydrolytically unstable, and unless particular care is taken, polymerization of the cyclic monomer leads to an insoluble cross-linked product.

The following procedure provides a polymer that is readily soluble in benzene under anhydrous conditions. This soluble polymer can be converted to hydrolytically stable products by treating it under anhydrous conditions with a variety of anions. A typical example is the reaction with trifluoroethoxyl anion.

176. Preparation of Phosphonitrilic Chloride (75)

$$NH_4Cl + PCl_5 \longrightarrow (PNCl_2)_3 + (PNCl_2)_4$$

To a solution of 450 g phosphorus pentachloride in 1200 mL dried sym-tetrachloroethane is added 140 g dry ammonium chloride. The mixture is refluxed for about 12 h, or until evolution of hydrogen chloride has ceased. A calcium chloride tube is then attached to the open end of the condenser, and the mixture is cooled. The residual ammonium chloride is filtered, and the solvent distilled under water pump vacuum, with the temperature not exceeding 50 to 60°C. The residual mass is then transferred to an open dish and allowed to cool. The oily material and traces of excess solvent are removed by suction on a funnel, and the product is then washed with a little 50% aqueous ethanol. The residual powder, which consists almost entirely of phosphonitrilic chloride trimer, is recrystallized from benzene for purification. Before polymerization, the monomer should be recrystallized an additional two times. The product melts at 114°C.

177. Polymerization of Phosphonitrilic Chloride (76)

$$(PNCl_2)_3 \longrightarrow [-PNCl_2-]_n$$

A total of 250 g phosphonitrilic chloride is purified by recrystallization from about 400 mL warm heptane (<75°C) after decolorization with activated charcoal. Then 116 g (0.33 mol) recrystallized phosphonitrilic chloride is evacuated in a contricted Pyrex tube, with intermittent melting at 114°C to remove occluded air; the tube is then vacuum sealed and immersed in a constant temperature bath at 250°C for 48 h. After this time, the product is a transparent, immoble rubbery material in which crystallization of residual trimer and other oligomers occurs slowly at 25°C. The tube is then wrapped in a towel and broken open. The polymer is cut into small pieces (0.25-in cubes), preferably in a stream of dry nitrogen, and is added to 500 mL dry benzene. It dissolves to a viscous, colorless solution after 24 to 48 h continuous agitation.

178. Preparation of a Modified Phosphonitrilic Chloride (76)

$$\left[-N=\underset{Cl}{\overset{Cl}{\underset{|}{\overset{|}{P}}}}-\right]_n \xrightarrow{CF_3CH_2ONa} \left[-N=\underset{OCH_2CF_3}{\overset{OCH_2CF_3}{\underset{|}{\overset{|}{P}}}}-\right]_n$$

The phosphonitrilic chloride polymer above (1.725 mol of $NPCl_2$) is dissolved in 100 mL dry benzene and treated with a solution of 3.45 mol sodium trifluoroethoxide in 1500 mL diethyl ether for 28 h at 57°C. The mixture is then neutralized to litmus with concentrated hydrochloric acid; and the precipitate is filtered, washed with methanol and then water, and dried. The product is a mixture of fully substituted oligomers and polymers. The high polymers are obtained by fractional precipitation of an acetone solution of the products into benzene to give 55 g poly(bis[trifluoroethoxy] phosphonitrile. The number average molecular weight is 90,000; the weight average by light scattering in ethyl trifluoroacetate is 1.7 million (±500,000); the intrinsic viscosity of the polymer in acetone at 30°C is 1.92.

8.10. CYCLIC POLYSULFIDES

8.10.1. Fluorinated Cyclic Polysulfides

Sulfur and tetrafluoroethylene (TFE) react in the vapor phase as shown in **8.4**. These monomers are easily ring opened to give copolymers of sulfur and TFE of high molecular weight.

$$F_2C=CF_2 + S_x \longrightarrow \text{cyclo-}(CF_2CF_2S_3) + \text{cyclo-}(CF_2CF_2S_4)$$

Schematic 8.4

179. Preparation of Tetrafluoro-1,2,3,4-Tetrathiane and Tetrafluoro-1,2,3-Trithiolane (77)

$$S_x + CF_2{=}CF_2 \longrightarrow$$

$$\text{(tetrafluoro-1,2,3-trithiolane)} + \text{(tetrafluoro-1,2,3,4-tetrathiane)} \longrightarrow$$

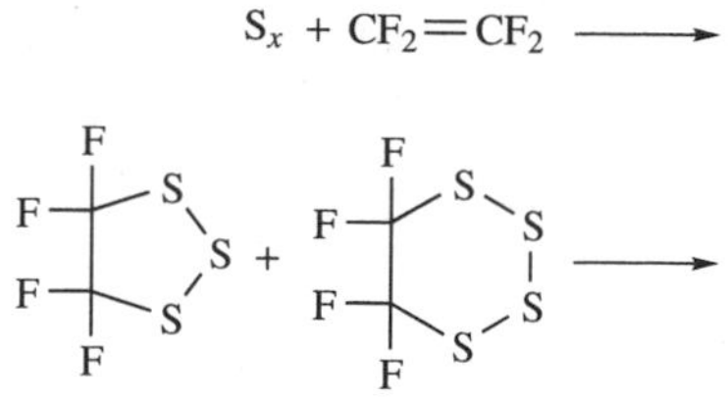

The preparation of cyclic polysulfides by reaction of TFE with the vapors of boiling sulfur at atmospheric pressure is carried out as follows. The preferred apparatus is a round-bottom glass reactor provided with an inlet tube to deliver the TFE into the sulfur vapors and an upright outlet neck (35 cm in length and 25 mm id) to serve also as a condenser for the sulfur vapors. A total of 900 g (28 g atoms) sulfur is placed in the reactor, blanketed with nitrogen, and heated to the refluxing point, about 445°C, at atmospheric pressure. Then 480 g (4.8 mol) TFE is passed through the sulfur vapors over a period of 4 h. Heating is regulated so that the temperature of the escaping reaction products at the head of the outlet tube is 250 to 300°C. The products are passed through an air-cooled downward condenser and collected in an acid washer receiver. Volatile products not condensed are collected in a trap cooled to −80°C. The volatile fraction consist of 60 g of a low boiling liquid, mainly thiocarbonyl fluoride (bp about −60°C, trifluorothioacetyl fluoride (bp about −25°C, bis(trifluoromethyl) disulfide and carbon disulfide (bp 29 to 31°C) for the azeotrope, and bis(trifluoromethyl)trisulfide (bp 84 to 85°C. The main product is 1098 g of a liquid containing some solid material. Rapid distillation of this condensate at <5 mm gives 813 g yellow oil. Fractionation through an acid-washed packed column into acid-washed receivers gives two products. The first is 97 g (10% yield) tetrafluoro-1,2,3-trithiolane (bp 26 to 32°C 15 mm), a yellow oil with a pronounced tendency to polymerize. The major product is 480 g (44% yield) of pale yellow tetrafluoro-1,2,3,4-tetrathiane (bp 59 to 61°C 15 mm) and mp 12.5°C, with a refractive index of 1.5447 at 25°C.

180. Polymerization of Tetrafluoro-1,2,3-Trithiolane (77)

$$\text{(tetrafluoro-1,2,3-trithiolane)} \longrightarrow [-CF_2CF_2SSS-]_n$$

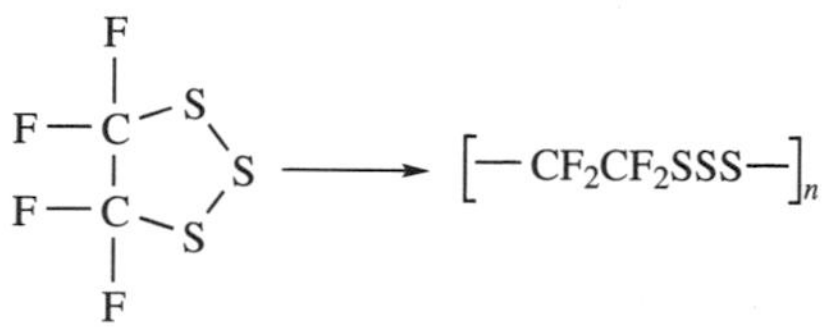

A solution of 164 g (0.84 mol) tetrafluoro-1,2,3-trithiolane in 210 mL of pentane is cooled to −80°C. To this solution is added a solution of 0.1 g trimethyl phosphite in 0.7 g pentane, and the mixture is allowed to stand for 1 h at −80°C.

The solvent is decanted, and the solid poly(tetrafluoro-1,2,3-trithiolane) that forms is dissolved in toluene and precipitated with pentane. A second precipitation, followed by washing with pentane yields 30 g (18%) of polymer with an inherent viscosity of 1.21 in 0.1% solution in toluene at 25°C. The polymer can be pressed to an opaque white film, which clarifies at 95 to 100°C.

181. Polymerization of Tetrafluoro-1,2,3,4-Tetrathiane (77)

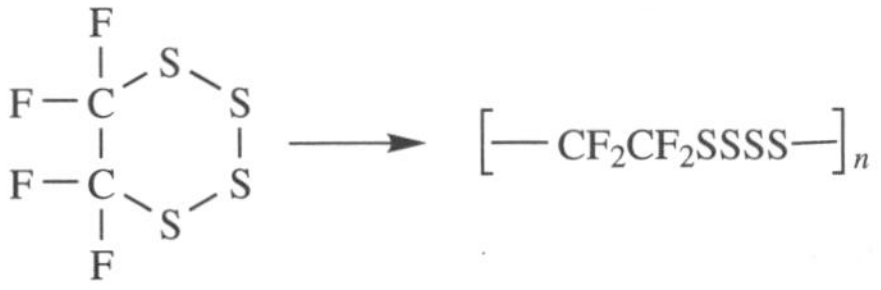

$$\text{(cyclic } F_2C{-}CF_2 \text{ with } S{-}S{-}S{-}S\text{)} \longrightarrow [-CF_2CF_2SSSS-]_n$$

Tetrafluoro-1,2,3,4-tetrathiane is polymerized by adding 200 g (0.88 mol) of the monomer over a 30-min period to 1200 mL of rapidly stirred acetonitrile maintained at −40°C. The precipitated polymer is isolated by filtration and washed successively with 300-mL portions of ethanol, ether, and pentane. The dried product weighs 170 g (85% yield). A 100-g portion of the polymer is dissolved in 400 mL chloroform and precipitated by pouring into a solution of 500 mL of vigorously stirred, cold (0°C) pentane. After washing well with pentane, the polymer is dried and weighs about 94 g (80%). The inherent viscosity in a 0.1% solution in toluene at 25°C is about 1.0. The crystalline melt temperature is 55 to 60°C. Tough, cold-drawable films can be obtained by pressing freshly prepared samples at 90 to 100°C.

8.10.2. Polymerization of Sulfur

Another inorganic cyclic monomer that opens to give linear polymer is sulfur (an eight-membered ring). This may be converted to high molecular weight in the order of 1.5 million. The melted sulfur can be drawn out into fiber with surprisingly good tensile properties, provided the melt is quenched rapidly in water. It will crystallize to give a typical fiber diagram. Unfortunately, the high polymer does not remain polymeric at room temperature, but reverts rapidly to the cyclic monomer and loses all its polymeric characteristics.

182. Polymerization of Sulfur (74, 78)

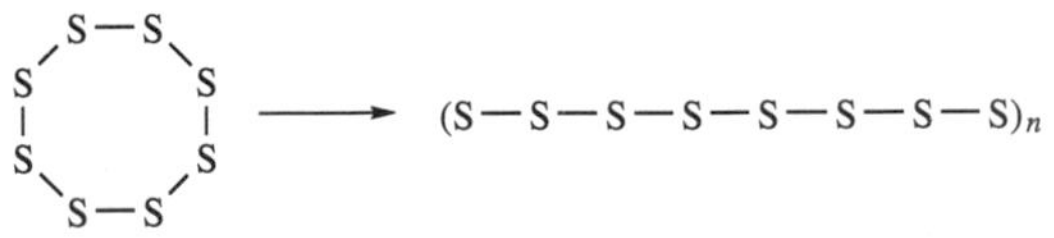

$$\text{cyclo-}S_8 \longrightarrow (S{-}S{-}S{-}S{-}S{-}S{-}S{-}S)_n$$

Ordinary rhombic sulfur is heated in a test tube gradually to 180°C. The rhombic material melts about 113°C, and at 180°C, the previously fluid material turns brown and becomes viscous. The viscosity reaches a maximum at about 187°C. Fibers may be drawn from the viscous mass at that temperature and after quenching have surprisingly good tensile strength.

As a variation on this experiment, a Carver press is heated to 180 to 187°C, and a lump of polymeric sulfur is prepared as above and quenched in ice water is immediately pressed into film. The film is removed from between the two sheets of aluminum used for pressing and quenched rapidly in an ice/water mixture. A tough, rubbery darkbrown film is obtained.

8.11. FREE-RADICAL RING-OPENING POLYMERIZATIONS

Although ionic ring-opening polymerizations of heterocyclic compounds are well-established (1, 2), free-radical ring openings are less common. Bailey et al. (2), however, demonstrated that ketene acetals, cyclic ketene aminals, cyclic vinyl ethers, unsaturated spiro orthocarbonates, and unsaturated spiro orthoesters all undergo free-radical ring-opening polymerizations. These can be copolymerized with a wide variety of vinyl monomers to introduce ester, amido, keto, or carbonate functions into the backbone of addition polymers. The free-radical ring-opening process is illustrated in **8.5**.

$$\xrightarrow{R^\bullet} -\!\!\left(CH_2\overset{O}{\overset{\|}{C}}-O-CH_2\underset{Ph}{\underset{|}{C}H}\right)_{\!n}-$$

Schematic 8.5

183. Preparation of 2-Chloromethyl-1,3-dioxepane (2)

A solution of 50 g (0.4 mol) chloroacetaldehyde dimethyl acetal and 36 g (0.4 mol) 1,4-butanediol is heated at 115 C with 0.5 g Dowex (H^+) resin in a 200-mL flask equipped with a 10-cm. Vigreux column. When the calculated amount of methanol (about 40 mL) has been collected over about 3 h, the resin is filtered, and the filtrate liquor fractionally distilled at 5 mm to give about 40 g of the dioxepane (bp 74 to 75°C at 6 mm).

184. Polymerization of 2-Methylene-1,3-dioxepane (2)

Polymerization of 2-Methylene-1,3-Dioxepane

$$=CH_2 \longrightarrow -CH_2-\overset{O}{\overset{\|}{C}}-O-CH_2-CH_2-CH_2-$$

A solution of 24 g of 2-chloromethyl-1,3-dioxepane and 60 g KOH in 40 mL hexadecene is heated at 130°C for 12 h. The product is distilled from the mixture under partial vacuum to give a liquid (bp 40 to 45°C at 5 mm). Further purification by redistillation over metallic sodium through a 10 mm Vigreux column gives about 12 g 2-methylene-1,3-dioxepane (bp 42 to 43°C at 5 mm).

In a dry 10 × 40-mm-thick-walled polymer tube flushed nitrogen is added 1.5 g of the above methylene dioxepane and 2 mol% of azoisobutyronitrile (AIBN) initiator. The tube is flushed with nitrogen, sealed, and heated in an oil bath at 50°C for about 10 h. The tube is then cooled and broken open. The polymeric product is extracted with chloroform. The solution was added to excess hexane to precipitate about 1 g of a waxy or rubbery product. The linear ester structure is confirmed by and nuclear magnetic resonance (NMR). Di-*t*-butyl peroxide can be used as initiator instead of the AIBN, but the temperature of polymerization must be raised to 125°C. Copolymerization of the above methylene dioxepane with vinyl monomers is readily accomplished.

184. Copolymerization of 2-Methylene-1,3-dioxepane with Styrene (2)

$$\text{(2-methylene-1,3-dioxepane)}=CH_2 + C_6H_5-CH{=}CH_2 \longrightarrow -\!\left[CH_2-\underset{\underset{O}{\|}}{C}-O-CH_2-CH_2-CH_2-\underset{\underset{C_6H_5}{|}}{C}-\right]_n$$

In a 10 × 40-mm-thick-walled glass polymer tube under nitrogen is added 1.2 g dioxepane, 1.09 g freshly distilled styrene (removes inhibitor), and 2 mol% of di-*t*-butyl peroxide. The tube is sealed under nitrogen and heated in an oil bath at 120°C for 36 h (**Shield**). The tube is removed from the bath, cooled, and broken open. The polymer is dissolved in chloroform (**toxic**), and the solution poured into excess methanol to precipitate the polymer. The polymer is filtered and dried in-vacuo at 70°C to give about 1.5 g of a white powdery copolymer. NMR and IR confirm the copolymer structure.

REFERENCES

1. K. J. Ivin and T. Saegusa, eds. *Ring Opening Polymerization*, 2 Vols., Elsevier Applied Science, London, 1984. D. J. Brunelle, *Ring Opening Polymerization*, Hanser, Manich, Germany, 1993
2. W. J. Bailey et al., *J. Poly. Sci. A*, **20**, 3021 (1982); *Polym. J.*, **17**, 85 (1985); and *Ind. Eng. Chem.*, **50**, 8 (1958).
3. H. K. Hall Jr. and P. Ykman, *Macromol. Rev.*, **11**, 1 (1976).
4. H. K. Hall Jr. et al., *J. Am. Chem. Soc.*, **80**, 6404–6428 (1958).
5. W. Sweeny and J. Zimmerman in *Encyclopedia of Polymer Science and Technology Polyamides*, Vol. 10, John Wiley & Sons, New York, 1909.
6. H. K. Hall Jr. *J. Am. Chem. Soc.*, **82**, 1209 (1960).
7. R. F. Moore, *Polymer*, **4**, 493 (1963). P. W. Allen, *Techniques of Polymer Characterization*, Academic Press, New York, 1959. A Weissberger, Ed. *Techniques of Organic Chemistry*, Vol. 1, Wiley-Interscience, New York, 1949.
8. O. Wichterle, et al., *Adv. Poly. Sci.*, **2**, 578 (1961). 2,739,959 (March 27, 1956) W. O. Ney Jr. and M. Crowther. U.S. Pat.

9. O. Wichterle et al., *Macromol. Chem.*, **35**, 174 (1960).
10. W. Memeger et al. *Macromolecules*, Vol. 29, No. 20, 6475 (1996); U.S. Pat. 5,399,662 (March 21, 1995), W. Memeger, to DuPont.
11. R. Schwesinger et al. *Angew. Chem. Int. Educ. Engl.*, **32**, 1361 (1993). et al., *J. Am. Chem. Soc.*, **115**, 5015 (1993).
12. D. D. Coffman et al., *J. Poly. Sci.*, **3**, 85 (1948).
13. U.S. Pat., 2,647,105 (July 28 1953), H. R. Mighton, to DuPont.
14. U.S. Pat., 2,638,463 (May 12 1953), W. O. Ney, W. R. Nummy and C. E. Barnes, to DuPont.
15. W. E. Hanford and R. M. Joyce, *J. Poly. Sci.*, **3**, 167 (1948).
16. P. H. Hermans et al., *J. Poly. Sci.*, **30**, 81 (1958).
17. H. K. Reimschuessel, *J. Poly. Sci. Macromol. Rev.*, **12**, 65 (1977).
18. U.S. Pat., 2638463 (May 12 1953), W. O. Ney et al., to DuPont.
19. R. Graf et al., *Angew. Chem. Int. Educ.*, **1**, 481 (1962).
20. E. Testa and L. Fontanella *Ann.*, **625**, 95 (1959).
21. Germ. Offen. 1911834 (1970) and 2142757 (1972), C. E. Barnes, H. Sekiguchi et al., *Eur. Poly. J.,* **10**, 1185 (1972). K. Dachs and E. Schwartz, *Nuova Chim.,* **48**, 43 (1972).
22. H. Rinke and E. Istel in-**Houben-Weyl** ed., page 118 Editor, Eugen Miller
23. A. D. Bliss et al., *J. Org. Chem.*, **28**, 3537 (1963).
24. U.S. Pat., 2,983,713 (May 9, 1961), W. H. Libby, to 3 M Co.
25. U.S. Pat., 3,216,980 (November 9, 1965 to 3 M Co. 1965), C. J. Berg.
26. M. Szwarc, *Adv. Poly. Sci.*, **4**, 1 (1965).
27. D. Coleman, *J. Chem. Soc.*, **1950**, 3222 (1950).
28. D. Colman and A. C. Farthing, *J. Chem. Soc.*, **1950**, 3218 (1950).
29. A. C. Farthing, *J. Chem. Soc.*, **1950**, 3213 (1950).
30. C. H. Bamford, A. Elliot and W. E. Hanby, *Synthetic Polypeptides*, Academic Press, New York, 1956. S. Sugai et al., J. *Poly. Sci. A*, **4**, 183 (1966).
31. U.S. Pat., 2,811,511 (October 29, 1957) T. Alderson, to DuPont.
32. E. E. Blaise and M. Montagne, *Compt. Rendu.*, **174**, 1173, 1553 (1922).
33. R. Thiebaut et al., *Ind. Plastiques Mod. (Paris)*, **14**, 13 (1962).
34. Can. Pat. 549347 (1957), R. J. W. Reynolds.
35. A. Mooradian and J. B. Cloke, *J. Am. Chem. Soc.*, **67**, 942 (1945).
36. E. Testa et al., *Annalen.*, **639**, 166 (1961).
37. R. Blume, *Macromolecular Synth.*, **7**, 39. U.S 347145 (Oct 7, 1969) to DuPont.
38. U.S. Pat. 3,367,921 (February 6, 1968), W. Sweeny and D. J. Casey, to DuPont.
39. U.S. Pat. 2,668,162 (Feb. 2, 1954), C. E. Lowe.
40. N. Spassky et al, *Macromol. Chem. Phys.*, **197**, 2627 (1996).
41. G. W. Coates and T. M. Ovitt, *J. Am. Chem. Soc.*, **121**, 4072 (1999).
42. G. L. Baker, M. R. Smith III, and C. P. Radano, *J. Am. Chem. Soc.*, **122**, 1552 (2000).
43. D. J. Brunelle et al., *J. Am. Chem. Soc.*, **12**, 2399 (1990) and *Poly. Prep.* **30**, 569 (1989).
44. Brit. Pat. 613280. (Nov. 24, 1948) to Marx & Clerk to Wingfoot Corp.

45. R. S. Bly et al., *J. Am. Chem. Soc.*, **44**, 2899 (1922).
46. T. Mukiyama et al., *Bull. Soc. Chem. Japan*, **35**, 687 (1962).
47. T. Mukiyama et al., *J. Org. Chem.*, **27**, 3337 (1962).
48. U.S. Pat., 2,856,370 (October 14 1958), E. L. Muetterties, to DuPont.
49. E. L. Wittbecker, paper presented at 129th meeting of American Chemical Society, Dallas, April, 1956.
50. E. L. Wittbecker, H. K. Hall, and T. W. Campbell, *J. Am. Chem. Soc.*, **82**, 1218 (1960).
51. V. Drägujan et al. *Olefin Metathesis and Ring Opening Polymerizations of Cyclic Olefins*, Wiley Interscience, New York, 1995. K. J. Ivin *Olefin metathesis*, Acadamic Press, London, 1983.
52. F. N. Hill, F. E. Bailey Jr., and J. T. Fitzpatrick, *Ind. Eng. Chem.*, **50**, 5 (1958).
53. K. L. Smith and R. Van Cleve, *Ind. Eng. Chem.*, **50**, 12 (1958).
54. U.S. Pat., 2,706,181 (April 12 1955), M. E. Pruitt and J. M. Baggett, to Dow Chemical Co.
55. U.S. Pat., 2,706,189 (April 12 1955) M. E. Pruitt and J. M. Baggett, to Dow Chemical Co.
56. U.S. Pat. 2,811,491 (October 1957), M. E. Pruitt and J. M. Baggett, to Dow Chemical Co.
57. U.S. Pat. 2,455,912 (December 14 1948) T. L. Cairns and R. M. Joyce, to DuPont.
58. K. T. Garty et al., *J. Poly. Sci.*, **1**, 85 (1963).
59. T. B. Gibb Jr. et al., *J. Poly. Sci. A*, **4**, 917 (1966).
60. A. Noshay and C. C. Price, *J. Poly. Sci.*, **34**, 165 (1959).
61. E. J. Vandenberg, *J. Poly. Sci.*, 47, 486 (1960).
62. J. B. Rose, *J. Chem. Soc.*, **1956**, 542, 546 (1956).
63. T. W. Campbell, *J. Org. Chem.*, **22**, 1029 (1957).
64. Y. Etienne, *Ind. Plastique Mod.*, **9**, 37 (1957).
65. V. S. Foldi and W. Sweeny, *Macromol. Chem.*, **72**, 208 (1964).
66. D. G. Hummel et al., *J. Am. Chem. Soc.*, **82**, 2928 (1960).
67. S. Searles Jr. and E. F. Lutz, *J. Am. Chem. Soc.*, **80**, 3168 (1958).
68. J. F. Brown Jr. et al., *J. Am. Chem. Soc.*, **86**, 1120 (1964).
69. G. Natta et al., *Angew. Chem. In. Educ.*, **3**, 723 (1964). Fr. Pat. 1,394,380.
70. W. Patnode and D. F. Wilcock, *J. Am. Chem. Soc.*, **68**, 358 (1946).
71. U.S. Pat. 2,634,252 (April 7 1953), E. L. Warrick to Dow Corning.
72. J. F. Brown Jr. et al., *J. Am. Chem. Soc.*, **86**, 1120 (1964).
73. J. F. Brown Jr. *J. Poly. Sci. C*, **1**, 83 (1963).
74. K. H. Meyer, *Natural and Synthetic High Polymers*, Wiley-Interscience, New York, 1950.
75. C. J. Brown, *J. Poly. Sci.*, **5**, 465 (1950).
76. H. R. Allcock and R. L. Kugel, *J. Am. Chem. Soc.*, **87**, 4216 (1965).
77. C. G. Krespan and W. R. Brasen, *J. Org. Chem.*, 27, 3995 (1962).
78. A. V. Tobolsky and A. Eisenberg, *J. Am. Chem. Soc.*, **81**, 780 (1959).

9

ADDITION POLYMERIZATION

9.1. GENERAL CONSIDERATIONS

Addition polymerization was initially discussed in general terms in chapter 1. Here we will add more detail and experimental examples. *Addition* is used because in most instances this polymerization process involves the joining, or addition, of successive monomers to the end of a chain without the need for elimination of a small by-product molecule as occurs in condensation polymerization. The term *chain growth* has also been applied to this format of polymerization. The application of these adjectival terms has been almost exclusively used for the coming together of unsaturated molecules, e.g., olefins and dienes, typically, but strictly speaking it would also seem to apply to ring-opening polymerizations, by which the cyclic monomers are also "added" to a growing chain without the need for elimination. The convention is to consider such cases separately, as special classes of polymerization, because their chemistry is formally more exotic than simply joining unsaturated monomers, although the mechanisms and catalysts used in the polymerization of unsaturates can be brilliantly exotic. We have elected to stay with the conventions, as evidenced by the presence chapter 8 on ring-opening and other "chain growth" systems that arbitrarily or otherwise seem worthy of a special consideration.

The simplest presentation of an addition polymerization is the following, involving an appropriate vinyl monomer:

$$CH_2{=}CHR \longrightarrow -(CH_2{-}CHR)-$$

Nothing is said of mechanism in the foregoing, a matter that is fairly vital to understanding and working in the field, because obviously not all such prototype monomers are going to be polymerizable through the same mechanism, and thus initiation process, or known at any given moment to even be polymerizable at all by any known procedure. Resolving how to polymerize a hitherto refractory addition monomer is one of the allures of the discipline. Vinylic and dienic monomers are among the oldest category in polymer chemistry and so have a prodigious history. References 1–5 should be consulted for some of this background, especially the mechanistic elements.

The classic mechanistic steps for addition polymerization are

1. Initiation:

$$\underset{}{X^*} + \underset{\text{I}}{CH_2{=}CHR} \longrightarrow \underset{\text{II}}{XCH_2{-}CHR^*}$$

2. Propagation:

$$\text{II} + n\text{I} \longrightarrow \underset{\text{III}}{X(CH_2{-}CHR)_{n+1}{}^*}$$

3. Termination:

$$\text{III} \longrightarrow \text{Inactive polymer chain}$$

The details of each of these steps are different for the four broad classes of initiators, and it is the latter that determines the character of the active center * as a free radical, cation, or anion, or the complex metallic catalysts termed Ziegler–Natta or coordination catalysts (4, 6).

The terms *initiator* and *catalyst* have fairly distinct, if hardly precise, meanings as applied to polymerization. An initiator starts the growth of a chain by becoming part of it. Growth continues at an active site in which the initiator no longer plays a role, as the above reaction sequences show. A catalyst starts the growth of many chains, usually, and participates in the addition of each monomer to a growing chain, i.e., the catalyst is literally the active site or an integral part of it. Free-radical polymerizations invariably require initiators, and coordination catalysts almost certainly are involved in each monomer placement. Anionic processes are almost always said to use appropriate initiators; but the situation is complicated by the fact that the anion may initiate and the counterion may guide the steric placement of each monomer at the active site. In radical polymerizations, chain growth moves away from the initiating entity; whereas in coordination systems, the chain effectively grows away from the active site. Anionic polymerization in a sense splits the difference, in that the initiating entity remains behind at the chain end while growth continues, but if the counterion is playing a steric role at the active site, chain growth resembles the coordination case, with the chain growing away from the sterically guiding ion. These characterizations are generalizations; they have some conceptual value in terms of the design and choice of initiators and catalysts.

Few monomers are polymerizable by more than one or two initiator or catalyst categories; and when they are, it is often not with equal effectiveness in terms of rate, molecular weight, and process conditions. The susceptibility of a monomer to each type of initiation is determined by its structure; and as in virtually all organic reactions, the influences are steric and electronic. Generally, the more hindered an unsaturated compound, the less likely it is to polymerize easily with itself or at all. Sometimes a monomer will copolymerize with certain others, yet be extremely difficult to homopolymerize. Some substituents around the unsaturated center(s) contribute to the stabilization of certain active centers but not others, so the monomer will polymerize under some kinds of initiation but not others. Styrene, for example, polymerizes well under free-radical and anionic influence but less so under cationic. Methyl methacrylate responds nicely to free-radical and anionic initiators but essentially not at all to cationic; isobutylene responds to the latter and practically not at all to the others. *N*-Vinylcarbazole polymerizes cationically and via free radicals. The degree of reactivity of a monomer can be intuitively assessed by considering the classical extent of resonance stabilization available to an active center. Styrene is likely to be more reactive in free-radical polymerization than vinyl chloride or vinyl acetate, although all three readily homopolymerize. But copolymerization among any two of the three becomes far more complex in terms of reactivity. The so-called Q-e scheme developed from copolymerization studies (5) illuminates nicely the varying reactivities of monomers.

It must be kept in mind that "to polymerize" in practical terms means that the initiation reaction occurs at acceptable rates and the subsequent propagation reaction is considerably faster than any of the competing termination reactions. The ultimate of the competition is the case in which no termination reactions occur at all, giving the remarkable event of living polymer chains, although it does not mean that the overall polymerization is particularly fast.

Termination processes (7) wherein polymer chains cease to have centers for active growth occur in many ways, depending on the specific monomers and polymerization type. Some of these are shown below.

9.1.1. Free-Radical Process

1. Combination of two radicals:

$$2 \sim CH_2CHR^\bullet \longrightarrow \sim CH_2CHRCHRCHCH_2 \sim$$

2. Disproportionation of two radicals:

$$2 \sim CH_2CHR^\bullet \longrightarrow \sim CH_2CH_2R + \sim CH_2CH{=}CHR$$

3. Chain transfer:

$$\sim CH_2CHR' + R'SH \longrightarrow \sim CH_2CHR + R'S'$$

$$R'S^\bullet + CH_2{=}CHR \longrightarrow R'SCH_2CHR^\bullet$$

$$\downarrow$$

continued propagation

Shown above is chain transfer to an added mercaptan; but transfer to monomer, solvent, and dead polymer chains atoms are other possibilities. Note that inhibition is the consequence when chain transfer terminates a chain, but the resulting transfer radical is largely incapable of effective reinitiation.

9.1.2. Anionic Processes

The principal mode of termination is abstraction of a hydrogen relatively acidic to the anion chain end; such proton sources are usually present as an impurity of the system. If the monomer can be initiated by the resulting transfer anion, chain transfer can be said to have occurred. The presence of small amounts of alcohol in the anionic polymerization of formaldehyde may be such a case.

$$\sim CH_2O^- + ROH \longrightarrow \sim CH_2OH + RO^-$$

Termination may also occur by addition of the propagating anion to any center susceptible to such attack, to an ester group, as in the intramolecular termination that occurs in the anionic polymerization of methyl methacrylate with phenyl magnesium bromide (8).

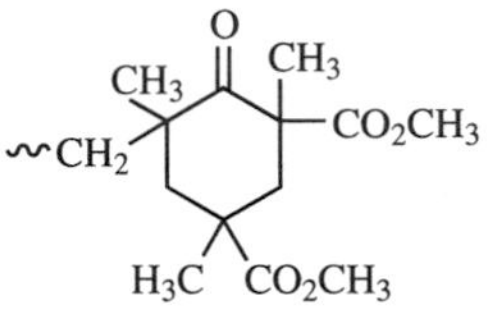

O
CH_3 CH_3
CO_2CH_3
$\sim CH_2$
H_3C CO_2CH_3

A remarkable aspect of anionic polymerization is that in the absence of built-in addition or transfer reactions (especially those due to impurities such as water), termination may not occur at all, so that the chain remains in a living state when monomer is consumed. On addition of more or different monomer, propagation continues (9–11).

9.1.3. Carbocationic Process

1. Elimination:

$$\underset{\text{I}}{\sim CH_2(CH_3)C^+A^-} \longrightarrow \sim CH_2C(CH_3){=}CH_2 + HA$$

2. Chain transfer to monomer:

$$\text{I} + CH_2{=}C(CH_3)_2 \longrightarrow \sim CH_2C(CH_3){=}CH_2 + \text{I}$$

3. Reaction with anion:

$$\sim CH_2CHPhH^+CF_3CO_2^- \longrightarrow \sim CH_2CPhH{-}OCOCF_3$$

The polymerization of the cyclic ether tetrahydrofuran (THF) by a carbocationic or, possibly more correctly, an oxonium ion process is an example of a living polymer with a positively charged active chain end that does not terminate under the proper conditions (12). THF polymerization is considered in chapter 8.

Termination processes in coordination polymerization are probably more clearly visualized than, say, the initiation and propagation steps. Evidence exists to show that, in at least some instances, the growing chains are long lived and might be considered as living polymers.

The kinetics of addition polymerization by all four processes (radical, anionic, cationic, coordination (Ziegler–Natta)) with a variety of monomers have been extensively studied, These are thoroughly discussed in books and publications and are essential to a good insight into the mechanisms involved. It is beyond the the purpose of this discussion to go further than the introductory treatment already accorded. The preparative examples of this chapter will embody in concrete experiment most of the principles involved.

9.2. CONJUGATED DIENES

Although most of the preceding section focused on vinyl compounds, essentially the same principles apply to conjugated dienes (13–16). The major difference between the two classes of unsaturates is in the greater structural variation possible in diene polymer, where in the simplest case — butadiene — 1,2-; 1,4-cis; and 1,4-trans additions are possible in each monomer added to the chain. With more comples dienes, e.g., isoprene and 1,3-pentadiene, even more variation can occur. A major accomplishment of synthetic polymer chemistry is the capability of achieving remarkably uniform monomer addition patterns through apt choices of initiator/catalyst and reaction conditions.

The foregoing makes clear that in conjugated dienes the double bonds usually react in concert. For 1,4-addition this is the obviously the case for valid organic chemical reasons. But in nonconjugated dienes, the double bonds react in most instances independently of each other. Glycol dimethacrylate produces a cross-linked resin, as an example. But exceptions exist, such as the cyclopolymerization of nonconjugated but appropriately structured dienes, as described in chapter 12.

9.3. COPOLYMERS

Copolymers are the result of putting two or more unsaturated monomers in a macromolecule. The objective is usually to achieve a modified set of physical properties, not achievable in economic fashion, or at all, from a homopolymer. For example, poly(vinyl chloride) can be made more easily soluble and lower softening by copolymerization with vinyl acetate, polyacrylonitrile more dyeable with vinyl pyridine, and polyisobutylene sulfur-vulcanizable with isoprene.

Copolymerization has often been seized on as a panacea for any adverse property of a homopolymer, but the reality is usually that one or more valuable properties of the homopolymer are compromised, e.g., a reduction of brittleness may also be accompanied by a decrease in tensile strength, chemical resistance, surface hardness, and melting point, not all or any of which may be tolerable. The synthetic polymer chemist must decide, usually in collaboration with nonchemists who must market the product, which losses and how much can be tolerated. The effect of copolymerization on polymer properties is perhaps seen fundamentally in its effects on crystallinity and glass-transition temperature T_g. Because the development of crystallinity is due to molecular symmetry and/or intermolecular association, it is not surprising that simple copolymerization should reduce or obliterate an ability of polymer chains to crystallize. Similarly, the T_g in amorphous and crystalline polymers is modified by copolymerization, but in a different way. The general tendency is for T_g in a copolymer to lie between the glass-transition temperatures of the parent homopolymers; and for random (further defined below), a weighted contribution of each parent is sometimes observed, according to the relationship

$$1/T_g = W_1/T_{g1} + W_2/T_{g2}$$

where T_{g1} and T_{g2} are in degrees absolute (K) and W_1 and W_2 are the weight fractions of the corresponding monomers in the copolymers. The problem of how to achieve coplymerization and still maintain the best of two or more worlds has been attacked through synthetic methods designed to prepare copolymers in four basically distinct categories. For monomers A and B these can be described as follows:

1. Random:

$$\sim\text{ABBAABBBBABAAB}\sim$$

2. Alternating:

$$\sim\text{ABABABABABABAB}\sim$$

3. Block:

$$\sim\text{AAAAABBBBBBBBBBAAAA}$$

4. Graft:

```
AAAAAAAAAAAAAAAAAA
      |      |      |
      B      B      B
      B      B      B
      B      B      B
      B      B      B
```

All of the above are idealized examples; in particular, random copolymers are impossible to truly represent, because, well, they are so random. Sequence length distribution for each monomer becomes an important concept in dealing with their structures.

9.3.1. Random

The preparation of random copolymers is the most superficially straightforward. Monomers A and B are mixed, and the catalyst is added. In most cases, A and B will copolymerize at different rates, so that the composition of a copolymer isolated before polymerization is complete will depend on the relative reactivity of A versus B. This has led to the concept of reactivity ratios (5), in which monomer pairs are compared in their reactivity toward each other. They are defined as the ratios of relative reaction rates of the monomers toward active sites on the growing chain ends. For the monomer pair M_1 and M_2, a growing chain end can be of two types and reaction is possible for each of them with either monomer, leading to four reaction rate constants. This is shown below for a free radical system, where the reactivity ratio concept has been applied.

$$\sim M_1\cdot \xrightarrow[k_{11}]{M_1} \sim M_1{-}M_1\cdot$$

$$\sim M_1\cdot \xrightarrow[k_{12}]{M_2} \sim M_1{-}M_2\cdot$$

$$\sim M_2\cdot \xrightarrow[k_{22}]{M_2} \sim M_2{-}M_2\cdot$$

$$\sim M_2\cdot \xrightarrow[k_{21}]{M_1} \sim M_2{-}M_1\cdot$$

The constant k_{11} is for reaction of terminal radical M_1 with monomer M_1, etc. The reactivity ratios, r_1 and r_2, are defined as

$$r_1 = k_{11}/k_{12}$$

$$r_2 = k_{22}/k_{21}$$

which is to say, r_1 and r_2 show the preference of a given radical toward its own monomer compared to the second monomer. A high value of r_1 means that M_1

prefers to react with M_1 and a low value means a preference for M_2. If r_1 is very high and r_2 is very low for a monomer pair, it will be difficult to obtain copolymers of uniform composition by simply charging both monomers all at once; rather, a gradual addition of the more reactive monomer will be necessary. An example is the staged copolymerization of vinyl chloride and vinyl acetate (prep. 191.)

Several methods are known for determining r_1 and r_2 valuations, and extensive compilations exist (5, 17). Once obtained, they can be used with the instantaneous copolymer compositions equation to estimate the monomer feed composition needed to get a certain copolymer composition, when the effect of conversion is taken into account. In one of its forms, the equation is:

$$F_1 = \frac{r_1 f_1^2 + f_1 f_2}{r_1 f_1^2 + 2 f_1 f_2 + r_2 f_2^2}$$

where f_1 is the mole fraction of M_1 in the copolymer being formed at any instant, f_1 and f_2 are mole fractions of monomers M_1 and M_2 in the feed at that instant, and r_n is the reactivity ratio for the appropriate monomer. Knowing each r value, each F can be calculated for various values of the two f values. With the two r values known and with given f values in the initial feed, the values of the two Fs can be obtained for various conversions of the two monomers. The computer allows rapid solution of the equation at selected small increments of conversion and the correspondingly altered values of the two fs at these instants.

If polymerization is allowed to go to completion from a single initial monomer feed, the copolymer will have an overall composition corresponding to the original ratio of the monomers, but the polymer composition formed toward the end will be richer in the slow component, the faster having been largely consumed. At any instant, the composition of polymer being formed is necessarily different from that being formed at any other instant, as described above, except for the situation in which r_1 and r_2 are both less than *1* and $f_1 = (1 - r_2)/(2 - r_1 - r_2)$ and $f_1 = f_2$ at every instant.

Frequently, a monomer that doesn't ordinarily homopolymerize will be found to copolymerize because of favorable electronic effects (see maleic anhydride with styrene, below) or because steric restraints are less in copolymerization than homopolymerization; r_2 in such a case may be zero. On the other hand, two easily homopolymerizable monomers will occasionally copolymerize with great difficulty or not at all. Such a case is that of vinyl acetate and styrene; a small amount of styrene will essentially inhibit polymerization. Because styrene is so reactive, it rapidly captures most of the radicals generated; but the styrene radical is too stable to add easily to the sluggishly reactive vinyl acetate. Thus polymerization dies aborning as styrene radicals are eventually terminated at a low degree of polymerization. Butadiene inhibits vinyl chloride in the same way.

9.3.2. Alternating Copolymers

Alternating copolymers represent a special case and require that each monomer be more reactive toward the other species than toward its own kind. Such copolymers result when r_1 and r_2 are both extremely low. Thus styrene radical has a great affinity for maleic anhydride, but

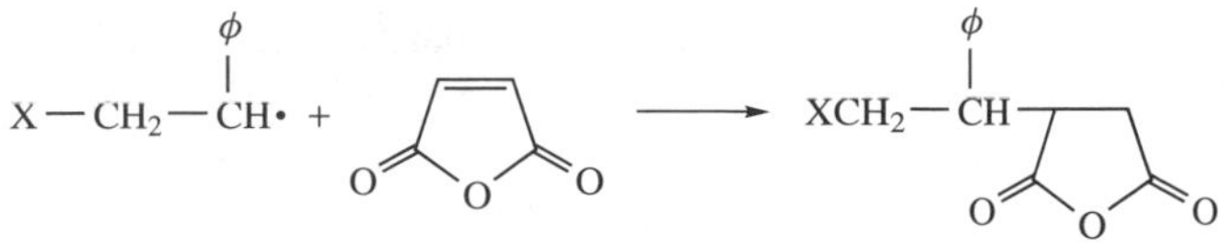

$$X-CH_2-\underset{}{\overset{\phi}{\overset{|}{C}}}H\cdot + \text{(maleic anhydride)} \longrightarrow XCH_2-\overset{\phi}{\overset{|}{C}}H-\text{(succinic anhydride ring)}$$

the maleic anhydride radical will not add to itself easily; hence a styrene molecule is added next to the growing chain, i.e., r_1 (for styrene) is low and r_2 (for maleic anhydride) is essentially zero in the presence of styrene.

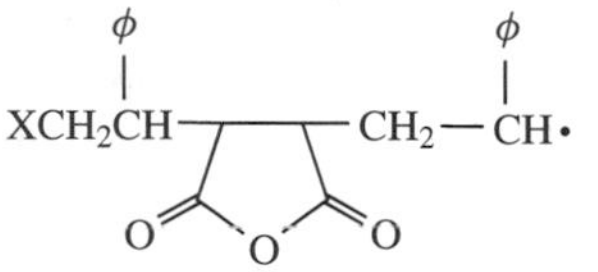

$$XCH_2\overset{\phi}{\overset{|}{C}}H-\text{(succinic anhydride ring)}-CH_2-\overset{\phi}{\overset{|}{C}}H\cdot$$

This alternation continues until the chain is terminated (5, 18).

Another particularly interesting and unusual case of alternation occurs during the copolymerization of terminal olefins with sulfur dioxide (19–23):

$$XCH_2-CH_2\cdot + SO_2 \longrightarrow XCH_2CH_2SO_2\cdot \longrightarrow XCH_2CH_2SO_2CH_2CH_2\cdot$$
$$\longrightarrow XCH_2CH_2SO_2CH_2CH_2SO_2\cdot,\ \text{etc.}$$

An olefin—SO_2 complex has been suggested as an intermediate (24, 25).

9.3.3. Block and Graft Copolymers

The synthesis and characterization of block and graft copolymers has been reviewed (26–29); and while special synthetic techniques have been worked out, in many cases it is difficult to prepare pure graft or block copolymers without contamination with homopolymers corresponding to both monomers. In all but a few favorable instances, removal of homopolymer by extraction or fractionation is difficult and time-consuming. In practical cases, the polymer chemist must design the block or graft system so that he or she can live with a mixture of products and yet derive benefit from the presence of the block or graft component. A case in point is the preparation of an acrylonitrile-butadiene-styrene (ABS) polymer (prep. 190), by copolymerizing styrene and acrylonitrile in the presence of polybutadiene. In this particularly complex system, the net result is a mixture of an essentially random styrene-acrylonitrile copolymer (the major constituent) and a graft of the same random copolymer to a backbone of polybutadiene; there is probably a small amount of ungrafted polybutadiene also (30).

There are practical reasons for attempting to make block and graft copolymers apart from the satisfying element of simply finding a new way to do so. They offer perhaps the best way of realizing a combination of the most desirable properties of two or more homopolymers with a minimum of loss in these properties. In the ABS case just noted, the stiffness, creep resistance, tensile strength, and chemical resistance of poly(styrene-co-acrylonitrile) are all maintained at an acceptable level, but combined with a useful part of the good impact strength and low-temperature properties of polybutadiene. Likewise, when propylene is polymerized by certain coordination catalysts with intermittent admission of ethylene to the system in the absence of propylene (31), a certain amount of block copolymer is formed containing separate propylene and ethylene segments. Only a few percent of total ethylene in block form serves to improve the impact strength over polypropylene, especially at low temperatures.

Another example is the block copolymer formed by butyl lithium–initiation of butadiene, or isoprene, which, when exhausted, is followed by addition of styrene, and again by diene (32, 33). This sandwich of diene-styrene-diene blocks exhibits fully elastic behavior without the need for vulcanization and is thus a thermoplastic rubber. The preparative method used is based on the living polymer technique developed originally by Szwarc (10, 11), which depends on the existence of a nonterminating anionic system. Examples are found in the experimental part of this chapter.

Grafting depends on creating active sites on the backbone of the substrate polymer. These may be radical in nature and originate from preirradiation of the substrate polymer or they may occur via chain transfer to the substrate during polymerization of the grafting monomer in its presence. The efficiency of grafting is usually much greater for the former method. Oxidation of carbon atoms in the backbone of poly A can lead to hydroperoxide groups that can, on heating in the presence of monomer B, initiate polymerization of B from poly A. This has been done with polypropylene as the backbone polymer. Grafting can also occur through reaction with functional groups, e.g., the addition of living polystyrene anion to a carbonyl in poly(methyl methacrylate); the presence of acryloyl chloride units in the substrate polymer allows grafting of a condensation polymer in a low-temperature system.

What block and graft copolymers seem to do is allow the essential homopolymer properties of each constituent to be developed. The interruption of the polymer chain by a comonomer and the consequent loss of certain properties are markedly reduced or absent; yet, the comonomer, as homopolymer, exerts its own effect on the system because its own properties are developed in the same way and are ineradicably bound into the total system by covalent linkages. But it should be understood that homopolymer properties are not fully maintained. The value of these copolymers lies in the attractive compromises they provide, in the best cases allowing wholly new classes of materials.

9.4. FREE-RADICAL POLYMERIZATION OF VINYL MONOMERS — EXPERIMENTAL METHODS

The number of examples that could be included in this section is quite large, because each of the many available monomers usually may be polymerized in a variety of ways (i.e., in bulk, suspension, and emulsion) and by a variety of catalysts that give free radicals. The monomers considered will be limited to those known to polymerize to high molecular weight readily, while satisfactory representative polymerization techniques will be described. No attempt will be made to consider exhaustively all methods for all monomers.

9.4.1. Monomers

The largest single factor in determining whether a polymerization is successful or not is the purity of the monomer used. It is absolutely essential that the material be pure. The ultimate test is successful polymerization. While the presence of small amounts of inhibiting substances may in some cases be overcome by the use of excess initiator, the only generally satisfactory, reproducible approach is to remove them.

9.4.2. Polymerization Systems (34)

Bulk polymerization is the simplest laboratory technique for converting monomer to polymer. In this method, catalyst is added to undiluted monomer and this mixture is carried through the polymerization cycle. This technique is useful because of its simplicity and because it makes possible the direct preparation of castings, since the monomer-catalyst mixture will polymerize to a solid shape controlled by the shape of the polymerization vessel. This method, however, suffers from certain disadvantages. For example, polymerizations that are exothermic are liable to form local hot spots within the polymerization with consequent charring of the product. Other difficulties will be apparent during later discussion. However, continuous bulk polymerization on a commercial scale is routine for some monomers (e.g., polystyrene, low density polyethylene).

To obviate some of the difficulties encountered with bulk polymerization, solution polymerization has been used to some extent. In this technique, the monomer is diluted by an inert liquid that makes control of the temperature of the reaction much simpler. This technique, if the solvent is chosen properly, will give a solution of the polymer ready for casting or spinning. However, many solvents have a deleterious effect on polymerization, since they may act as chain transfer agents, thus lowering the molecular weight of the polymer. Also, the last traces of solvents are sometimes quite difficult to remove.

Another technique that is used to compensate for the poor features of bulk polymerization is suspension polymerization. In suspension polymerization, monomer is suspended rather than dissolved in an inert liquid, usually water; the net effect is a large number of bulk polymerizations, one in each suspended monomer droplet. The initiator used must be soluble in the monomer droplets. Suspending agents such as gelatin, methyl cellulose, and other water-soluble polymers may be used to keep the monomer in a state of suspension. Mechanical agitation is required to maintain the liquid in suspension; and at the end of the polymerization the product is ordinarily obtained as a fine, granular product that is easily filtered and handled. For polymers that are soluble in their monomer, a stop in stirring during polymerization and excessive or insufficient stirring may lead to agglomeration of the droplets when they are in a partially polymerized, tacky state. Suspension polymerization is not satisfactory for inherently tacky polymers such as elastomers, because agglomeration is virtually inevitable with such materials. An insufficiency of suspending agent will also permit agglomeration, even for polymers insoluble in their monomer, such as poly(vinyl chloride). On finishing the polymerization, it is sometimes necessary to wash and dry the polymer thoroughly to remove traces of the suspension stabilizer and the reaction medium, usually water.

When the monomer is significantly soluble in water, initiation can take place in the aqueous phase using water-soluble initiators. An example is acrylonitrile (soluble in water to the extent of about 8%) polymerized by potassium persulfate. Initiation occurs in solution, but polymer quickly separates as a solid and continued polymerization appears to occur in the latter phase. This is often termed slurry *polymerization*.

The last technique to be considered is emulsion polymerization (35, 36). In this system, water is again used as a carrier. However, an emulsifying agent such as a synthetic detergent is added. The emulsifier forms micelles that solubilize a certain quantity of monomer, the majority of which is dispersed in small droplets. The water-soluble initiator radicals penetrate monomer-rich micelles and initiate polymerization. The micelles serve as the locus of polymerization of more monomer as the latter diffuses to these sites, the monomer droplets thus acting as a reservoir for this purpose. The micelles dissipate as the polymer particles grow and continue as polymerization sites. The emulsifier then is adsorbed on the surface of these particles. Because fine (0.1 to 1.0 μ), stably suspended (emulsified) particles are produced, the system is well adapted for making tacky polymers, which are thus prevented from agglomerization. The kinetics of emulsion polymerization differ from those in suspension (1).

One of the big advantages of emulsion polymerization is that it gives a fluid system in which temperature control can be achieved. Rapid polymerization to very high molecular weights in easily achieved. Laboratory workup of emulsions is usually via coagulation and filtration, which can be tedious. Commercially, emulsions of nontacky polymers are often spray dried, giving a fine powder with the emulsifier retained on the particles.

9.4.3. Initiators

There are a number of different techniques for initiating polymerization of vinyl monomers, which depend on the generation of free radicals. The simplest method is thermal polymerization in which free radicals are developed in the monomer simply by heating. About the only monomer with which this technique is satisfactory is styrene. In other cases it is necessary to add in small proportions an agent that will generate radicals in situ.

The following compounds have been widely used, but as in any field of science newer materials have steadily appeared that have advantages over the older ones in certain applications. Veterans are benzoyl peroxide; α, α'-azodiisobutyronitrile (special versions having other than butyl groups are now available); *t*-butylhydroperoxide; cumene hydroperoxide, and di-*t*-butylperoxide. Newer examples are the entire class of peroxydicarbonates based on alcohols of different chain lengths and structures, e.g. $(ROCO_2)_2$, usually where R isn't less than a C_4; peroxy esters of neoacids of differing acid and alcohol structures, e.g., *t*-amyl peroxyneodecanoate; and mixed acyl-carboxylic peroxides, e.g., RSO_3OOCOR. Many factors influence the choice of initiators for a system, but a central one is the desired rate of polymerization at the desired temperature of polymerization. And into this latter enters the ability of the process to remove the heat of polymerization to control the temperature. It is the half-life of the initiator at a given temperature and in a specific organic medium that indicates its position for potential application. On example showing these principles is the suspension polymerization of vinyl chloride later in this chapter. Literally dozens of initiators are likely to be available for consideration, either commercially or on a developmental basis.

These agents are generally used in bulk or suspension polymerization, for which organic solubility is needed. In aqueous media, it is also possible to use certain inorganic oxidizing agents. Among those that are popular are hydrogen peroxide, sodium perborate, and various persulfates. The effectiveness of the radical catalyst can, in many cases, be enhanced greatly by the inclusion of a reducing agent, such as ferrous ion and other modifiers. Thus the important redox type of reduction-activated polymerization recipe contains a number of ingredients, each added for a specific purpose (15). For details on kinetics and mechanism of free-radical catalyzed polymerizations, the reader should refer to a standard text on the subject.

It should be noted that the literature, in particular the patent literature, is replete with additional examples of compounds suitable for use as either sources of free radicals or as reducing agents for use in redox systems. There is little point in attempting to list all of these. Only those that will actually be used in the preparation of polymer will be mentioned, and those in the appropriate place.

9.5. FREE-RADICAL CATALYZED POLYMERIZATION OF MONOSUBSTITUTED ETHYLENES

Low-density polyethylene (LDPE) by free-radical initiation was first prepared in 1933 (37). Its preparation requires high pressures and temperatures, well beyond the abilities of all but laboratories equipped for just such purposes. However, in the age of modern coordination catalysts, the synthesis of LDPE of varying degrees of branching of controlled lengths is done at moderate temperatures and pressures, accessible to most ordinary laboratories. In fact, polyethylene of virtually any intermediate density between the highest available, about 0.965 g/mL, to the lowest of wide utility, 0.915 g/mL, can be made by these metallic catalyst systems. The latter are covered in a later section.

Higher aliphatic 1-olefins such as propylene and 1-butene do not polymerize to high molecular weight with free radicals. Coordination catalysts are required. The most common hydrocarbon type known to polymerize with free radicals (aside from ethylene) is styrene and ring-substituted derivatives, as well as other vinyl aromatics such as vinylnaphthalene. Styrene homopolymerizes under a wide variety of conditions (38). A number of these will now be considered.

185. The Thermal Polymerization of Styrene in Bulk (39)

$$C_6H_5CH{=}CH_2 \longrightarrow \left[-\underset{}{\overset{C_6H_5}{\overset{|}{C}H}}-CH_2-\right]_n$$

Pure styrene monomer polymerizes thermally without added initiator at a rate that increases rapidly with increasing temperature. Polymerization at room temperature may require months or years. However, at 150°C polymerization is over in a short time. The temperature chosen for the thermal polymerization of the styrene should be somewhere in between these extremes because, although lower temperature requires much longer period of time, it also produces much higher molecular weight.

Preparation of polystyrene with a molecular weight of about 150,000 may be carried out at 125°C. A polymer tube or beverage bottle is charged with approximately 50 g styrene monomer and is flushed and sealed under nitrogen. The container is immersed in a heating bath at approximately 125°C and allowed to remain at this temperature for approximately 24 h. Under these conditions better than 90% of the monomer will be converted to polymer. For complete conversion to polymer, it is necessary to allow the polymerization to run for 7 days or more and then finish the polymerization by heating at 150°C for an additional 2 days. This will reduce the content of volatile material to less than 1%. However, for a volatile content of less than 10%, 24 h is all the time that is required at 125°C.

If polymerization is run at 102°C for 15 h, the conversion is about 40%, and the inherent viscosity is about 1.5 (toluene, 30°C, 0.5 g/100 mL).

Heat transfer problems can be serious if the polymerization is scaled up. The polymerization may become violent, so a larger scale is not recommended.

If desired, the polymer may be purified by first grinding to a small particle size in, for example, a Wiley mill, then dissolving in benzene. The benzene solution is poured into methyl alcohol and agitated vigorously in a high-speed mixer to precipitate the polymer in a finely divided condition. The solid polymer is filtered and dried in a vacuum at 110°C.

186. Bulk Polymerization of Styrene with Peroxide Catalysis (39)

To 45 g freshly distilled styrene contained in a suitable vessel such as a polymer tube, flask, or screw-cap bottle is added 1.0 g benzoyl peroxide. This mixture is heated at 50°C for approximately 3 days. At the end of this period, a solid plug of polymer is obtained that should have an inherent viscosity of about 0.4

(0.5 g/100 mL in benzene, 25°C). The rate of polymerization can be increased markedly by carrying out the reaction at temperatures higher than 50°C, but at sacrifice of molecular weight; by reducing the initiator concentration to one half or less, a slower rate but substantially higher molecular weight will result at a given temperature, because rate is proportional to the square root of initiator concentration and degree of polymerization is inversely proportional thereto. With 0.25 g initiator, polymer conversion in 2 days is about 30%, and the inherent viscosity is about 1.0.

Polymerization mixtures of this type are suitable for preparing castings or embedding small objects in a polystyrene matrix. For example, polymerization of the mixture of styrene and benzoyl peroxide indicated above may be carried to the point where the polymerized mixture is so viscous it is difficult to pour. It may be then transferred to a mold, such as a square box, and the object to be embedded is immersed in the viscous mixture. Polymerization is then continued at elevated temperatures until the polystyrene solidifies around the object to be imbedded. It is possible to suspend the object in the viscous mixture if it has a tendency to sink by use of a fine thread or wire that may be removed when the polystyrene reaches a gel-like state. Alternatively, the casting may be built up in layers. Half of the mold is cast and hardened. The object to be embedded is laid down, the remainder of the viscous prepolymer added, and polymerization completed. Any bubbles that occur in the final casting may be removed simply by drilling a hole with a fine drill and injecting more styrene-catalyst mixture into the bubble to fill it. Polymerization is then continued to polymerize the styrene that was added to the air pockets.

Polystyrene molds can be made easily, again from the viscous polymerizing mixture such as described above. It is poured into the object to be reproduced and polymerization continued until the polystyrene is solid. The polymer mass will, of course, take on the outline of the surrounding vessel. Polystyrene does not adhere to glass; a coating of methyl cellulose or a silicone spray may be applied to objects that stick to polystyrene to facilitate separation.

It is not essential to use benzoyl peroxide for the polymerization of styrene in bulk; other peroxide or azo-type catalysts can be used. Techniques would be essentially the same as those described for benzoyl peroxide with the added precaution that some of the azo compounds may be more active toward producing free radicals at low temperatures.

The polymerization of styrene in a solvent offers little advantage (other than improved heat transfer) over the bulk polymerizations described previously, because many solvents tend to react with the growing polymer chain, limiting the molecular weight obtainable. Furthermore, rather dilute solutions must be prepared, otherwise the viscosity of the polymer solution becomes so great that manipulation becomes a problem. The principal utility of a solution method where styrene is concerned is in copolymerization with other monomers (e.g., acrylate esters and acrylic acid) to make surface coatings that can be applied directly from the solution in which they are made.

187. Emulsion Polymerization of Styrene with Persulfate (18, 40)

For the simple laboratory polymerization of styrene in an emulsion, the following experiment is quite satisfactory. In a soda pop or beer bottle is placed 100 g water, 0.05 g potassium persulfate, 0.05 g sodium hydrogen phosphate, and 1.0 g sodium laurylsulfate. When this mixture has dissolved, 50 g styrene is added. Nitrogen is bubbled through the mixture to replace the air and disperse the styrene. The nitrogen tube is removed, and the bottle is capped and sealed. The bottle is wrapped with some wire screen to prevent serious damage in the event the polymerization gets out of control and is maintained with intermittent agitation at 70°C for 2 h, then at 95°C for 2 h. The polymer latex so produced is precipitated by adding alum solution and boiling of the resulting mixture. Polystyrene is separated by filtration, washed, and handled in the usual way. Instead of a bottle, a three-necked flask with stirrer, condenser, and nitrogen inlet for blanketing the mixture, is quite satisfactory.

188. Suspension Polymerization of Styrene (41)

The following ingredients are added to a 2-L round-bottom flask equipped with a mechanical stirrer and a condenser and maintained at a temperature of about 80°C: 500 mL deaerated water, 0.1 g sodium laurylsulfate, 1.5 g sodium polyacrylate, 5 g sodium sulfate, 150 g styrene, 1.5 g stearic acid, and 0.7 g benzoyl peroxide. This mixture is stirred vigorously and maintained at 80°C for 12 to 24 h. At the end of this time, the beads of polystyrene are filtered, washed with water, and dried in an oven at 60°C in vacuum. Conversion should be complete. It is necessary that agitation be effective until the beads attain a solid consistency, otherwise the droplets will tend to fuse together. For this reason, it is generally necessary to add a small amount of a suspending agent. Gelatin or sodium polymethacrylate have been used, but other kinds, such as cellulosics, also work.

The brittleness of unmodified polystyrene constitutes a major deficiency in its performance. To correct this problem, medium- and high-impact polystyrenes are made by incorporating rubber into the hard polystyrene. This flexible component has the effect of increasing the ability of the composite to absorb rapidly applied stresses, as in impact. The rubber-modified resin is usually made commercially either by mechanically blending polystyrene with the rubber (42) as on a mill or in a Banbury mixer, or by polymerizing styrene in the presence of the rubber (43) to give, at least in part, a graft of polystyrene onto the rubber. Styrene-butadiene or acrylonitrile-butadiene rubbers are often used in both the polyblend and graft types, from about 5 to 15% by weight of the hard polymer. Such materials are opaque in contrast to the clear polystyrene. Some high styrene (>50%)-butadiene copolymers are also made, but are not generally classed with the impact polystyrenes.

A variation of the high-impact polystyrene resins has achieved considerable stature because of the unusual combination of stiffness, tensile strength, and impact resistance obtained. These are the ABS resins, which are in the category of "engineering" plastics because of the excellent resistance to creep combined with the other properties noted. As with high-impact polystyrenes, ABS resins are either of the polyblend or of the graft type. The former is made by mixing rubber and hard polymer, e.g., a styrene-acrylonitrile (70/30 w/w) copolymer with a butadiene-styrene (75/25 w/w) copolymer, either by coagulation of a mixed latex of the two or by hot-milling the two materials. The graft resin is made by copolymerizing styrene and acrylonitrile (again, at about the 70/30 level) in a preformed latex of polybutadiene or styrene-butadiene (85/15). The final mixture is coagulated to give a mixture of free poly(styrene-co-acrylonitrile) resin and the graft of this resin onto the rubber. The compositions cited above are typical, but there is a great deal of variation possible to give a spectrum of products with different stiffness values, tensile and impact strengths, and hardnesses. α-Methylstyrene in place of styrene is said to impart greater resistance to deformation in heat (44), while replacement of the acrylonitrile by methyl methacrylate gives a transparent graft polymer instead of the conventionally opaque. In addition, blends of styrene-acrylonitrile copolymer, a hard resin, are sometimes made with a latex of styrene-acrylonitrile copolymer/grafted onto poly(butadiene) as prepared in the following sequence.

In the following examples, a latex of polybutadiene is prepared. This polymer has a relatively scrambled structure (cis- and trans-1,4-, and 1,2- units) to the stereoregular polybutadienes prepared in other examples. However, the lower the temperature of polymerization, the more regular the structure of the polybutadiene, trans-1,4- increasing at the expense of 1,2-. The impact properties of the final polymeric composition are apparently better when the butadiene is polymerized at lower temperatures.

189. Preparation of a Polybutadiene Latex (45)

$$CH_2{=}CH{-}CH{=}CH_2 \longrightarrow -[CH_2{-}CH{=}CH{-}CH_2][CH_2{-}CH(CH{=}CH_2)]-$$

To an 8-oz beverage bottle is charged 150 mL deaerated, distilled water, 77 g polymerization-grade butadiene, 7.5 g carbon tetrachloride, 2.3 g sodium oleate (technical grade is satisfactory), and 0.4 g potassium persulfate. About 2 g butadiene is allowed to boil from the bottle to remove air, the bottle is sealed with a neoprene septum, and a metal cap with a hole drilled in it. It is tumbled at 50°C for 65 h. The conversion to polymer is about 69%. The solids content can be estimated from the difference in the weight of the bottle before and after discharge of unreacted butadiene. The latex should be stripped of residual monomer by passing nitrogen through it.

190. Preparation of an ABS Polymer (45)

$$\text{Polybutadiene} + CH_2{=}CHPh + CH_2{=}CHCN \longrightarrow$$

$$[\text{Polybutadiene}]-[(CH_2CHPh)(CH_2CHCN)]- + -[(CH_2CHPh)(CH_2CHCN)]-$$

Sufficient quantity of the latex prepared in the previous example is used to give 21 g polybutadiene. This is charged to a beverage bottle along with 32 g inhibitor-free styrene (twice washed with 10% aqueous sodium hydroxide, once with distilled water; drying is not essential), 16 g acrylonitrile (inhibitor removed by passage through a silica gel column), 22 g deaerated distilled water, 0.7 g sodium oleate, and 0.1 g potassium persulfate. The bottle is then tumbled at 40°C for 26 h. (Since none of the monomers is a gas at the temperature used, a three-necked flask equipped with a stirrer condenser, and nitrogen inlet may be used instead of a sealed bottle.) The latex can be coagulated by addition of a salt or alum solution; the polymer is washed first in methanol to remove residual monomers, then several times in distilled water to remove inorganic impurities. The conversion of the styrene and acrylonitrile together is about 75%. The polymer can be compression molded to give opaque, tough specimens. Larger quantities than that prepared here are usually necessary for injection molding.

A wide variety of substituted styrenes has been made and polymerized.

An interesting example of an alternating copolymer involving vinyl monomers in which the other monomer is inorganic, is the propylene-sulfur dioxide copolymer. Polymerization probably occurs through a 1 : 1 complex of monomers and is the exceptional case where higher α-olefins polymerize by means of free radicals. Most α-olefins undergo SO_2 copolymerization, but so do many olefins not so conventionally described; styrene, vinyl chloride, and cycopentadiene are among them.

$$CH_2{=}\underset{\displaystyle CH_3}{\overset{|}{CH}} + SO_2 \longrightarrow \left[-CH-\underset{\displaystyle CH_3}{\overset{|}{CH}}-SO_2-\right]_n$$

This experiment requires the use of high-pressure equipment. To a 1-L rocker bomb is added 0.5 g α, α'-azobis(α, γ-dimethylvaleronitrile). The bomb is pressure tested with nitrogen at 400 psi to detect leaks and then evacuated to less than 1 mm for 3 to 4 h. The bomb is chilled in a dry ice–acetone mixture and 42 g propylene and 240 g sulfur dioxide (99% purity) are distilled in. The bomb is sealed. Polymerization is allowed to proceed at 40 to 45°C for 8 h. The product, which is obtained in a yield of approximately 100 g, is removed from the bomb and washed twice with alcohol in a high-speed mixer. The dry polymer is ground in a Wiley mill, or similar device, to pass through a 20-mesh screen and rewashed with alcohol. It is dried at 80°C overnight. The yield is 96 g of a hard white product that has an inherent viscosity of 3.3 as measured in concentrated sulfuric acid at room temperature. The product has a polymer melt temperature (PMT) of 300°C. At this temperature, it decomposes into monomer.

Of the vinyl halides, only vinyl fluoride and vinyl chloride (46) have given useful high polymers. Poly(vinyl fluoride), a highly crystalline moldable polymer, melts about 100°C higher than polyethylene. With ordinary free-radical catalysts, rather high temperatures and pressure are required (47–49) for the polymerization of vinyl fluoride. It can, however, be polymerized under pressure to a high molecular weight product by radiation at room temperature (50).

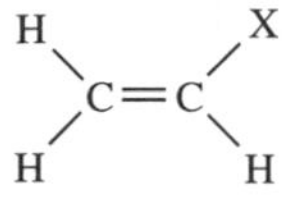

$$\begin{matrix} H & & & & X \\ & \diagdown & & \diagup & \\ & & C{=}C & & \\ & \diagup & & \diagdown & \\ H & & & & H \end{matrix}$$

where X = F, Cl, Br, I.

It is worth noting that some of the chemical reactions undoubtedly occurring in this polymerization system lead to the thermal sensitivity of PVC. Chain transfer from radical chain to monomer is thought to lead to allylic chlorine atoms, for example.

$$\sim\!\sim CH_2CHClCH_2\dot{C}HCl\cdot + CH_2{-}CHCl \longrightarrow \sim\!\sim CH_2{-}CHCl{-}CH{=}CHCl + CH_3{-}\dot{C}HCl$$

Tertiary chloride is understood to arise by radical-induced branching at the −CHCl− carbon, giving rise to heat-sensitive sites. It is the splitting out of HCl that occurs, thermally, from these labile groupings. Each loss of HCl produces a double bond and thus another allylic chloride, leading to repeated HCl eliminations. The conjugated double bonds that so build up are considered to be responsible for the ever deepening color, from yellow to brown to black as thermal assault continues. This sequence is particularly severe in rigid PVC, where diluting plasticizer is absent and higher temperatures are needed for thermal processing. But heat stabilizers have been developed to retard this problem; with proper temperature control of the processing operation (be it extrusion, molding, or others) and adequate stabilizers, the problem can essentially be obviated. Absent stabilizers, PVC may have been doomed to a life as a laboratory curiousity.

The polymerization example that follows is not intended for the unprepared laboratory or experimenter. Rather, it is meant to be instructive in the sense of demonstrating what is done in practice in the instance of an important commercial material, and in how its polymerization kinetics and its important properties are controlled through the operating variables. We are indebted to T. J. Martin (Condea Vista Co., Austin, Tex.) for providing the detailed exposition of the subject that makes this presentation possible.

The suspension process is by far the most important commercial method for making PVC. This involves dispersing VCM in water in a ratio of about 1.2 (w/w, H_2O/VCM). A so-called suspending agent is initially dissolved in the water and the VCM is pressured into the reactor. (The suspension process should not be confused with the emulsion process, discussed later.) A free radical initiator, dissolved in a hydrocarbon such as hexane, is then usually also pressured into the stirred reactor. The initiator dissolves in the monomer droplets. The suspending agents commonly used are primarily alkylated and/or alkoxylated cellulose polymers, e.g., a methyl hydroxypropyl cellulose. Poly(vinyl alcohols), from varying degrees of saponification of poly(vinyl acetates), are another useful suspending agent class. Between the basic polymer types, their molecular weights and their kinds and degrees of substitution or hydrolysis (for the polyalcohols), researchers have developed a host of special formulations used commercially. The choice of suspending agent influences the particle size of the resulting PVC, but in combination with the reactor mechanics and charge ratio. That is, the agitator speed and diameter relative to the reactor diameter, the height/diameter ratio of the liquid column in the reactor, and the nature of the reactor baffles are highly interactive with the suspending agent in determining the nature of the particle character of the product. The latter in turn is one factor that determines the processibility of the PVC in the hands of those who process it further by extrusion, calendering, or molding into finished products of commercial importance.

After all the ingredients are charged, heat is applied via hot water in the reactor jacket or, in a small laboratory autoclave, via a heating mantle or oil bath. Commercially, hot water is often used in making the initial suspension to shorten the reactor heat-up time. In any event, each VCM droplet, in the ideal case, contains an equal concentration of initiator. Each droplet becomes an individual site of polymerization. PVC is not very soluble in its monomer, so fine micrograins of polymer soon begin to precipitate after the temperature has reached a level to dissociate some initiator to its radical fragments. These micrograins begin to coalesce, initially at the VCM side of the water-monomer interface, but also in the body of the VCM liquid, to form the particles that underlie the internal structure of the final individual particles; the ultimate particles having, say a 140-μm diameter, are themselves composed of coalesced, but still discretely evident, finer particles. Until the individual droplets acquire mechanical strength via solid buildup, especially at the surface, droplet-droplet collisions and coalescence can occur. A purpose of the suspending agent is to prevent these coalescences from getting out of hand, possibly provoking the ultimate disaster — one particle occupying much of the volume of the reactor and requiring mining procedures to remove. (The occurrence of such ultimate disasters may be apochryphyl, although much feared; exceedingly coarse, and thus useless, particle size distributions are certainly known.)

The kinetics of VCM polymerization exemplify an important possible event whenever a polymer is insoluble in its monomer. In the case of PVC, the polymer is largely insoluble in monomer but remains swollen by it as the polymer separates. This changes the relative rates of the three classic reactions in free-radical polymerization. The rate of termination slows down relative to that of initiation, the reason being that the rates of combination and disproportionation between radical-ended polymer chains, which are important chain-terminating reactions, begin to slow down as these high molecular weight chains become immobilized in the viscous, monomer-swollen polymer phase. But while chain-end interactions bringing about termination are slowed, the effect of viscosity on monomer is not nearly so great. Thus monomer can diffuse to radical chain ends with little difficulty. The chain growth reaction speeds up as termination is slowed and the heat release in the polymerization locus, i.e., each droplet, is markedly speeded up. This phenomenon has been termed the *Tromsdorf effect*, after its definer. Temperature control of the polymerizing batch becomes difficult unless a great cooling capacity can be applied.

This peak-load cooling demand can raise plant investment and operating costs significantly, yet cooling is essential because the temperature of the polymerization is the single dominating factor in controlling the average molecular weight of the polymer being formed. Initiator concentration is unimportant, for practical purposes, because its effect is so small relative to chain transfer to monomer and the Tromsdof effect. The rate of the polymerization is in contrast highly influenced by the initiator concentration at any selected polymerization temperature. The industrial manufacturer and the laboratory researcher face a dilemma: Both want the shortest polymerization time possible at the chosen temperature to make efficient use of reactor time, i.e., to get the most batches possible per working hour of reactor time. Using high concentrations of initiator is the obvious answer; but if the batch temperature cannot be controlled, the polymer will not meet the desired average molecular weight.

The curve of heat release with time increases gradually and steadily from the beginning, but at about 60 to 70% monomer conversion it begins to accelerate steeply. A plot of calories/minute/minute vs. time shows the inefficiency of the process, in that something over half the total cooling needed is demanded in the roughly the last 10 to 15% of the intended fraction of monomer to be converted to polymer, constituting the famous "heat kick" characteristic of VCM polymerization.

The long-standing initiator of choice for VCM is lauroyl peroxide. For many years its half-life was about the best that could be obtained for polymerizing this monomer; but even then it created a slow initiation at the beginning, which became thunderously fast as the termination rate fell off. To get around the problem, peroxydicarbonate esters were developed with shorter half-lives. These and other "fast catalysts," e.g., peroxyneodecanoates, gave faster early initiation, consuming more monomer earlier in the process than occurred with lauroyl peroxide, with the result that the last part of the polymerization had a much reduced heat kick. This meant that the curve of heat evolution (and cooling demand) was more nearly "squared," i.e., heat evolved more nearly linearly, thus diminishing the great heat kick near the end. And of enormous practical importance as a consequence, the polymerization time for a given reactor batch could be significantly shortened.

It is not uncommon to find mixtures of initiators used, their half-lives of decomposition being matched to the polymerization temperature desired, resulting in profound plant capacity enhancement by shortening the polymerization time. One of these should have a relatively short half-life and thus be capable of rapid radical production at the start; a second initiator of longer half-life of slower radical generation rate can be charged at the beginning also. The first (front-end) initiator is charged at a concentration so that it burns out relatively early in the reaction cycle, while the second (back-end) continues at a slower rate to bring the batch to its desired conversion, giving an even better "squaring" of the curve of heat release. Although it may rarely be so ideal in practice as the latter term implies, the technique is often used because it permits much more efficient cooling usage. An example of such a combination is α-cumyl peroxyneodecanoate and *t*-butyl peroxyneodecanoate.

A process was developed in the 1960s by the French that was termed *mass polymerization*. This method used VCM as the sole fluid phase. It was a kind of suspension process without water or suspending agents; excess monomer was the medium of polymerization and the carrier of the solid polymer generated. This was done by generating a small quantity of polymer 'seeds,' effectively the micrograins formed in the early part of suspension polymerization, in a small volume of VCM, then adding this charge of seeds in liquid VCM to a horizontal, rotating reactor containing the majority of the monomer and the initiator(s). The polymer is distinctive in that it has less of a solid surface layer, but rather is more porous and open at the surface. This openness tends to speed up fusion of the polymer in subsequent fabrication as well as making it more absorptive toward plasticizers (discussed later). The economics of the mass vs. suspension processes has been much debated; both are used today, but suspension heavily predominates.

Suspension and bulk polymerized PVCs are usually made at temperatures from about 43°C to 65°C to create very high and very low molecular weight polymers, respectively. The high molecular weight resins are generally always more porous and able to readily absorb plasticizers (see below) than the low molecular weight. Porosity can be closely controlled by limiting or extending the conversion of monomer. The most plastizer-absorptive high molecular weight materials may be purposely terminated at, e.g., 75% conversion; but more typical resins, used in pipe extrusion and clear film, are terminated at 85%. Resins intended for use in nonplasticized end products may be permitted to reach even higher conversions. Termination may be achieved by simply rapidly stripping the remaining monomer from the reactor when the desired conversion is reached, determinable by noting a drop in reactor pressure to a desired level or by counting calories evolved, which is more sensitive but more difficult. Chain-terminating agents such as bisphenol-A can be added; small amounts of styrene and methyl methacrylate also function in this role by forming a terminal or transfer radical that will not propagate easily with vinyl chloride, though these would not be practical where VCM is to be recovered and recycled.

It is worth noting that suspension and bulk processes for PVC are unique among the large-volume plastics from addition polymerization of unsaturated monomers in that they still require batch as opposed to continuous methods. This anomalous state of affairs derives from the microdynamics of the polymerization of the monomer. Each finished polymer particle is the consequence of an individual polymerization history that must be as alike among particles to the highest degree possible and must be completed in essentially the same time frame as its multitude of bretheren for the assemblage to possess the correct physical properties, especially particle porosity. These factors do not intrude in anywhere near the same degree in the processes for the other major plastics, which is most of them, made by continuous processes. The principles governing PVC have broken all attempts to develop a continuous PVC process over several decades of trying. Though one learns never to say never, this problem may outrun Fermat's last theorem in waiting for breakthrough.

Preparation 191 is an example of polymerization at 58°C under conditions that produce an extrusion-grade resin (polymer) suitable for the fabrication of pipe, typically usable in assembling potable water or drain, waste, and vent systems.

191. The Suspension Polymerization of Vinyl Chloride

All traces of polymer scale are removed from the walls, agitator blades, and baffles of a 2-L, 316 stainless-steel reactor using THF. Then 15.7 g of a 2% hydroxypropyl methyl cellulose solution (0.08 parts suspending agent per hundred parts monomer; phm) is mixed with 491.3 g (125 phm) deionized water and added to the reactor. The reactor should be equipped with two baffles, each about 0.25 in wide and a similar distance from the reactor wall, the reactor top (containing the drive shaft and agitator blade) is bolted down, and the drive belt put in place. The reactor is evacuated to −14.7 psig at room temperature to remove noncondensable oxygen and nitrogen (the former would seriously inhibit the polymerization); the VCM supply cylinder is then heated to give 60 psig pressure, and 393 g (a volume of about 470 mL at reaction temperature) is pressured into the reactor. All lines are then closed, agitation is set to 800 rpm and the reactor contents heated to 135°C. When at this temperature, 0.48 mL of a 75% solution of *t*-amyl peroxyneodecanoate in hexane, or similar inert, volatile, and monomer-soluble solvent, is added using a high-pressure syringe with an 18-guage needle capable of reaching below the liquid's level. By injecting initiator at batch temperature, polymerization does not take place at any temperature below the one desired, thus providing better molecular weight uniformity. The batch temperature is controlled at ±0.1°C for about 222 min, when the pressure of the batch has dropped from 125 psig, which prevails as long as any free liquid monomer phase exists, to 105 psig, corresponding to about 85% conversion of VCM. At this point, the dump valve of the reactor is opened; the slurry is then transferred to a vented bucket, from which the unpolymerized monomer is allowed to evaporate over a period of about 1 h. (The initial pressure drop occurs at about 73% conversion; at that point, the residual monomer is present only dissolved in the swollen polymer particles rather than as a liquid phase.) The product PVC is filtered free via a Buchner funnel and oven dried at about 38°C overnight. Polymer yield is 333 g, including a small amount of scale on the reactor, agitator, and baffle surfaces. Obviously, the venting of monomer and filtration of slurry must be done in efficient fume hoods and the monomer vapor recovered before the hood air is vented, all in keeping with the need of having a laboratory designed to handle vinyl chloride.

The polymerization in this example is stopped by quickly venting the slurry and residual monomer, largely starving any further polymerization. When such quick termination is not possible, as may occur in cicumstances in which the slurry must be held in the reactor before dumping, terminating agents can be added, as dicussed earlier, usually fatty alcohol esters of thioglycolic acid. These act by reaction of the thiol hydrogen with chain end radicals (or initiator radicals, if still forming) to give a thiol radical, $-S\bullet$, which is too stable to effectively initiate new chains.

The average molecular weight is determined almost solely by the polymerization temperature. In the above example, the inherent viscosity (0.5% in cyclohexanone, ASTM D1243-79) should be 0.87 to 0.93, though with careful controls and technique it should be possible to achieve ± 0.05 unit variation between batches. Various viscosity indexes can be used as a surrogate for the molecular weight. In practice, the one just cited is not more useful than another viscosity index that has been determined by experience to produce the desired physical properties in finished products. Other physical characteristics depend highly on the polymerization environment, e.g., the particle porosity, cold plasticizer absorption (affected by degree of conversion and polymerization; lower temperatures and conversions both contributing to higher porosity and hence absorption of liquids), the bulk density of granular polymer and the particle size mean and distribution (both being a strong consequence of agitation, baffling, and reactor geometry). In this example, the polymer could be expected to have properties similar to these: mean particle size, 150 to 165 μm (by screening or laser light scattering, the latter now almost as routine as the first); bulk density, 0.50 to 0.60 g/mL (ASTM D1895-69); and the cold plasticizers absorption, 17 to 22%. In contrast, from polymerization at 61–67°C the inherent viscosity would be 0.50 to 0.85; the mean particle size, 130 to 150 μm; the cold plasticizer absorption, 17%; and the bulk density, 0.50 to 0.60.

The use of plasticizers with PVC has been referred to several times. Plasticizers are relatively smaller molecules, almost all of them liquids, that in effect solvate PVC, converting it to a softer, flexible mass after heating. The resulting products from plasticized, or flexible, PVC are ubiquitous in the form of electrical cable insulation, upholstery in a form that resembles leather (without the breathability), shoes, luggage, clear film, and many others. Most plasticizers are phthalate esters of middle-length alcohols; popular ones are branched C_8-, branched C_{13}-, and linear C_{6-10} alcohols along with liquid aliphatic polyesters for higher temperature uses. Choices are made on the basis of the needed properties in the end uses and the economics of the materials. Other acids used with some of these alcohols are adipic and phosphoric, the latter mainly used as phenolic esters for flame retardancy. The choices are many and the products multiple, accounting for about half the total use of PVC. Absent plasticizers, PVC is important for stiff, high-modulus products, of which pipe of a considerable range of diameters and uses is the most important.

Vinyl chloride can also be polymerized by emulsion methods. Emulsion PVC (dispersion, plastisol, or paste resin, as it is more often termed in the PVC industry) is used to make viscous, nonsettling dispersions of resin in plasticizer, with stabilizers and other additives added. Called pastes or plastisols, these compositions are used in manufacturing finished product forms not obtainable by other more conventional methods, such as extrusion or injection molding: pastes can be converted to balls; used to coat tool handles; doctored onto a web of fabric to provide a decorative or protective coating; and made into thin, protective medical gloves by dipping a hand-form into a paste followed by heating to fuse the resin-plasticizer combination.

As a way to demonstrate emulsion polymerization techniques, vinyl chloride serves no better than, say, acrylonitrile or styrene as they appear in this chapter but offers considerably greater operational difficulties in complying with handling regulations. For this reason, no example using vinyl chloride is given. But there is another reason. In practical terms, the method of recovery of PVC from its emulsion determines its application properties and overall utility at least as strongly as the polymerization itself. It is customary in industrial manufacture to spray dry the emulsion taken from the polymerizer, followed by grinding to break up clumps. These extremely critical steps cannot in any reasonable way be imitated by a benchtop laboratory method, which typically coagulates the polymer, filters it (with great slowness, usually), and dries it in an oven or that simply evaporates it to dryness. The resin from spray drying is usually a mixture of hollow spheres of PVC from 1 to 20 μ in diameter, and fragments of such spheres are broken in drying and handling. PVC's properties depend critically on the balance of physical state and chemical surface composition.

Rigid PVC (i.e., no plasticizer) can be improved in its resistance to deformation on exposure to heat by chlorination of the polymer in such a way that 95% or more of the vinyl chloride monomer units are believed to be converted to 1,2-dichloroethylene units, –CHClCHCl–. (The latter monomer has not yet been successfully homopolymerized to a high molecular weight.) The resulting product, termed chlorinated PVC (CPVC) can have a T_g of over 100°C vs. 75°C for unmodified, rigid PVC. CPVC also appears to have a better heat stability than PVC itself probably via chlorination of any double bonds.

The avoidance of 1,1-dichloroethylene monomer units in the present case is believed due to the maintenance of a sufficiently high chlorine concentration at the locus of the light-catalyzed reaction. Two factors determine the course of the reaction: dissolved chlorine content and level of photochemical activation. Moderate levels of photochemical activation can easily outstrip the ability to dissolve chlorine in the system; thus chlorine starvation must be avoided.

The starting PVC must be of a high molecular weight and have a relatively large granular form (suspension resin) with high porosity. Chloroform is used as a swelling agent and is necessary to obtain a superior chlorination product.

192. Chlorinated PVC (51)

$$-\!\!\!\left(CH_2-\underset{\underset{Cl}{|}}{CH}\right)\!\!\!- \xrightarrow{Cl_2} -\!\!\!\left(\underset{\underset{Cl}{|}}{CH}-\underset{\underset{Cl}{|}}{CH}\right)\!\!\!-$$

To a creased, 1-l three-necked flask equipped with a highly efficient condenser, stirrer, and gas inlet tube extending nearly to the bottom of the flask and preferably having a fritted glass tip is charged 160 g of a granular high porosity PVC such as Geon 101 EP (B. F. Goodrich Co.) 320 mL concentrated (37%) hydrochloric acid, 275 mL distilled water, and 60 mL chloroform. A 100-W mercury are lamp (A4 H 100 type) is positioned adjacent to the flask; it is preferred, however, to insert a lamp into the flask. The flask is immersed in a water bath held at a temperature of 55°C. The mixture is stirred extremely rapidly, and chlorine is admitted at a rate of 1375 mL/min, as determined by a rotameter calibrated for this purpose. The reaction mixture should evidence the presence of chlorine by a greenish yellow color throughout the addition. If this does not occur, the chlorine rate should be increased. After 7.5 h, the reaction is stopped, and the slurry filtered. The solid polymer is washed with dilute bicarbonate to neutralize the residual acid, then with water to free it of salts. The chlorine content should be about 71.9%; theoretical for poly(vinyl dichloride) is 72.3%. The T_g is well over 100°C, and the intrinsic viscosity is about 1.24, which is somewhat higher than the starting polymer.

Acrylonitrile polymerizes with the greatest of ease with a wide variety of catalysts, both free radical and anionic (chapter 8). Solution, suspension, and emulsion techniques have been successfully employed, but the polymerization is too exothermic to make bulk polymerization satisfactory.

193. Polymerization of Acrylonitrile in a Slurry (52)

$$CH_2{=}CHCN \longrightarrow \left[CH_2{-}CHCN\right]_n$$

Commercial acrylonitrile contains an inhibitor (usually a hydroquinone derivative) that is best removed before polymerization. Immediately before use, commercial acrylonitrile is passed through a column of silica gel about 24 in long and 1 in wide. It is maintained under a slight head of nitrogen and is collected at the bottom of the column and used immediately. If it is retained without the inhibitor, it may polymerize.

A 500-mL three-necked flask is equipped with a stirrer, condenser, and nitrogen inlet and surrounded by a constant temperature bath maintained at about 40°C. Then 300 mL water, which has previously been deaerated by boiling for 10 min, and 22 g purified acrylonitrile are placed in the flask and stirred gently for 10 min to allow the mixture to come to bath temperature. The initiator is now added. It is composed of 0.3 g potassium persulfate dissolved in 10 mL water, followed after 1 min by 0.15 g sodium bisulfite, also in 10 mL water. Almost immediately, the colorless, aqueous solution becomes somewhat opalescent, and white polymer begins to precipitate. After 3 h, the polymerization should be complete. The solid product is filtered, washed with water, and dried in a vacuum over at 60°C overnight. The yield of polymer is 80 to 90%. The inherent viscosity as measured in DMF will be about 2.0 (0.5%, 25°C).

193a. Preparation of High Molecular weight Polyacrylonitrile at Low Conversions

In a two liter three-necked flask equipped with a Dry-Ice condenser, a paddle stirrer and thermometer, was added 146 g of air -free deionized water (the water was boiled to eliminate air and then cooled under a stream of air-free nitrogen), 342 g of 'Spectro' grade dimethylsulfoxide and 273 g of distilled acrylonitrile. A slow stream of nitrogen was bubbled through the solution and stirring started. The flask and content were heated to an internal temperature of 72°C using a heated water bath. 0.4 g α, α'-azodiisobutyronitrile catalyst was added all at once and the reaction temperature was maintained at 72–74°C by heating or cooling via the bath. After 30 minutes the flask contents was a thick gelatinous slurry. Thioglycollic acid (2 ml) was added with vigorous stirring, followed by 600 ml. distilled water. The precipitated solid was then filtered, washed with 1 liter water, then acetone and dried under vacuum at 65°C. Yield of polymer was 57 g. Single point intrinsic viscosity at 0.1% conc. in dimethylacetamide/LiBr solution was 7.5. Weight average molecular weight was about 400,000. Dispersivity measured by GPC was 4.

194. Emulsion Polymerization of Acrylonitrile (53)

$$CH_2{=}CHCN \longrightarrow [-CH_2-\underset{\displaystyle |}{\overset{CN}{\overset{|}{C}}}H-]_n$$

A 500-mL three-necked, round-bottom flask is fitted with a nitrogen inlet, a stirrer, and a reflux condenser. The flask is thermostatted in a bath at about 35°C and flushed for 15 min with nitrogen. Then 120 mL freshly boiled distilled water is added, stirring is started, and the nitrogen flow is reduced to a slow rate over the surface. To this flask is then added, in order, 2 g sodium lauryl sulfate, 80 g acrylonitrile freed of inhibitor by the method described in the previous experiment, 0.1 g potassium persulfate, and 0.033 g sodium bisulfite. Evidence that the polymerization has started is the appearance of a milkiness, usually in about 5 to 20 min. If the milkiness does not appear within about 1 h, an additional amount of initiator and activator may be added. Once begun, polymerization is usually complete in 2 to 3 h. However, a small additional yield may be obtained by allowing it to stand overnight. A nearly quantitative yield of polymer is obtained as a stable dispersion. The particles are nearly spherical with a diameter of approximately 0.1 μ. The polymer is isolated by pouring the dispersion into approximately 500 mL water, then slowly adding salt with stirring to coagulate the emulsion. The product is collected by filtration. It is washed with water and dried in air at room temperature. The molecular weight of the polymer prepared in this manner is extremely high, the inherent viscosity usually being of the order of 10.5 as measured in DMF (0.5%, 30°C).

Polyacrylonitrile homopolymer is resistant toward most chemical reagents. It is, however, degraded by alkali and by heat. A most interesting transformation occurs when an acrylonitrile fiber is **heated** in a controlled fashion (54) at 200°C and over. The originally white, flammable yarn or fabric turns black and becomes fireproof without losing its identity as a fiber. The density of the polymer increases from 1.17 to 1.60, while the tensile strength of the yarn decreases, perhaps as much as 50%, though it is still equivalent or superior to many commercial yarns. It has been suggested (54, 55) that this product is produced in part by the following transformation; although this is a simple illustration of a more complex processor.

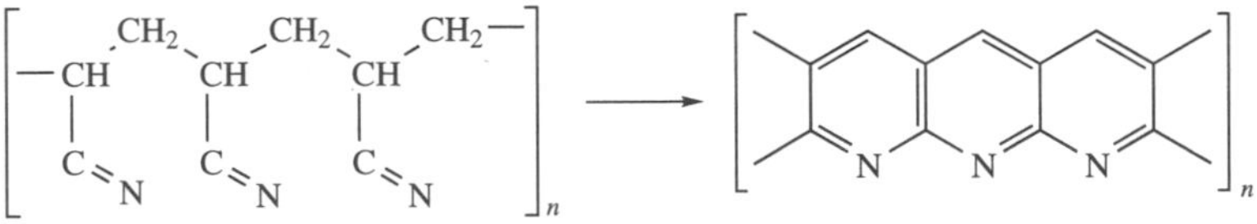

$$\left[-\underset{\underset{C\equiv N}{|}}{CH}-CH_2-\underset{\underset{C\equiv N}{|}}{CH}-CH_2-\underset{\underset{C\equiv N}{|}}{CH}-CH_2-\right]_n \longrightarrow \left[\text{fused ring ladder structure with three ring N atoms}\right]_n$$

195. Thermal Condensation of Polyacrylonitrile (54, 56)

For this preparation, almost any type of a forced-draft oven capable of operating at temperatures in the range of 250°C or more can be used successfully. The condensation product may be prepared in the form of either fabric or yarn. However, it is more convenient to use a portion of fabric made, for example, from Orlon acrylic fiber. The fabric is placed in a forced-draft oven and heated in air for a period of time suitable for converting the polyacrylonitrile to the black modification. The time varies; at 220 to 240°C, the time required is 4 to 6 h. It should be noted that the conversion of polyacrylonitrile to the black modification is an extremely exothermic reaction, so that the material being treated should not be wadded together or compressed, because this will tend to give a charred product from its own heat of reaction. It should be noted also that oxygen is necessary for the conversion. The optimum oxygen content fortunately is about 20%, so air is the most satisfactory medium in which to carry out this conversion. Black fabric produced as described has a remarkable resistance to fire. A sample of the fabric can be held directly in a flame. The only effect will be that the sample will glow around the edges. If left in the flame for any length of time, the fabric will lose its strength completely. However, under no conditions will it show signs of burning.

Another chemical reaction that may be carried out on poly-acrylonitrile is the reaction with hydroxylamine (57). The resulting polyamideoxime is formed without cleavage of the backbone, hence has the same degree of polymerization as the parent polymer. The product is film and fiber forming.

196. Conversion of Polyacrylonitrile to Polyacrylamideoxime (57)

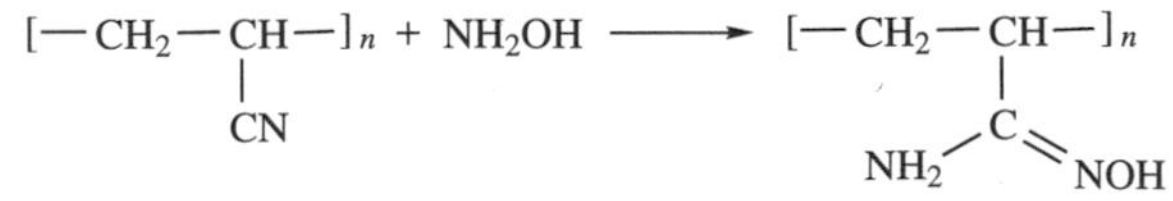

$$[-CH_2-\underset{\underset{CN}{|}}{CH}-]_n + NH_2OH \longrightarrow [-CH_2-\underset{\underset{NH_2-C=NOH}{|}}{CH}-]_n$$

A 1-L three-necked flask containing 300 mL dimethylformamide and 50 g polyacrylonitrile is equipped with a mechanical stirrer and maintained at a temperature of 75°C by an external water bath. Then 20 g hydroxylamine hydrochloride and 15 g anhydrous pulverized sodium carbonate are added to the flask, and the resulting mixture is stirred and heated for 3 h. At the end of this time, the polymer solution is filtered and precipitated into methanol; the product washed with methanol and dried. The nitrogen content will vary between 23 and 25%. This is not too different from the nitrogen content of pure polyacrylonitrile, suggesting that some hydrolysis of amideoxime groups has occurred. That some chemical reaction has taken place on the polymer is evidenced by the fact that although the polymer remains in solution during the reaction, the precipitated and dried product no longer will dissolve in dimethyl formamide. However, it will dissolve in dilute hydrochloric acid and dilute sodium hydroxide. The dilute hydrochloric acid solution gives a deep red to violet color with ferric chloride, indicating the formation of a ferric chelate of the amideoxime group.

Vinyl acetate may be polymerized by a variety of free radical catalysts. Several typical systems are given.

197. Solution Polymerization of Vinyl Acetate in Benzene

$$CH_2{=}CHOCOCH_3 \longrightarrow [-CH_2-\underset{|}{\overset{OCOCH_3}{CH}}]_n$$

A 2-L three-necked flask equipped with a stirrer, condenser, nitrogen inlet, and thermometer is charged with 200 mL vinyl acetate and 300 mL dry benzene. The vinyl acetate is purified by distillation, then passed through a silica column just before use. The mixture is heated to reflux under nitrogen, and about 0.2 g α, α'-azodiisobutyronitrile is added. Polymerization is allowed to proceed to reflux temperature for about 2 h, and the mixture is treated with steam to remove unreacted monomer and solvents. The polymer is broken up, filtered, then dried. This polymer has an inherent viscosity in chloroform of 1 (0.5% at 25°C).

198. Emulsion Polymerization of Vinyl Acetate (58)

To a 500-mL multinecked flask or resin kettle equipped with stirrer, reflux condenser, dropping funnel with pressure-equalizing side arm, thermometer, and nitrogen inlet is charged 70 mL water, 1.5 g polyethoxylated lauryl alcohol, 5.0 g sodium tridecylethoxylate sulfate, and 1.0 g of a protective colloid such as Cellosize WP-09. The mixture is stirred and heated to 70°C to effect solution as nitrogen is bubbled through it to remove the air. The solution is then cooled to 50°C, and 0.05 g sodium carbonate (to act as a buffer) and 0.15 g potassium persulfate are added. Then, 10 g distilled vinyl acetate is added to the flask, and the mixture is stirred and refluxed (about 65°C). The polymerization begins, and the temperature is raised to 80 to 85°C at a rate governed by the maintenance of a mild reflux. When the temperature reaches 85°C, 90 g distilled vinyl acetate is added from the dropping funnel at a rate that will maintain reflux. After each fourth of the vinyl acetate in the funnel has been added, 1 mL of a 5% potassium persulfate solution (in deaerated water) is added to the flask. When the last monomer and initiator solution has been added, the temperature is raised to 90°C for 1 h. The emulsion is cooled to room temperature and enough of a 5% solution of sodium carbonate is added to adjust the pH to 4.5–5.5. The creamy emulsion has a solids content of about 55%. Such a product is best used in its emulsion form, e.g., for surface coatings, rather than attempting to isolate the polymer for subsequent use.

199. Polymerization of Vinyl Acetate (59)

$$CH_2{=}CHOCOCH_3 \longrightarrow \left[-CH_2-\underset{\displaystyle |}{CH}- \right]_n \quad (\text{with } OCOCH_3 \text{ on } CH)$$

In a 3-L three-necked flask equipped with a stirrer, an efficient condenser, and a nitrogen inlet is placed 300 g pure vinyl acetate, 800 mL water, 3 g commercial sodium dodecyl sulfate detergent, 3 g sodium dihydrogen phosphate monohydrate, and 1.5 g α, α'-azodiisobutyronitrile. The vessel is swept with nitrogen to remove the air, and the mixture is maintained with stirring at a temperature of about 40°C for 17 h. The reaction mixture is then treated with steam to remove unpolymerized vinyl acetate and cooled; then the solid polymer is filtered from the aqueous phase, washed with water, and dried. The polyvinyl acetate is obtained in the form of tiny spheres or granules that are easily handled. The reaction time may be decreased by increasing the amount of catalyst.

High molecular weight poly(vinyl acetate) is a clear, glassy solid soluble in many organic solvents. It softens at a fairly low temperature and discolors above 200°C. It is amorphous.

Poly(vinyl acetate) is readily hydrolyzed in alcohol solution to poly(vinyl alcohol) (PVA) the polymer of the unknown vinyl alcohol, or acetaldehyde enol. This polymer is less readily soluble in organic media, but dissolves in water. The bulky acetoxy group has been replaced by the smaller —OH, so the polymer can now crystallize; fibers with high crystallinity and orientation can be obtained. Poly(vinyl alcohol) of various degrees of hydrolysis and of several different molecular weight ranges are available commercially.

200. Hydrolysis of Poly(vinyl Acetate) (60–63)

$$[CH_2-\overset{\displaystyle OCOCH_3}{\overset{|}{CH}}-]_n \longrightarrow [CH_2-\overset{\displaystyle OH}{\overset{|}{CH}}-]_n$$

A total of 50 g high molecular weight poly(vinyl acetate) is dissolved in about 500 ml boiling methanol in a 1-L flask equipped with a condenser, with drying tube, pressure-equalizing dropping funnel, and a mechanical stirrer. All due precautions are observed to maintain anhydrous conditions. Then 5% sodium methoxide in methyl alcohol is added to the stirred refluxing polymer solution in 5- to 10-mL portions at intervals of 5 min. Approximately 25 to 30 mL of the solution is sufficient to catalyze the methanolysis. The reaction, once begun, proceeds rapidly with the precipitation of the poly(vinyl alcohol), which is insoluble in methanol. If the hydrolysis proceeds at too great a rate, the reaction may be moderated by the addition of 100 to 200 mL methanol and by external cooling. After the reaction has subsided, refluxing is continued for about 3 min, and the mixture is filtered. If the product is gel-like, it may be broken up by vigorous agitation in a home blender with cold methanol. The product is filtered and washed several times with alcohol and then dried.

Alternatively, the hydrolysis of poly(vinyl acetate) may be carried out in acidic media. For example, 30 g poly(vinyl acetate), 100 g water, and 1 g 95% sulfuric acid is heated and stirred for 6 to 8 h. at 95°C, or until a clear solution is obtained. Steam is then passed through the solution to remove acetic acid and complete the hydrolysis. The poly(vinyl alcohol) is isolated by precipitation with concentrated salt solution or the aqueous acid solution may be used directly, for example in the preparation of poly(vinyl butyral) described below.

201. Films and Fibers of Poly(vinyl Alcohol) (64, 65)

The preparation of poly(vinyl alcohol) films may be carried out by dissolving polymer in solvent to a solid content of 15 to 20% and casting the solution on a glass plate or a polished metal surface with a doctor knife (see chapter 2) of appropriate clearance. Poly(vinyl alcohol) is hygroscopic. Therefore, drying of the film cast from water is relatively difficult and slow. It is preferable to use a solvent such as alcohol-water (30–70) for this purpose. It is also preferable to carry out the drying of the film in a closed container in a slow stream of dry air.

It is interesting to note that poly(vinyl alcohol) film will react with iodine very much in the same manner as starch to form a sorption complex. If these films are stretched, the complex is oriented to a structure that is light polarizing. This is the basis of many of the polarizing filters now in use (66).

Fibers of poly(vinyl alcohol) can be prepared rather simply by extrusion of aqueous solutions into precipitating baths consisting of aqueous solutions of salts in a high concentration or into an organic nonsolvent. The fiber is then further processed by stretching in the usual manner. Such fibers are sensitive to cold water and usually will dissolve in hot water. These fibers are less soluble in water when under tension than when relaxed. The following represents a typical preparation of a poly(vinyl alcohol) fiber.

A 20% solution of high molecular weight poly(vinyl alcohol) in water is prepared at a temperature of about 73°C. It is placed in a spinning apparatus, such as the ones described in chapter 2, maintained at this temperature; and extruded by a mechanical pump through a spinneret into a 40% aqueous solution of ammonium sulfate at about 50°C. The rate of extrusion should be such that the filaments are completely coagulated and capable of supporting their own weight when removed at the other end of the bath. The fibers are wound up on a mechanical windup, and the yarn is washed on the bobbin in 50% acetone-water mixture. Finally, the yarn is treated with 95% acetone-water, and the yarn is then allowed to dry overnight at room temperature. The yarn is then stretched to its maximum degree at 75 to 100°C and wound onto a cone.

The techniques of precipitation of poly(vinyl alcohol) solution in strong salt solutions may also be applied to the preparation of poly-(vinyl alcohol) film.

Thus a 50% aqueous solution of poly(vinyl alcohol) at 75°C to which has been added approximately 0.01% of sodium dioctylsulfosuccinate is cast on a glass plate using a doctor knife of approximately 0.004 in clearance. The plate and film are then immersed in a saturated aqueous solution of sodium sulfate or of ammonium sulfate and maintained at a temperature of 40 to 50°C. The solution is allowed to coagulate for 5 min or more; then the plate is removed and immersed in a 50% aqueous acetone solution. After 3 h, the film is removed from the plate and washed with fresh portions of the aqueous acetone. It is finally washed with pure acetone. The product is a strong self-supporting film of poly(vinyl alcohol).

It is also possible to use organic media for the precipitation of poly(vinyl alcohol) solutions in fiber or film form. Thus a solution of 14% by weight poly(vinyl alcohol) in water at 60°C is extruded into a precipitating bath consisting of 94% acetone and 6% water at about 30°C. The polymer solution is precipitated in the form of a thread that is wound up at a rate sufficiently slow so that the thread, on issuing from the bath is completely self-supporting. The yarn is then stretched approximately five times its former length in a mixture of 50% diethylene glycol monomethyl ether and 50% water at about 30°C. The yarn so obtained should have tenacities of better than 2 g/denier.

Poly(vinyl alcohol) with its free hydroxyl groups offers considerable latitude for chemical transformation. Thus acetylation converts it back to the parent poly(vinyl acetate) while, for example, butyric anhydride, will produce poly(vinyl butyrate).

202. Preparation of Poly(vinyl butyrate) by Esterification

$$\left[-CH_2-\underset{\displaystyle OH}{\underset{|}{CH}}-\right]_n \longrightarrow \left[-CH_2-\underset{\displaystyle OCOCH_2CH_2CH_3}{\underset{|}{CH}}-\right]_n$$

A total of 10 g poly(vinyl alcohol) is refluxed with a mixture of 50 mL pyridine and 50 mL butyric anhydride until a clear, homogeneous solution results. The product is precipitated by pouring into water. The solid polymer is filtered, washed thoroughly, and dried. To obtain a pure product, it is preferable to dissolve the polymer in a solvent such as methanol and reprecipitate by pouring into a home blender containing water agitated at a high rate of speed. The slurry is filtered and the polymer dried. It exhibits properties similar to those of poly(vinyl acetate), except that it is somewhat less brittle at lower temperatures.

Because poly(vinyl alcohol) is a 1,3 glycol, it forms cyclic acetals with aldehydes. For example, water insoluble fibers may be prepared by extrusion of aqueous poly(vinyl alcohol) into aqueous formaldehyde. Acetalization occurs, together with an occasional cross-link, rendering the fiber insoluble (67).

Other aldehydes can be substituted for formaldehyde, giving a whole family of polymers, each with different characteristics with respect to melting point, stiffness, solubility, etc. Furthermore, acetalization may be carried out on the finished yarn, or film. For example, a film such as was made in an earlier preparation, soaked in an aqueous solution of benzaldehyde (10%) and phosphoric acid (1%) in hot ethanol will pick up a substantial weight of benzaldehyde in an acetal structure, during the course of several hours.

An important plastic that finds use in safety glass is poly(vinyl butyral), made from poly(vinyl alcohol) and butyraldehyde.

203. The Preparation of Poly(vinyl Butyral) (68)

$$\left[-CH_2-\underset{\displaystyle OH}{\underset{|}{C}H}-CH_2-\underset{\displaystyle OH}{\underset{|}{C}H}- \right]_n \xrightarrow{C_3H_7CHO} \left[-CH_2- \text{(1,3-dioxane ring, O–CH(}C_3H_7\text{)–O)} - \right]_n$$

A solution of 100 g poly(vinyl alcohol), 80 ml methyl alcohol, and 0.3 sulfuric acid in 820 mL water is prepared by mixing the ingredients in warming in a vessel equipped with a mechanical stirrer. To the agitated solution is added 80 g butyraldehyde. Then 300 g of this solution is placed in a 2-L three-necked flask equipped with mechanical stirrer and a condenser. Then 80 g butyraldehyde is added with vigorous stirring, followed by the remainder of the poly(vinyl alcohol) solution over a period of about 20 min. During this time, the internal temperature should rise to about 70°C. At the conclusion of this period, 600 mL hot water at approximately 70°C is added over a period of 15 to 20 min. The resulting mixture is agitated for an additional 10 min, then 3 g concentrated sulfuric acid dissolved in 25 mL water is added. The reaction mixture is allowed to stir for an additional 1 h; then the resin is filtered and washed repeatedly with water. A product is obtained, if the agitation is satisfactory, of a particle size that is appropriate for easy handling and filtering.

The poly(vinyl butyral) prepared in this way must be washed thoroughly with dilute alkali to remove the last traces of acid, which would catalyze decomposition.

The major use of poly(vinyl butyral) is in the preparation of safety glass in which a thin layer of plasticized polymer is sandwiched between two sheets of glass. The poly(vinyl butyral) has outstanding adherence to the glass. It is elastic and tough and serves admirably for this purpose.

To prepare a plasticized poly(vinyl butyral), 100 g polymer is mixed with a plasticizer such as triethylene glycol di-2-ethylbutyrate. The polymer and plasticizer are blended together with enough ethanol to form a plastic mass. This material can be spread or rolled into a sheet and the solvent allowed to evaporate. The resulting product may be dusted with talc to decrease the tackiness of the material, if it is to be stored. To prepare a laminate with two pieces of glass, a sheet of the butyral is washed and placed between carefully cleaned pieces of glass. The seal is effected by heating in a Carver press or similar source of heat at temperatures of 150 to 175°C and moderate pressures. For the preparation of commercial glass laminates, it is necessary to finish the process by heating at higher temperatures and pressures. Specialized equipment, however, is necessary.

Other groups reactive toward secondary hydroxyls will also transform poly(vinyl alcohol). For example, cyanoethylation occurs readily, without chain cleavage.

204. Cyanoethylation of Poly(vinyl Alcohol) (69)

$$\left[\begin{matrix} CH_2-CH \\ \quad\ \ | \\ \quad\ \ OH \end{matrix}\right]_n + CH_2{=}CH-CN \longrightarrow \left[\begin{matrix} -CH-CH- \\ \qquad\ | \\ \qquad\ OCH_2CH_2CN \end{matrix}\right]_n$$

A slurry of 45 g high-viscosity poly(vinyl alcohol), 265 g acrylonitrile, and 5 g of a 5% aqueous solution of sodium hydroxide is placed in a 500-ml three-necked flask fitted with a mechanical stirrer and a reflux condenser. The mixture is heated externally to reflux. After about 30 mins the poly(vinyl alcohol) begins to go into solution and forms a gel. After another 15 min at reflux, the external heat is discontinued, and the reaction mixture is cooled and then neutralized with glacial acetic acid. The resulting viscous, light tan product, which consists of a solution of polymer in acrylonitrile, is poured into about 2-L diethyl ether, and a taffy-like precipitate is obtained. This precipitate is dissolved in 300 mL acetone and again precipitated by pouring into diethyl ether. The product is then dried in a vacuum over phosphorus pentoxide. The yield is about 95 g.

Poly(vinyl acetate) from free-radical polymerization differs little in tacticity regardless of the circumstances under which it is prepared; an essentially atactic placement occurs from polymerization with an azo initiator at room temperature and above as well as by triethylborane/oxygen at −78°C. (70). PVA derived by hydrolysis of the acetate is consequently also atactic in structure, although a high level of crystallinity can be developed. Differences in properties of PVA's derived from poly(vinyl acetates) prepared in varying ways have been reported, but these differences are believed due to variances in molecular weight and chain branching rather than tacticity changes.

PVA of syndiotactic-enriched structure can be prepared from poly(vinyl trifluoroacetate) that is made by radical initiation at −78°C (71). Isotactic-enriched PVA results from the cleavage of poly(vinyl-*t*-butyl ether) made at −78°C in a cationic polymerization (72, 73) and from debenzylation of poly(vinyl benzyl ether) cationically polymerized at −78°C (74). It has also been reported that polymerization of vinyl trimethylsilyl ether at −78°C with $SnCl_4$ or Et_2AlCl as catalyst gives, after cleavage of the trimethylsilyl groups, a predominantly isotactic or syndiotactic PVA, the former in less polar solvents (toluene), the latter in polar ones (methylene chloride) (75). While the question of tacticity assignment in PVA may not be fully resolved, it appears that in none of these cases is the stereoregularity overwhelmingly of one type, as it is with, say, isotactic polypropylene. The vinyl trimethylsilyl ether route to both isotacticity and syndiotacticity may, at the present, achieve the highest levels of tactic purities. Isotactic, heterotactic, and syndiotactic triads of 86/10/4 and 6/40/54 have been reported (75).

There is apparently no increase in crystallinity in PVA with an increase in stereoregularity. The OH and H are evidently quite interchangeable in the crystal lattice development of this highly polar polymer. There is an influence of stereoregularity on other properties, however. Both syndiotactic and isotactic PVA are more water resistant than the atactic (76).

205. Preparation of Mainly Isotactic Poly(vinyl Alcohol) from Poly(vinyl-t-butyl Ether) (73)

$$CH_2{=}CH{-}O{-}C(CH_3)_3 \longrightarrow \left[CH_2{-}CH(O{-}C(CH_3)_3)\right]$$

I

$$\text{I} + HBr \longrightarrow \left[CH_2{-}CH(OH)\right] + (CH_3)_3CBr$$

A 200-mL polymerization vessel and stirrer apparatus are baked dry at 120°C and assembled in a stream of nitrogen to maintain dryness on cooling. To the nitrogen-purged vessel is added 100 mL dry toluene and a solution of 0.034 g distilled, boron trifluoride etherate (2.4×10^{-4} mol) in 5 mL dry toluene. The mixture is cooled to −87°C in dry ice-acetone and 5 g freshly distilled (from sodium) and previously chilled vinyl-*t*-butyl ether (bp 76 to 78°C) is added by syringe through a serum cap. The polymerization occurs in a homogeneous manner and is completed within 1 h. The polymer is precipitated in methanol containing a small amount of aqueous ammonia, washed thoroughly in methanol, and dried in vacuum at 50°C. The polymer should show moderate crystallinity by x-ray diffraction. The intrinsic viscosity is about 0.7 in benzene at 30°C.

To prepare mainly isotactic poly(vinyl alcohol), 1 g of the above poly(vinyl-*t*-butyl ether) dissolved in 100 mL dry toluene in a 200 mL round-bottom flask, with a gas inlet and outlet protected by a drying tube, is cooled to 0°C, and gaseous HBr is passed through the solution for 15 min. The precipitated polymer is filtered, washed thoroughly with methanol containing a small amount of aqueous ammonia, and dried in vacuum at 50°C. Ether cleavage is essentially complete. The polymer has an excess of isotactic diads or triads by NMR examination of a deuterium oxide solution (72); the ratio of percent isotactic/heterotactic/syndiotactic triads is about 49/38/13.

206. Preparation of Mainly Syndiotactic Poly(vinyl Acetate) from Poly(vinyl Trifluoroacetate) (71)

$$CF_3-\overset{\overset{O}{\|}}{C}-OCH=CH_2 \longrightarrow \left[CH_2-\underset{\underset{\underset{\underset{CF_3}{|}}{C=O}}{|}}{\underset{O}{|}}{CH}\right] \quad \text{I}$$

$$\text{I} \longrightarrow \left[CH_2-\underset{OH}{\underset{|}{CH}}\right]$$

A 500-mL three-necked round-bottom flask is fitted with a gas inlet, a stirrer, and a gas outlet protected with a drying tube. The flask is flamed under argon, and a positive pressure of argon's maintained. The flask is cooled to −78°C, and 66 g redistilled vinyl trifluoroacetate (bp 39°C) (Peninsular Chemical Research) and 18 mL *n*-heptane are added. To the cold solution is added 1 mL tri-*n*-butylborane in 5 mL *n*-heptane by syringe. At 10-min intervals, five 10-mL portions of dry air are added also by syringe. The polymerization is allowed to continue for 16 h at −78°C. The polymerization mixture is placed in a home blender and stirred with 300 mL cyclohexane. The precipitated polymer is filtered, washed with 300 mL cyclohexane, and dried in a vacuum over at 50°C. The dried poly(vinyl trifluoroacetate) weighs about 38 g and has an intrinsic viscosity of about 2.2 in methyl ethyl ketone at 30°C.

Then 1 g of this polymer is dissolved in 10 mL acetone and added to 240 mL of a 50 : 50 ammonium hydroxide-methyl alcohol mixture with rapid stirring. The solution is stirred rapidly for 3 h, filtered, washed with acetone, and dried. PVA is obtained, which is free of trifluoroacetate groups as determined from IR and NMR spectra and analyses for fluorine. The percentages by NMR are 11/46/43 for isotactic-/heterotactic-/syndiotactic triads.

Acrylic acid, its salts, amides, esters, and acid chlorides can be polymerized.

The preparation of a vinyl polymer of a high degree of reactivity is made possible by the availability of acryloyl chloride (77), which can be polymerized to a polymeric acid chloride. It has the expected high degree of sensitivity toward hydroxylic reagents and amines and can be prepared only in anhydrous media and handled only in a dry box. Homopolymerization is not particularly easy. However, minor proportions of acryloyl chloride can be copolymerized (78) with other vinyl monomers. Cross-linking reactions, such as treatment with a diamine can then be carried out to give insoluble, infusible products with improved tensile properties.

Acrylamide will undergo vinyl polymerization via free radical catalysis to give a polymer with pendant carboxamide groups (see chapter 12, for a Michael addition-type polymerization of acrylamide.) Water-soluble polymers can be prepared in an aqueous system, as in the following example.

207. Solution Polymerization of Acrylamide (79)

$$CH_2{=}CH{-}\overset{\overset{\displaystyle O}{\|}}{C}{-}NH_2 \longrightarrow \left[{-}CH_2{-}\underset{\underset{\displaystyle NH_2}{|}}{\underset{\displaystyle C{=}O}{\underset{|}{CH}}}{-} \right]_n$$

Acrylamide is toxic. Use care in handling.

In a 1-L three-necked flask equipped with stirrer, gas inlet, thermometer, and condenser are placed 51.8 g acrylamide and 414.7 g distilled water. The acrylamide solution is stirred and heated to 68°C under a rapid stream of carbon dioxide. Then 7.7 g isopropyl alcohol and 0.096 g potassium persulfate are added. The temperature of the reaction rises to 75 to 80°C., where it is maintained by a heating bath for 2 h. The product is obtained in a clear, colorless solution with a very high viscosity. The polymer can be precipitated in methanol, washed well with methanol, and dried in a vacuum at 50°C. The inherent viscosity is about 1.0 (1N solution of sodium nitrate, 0.5% polymer concentration, 30°C). The relationship of intrinsic viscosity to molecular weight is

$$[\eta] = 3.73 \times 10^4 M^{0.66}$$

where M is weight average molecular weight.

The polymerization of acrylate esters can be carried out readily. The acrylate esters are rubbery polymers exhibiting a decreasing T_g with increasing chain length of the alkyl portion up to C_8, after which the T_g increases. T_g of the methyl ester is around 0°C, of the octyl, −65°C. Hexadecyl acrylate forms a polymer with a T_g of about 35°C. Consequently, acrylate polymers are often made in emulsion. The polymerization of methyl acrylate is typical of the preparation of the acrylate esters. An extensive series of poly(acrylate esters) has been prepared and reported in the literature (80).

208. The Polymerization of Methyl Acrylate (80)

$$CH_2{=}CHCO_2CH_3 \longrightarrow \left[{-}CH_2{-}\underset{\underset{\displaystyle CO_2CH_3}{|}}{CH}{-} \right]_n$$

A three-necked flask is fitted with a stirrer, a reflux condenser, and a thermometer. The flask is charged with 400 mL water, 2 g Triton 720, which is an alkyl arylether sulfonate, 2 to 4 g Tergitol paste, which is sodium 2-methyl-7-ethyldecyl-4-sulfate penetrant, and 0.01 g ammonium persulfate. The solution is stirred slowly and 200 g distilled methyl acrylate is added. Heat may be applied to the reaction vessel to induce polymerization. If polymerization does not start within 10 min after refluxing has occurred, additional ammonium persulfate may be added. If excessive quantities are required the monomer is not of sufficient purity. Once initiated, the polymerization usually proceeds at a rate sufficient to cause refluxing without external heating for 15 to 30 min. After about 30 min, heat is applied, and the refluxing temperature is allowed to rise until it is about 95°C, at which point the polymerization may be considered to be complete. The resulting polymer emulsion is steam distilled for 15 to 30 min to remove excess monomer and is run slowly into twice its volume of hot 5% sodium chloride solution. The polymer is precipitated in the form of discrete particles that are filtered and washed with hot water until free of salt. It is then air dried. The yield is quite good in all cases.

Higher alkyl acrylates may be polymerized using the identical recipe described here. The properties of a whole series of such esters are available (80).

A number of *N*-vinyl lactams have been prepared by the reaction of acetylene with the lactam. The most important of these is *N*-vinyl pyrrolidone (81 to 84).

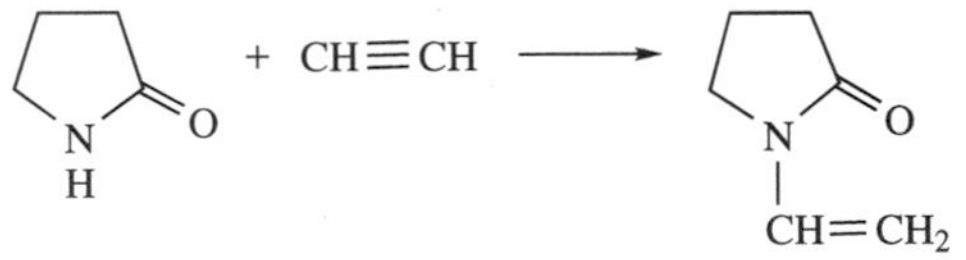

N O H + $CH{\equiv}CH$ ⟶ N O $CH{=}CH_2$

This monomer polymerizes readily with most free-radical catalysts, under a variety of conditions to give a water-soluble high polymer. This material has been used as a blood plasma substitute.

209. Polymerization of N-Vinyl Pyrrolidone (81–84)

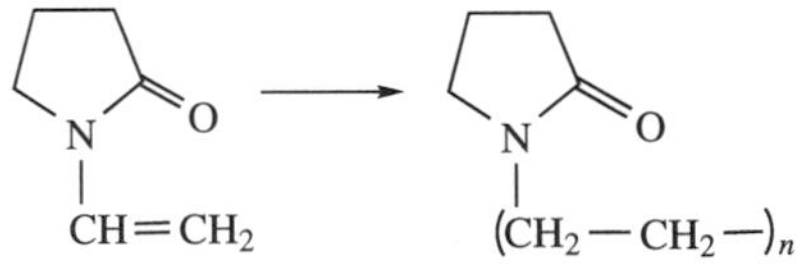

N O $CH{=}CH_2$ ⟶ N O $(CH_2{-}CH_2{-})_n$

A mixture of 30 g distilled *N*-vinyl pyrrolidone and a solution of 40 g neutral potassium sulfite in 200 mL water is stirred vigorously in an atmosphere of nitrogen for 24 h at a temperature of 35 to 40°C. At the end of 24 h the polymerization product, which is a viscous solution, is decanted into a dish and evaporated under a stream of dry air or nitrogen on a steam bath with stirring. The product, poly(vinyl pyrrolidone), is obtained in good yield in the form of a clear, horn-like solid mass mixed with potassium salts. It will dissolve in water to give a viscous solution. The polymer can be separated from the potassium salts by extraction into alcohol, or by dialysis of an aqueous solution in, for example, a cellophane bag, against running water.

Some unsaturated ketones and aldehydes will polymerize with peroxide catalysts. Methyl vinyl ketone and methyl isopropenyl ketone give products with structural similarity to poly(methyl acrylate) and poly(methyl methacrylate), respectively. They are not easy to achieve at high molecular weight, are costly monomers and show less attractive properties than their ester counterparts (85). Acrolein is also polymerizeable in a water solution using a silver nitrate catalyst system (86, 87), but the resulting polyaldehyde forms a hydrated structure thought to involve one water molecule added to two closely placed –CHO along the chain.

Pendant aldehyde groups can be attached to a polymer chain by hydroformylation (oxo reaction) of the double bonds in an unsaturated polymer, such as polybutadiene or poly(butadiene-co-styrene) (88a).

$$-CH{=}CH- \; + CO + H_2 \xrightarrow{[CO(CO_4)]_2} -CH_2-\underset{\displaystyle CHO}{\underset{|}{CH}}-$$

High conversion to aldehyde gives a polymer with a strong tendency to gel; this can be avoided by inclusion of an agent that reacts rapidly enough to convert aldehyde to acetal, such as 2,3-butanediol.

High pressures (2400 psi at room temperature charge) and temperatures (125 to 180°C) are required.

210. Suspension Polymerization of Methyl Methacrylate

$$CH_2{=}C(CH_3)CO_2CH_3 \longrightarrow \left[-CH_2-\underset{\displaystyle CO_2CH_3}{\underset{|}{\overset{\displaystyle CH_3}{\overset{|}{C}}}}-\right]_n$$

A mixture of 200 g monomeric methyl methacrylate (freshly distilled) with 2.5 g α, α'-azodiisobutyronitrile, 40 mL of a 5% aqueous poly(methacrylic acid) solution, 20 g disodium hydrogen phosphate dodecahydrate, and 400 mL water is placed in a 1-L three-necked flask mounted on a steam bath and equipped with a nitrogen inlet, an efficient stirrer, a thermometer, and a reflux condenser. The reaction mixture is stirred vigorously and heated to boiling in a steam bath, under an atmosphere of nitrogen. The initial reflux temperature will be about 82°C. As polymerization continues, the temperature will rise to 93°C. At this temperature, polymerization is essentially complete. The granular polymeric methyl methacrylate is filtered, washed in water, and dried. The total time required for the polymerization is approximately 20 min.

Alternatively, polymerization may be carried out without added initiator. This is not a true thermal polymerization, but is initiated by adventitious impurities.

Highly purified monomer will not polymerize under these conditions. In a 1-L three-necked flask equipped with condenser and stirrer is placed 100 g methyl methacrylate and 200 mL water containing about 4 g poly(vinyl alcohol) dispersing agent. This mixture is stirred vigorously and heated at about 80°C for approximately 40 min. The internal temperature begins to rise at this point and will reach a maximum of 85°C. At this point, the mixture is cooled to 60°C by the addition of cold water to the flask. The granules of polymer are separated, washed with water, and dried at 100°C. The poly(methyl methacrylate) may be converted to molded objects.

Methyl methacrylate may be safely bulk polymerized under mild conditions. In this way it is possible to prepare castings by in situ polymerization at 40°C.

211. Bulk Polymerization of Methyl Methacrylate (89)

$$CH_2{=}\underset{CO_2CH_3}{\overset{CH_3}{\overset{|}{\underset{|}{C}}}} \longrightarrow \left[-CH_2-\underset{CO_2CH_3}{\overset{CH_3}{\overset{|}{\underset{|}{C}}}}- \right]_n$$

A total of 100 g freshly distilled monomeric methyl methacrylate is mixed with 3 g high molecular weight poly(methyl methacrylate) thickener, 0.007 g methacrylic acid, and 0.05 g α, α'-azodiisobutyronitrile. The viscous mixture is maintained in an oven at about 40°C, under which conditions it will polymerize in about 25 to 30 h to a clear, solid block in the shape of the polymerization vessel. For example, the viscous mixture may be poured between two glass plates that are separated by a compressible gasket. If this assembly is maintained in the oven at about 40°C, as previously described, for 25 to 30 h, a clear sheet of poly(methyl methacrylate) is obtained. It may be separated from the glass-retaining plates by raising the temperature to 95°C.

Poly(methyl methacrylate) has a T_g of about 105°C. Stereoregular forms of the polymer are considered further later in this chapter. Objects may be embedded in poly(methyl methacrylate) using the same techniques given for polystyrene.

If higher esters of methacrylic acid are polymerized, the resulting materials will be lower in softening temperatures than methyl methacrylate, e.g., butyl methacrylate leads to a polymer softening at around 30°C, while the n-amyl ester polymer is essentially elastomeric with a T_g below room temperature (89, 90). Methacrylonitrile is polymerizeable in much the same way as acrylonitrile to give a lower softening product, one that can be molded at 135 to 155°C, though with discoloration likely. Methyl α-chloroacrylate is a potent lachrymator but can be polymerized, if anyone feels obligated to do so, just as methyl methacrylate, but with a softening temperature about 40°C higher than the latter, the chloro-analogue is more scratch resistant, but discolors easily.

Multitudes of other methacrylic acid derivatives have been polymerized, but only a few have seen any commercial application. Among the compounds examined have been glycol dimethacrylate (91), polymerized as methylmethacrylate but producing a cross-linked polymer used in dental fillings at one time. Another is methacrylonitrile (92), whose polymer offered no benefits and some disadvantages (discoloration) relative to acrylonitrile. Chlorine-containing analogues have also surfaced, e.g., methyl α-chloroacrylate (93) and α-chloroacrylonitrile (94).

Vinylidene chloride is a low boiling (32°C) liquid that polymerizes readily. Bulk polymerization may be carried out at 40 to 50°C in sealed tubes but should not be done on a large scale, because the polymerization may get out of control if efficient dissipation of heat is not possible. It is best polymerized in an emulsion. The polymer may be molded, but the molding temperature (about 200°C) is rather high and some decomposition occurs. Vinylidene chloride is found most frequently in copolymers. Some are described in a later section.

Caution! Vinylidene chloride monomer tends to form peroxides and phosgene in contact with air, giving a mixture that may explode on heating (95). Those planning to work with this monomer should acquaint themselves fully with the hazards (96).

212. Polymerization of Vinylidene Chloride (97)

$$CH_2{=}CCl_2 \longrightarrow [-CH_2-CCl_2-]_n$$

In a 1-L three-necked flask equipped with a nitrogen inlet, a condenser, and a stirrer is placed 100 g pure vinylidene chloride, 300 mL of an aqueous solution containing 3 g ammonium persulfate, 1 g sodium hydroxide, 1.5 g sodium thiosulfate, and 3 g of a detergent such as sodium lauryl sulfate. The air in the reaction vessel is displaced by nitrogen, and the temperature is maintained at 30°C with stirring. After about 6 h, polymerization is essentially complete and a polymer emulsion is obtained. This is removed from the reaction vessel, and the polymer is precipitated by the addition of 100 to 150 mL saturated salt solution with stirring. The easily filterable, finely divided white powder is removed, washed with water, and dried. The yield is approximately 85 g. T_m is about 160°C.

In view of the low boiling point of vinylidene chloride (32°C) a nitrogen sweep should not be used. Instead, the reaction vessel should be kept under a slight positive pressure of nitrogen. If desired, steel pressure vessels can be used for the polymerization.

Vinylidene fluoride may also be polymerized to a crystalline polymer remarked for its thermal stability. Although easier to fabricate than poly(tetrafluoroethylene), it is not at all an equal to the latter, nor has it captured the immensely widespread group of applications of the latter or the accompanying production volume. Polymerization of vinylidene fluoride must be done under rather high-pressure conditions, although with di-*t*-butyl peroxide this need is kept to a minimum. Still, an autoclave and all due precautions in its use are required.

213. Polymerization of Vinylidene Fluoride (98)

$$CH_2{=}CF_2 \longrightarrow \left[CH_2{-}CF_2 \right]$$

A 300-mL stainless-steel autoclave is charged with 100 mL deionized and deoxygenated water and 0.8 g di-*t*-butyl peroxide. The autoclave is evacuated, cooled in liquid nitrogen, and then charged with 35 g vinylidene fluoride by gaseous transfer in vacuo. The autoclave is then sealed and placed in an electrical heating jacket mounted in a shaking apparatus and held at 122 to 124°C, for 1 to 5 h. A maximum pressure of about 800 psi is reached after the first 2 h of heating, and the pressure then decreases as the polymerization proceeds.

After the reaction period, the autoclave is cooled, vented, and opened. The contents consist of precipitated poly(vinylidene fluoride) suspended in a liquid phase having a pH of 2.5. The polymer is vacuum filtered and washed on the filter funnel with methanol, then washed 10 times with distilled water, and given a final wash with methanol. It is then dried in a vacuum oven at 102°C. The washed and dried polymer weighs 29 g, representing an 83% conversion of the monomer to polymer. The polymer, which melts at 160 to 165°C, has a high molecular weight, as indicated by its plasticity number of 3020. The *plasticity number* is an empirical index that indicates relative molecular weight of vinylidene fluoride polymers. Because of the difficulty of obtaining a true solution of the polymer, absolute molecular weight determinations have not been possible to obtain. The plasticity number is the area in square millimeters of a plaque made by placing 0.5 g polymer powder piled in a cone between the platens of a Carver press heated at 225°C. The platens are brought together to compress the powder under slight pressure (less than 50 psi) between the heated platens, and the powder is preheated in this manner at 225°C for 30 s. A pressure of 2500 psi is then applied for 60 s at a platen temperature of 225°C. The greater the area of the plaque so produced, the lower the molecular weight of the polymer, and conversely. Thus comparison of different preparations of the polymer is possible.

The polymer likewise has excellent thermal stability. It does not decompose when heated to 320°C and shows substantially no discoloration when exposed in a circulating air oven for 100 h at 200°C. It has good low-temperature properties as shown in a test in which the polymer is flexed 180° over a 0.125-in mandrel at −70°C, without breaking.

9.6. THE FREE-RADICAL POLYMERIZATION OF OTHER DISUBSTITUTED, TRISUBSTITUTED AND TETRASUBSTITUTED OLEFINS

In sharp contrast to the number of mono- and 1,1-disubstituted olefins known to homopolymerize by a free-radical mechanism, there are only very few 1,2-disubstituted olefins that give high polymer. It is interesting to note that most of the 1,2-disubstituted olefins that polymerize well (including those that polymerize by other mechanisms; cf. acenaphthylene) are unsaturated 5-membered cyclic compounds.

An interesting 1,2-olefin is vinylene carbonate, first reported in the literature by Newman and Addor (99, 100). It is prepared by chlorination of ethylene carbonate, followed by dehydrohalogenation. Polymerization is carried out with benzoyl peroxide or an azo catalyst (101).

It is essential that the monomer be very pure if high molecular weight polymer is to be obtained. A thermal treatment has been claimed (102), but treatment with sodium borohydride is preferred.

214. Preparation of Chloroethylene Carbonate (101)

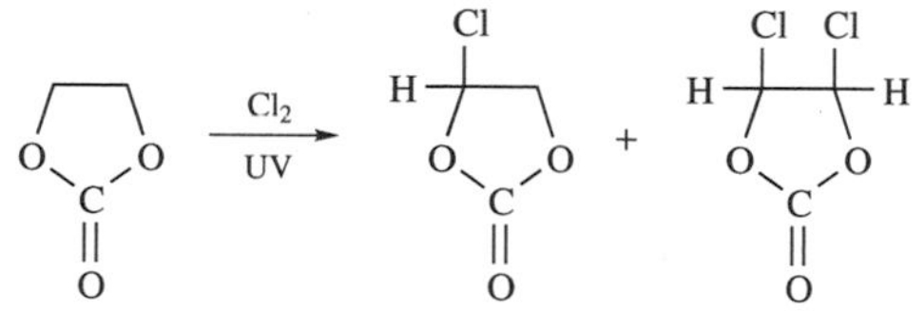

A total of 500 g (5.7 mol) ethylene carbonate and 1-L carbon tetrachloride are mixed in a 2-L three-necked flask equipped with a uv lamp in a quartz jacket immersed in the flask, a gas inlet tube, and an efficient condenser. The two-phase system is heated by use of a heating mantle until refluxing occurs. The heat is then turned off, the uv lamp is turned on, and chlorine from a cylinder is introduced at a rate sufficient to maintain vigorous refluxing. The ethylene carbonate–rich phase gradually disappears and a homogeneous solution forms. Chlorination is continued beyond this point until the total weight of chlorine added is about 600 g (8.5 mol). This requires 3 to 5 h. The product is isolated by fractional distillation through an efficient column. After removal of the solvent and a low-boiling solid impurity, 114 g (37%) 1,2-dichloroethylene carbonate (bp 91°C/30 mm; n_D^{25} 1.4606) and 420 to 475 g (60 to 68%) chloroethylene carbonate (bp 102°C/8 mm; n_D^{25} 1.4525) are isolated.

215. Preparation of Vinylene Carbonate (101)

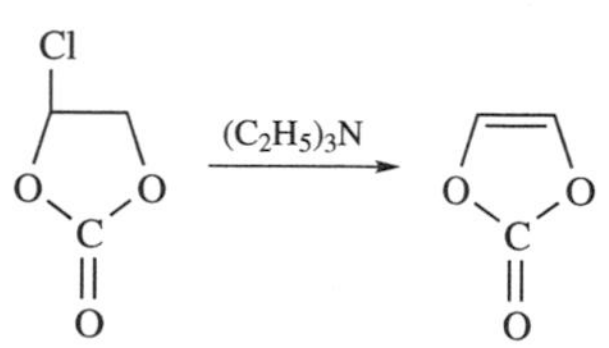

The 450 g (3.65 mol) chloroethylene carbonate from the preceding preparation, 450 mL anhydrous ether, and 4 g di-*tert-butyl-p*-cresol (to prevent adventitious polymerization of the monomer) are added to a 2-L three-necked flask equipped with a precision-ground stirrer, dropping funnel, and reflux condenser to which a drying tube filled with Drierite is attached. Then 560 mL (4.1 mol) triethylamine is added slowly over about 4 h to the stirred refluxing solution (a temperature of about 45°C should be maintained). Gentle refluxing is maintained for 2 days. A copious precipitate of amine salt forms, and the color of the solution becomes dark brown. The precipitate is collected and washed four times, each time with 400 mL of a mixture of 50/50 vol. % benzene and ether. The second and third washings are carried out by slurrying the solid precipitate with the solvent mixture in a beaker. The filtrate and washings are combined, and most of the ether and some of the benzene are removed by simple distillation. Distillation of the remainder of the solution at reduced pressure through an efficient column yields 200 to 230 g (63 to 73%) of vinylene carbonate (bp 74°C/30 mm). This material rapidly colors (brown) on standing. It is further purified by refluxing for 1 h over 1.5% by weight sodium borohydride and then distilling. A second treatment with sodium borohydride is recommended to obtain a color-stable product (n_D^{25}, 1.4185; mp 20.5°C).

216. Polymerization of Vinylene Carbonate (101)

[Reaction scheme: vinylene carbonate (cyclic O–C(=O)–O ring with C=C) —AIBN→ poly(vinylene carbonate), repeat unit in brackets with subscript n]

A thick-walled test tube is necked down near the top and, after cooling, 0.01 g azobiisobutyronitrile (AIBN) is introduced. By use of a hypodermic syringe, 5 mL sodium borohydride-treated vinylene carbonate is added. The tube is cooled in ice water to freeze the monomer and evacuated through a stopcock to 1 mm pressure. The stopcock is closed, and the monomer is then degassed by melting and refreezing. The evacuation and degassing procedures are then repeated, after which the system is sealed in vacuo. The sealed tube is placed in a bath thermostated at 60 to 65°C. Polymerization takes place slowly to give a clear solid resin in 18 to 72 h. The tube is broken, and the tough plug of polymer is dissolved in 50 mL DMF at room temperature and reprecipitated as a white fibrous solid by adding this solution slowly with stirring to 200 mL methanol. The polymer is collected by filtration and washed repeatedly by slurrying with methanol until the filtrate is nearly clear. The yield of polymer is 3.7 to 5.6 g (57 to 87%); the inherent viscosity at 30°C of a 0.5% solution of polymer in *N,N*-dimethylformamide is 2.0 to 3.5.

Vinylene carbonate polymer can be hydrolyzed to polyhydroxy-methylene, an intractable polymer. Hydrolysis is best carried out on a finished product, e.g., a cast film of poly(vinylene carbonate) to give an orientable film of polyhydroxymethylene.

217. Preparation of Polyhydroxymethylene (101)

[Reaction scheme: poly(vinylene carbonate) $\xrightarrow[CH_3OH]{NaOCH_3}$ $(-CH(OH)-)_n$]

Poly(vinylene carbonate) is dissolved in DMF to form a 10% solution. This is cast as a 10-mL film on a glass plate by use of a doctor knife. After drying overnight at room temperature, the clear film is removed from the plate and hydrolyzed by suspending it in a 1% sodium methoxide solution in methanol in a covered beaker. Hydrolysis to clear, but crinkled, films of polyhydroxymethylene is complete after 24 h at 50 to 60°C or after 3 to 5 days at room temperature. The progress of hydrolysis may be followed conveniently by noting the disappearance of the carbonyl bond at 5.5 μ in the ir spectrum.

The films of polyhydroxymethylene are stiff and brittle when thoroughly dry, but become limper and tougher in moist air. They are insoluble even in boiling water and retain a moderate amount of strength when wet. The wet film can be cut into narrow strips (about 5 mm wide), and these can be oriented by drawing them quickly over a rod heated at 200°C. Such films are quite strong and stiff.

Surprisingly, the rather highly hindered phenyl vinylene carbonate can be polymerized in bulk to a modest molecular weight (e.g., degree of polymerization of about 80). The homopolymer has a high T_g of 213°C by differential thermal analysis (DTA).

Other 1,2-disubstituted monomers of a similar structure will, in some cases, polymerize or at least copolymerize. These include *N*-substituted maleimides (103–105), and maleic anhydride (106). In addition, the noncyclic maleonitrile will form copolymers (107).

Tetrasubstituted monomers known to polymerize satisfactorily include chlorotrifluoroethylene(I) (102–110) and tetrafluoroethylene (II) (111–113).

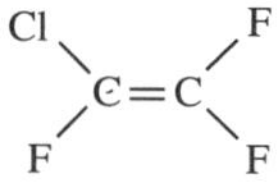

$$\mathrm{(Cl)(F)C{=}CF_2} \quad \text{(I)} \qquad \mathrm{F_2C{=}CF_2} \quad \text{(II)}$$

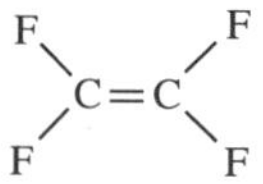

These compounds polymerize only under pressure and are dangerous to handle; therefore, the polymerizations should not be attempted except by those experienced in the handling of these materials.

9.7. CATIONIC POLYMERIZATION OF VINYL COMPOUNDS

Cationic polymerization of vinyl compounds is not nearly so applicable as is free-radical polymerization. The growing chain has a terminal carbonium ion together with its counter ion, and polymerization is ordinarily much more rapid and vigorous. Cationic polymerizations are usually initiated at low temperatures to suppress undesirable chain-terminating side reactions. The choice of solvent, catalyst concentration, cocatalyst, etc., is extremely important. It is, of course, essential that the solvent, if used, should be almost completely unreactive. The theoretical implications of general cationic catalysis are discussed at length in the literature (2).

Advances have been made in developing "living carbocationic" polymerizations. This work has been pioneered by Higashimura (114), who reported living polymerization of vinyl ethers using HI/I initiating systems, and by Kennedy et al (115) who described the living polymerization of isobutene using cumyl ether-BCl_3 systems. These studies recognized that chain transfer in carbocationic systems could be controlled by using a large counterion and by eliminating spurious initiation from protons generated from adventitious water. Living carbocationic polymerizations of *p*-chlorostyrene, for example, were achieved using a 2-chloro-2,4,4-trimethylpentane/titanium tetrachloride initiating system in the presence of dimethylacetamide as an electron donor and 2,6-di-*t*-butylpyridine as a proton trap in methylene chloride/cyclohexane solvent at −80°C. The living nature of the polymerization was confirmed by the linear plot of M_n vs weight of polymer formed and by the plot passing through the origin. Molecular weight distributions were narrow $M_w/M_n = 1.2$ to 1.9. Polymerizations of vinyl ethers using the same principles gave $M_w/M_n = 1.2$.

9.7.1. Conventional Cationic Polymerizations

The following experimental examples illustrate some of the techniques used in the polymerization of several vinyl compounds. It will be noted that the procedures are quite different from those used in the polymerization of vinyl monomers by typical free-radical recipes.

One of the few monomers that responds satisfactorily to cationic and to free-radical catalysis is styrene, although it is admittedly difficult to get a product of as high molecular weight with cationic catalysis as is possible with free radical.

Substitution of styrene with an α-methyl group gives a monomer that will polymerize readily with typical cationic catalysts. Again, it is necessary to operate at very low temperatures to obtain high molecular weight. It appears that, in general, for cationic catalysis to occur, a cocatalyst is necessary that is capable of generating a proton or carbonium ion by reaction with what is customarily designated as the catalyst. The simplest cocatalyst is water, which is often adventitiously present. For example, with boron trifluoride,

$$BF_3 + H_2O \longrightarrow H^{\oplus} + \overset{\ominus}{B}F_3OH$$

water provides a proton as the initiating fragment. The unavoidable presence of small quantities of water almost certainly provides the cocatalyst for the next preparative example, which is a convenient cationic polymerization of α-methylstyrene.

218. Preparation of Cationic Polymerization of Methylstyrene (116)

Toluene is dried over calcium hydride and used directly. α-Methylstyrene is purified by first washing with 5% aqueous sodium hydroxide twice and with distilled water three times; it is then dried over calcium hydride. It should be distilled before use (bp 52°C/10 mm), taking a heart cut. To a dry, nitrogen-flushed, 8-oz beverage bottle is charged 50 mL toluene and 10 ml α-methylstyrene, using a nitrogen sweep during addition. The bottle is closed with a rubber septum and crown cap with a hole for catalyst addition. The bottle is immersed in dry ice-acetone.

Boron trifluoride (BF_3), gas is flushed through a 1-ft length of Tygon tubing attached directly to a cylinder of the gas; the end of the tubing should be vented through a hypodermic needle to ensure high gas velocity at the exit and no possible backup of air. When the tubing is thoroughly flushed, a dry hypodermic syringe is inserted through the tubing, and 10 mL BF_3 is withdrawn; it is injected directly into the bottle whose contents are at −78°C. The bottle is swirled gently and cooling is maintained. Polymerization occurs very quickly, but the bottle is allowed to stand at −78°C for 5 h. The contents are poured with stirring into about 400 mL methanol in a blender, filtered, washed three times with methanol in the blender and dried at 60°C in vacuum. About 7 g polymer is obtained, which softens at 175 to 200°C. The inherent viscosity in toluene (0.1 g/100 mL toluene, 25°C) should be 0.85 to 1.0.

Probably the most important monomer that is polymerized commercially by a conventional cationic catalyst is isobutylene (117). It polymerizes easily at low temperatures to high molecular weight polymers, which are relatively soft and rubbery. Isobutylene is not affected by free-radical catalysts. However, the cationic polymerization of carefully purified monomer may occur with almost explosive violence, hence it is ordinarily necessary to moderate this reaction by using very low boiling diluents such as methyl chloride, or ethylene. The following represents a typical technique for polymerizing isobutylene with a cationic catalyst.

219. Polymerization of Isobutylene with Cationic Catalyst

$$CH_2{=}C(CH_3)_2 \longrightarrow \left[-CH_2-C(CH_3)_2- \right]_n$$

A mixture of 10 g isobutylene with approximately 5 g anhydrous Dry ice is stirred at −80°C and approximately 10 mL boron fluoride gas is introduced via syringe (see prep. 218). Polymerization is initiated almost at once, and polymer is produced; the heat of reaction is dissipated by the Dry Ice present in the reaction vessel. The product is a high molecular weight, clear, nontacky, elastomeric material. Use of this technique requires Dry Ice from which all excess moisture is absent, because gross moisture will inhibit the activity of the catalyst (118). Taking the center from a large piece of Dry Ice should be satisfactory.

Another method (119) of dissipating heat and moderating violence of the polymerization is to use a low-boiling diluent. For example, a mixture of 20 parts of pure isobutylene and 80 parts of ethane (bp −88°C) or ethylene (bp −104°C) is cooled in a liquid nitrogen bath (−196°C) and treated with 0.2 parts of boron fluoride gas. Polymerization occurs rapidly, and the product is of high molecular weight. The mixture of polymer and solvents is allowed to warm to room temperature and is dried in a vacuum oven at 50°C. The product is a chunk of clear, rubber-like plastic.

Two other monomers that polymerize with Lewis acids are *N*-vinyl carbazole and acenaphthylene. The following techniques illustrate preparation of polymers from these monomers.

220. The Polymerization of N-Vinyl Carbazole (120, 121)

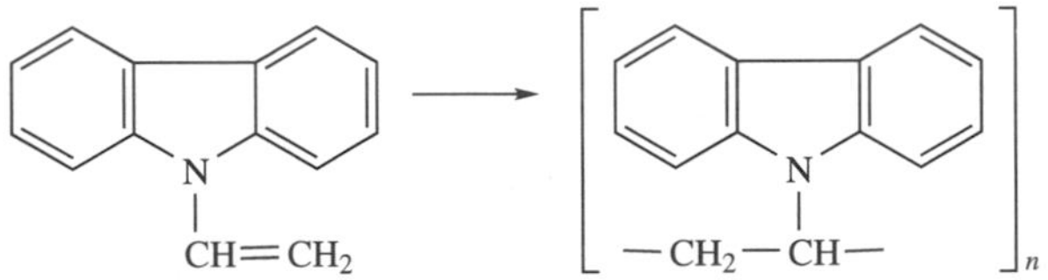

N

$CH{=}CH_2$ $\longrightarrow$ $\left[-CH_2-CH- \right]_n$ (N)

A total of 50 g freshly distilled *N*-vinyl carbazole and 150 mL methylene chloride in a 1-L three-necked flask equipped with a stirrer, an inlet tube, and an outlet tube is cooled in a Dry Ice bath to −60°C with exclusion of atmospheric moisture by means of suitably placed drying tubes. The solution is stirred rapidly, and 0.1 ml boron fluoride etherate solution is added by means of a hypodermic syringe through the inlet tube, which may be capped by a serum type stopper (chapter 2). Polymerization is initiated almost immediately, and the temperature of the reaction mixture rises. After approximately 5 min, the gel-like reaction mixture is treated with 1 ml concentrated aqueous ammonia to neutralize the catalyst present, and the polymeric solution is coagulated by stirring with methyl alcohol in a high-speed mixer. The polymer is filtered and is obtained as a white mass in a yield of 80 to 85%.

It is possible to carry out this experiment at room temperature or even at higher temperatures, but lower molecular weight polymer is always obtained. Other solvents may be used, such as toluene and carbon tetrachloride, but they are all inferior to methylene chloride in that lower molecular weight products are invariably obtained.

The polymerization of *N*-vinyl carbazole will take place under a wide variety of conditions. For example, highly purified monomer may be heated in the absence of catalyst at temperatures of 85 to 120°C to give a nearly color-less clear product similar in appearance to polystyrene. Again it must be emphasized that it is essential for the monomer to be very pure, otherwise high molecular weight material will not be obtained. The monomer should be distilled, or recrystallized from a suitable solvent such as methyl alcohol or cyclohexane. Poly(vinyl carbazole) can be molded at temperatures of 210 to 270°C to sheets that are clear and stiff. The polymer is soluble in chloroform, trichloroethylene, aromatic hydrocarbons, etc. The polymer has excellent electrical properties, but has not found widespread use in this field, mainly because of the high cost of the monomer.

221. Polymerization of Acenaphthylene (122)

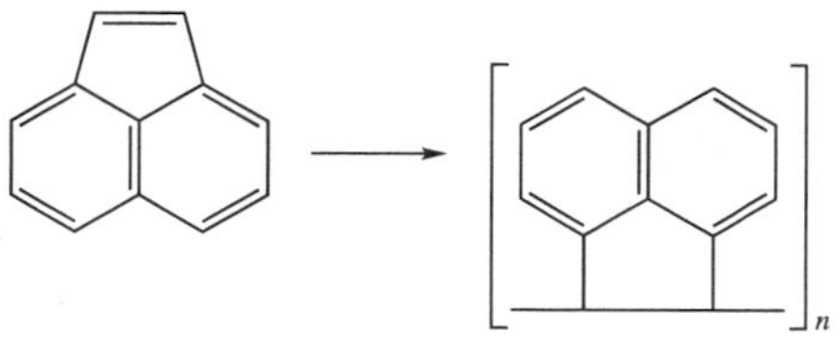

A solution of about 9 g of the hydrocarbon dissolved in ether at −50°C is treated briefly with a slow stream of boron trifluoride gas. The solution is then allowed to warm to 25°C. After 4 h, the precipitated polymer is removed by filtration, dissolved in benzene, and reprecipitated with methyl alcohol. The polymer has a molecular weight in the range of 183,000 to 341,000.

Alternatively, polymerization may be carried out by treating a solution of 50 g acenaphthylene in 140 mL chlorobenzene maintained at −50 to −20°C with a very slow stream of boron trifluoride gas for about 30 min. The polymer is obtained from the chlorobenzene by precipitation with alcohol. It is dissolved in benzene and reprecipitated with methanol. The yield is approximately 37 g of very high molecular weight material.

Polymerization at a higher temperature is also possible (123): a solution of 1 g acenaphthylene in 10 mL dry benzene is placed in a dry test tube under exclusion of moisture but not necessarily of air. It is kept at 40°C and injected with 0.15 mL redistilled boron trifluoride diethyl etherate (bp 125 to 126°C). After several hours, the darkened solution is poured into methanol; the recovered polymer is purified as above. It has an intrinsic viscosity of about 1.25 in benzene.

Acenaphthylene can be recrystallized several times from ethanol to improve its purity. From some sources, the original state is sufficient for acceptable polymerization.

Acenaphthylene can also be thermally polymerized in bulk by heating slightly above its melting point (e.g., 95°C; the mp is 92.5°C) for several hours (124). Fractionation of polymer prepared in this way has given the following intrinsic viscosity-molecular weight relationship for benzene at 25°C and is believed to be valid over the molecular weight range of 20,000 to 1,000,000:

$$[\eta] = 3.04 \times 10^{-2} M^{0.594}$$

Another group of vinyl compounds that is usually susceptible as a class to cationic polymerization is the vinyl ethers. Vinyl isobutyl ether has probably been more closely examined than any other member of the family, a fact owing in large measure to its early stereoregular polymerization using boron trifluoride etherate at low temperature and the subsequent question of the mechanism of occurrence and the physical state in which it takes place (125–132). Much of the mechanistic controversy centered on whether a heterogeneous surface was necessary for the stereoregular polymerization, because Schildknecht's original discovery of this phenomenon involved polymerization at the interface of the catalyst droplet and the monomer solution in a slow, proliferous fashion. Yet, it was not certain whether polymerization occurred in a truly heterogeneous system or in a homogeneous solution. It was finally shown that stereoregular polymer (at least, relatively so) could indeed form in a homogeneous solution (126, 131), and even in the heterogeneous system, the polymerization occurs in a homogeneous phase.

Polymerization with boron trifluoride etherate in the manner necessary to give relatively crystalline polymer occurs slowly. In contrast, boron trifluoride gas at low temperature in propane solvent causes polymerization to occur literally in a flash, and rubbery, amorphous polymer results. The crystalline polymer prepared in the former method has an isotactic chain structure. The concept and terminology of tacticity was developed initially around the crystalline polyolefins and later was extended in a fruitful manner to other polymer classes including the poly(vinyl ethers) particularly isotactic poly(vinyl isobutyl ether), despite the fact that the latter had been prepared well before the stereoregular polymerization of olefins (125).

222. Preparation of a Relatively Crystalline, Isotactic Poly(vinyl Isobutyl Ether)

$$CH_2{=}CH{-}O{-}CH_2CH(CH_3)_2 \longrightarrow \left[CH_2{-}\underset{\substack{| \\ O \\ | \\ CH_2 \\ | \\ CH(CH_3)_2}}{CH} \right]$$

This procedure (131) involves essentially anhydrous conditions, and the polymèrization is fully homogeneous through its initial stage (i.e., about 35% conversion); it then enters a clear, gel-like condition. The reaction is optically clear at all times. Polymer isolated at low conversion is slightly less stereoregular than that which has proceeded through the clear gel phase.

Materials are readied as follows: Toluene is distilled from calcium hydride. *n*-Hexane is shaken successively with acidic and alkaline potassium permanganate solutions, then washed with water, and finally distilled from calcium hydride. Isobutyl vinyl ether is shaken three times with 5% aqueous potassium hydroxide and dried by fractional distillation from sodium. At least 10% of forerun and tails are discarded when using commercial vinyl ether. To achieve a homogeneous polymerization, the boron trifluoride etherate catalyst should also be distilled from calcium hydride; inhomogeneity due to boron trifluoride hydrate present in the catalyst, which may be insoluble in the polymerization medium, is thus avoided.

A convenient reaction vessel (250 mL capacity) is a three-necked elongated flask equipped with a stirrer and a rubber serum cap stopper (Fig. 9.1). It is flamed out under nitrogen. Nitrogen inlet and outlet can be achieved with insertion of hypodermic needles through the serum caps. It is charged, after cooling to room

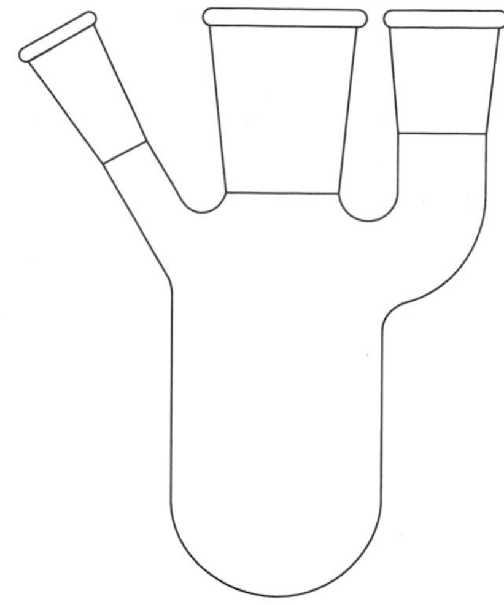

Figure 9.1. Elongated flask for exothermic polymerizations.

temperature, with a 90 ml of a 33/67 vol. % mixture of *n*-hexane and toluene, and 2.85 g isobutyl vinyl ether (0.0285 mol). The mixture is cooled to $-78°C$ in dry ice-acetone, and 0.051 g boron trifluoride etherate (3.6×10^{-4} mol) in a standard solution in about 10 mL hexane-toluene, previously chilled in dry ice–acetone, is added by syringe. The mixture remains stirrable to about 35% conversion, at which it gels and stirring is virtually impossible. (Before gelation at least, molecular weight increases with conversion, indicating a relatively slow propagation step.) After 90 min, the mixture is scooped into 400 mL isopropanol in a blender and washed thoroughly three times with 300-mL portions of isopropanol. It is dried at 50°C in a nitrogen stream under vacuum to give upward of 80% conversion. The inherent viscosity in toluene (0.1 g/100 ml) at 25°C is 1.0 or more.

An alternate procedure involving heterogeneous catalyst droplets requires no care to maintain anaerobic and anhydrous conditions, although monomer purity is important and use of distilled boron trifluoride etherate is desirable. A stirred reaction flask can be used, but stirring is not essential, and in fact a simple open flask can be used. The vessel is immersed in dry ice-acetone and 40 mL propane is condensed in, followed by 10 mL vinyl isobutyl ether. Three drops of catalyst ($BF_3 \cdot Et_2O$) is added. The polymerization occurs around the insoluble droplets of catalyst. After about 30 min, three more drops of catalyst is added, and reaction allowed to proceed for 90 min. The product is washed in methanol to remove catalyst and monomer and dried in a current of nitrogen in a vacuum oven at 50°C to give about 6 g polymer with an inherent viscosity of about 1.5 (toluene, as above).

The crystalline melting point of polymers prepared in each of the foregoing procedures is 90°C or more and usually is over a fair range, indicating the presence of some noncrystallizable polymer. The residue from extraction of the polymer with heptane, benzene, and tetrahydrofuran can have a T_m as high as 165°C.

223. Preparation of Amorphous Poly(vinyl Isobutyl Ether)

A total of 30 mL liquid propane is condensed into an open three-necked flask cooled in dry ice-acetone and containing 10 g pure vinyl isobutyl ether and in which is suspended a thermometer. The mixture is allowed to warm upto −60°C, and a stream of BF_3 gas is passed rapidly through the mixture. Reaction occurs almost instantaneously, and propane gas evolves rapidly from the heat of reaction. The white, rubbery solid is thrown all over, and possibly out of, the flask. The remaining propane is evaporated. The polymer should have an inherent viscosity of at least 1.0 (toluene, 0.1 g/100 mL, 25°C). The propane gas boiled out during the flash polymerization constitutes a flammability hazard, so the reaction should be conducted in a good hood.

One of the most remarkable developments in synthetic polymer chemistry has been the observation of asymmetric selection, i.e., the more favored polymerization of one optical antipode from a racemic mixture of *d*- and *l*-isomers. The polymerization of *d,l*-propylene oxide by a catalyst made up of diethyl zinc and either D- or L-diethyl glutamate is an example, each ester producing a predominant polymerization of one of the oxide enantiomers and hence a polymer with optical activity (133, 134). The stereoregular polymerization of optically pure *d*- or *l*-propylene oxide to a crystalline optically active polymer is also known, but is a different phenomenon.

Racemic α-olefins with an asymmetric carbon adjacent to the double bond have also been selectively polymerized with coordination catalysts using zinc alkyls having an asymmetric component, giving polymers of about 15% optical purity (135). In the olefin case, optical activity is due to an excess of monomers with one asymmetric side chain carbon over the other.

Related to asymmetric selection is the generation of an optically active polymer from a monomer having no asymmetric center. 1,3-Pentadiene, whose polymerization to various polytactic species is described in this chapter, has been converted to polymers showing optical activity by means of coordination catalysts where either constituent contains asymmetric groupings, e.g., (+) tris (2-methylbutyl) aluminum or (−) titanium tetramenthoxide (136). Optical activity in these polymers results from a difference in the immediately adjacent groups of a main-chain carbon atom, rather than a side-chain carbon.

The following experiment—the polymerization of benzofuran to a polymer with optical activity—is a case of a monomer that is not asymmetric in the sense of having optical antipodes, but an optical asymmetry is generated in the backbone chain by formation of the *threodiisotatic* or *erythro* diisotactic polymer. There is no internal compensation of the truly asymmetric carbons formed. Polymerization is cationic in nature, evidently, with an asymmetric counterion contributing to stereoselective monomer placement. Benzofuran is also of interest as an example of a polymerizable cyclic vinyl ether. The field of optically active addition polymers has been fully reviewed by Pino (137) and of optically active polymers generally by Schulz and Kaiser (134).

224. Preparation of an Optically Active Polybenzofuran (138, 139)

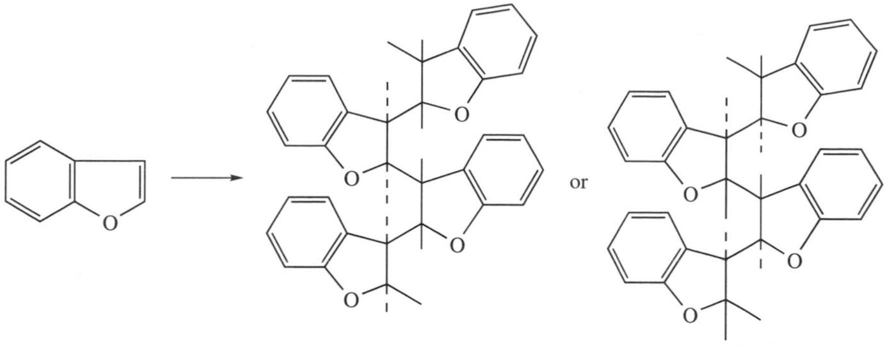

or

(*a*) *erythro*-diisotactic (*b*) *threo*-diisotactic

The above formulas portray in Fischer projection the two possible optically active forms of polybenzofuran. It is not known which is formed in this example. It is certain that the polymer obtained is not optically pure, so that units with asymmetric carbons are in a majority, but are not exclusive.

Benzofuran is fractionated at 60 to 70 mm until a chromatographically pure material is obtained. Toluene is washed with concentrated sulfuric acid and then with water; it is percolated through an active alumina column, distilled, refluxed over sodium for 20 h, and finally distilled directly into the reaction vessel as needed. Aluminum chloride is sublimed under nitrogen and collected in vials that are flame sealed. The amount of aluminum chloride used is always in excess of the β-phenylalanine component and is not severely critical. β-Phenylalanine of either optical persuasion is recrystallized several times from water and thoroughly dried at 60°C under vacuum. (+) β-Phenylalanine produces a polymer with (+) rotation; the reverse is true for the (−) amino acid. The pure β-phenylalanine is also sealed into vials. Aluminum chloride and amino acid vials are placed in a reaction vessel that has been flamed dry and purged with nitrogen and is fixed for receiving toluene via distillation, and a siphon for transfer of the catalyst solution to the polymerization reactor. The vessel is held under a high vacuum for several hours before toluene is distilled in. About 100 ml toluene is used for catalyst component quantities of about 0.067 g (+) or (−) β-phenylalanine (0.405 mmol) and 0.324 g aluminum chloride (2.43 mmol). The reactor can be basically a pearshaped flask; a Trubore stirrer shaft in the center neck can be used to crush the thin-walled vial seals, then used without the stirring blade to agitate the contents. After the glass vials are broken, the reactants are stirred at 20°C for 1 h, and the excess aluminum chloride is then allowed to settle until a clear supernatant solution results (about 3 h).

The clear solution containing the active catalyst has an aluminum chloride/β-phenylalanine mole ratio of about 3. It is siphoned with care into the reaction vessel, which can be a 500-mL Erlenmeyer or round-bottom flask fixed with a thermometer and addition funnel and a magnetic stirring bar. This flask and funnel can be connected by a pressure-equalizing tube on the funnel; in this way the system can be effectively closed. A three-way stop-cock on the pressure-equalizing side arm is convenient for applying vacuum for siphoning the catalyst solution. The assembly is, of course, thoroughly dried and flushed with nitrogen. The dropping funnel is modified by a surrounding jacket in which will be placed dry ice-acetone coolant. The funnel can be filled initially in a dry box before final assembly of the system. The funnel contains a solution of 4 g benzofuran in 100 mL dry toluene, which is cooled to −75°C. The catalyst solution is also cooled to −75°C, and the monomer solution is added with stirring. The polymer formed in the early stages of polymerization has a higher optical rotation (benzene solution, ca. 0.25%, sodium line, room temperature) than that formed later, due, it is believed, to termination reactions that tend to racemize the counterion. Hence, the sooner the reaction is terminated, the greater the specific rotation of the product. Samples can be withdrawn and quenched in methanol at intervals and the decrease in rotation noted. For convenience, the entire reaction can be terminated in about 30 min to give about 3 g polymer, after thorough washing in methanol and drying. The intrinsic viscosity is about 1.0 (in toluene, 30°C) and $[\alpha]_D$ is about ±50°C, depending on which isomer of β-phenylalanine is used. The polymer resists racemization in decalin solution at 200°C over several hours. The optical purity of the polymer is not known. Because the polymer is amorphous, a crystallographic determination of its exact steric structure has not been possible.

α- and β-Naphthofuran (140) have also been converted to optically active polymers under conditions similar to those given here. Polymers from both monomers have shown crystallinity on annealing, but films of the crystalline polymers could not be oriented, and x-ray fiber diagrams could not be obtained for structure analysis.

9.8. ANIONIC POLYMERIZATION OF VINYL COMPOUNDS

One of the most scientifically rewarding areas of addition polymerization is that of anionic initiation. The number of anionically polymerizable monomers is not great, but a detailed study has been made in some instances. The subject has been reviewed generally (141) and in some of its specific areas (142–146).

Polymerizations by free alkali metals are included in this category, because a free-radical propagation is apparently not involved, as witness the ready polymerization of α-methylstyrene by metallic potassium. The initiator is probably

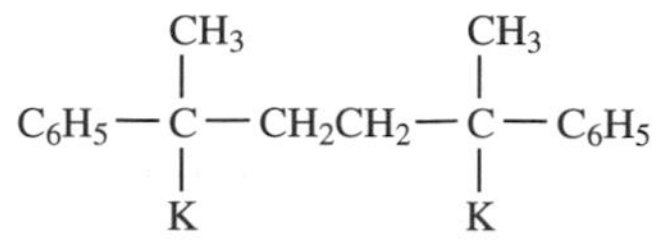

$$C_6H_5-\underset{K}{\overset{CH_3}{C}}-CH_2CH_2-\underset{K}{\overset{CH_3}{C}}-C_6H_5$$

225. Polymerization of α-Methylstyrene by Metallic Potassium (145)

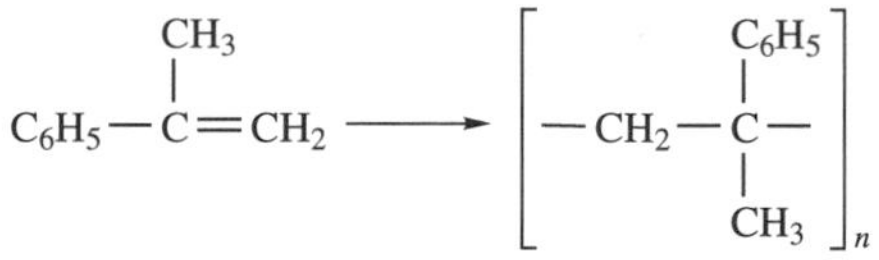

$$C_6H_5-\overset{CH_3}{C}=CH_2 \longrightarrow \left[-CH_2-\underset{CH_3}{\overset{C_6H_5}{C}}-\right]_n$$

In a glass-walled polymer tube of at least 150 mL content are placed 72.5 g freshly distilled α-methylstyrene together with about 2 g metallic potassium (**Danger!**) in the form of wire or pellets. The mixture is sealed under vacuum and agitated at 15°C for about 12 h. An induction period of about 7 h is noted followed by a polymerization period of about 5 h. The reaction mixture is removed from the tube, and the viscous liquid is separated from metallic potassium by filtration. Unpolymerized α-methylstyrene is removed from the mixture by heating at 190°C/1 mm, or by blowing with steam, after removal of the potassium. The product remaining in the flask consists of 35 to 40 g polymeric α-methylstyrene, with an inherent viscosity of 0.7 to 0.8 as measured in toluene (0.5%, 25°C).

α-Methylstyrene may be polymerized using similar conditions with metallic sodium, metallic lithium, or alloys of the alkali metals. The reaction with metallic sodium is considerably slower than with potassium, while lithium is intermediate.

Catalysts for the anionic polymerization of vinyls are ordinarily salts of very weak bases. The classic example is the polymerization of methacrylonitrile by Grignard reagents, or triphenylmethyl sodium, as described by Beaman (146).

226. Polymerization of Methacrylonitrile with Sodium in Liquid Ammonia (146)

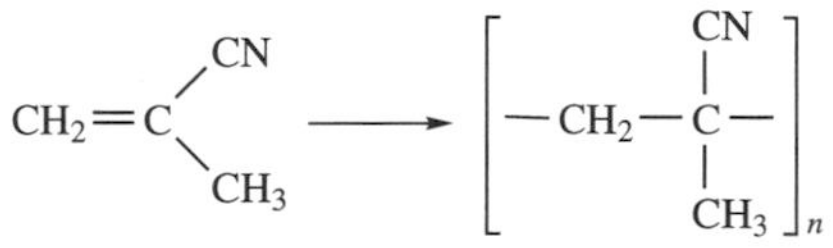

$$CH_2=C\begin{matrix}CN\\CH_3\end{matrix} \longrightarrow \left[-CH_2-\underset{CH_3}{\overset{CN}{C}}-\right]_n$$

A 500-mL round-bottom, three-necked flask equipped with a dry ice condenser, a stirrer, and a gas inlet is cooled in a dry ice bath, and about 100 mL anhydrous liquid ammonia is condensed into the flask. Approximately 0.4 g metallic sodium is now introduced into the liquid ammonia and allowed to dissolve to give the intense blue solution characteristic of sodium dissolved in liquid ammonia. To this solution, maintained at −75°C, is added 15 g freshly distilled α-methacrylonitrile. The blue color is discharged almost instantaneously. After approximately 15 min, 2 g solid ammonium chloride is added to decompose any organometallic compounds present, and the ammonia is allowed to evaporate. The residual solid polymer is washed with water to remove inorganic salts, filtered, and dried. The yield of polymer is quantitative, the inherent viscosity in the neighborhood of 0.8 as measured in dimethylformamide solution (0.5%, 25°C).

Methacrylonitrile can also be polymerized to a crystalline, predominantly isotactic polymer at room temperature and higher (e.g., 70°C) by the action of catalysts such as diethylmagnesium, di(piperamido)magnesium (147), and other complex organometallic compounds (148, 149).

*227. Stereospecific Polymerization of Methacrylonitrile (147)**

$$CH_2{=}C(CH_3){-}CN \longrightarrow \left[{-}CH_2{-}C(CN)(CH_3){-} \right]$$

Catalyst preparation (diethylmagnesium)

$$2EtMgBr \rightleftharpoons Et_2Mg + MgBr_2$$

$$Et_2Mg + Et_aAl \longrightarrow Mg(AlEt_4)_2$$

A 1-L three-necked flask equipped with a sealed mechanical stirrer, a reflux condenser, and a pressure-equalized dropping funnel is arranged for carrying out a reaction in an atmosphere of nitrogen by fitting into the top of the condenser a T-tube attached to a low-pressure supply of nitrogen and to a mercury bubbler. In the flask is placed 50 g (2.05 g atoms) of fine magnesium turnings and 250 mL absolute ether, which is freshly distilled over CaH_2 and under nitrogen immediately before use. After the vessel is purged with nitrogen the Grignard reaction is started in the usual way by introducing 5 mL of a solution of 250 g (175 mL, 2.3 mol) ethyl bromide in 250 mL freshly distilled absolute ether. The reaction, which starts at once, is maintained by gradually adding the remainder of ethyl bromide-ether solution. When the spontaneous reaction subsides, the mixture is heated gently under reflux with stirring for about 30 min. After the preparation of ethylmagnesium bromide is complete, the flask is cooled and then 193 g anhydrous dioxane is introduced dropwise through a dropping funnel with stirring under nitrogen atmosphere. A white precipitate is formed by introducing the dioxane to the Grignard solution. After the precipitate has settled down, the supernatant transparent liquid is withdrawn by a syringe and is transferred into another vessel which has been flushed with dry nitrogen gas. To the residual precipitate is added about 200 mL absolute ether with stirring, then is allowed to stand several hours. Supernatant liquid is again withdrawn by a syringe and added to the ether solution previously

* We are grateful to Dr. Yosushi Joh, Mitsubishi Rayon Co., for making available this synthesis in full detail before publication.

transferred in the same fashion. This procedure is repeated several times. The sum of the ethereal solutions is then concentrated under nitrogen until the white crystals of diethylmagnesium-dioxane complex is formed. After almost all the ether is removed, the vessel is connected to the vacuum pump to remove all traces of ether, then it is gradually warmed up to about 100°C to remove the complexed dioxane at 3 mm Hg. To remove the complexed dioxane completely from diethylmagnesium, more than 10 h are necessary under these conditions. After this procedure is over, nitrogen gas is introduced to the vessel to atmospheric pressure and the vessel is placed in a dry box filled with nitrogen. The crystals of diethylmagnesium are pulverized into uniform and finely divided form under nitrogen. The yield of diethylmagnesium is about 70%.

Then 0.12 mol diethylmagnesium is placed in a vessel that has been purged with dry nitrogen gas; 30 mL dry toluene is introduced. To this suspension, 0.2 mol $AlEt_3$ is added dropwise at room temperature. The mixture is warmed to and held at 70°C for 30 min with occasional shaking. After 30 min, almost all the Et_2Mg dissolves in the solvent. After cooling, the supernatant liquid is withdrawn by a syringe and transferred into another nitrogen flushed vessel.

The precipitate, which is composed of unreacted Et_2Mg and decomposition products derived from trace contaminants in the solvent, is washed several times with toluene and the clear, supernatant liquid is added to the supernatant liquid removed previously; the total volume should be 100 mL. This gives a catalyst solution of $Mg[AlEt_4]_2$ with a concentration of 0.001 mol/mL. Concentration of the catalyst can be checked by analysis for aluminum or magnesium which should have the indicated ratio of the two.

227.A. Polymerization

The reaction vessel is a 300-mL three-necked flask equipped with a mechanical stirrer, reflux condenser, and inlet and outlet tube for nitrogen gas. One neck of the flask is covered with rubber serum bottle cap. After the flask is flushed with dry nitrogen gas, 270 mL anisole, which has been distilled over CaH_2 immediately before use, is introduced with syringe through the serum cap. An amount of the catalyst solution calculated to give 0.5 g $Mg[AlEt_4]_2$ is then added under nitrogen. After the solution is maintained at constant temperature of 70°C, the polymerization is started by injecting 30 mL purified methacrylonitrile. On addition of the monomer, the simultaneous appearance of a deep red color that remains during the polymerization is observed. A positive nitrogen pressure is maintained during the polymerization. After 2 h, crude polymer is isolated by pouring the reaction product into 1000 mL methanol containing a small amount (usually 2 to 3%) of hydrochloric acid and then allowing the mixture to stand at room temperature overnight. Precipitated white polymer is collected, filtered, washed several times with methanol, and dried in vacuo. Thus 10.9 g polymethacrylonitrile is obtained. When the polymer is extracted with acetone in a Soxhlet extractor for 24 h, 79.7% of the polymer is isolated as an insoluble fraction, which gives a typical crystalline x-ray diagram. The intrinsic viscosity of the crystalline polymer can be measured in dichloroacetic acid at 30°C. The acetone-insoluble fraction obtained by this procedure has an intrinsic viscosity of about 4.7.

If one wishes to isolate a higher crystalline fraction, the following procedure is recommended: 1 g of the acetone-insoluble fraction of the polymer is placed in a flask, and 100 mL dimethyl formamide (or dimethylsulfoxide) is introduced. The flask is then warmed up to 70°C with stirring. After 2 h, the polymer is extremely swollen. The swollen polymer can be isolated by centrifuge. The white precipitate, which is obtained by addition of methanol to the swollen polymer, is washed with a sufficient amount of methanol, dried, and weighed; 0.7 g highly crystalline material can be isolated. The dimethylformamide soluble part can be also isolated by concentration of the solution followed by precipitating with methanol. This fraction gives an x-ray diagram of low crystallinity. Films can be cast at room temperature from trifluoroacetic acid; these can be oriented at 90°C after removing residual solvent by immersion in boiling water for 2 h. The intrinsic viscosity-molecular weight relationship determined for the acetone-insoluble polymer in Cl_2CHCO_2H is

$$[\eta] = 2.27 \times 10^{-4} M^{0.75}$$

In the anionic polymerization of styrene, the reactive chain propagating species, $[R{-}CH_2{-}CHPh]^{\ominus}M^{\oplus}$ is a substituted benzyl anion, which is stable indefinitely in an inert environment. The macromolecular anions have been called "living polymers" because they will resume chain growth in the presence of additional monomer (150). If an anionically polymerizable monomer other than styrene is added, block copolymers may be formed. Reaction with reagents such as CO_2 or SO_2 will give, after hydrolysis, macromolecular acids.

In the butyllithium-initiated polymerization of styrene, each mole of butyllithium generates one chain-propagating anion. Polymer molecular weight therefore is determined only by the quantities of monomer and initiator, according to the equation

$$M = \frac{\text{grams monomer}}{\text{moles of initiator}}$$

Polymers with predetermined molecular weights thus may be prepared by adjusting the quantities of monomer and initiator, providing that initiation is rapid with respect to propagation.

Because in any homopolymerization all macromolecular anions have equivalent reactivities, chain growth and chain length (molecular weight) distribution are governed by probability considerations. Theory (151) shows that the idealized polymer will have chain lengths that conform to a Poisson distribution, as given by the formula

$$X_{(P)} = \frac{n^{P-1} e^{-n}}{(P-1)!}$$

where n is the number of moles of monomer consumed per active chain end, P is the number of monomer units in a given chain, and $X_{(P)}$ is the mole fraction of the chains containing P monomer units. Polymers having a Poisson molecular weight distribution have a M_w/M_n of unity.

Any reactive impurities, such as protonic compounds, carbonyl compounds, halogen compounds, oxidizing agents, etc., can kill some of the growing chains, thereby broadening the molecular weight distribution. Elegant vacuum system techniques have been developed to attain the scrupulous purities necessary to prepare polymers with M_w/M_n approximating 1 (150, 152, 153). But by working in an inert atmosphere one can make quite good polymers using standard glassware (153, 154). The key to the latter procedure is that the deleterious impurities are removed from the polymerization medium and from the monomer by titration before polymerization (under conditions in which polymerization is slow) with the reactive anion, which constitutes the initiator. Because initiation by BuLi in benzene occurs more slowly than propagation, the molecular weight distribution tends to be broader than when initiation is faster than propagation. However, the problem is minimized when quite high molecular weight polymer is synthesized and when part of the initiation process occurs with only a small amount of monomer present, both of which conditions prevail in the present case.

228. Polystyrene of Predetermined Molecular Weight and Narrow Molecular Weight Distribution by Anionic Polymerization with Butyllithium (116)*

The benzene purification system is illustrated in Figure 9.2. The argon scrubber A and the solvent reservoir B are assembled ahead of time, and flushed with argon. Dry benzene (about 150 and 1500 mL, respectively), 1.6M butyllithium (20 mL to each) and styrene (2 mL to each) are then added, and the exit of B is stoppered until reaction flask C is attached. Over several hours both solutions gradually attain the characteristic yellow-orange to red color of the $R{-}CH_2CHPh^-$ anion. All subsequent operations are performed with a stream of argon moving through the system. A 1-L flask C (necks and outlets shown) is attached via the condenser (at 2). C is also fitted with a Teflon-covered stirring bar, a thermometer (at 3) a 125-mL graduated addition funnel with a Teflon stopcock (at 4), a serum cap (at 6), and an exit tube (at 7) leading to a mineral oil U bubbler, D. C and the addition

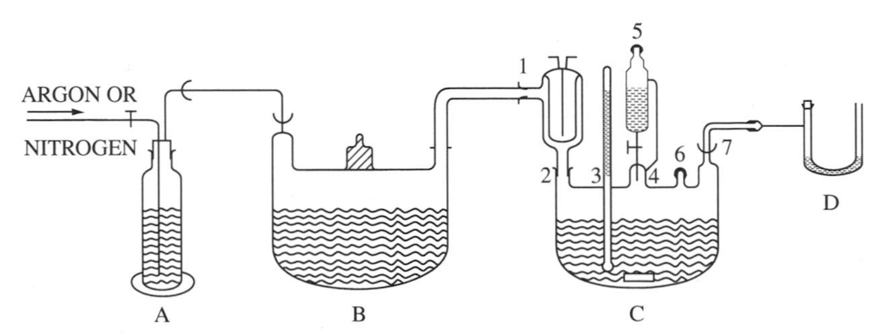

Figure 9.2. Apparatus for solvent purification and distillation.

* Our thanks are due Dr. Donald B. Miller, for this preparation.

funnel are flamed and cooled under argon; the addition funnel is then sealed with a serum cap (at 5). The tube connecting the exit 7 to D must be arranged so that any condensing vapors will not drain back into C. About 600 mL benzene is distilled from B to C. During the latter stages of the distillation, 125 mL benzene is transferred from C to the addition funnel with a 100 mL syringe with a 6-in needle (via caps at 6 and 5). When the distillation is finished, B is brought to room temperature, removed from the system, stoppered, and set aside for future use so long as the anion survives. C is quickly reattached to the argon supply at 7 and exit bubbler D is attached at 1 (Fig. 9.3).

The initiator solution is prepared by adding 0.81 mL 1.6*M* butyllithium and 0.4 mL styrene to the addition funnel with syringes, giving an initiator concentration of about 0.01 mEq/mL. Warming with a heat gun will hasten the formation of the orange-red anion. The benzene in C is heated to reflux and titrated with the initiator solution until a yellow color persists. The refluxing and titration should continue for about 2 h and may consume 10 to 30 mL initiator solution. After reactor C is cooled to 5 to 10°C, addition of 50 g styrene (freshly distilled from calcium hydride) with a syringe via serum cap at 6 dissipates the yellow color. Impurities in the monomer are titrated at 5 to 10°C by adding initiator until a yellow color persists for 0.5 h. (This will consume about 10 mL initiator.) After C is brought to 50°C with a warm water bath, 5.0 mL initiator is added. In 5 min, a 3 to 5°C exotherm will be observed; within 30 min, the solution will be quite viscous; and after 1 h, stirring is difficult. The temperature is kept at about 50°C for 6 h with a heat lamp, then polymerization is terminated by adding a few drops of isopropanol. The polymer is precipitated by slowly adding the viscous solution to 2 to 3-L methanol stirring in a 1-gal blender (or in four batches in a 1-L blender). Conversion is quantitative. The intrinsic viscosity in benzene at 25°C is about 2.8, corresponding to a molecular weight of about 10^6. The M_w/M_n ratio is 1.3 or less. If 50 mL initiator is used, polymerization may be so fast that cooling is necessary. The resulting polymer has a molecular weight of about 10^5 and as M_w/M_n ratio of <1.1.

Another effective type of catalyst for anionic polymerization is sodium naphthalene, or sodium benzophenone, which is an ion radical, but appears to

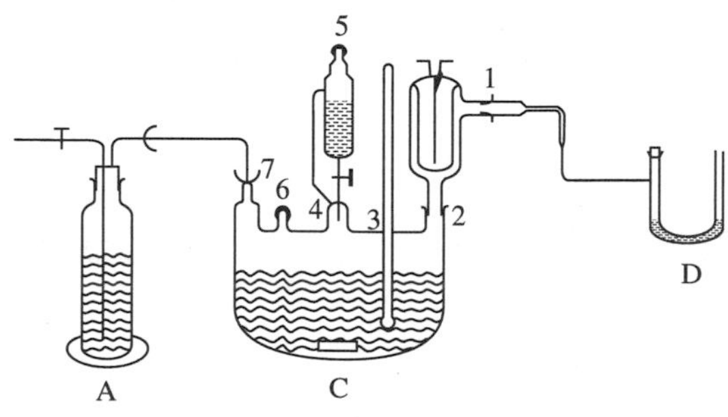

Figure 9.3. Apparatus for monomer polymerization.

initiate anionic polymerization (9–11, 150) as follows:

$$(C_6H_5)_2\dot{C}{-}O^- + CH_2{=}C\begin{smallmatrix}X\\Y\end{smallmatrix} \longrightarrow \phi_2CO + \cdot CH_2{-}C\begin{smallmatrix}X\\Y\end{smallmatrix}\ (-)$$

The radicals dimerize and the chain grows anionically in both directions, until terminated. Catalysts of this type will polymerize styrene and acrylonitrile to high molecular weight products, provided adequate precautions are taken to maintain purity of reagents.

229. Preparation of Lithium and Sodium Naphthalene Catalysts (155, 156)

$$\text{Naphthalene} + Li \longrightarrow \text{(naphthalene)}(-)\ Li^+$$

Lithium naphthalene catalyst is prepared in the following manner. A three-necked reaction flask of suitable size is equipped with a gas inlet, a stirrer, a stopper, and a means for allowing the inert gas to exit (which is protected by a drying tube). The flask is flamed out under nitrogen or, better, argon and 50 mL tetrahydrofuran, purified by distillation from lithium aluminum hydride, is added followed by 15 g resublimed naphthalene. To this solution at room temperature is added 1.5 g metallic lithium in small pieces, or as a lithium dispersion. The reaction begins almost at once as evidenced by the appearance of the dark greenish black color of lithium naphthalene. The reaction proceeds so rapidly that it is exothermic and becomes warm to the touch, and if lithium dispersion is used external cooling may be necessary. After 2 h of stirring, the reaction is considered to be complete. A 3-mL aliquot may be withdrawn at this point and quenched in methanol. The titer is then determined with standard hydrochloric acid. The solution should ordinarily contain approximately 1.6 meq of base/mL when prepared in this manner.

Sodium naphthalene is prepared in an identical manner to the lithium naphthalene. Again the solution is standardized by titration of a quenched sample with dilute hydrochloric acid.

230. Anionic Polymerization of Styrene with a Sodium-Naphthalene Initiator (9–11, 150)

$$C_6H_5CH{=}CH_2 \longrightarrow (-\underset{C_6H_5}{CH}-CH_2)_m(CH_2-\underset{C_6H_5}{CH}-)_n$$

To successfully polymerize styrene to high molecular weight polymer with an anionic initiator, it is necessary that all liquid reagents be distilled and maintained under nitrogen until required. The nitrogen used in flushing the equipment under which the reagents are to be stored should be previously purified by passing through silica gel. All solid reagents used should be dried in desiccators for at least 1 week. All glass equipment should be flamed out under dry nitrogen after assembly and immediately before use. All reaction vessels should be maintained under a positive nitrogen pressure.

A 100-mL three-necked flask fitted with a stirrer, a nitrogen inlet, and a side arm for the introduction of reagents is assembled. To the flask is added 10 mL styrene, 50 mL glycol dimethyl ether purified by stirring with sodium dispersion for several hours, followed by distillation at atmospheric pressure under nitrogen. This solution is cooled to −70°C by an external dry ice bath, and 50 mg sodium naphthalene catalyst is added. The polymerization proceeds rapidly, and the solution assumes a bright red color, due to the styrene anion. This color persists for long periods of time, presumably indefinitely if reactive agents are excluded. The polymerization is quenched by the addition of some alcohol, which causes immediate disappearance of the color. The mixture is allowed to warm to room temperature and the polymer is filtered, mixed with alcohol, and dried. The yield is quantitative; the inherent viscosity will run between 1.0 and 1.5 in toluene at 0.5% concentration.

231. Preparation of a Polystyrene with Carboxyl Ends (9–11, 150, 157)

$$C_6H_5CH{=}CH_2 \longrightarrow HO_2C(-\underset{\displaystyle C_6H_5}{\overset{|}{CH}}-CH_2)_m(CH_2-\underset{\displaystyle C_6H_5}{\overset{|}{CH}}-)_nCO_2H$$

Polymerization is carried out in a three-necked reaction vessel, preferably a creased flask of the type originated by Morton and co-workers (158). This flask is equipped with a nitrogen flush, a stirrer, stopper, and a means for the nitrogen to escape. The flask is flamed under nitrogen before adding the reagents. To the flask is then added 250 mL purified tetrahydrofuran. To make certain that the tetrahydrofuran is completely free of impurity, a few drops of the lithium naphthalene catalyst is added until the greenish black color of the organometallic reagent persists. At this point, a quantity of solution containing 21.9 mEq catalyst is added. The flask is then immersed in a dry ice bath and cooled until a temperature of about −77°C is reached. At this point, 150 meq, (13.6 g) styrene is added via a hypodermic syringe inserted through a serum type rubber stopper (chapter 2) in one of the necks of the flask. The green-black color of the lithium naphthalene catalyst is completely changed to the dark red color of the benzyl anion. The relatively nonviscous solution is allowed to come to room temperature when 35 g dry, solid carbon dioxide is quickly added. The dark red solution immediately becomes colorless and considerably more viscous. After the solid carbon dioxide has all evaporated, dilute hydrochloric acid is added to liberate the macrodicarboxylic acid. The mixture is precipitated with methanol and the solid filtered and washed repeatedly with water and alcohol until free of salts and mineral acid. The solid is dried overnight in an 80°C oven. The yield is approximately 90%. On the basis of the suggested ratio of ingredients, the degree of polymerization should be approximately 30. Experimentally, the value will not differ from this calculated figure appreciably. If a higher or lower degree of polymerization is desired, this may be obtained by varying the ratio of lithium naphthalene to monomeric styrene in the initial charge.

Again it should be stressed that in experiments of this type it is essential to maintain all glass equipment and all reagents absolutely free of impurities that might tend to either inactivate the catalyst or act as chain terminators. When due precaution is taken, no problems are encountered in preparing living polystyrene molecules, which can be terminated with carboxyl groups in the manner indicated above. It is possible to vary the molecular weight of the material obtained over a fairly wide range. The difunctionality of the resulting product is almost theoretical.

Acrylonitrile may be polymerized in essentially the same manner as described above for styrene with sodium naphthalene or sodium benzophenone. With acrylonitrile, it is convenient to use dimethylformamide as a reaction medium, in which case the polymer obtained remains in solution and may be used directly for further applications. A variation on this anionic polymerization technique uses a salt of a somewhat stronger (but still weak) acid, namely, sodium cyanide. This salt in dimethylformamide at low temperatures acts as an efficient anionic chain initiator for the polymerization of acrylonitrile.

232. Anionic Polymerization of Acrylonitrile (159)

$$CH_2{=}CHCN \longrightarrow [-CH_2-\underset{\displaystyle |}{\overset{\displaystyle CN}{C}}H-]_n$$

A 250-mL, three-necked, round-bottom flask is fitted with a stirrer, an inlet tube for the introduction of nitrogen, and an outlet tube. As in the previous experiments, the nitrogen must be dried by passing it through silica gel and the equipment should be flamed out under nitrogen immediately before use. In the flask is placed 60 mL freshly distilled dimethylformamide and 10 mL acrylonitrile recently distilled under nitrogen. The flask with its contents is immersed in a cooling bath consisting of dry ice and alcohol, and the temperature is lowered to about −50°C. The initiator, 2 mL of a saturated solution of anhydrous sodium cyanide in dry dimethylformamide, is rapidly introduced by means of a hypodermic syringe inserted in a serum-type rubber stopper. Sodium cyanide can be dried by storing in a vacuum desiccator over silica gel for several days before use. A saturated solution of this salt in dimethylformamide contains somewhat less than 1 g cyanide in 100 g dimethylformamide.

Within a few seconds of adding the initiator, the temperature of the reaction mixture will rise rapidly, and the solution may become so viscous it may climb up the stirrer shaft. The contents of the flask is allowed to stir for about 30 min in the cooling bath. At the end of this reaction time, 5 mL of 3% sulfuric acid in dimethylformamide is added to destroy unreacted initiator and to adjust the acidity of the mixture to a value of pH of 7 or less. The polymer may be isolated by precipitation in water, or the solution may be used directly for other purposes.

The yield is quantitative; the inherent viscosity measured in dimethylformamide is 2 to 3 (0.5%, 30°C).

Poly(methyl methacrylate) has been prepared in basically three stereoregular forms (160). Type I is highly syndiotactic and can be obtained from free-radical initiation at low temperatures (e.g., −40°C) and by anionic catalysis in ether solvents. Type II is highly isotactic and is prepared with an anionic catalyst (fluorenyllithium or Grignard reagent) in a hydrocarbon solvent at −60°C. Type III is a stereoblock composition of isotactic and syndiotactic segments and is the product from low temperature anionic polymerization in ether toluene mixtures. All three are crystallizable, the isotactic being the most readily so. There are distinct differences in T_m and T_g values, which are listed below along with the latter value for conventional poly(methyl methacrylate). It is interesting to note that even the conventional polymer has considerable syndiotactic stereoregularity, and that a truly statistically random polymer may not be known as yet.

	T_g	T_m	Density
Type I syndiotactic	115	200	1.19
Type II isotactic	45	160	1.22
Type III stereoblock	60–95	170–190	1.20–1.22
Conventional	104	—	1.188

Stereoregular acrylate polymers have also been characterized. Isotactic and syndiotactic (161, 162) forms of poly(isopropyl acrylate) have been prepared by initiation with Grignard-in-ether at −70°C.

233. Preparation of Isotactic Poly(methyl methacrylate) (163)

$$CH_2{=}\overset{\displaystyle CH_3}{\overset{|}{C}}{-}CO_2CH_3 \longrightarrow \left[CH_2 \diagdown \underset{CH_3\ \ \ CO_2CH_3}{C} \diagup CH_2 \diagdown \underset{CH_3\ \ \ CO_2CH_3}{C} \right]$$

9-Fluoroenyllithium is prepared as a 0.22*M* suspension by refluxing a slight molar excess of 10.4 g (0.062 mol) recrystallized fluorene with (3.52 g, 0.055 mol) butyllithium in enough toluene (dried over calcium hydride) to give 250 mL total solvent when that used in adding the butyllithium is included. Butyllithium should be used from a fresh bottle of commercial material in hexane or similar solvent and is added by syringe to the fluorene in a 500-mL three-necked flask equipped with condenser, nitrogen inlet, and serum cap closure. The apparatus should be flamed out and cooled under dry, oxygen-free nitrogen. The product separates as a bright orange, finely divided solid.

Methyl methacrylate is freed of inhibitor by low pressure distillation and stored at −20°C under nitrogen.

A solution of 50.0 g (0.5 mol) methyl methacrylate in 750 mL of anhydrous toluene is placed in a 1-L three-necked flask equipped with a mechanical stirrer, thermometer, and inlet and outlet tubes for nitrogen gas. One neck of the flask is covered by a rubber serum cap. The flask is flushed with dry nitrogen, and the contents cooled to −60°C. A 30-mL portion of the 0.22*M* suspension of 9-fluorenyllithium in toluene is added by means of a hypodermic syringe inserted through the rubber cap.

The temperature of the reaction mixture is maintained at −40 ± 6°C. for 1 h; during this period the solution becomes extremely viscous.

A small amount of methanol is added, and the reaction mixture is permitted to warm to room temperature. The polymer is precipitated by pouring the solution into 7 L vigorously agitated methanol. The swollen solid that separates is dried and dissolved in 800 mL acetone, and the solution is filtered. Precipitation of the polymer by pouring into 7 L water produces, after drying, about 41 g (83%) poly(methyl methacrylate).

A 5% solution of this polymer in di-*n*-propyl ketone, prepared at reflux temperature, is cooled to 30°C and held at that temperature for 48 h. Centrifugation of the mixture and removal of the supernatant liquid produces a fraction amounting to 32% of the original polymer. This material displays the crystalline x-ray diffraction characteristics typical of type II poly(methyl methacrylate). The intrinsic viscosity is determined in chloroform at 25°C, and the viscosity average molecular weight from the equation:

$$[\eta] = 4.8 \times 10^{-5}\overline{M}_v^{0.80}$$

An arbitrary ir parameter, the J value, is used to characterize stereoregularity. A clear, uniform film is cast on a silver chloride plate from a 3 to 5% solution of the polymer in benzene. The thickness (about 10 μ) should be such that the air-dried film transmits between 2 and 15% of the ir radiation at the intense band near 8.0 ± 0.2 μ. The plate bearing the film is heated at 135°C for 45 min to remove the last traces of benzene; no peaked absorption at 14.8 μ should remain.

The spectrum of the film is scanned at 5 to 15.5 μ in a Perkin-Elmer Model 21 recording ir spectrometer at standard scanning conditions. A 100% transmission line is drawn between 5.4 and 14.5 μ, and the absorbances at 6.75, 7.2, and 10.1 μ are calculated by use of this baseline. The absorbance at 9.3 μ is determined, a line connecting the points of minimum absorbance on either side of the 9.3 μ maximum being used as a baseline.

Two subparameters, J_1 and J_2, are calculated from these absorbances by means of the equations

$$J_1 = [179A_{9.3}/A_{10.1}] + 27$$

$$J_2 = [81.4A_{6.75}/A_{7.20}] - 43$$

The J value is the arithmetic average of J_1 and J_2. The lower the J value, the higher the isotacticity; high J values mean high syndiotacticity. In the present case, the J value of the polymer should be 30 to 35, indicating high isotacticity.

234. Preparation of Syndiotactic Poly(methyl methacrylate) (163)

$$CH_2{=}C(CH_3){-}CO_2CH_3 \longrightarrow \left[CH_2{-}C(CH_3)(CO_2CH_3){-}CH_2{-}C(CO_2CH_3)(CH_3) \right]$$

Tetrahydrofuran is dried by refluxing 24 h over calcium hydride; then 500 mL is distilled into, or otherwise transferred under anhydrous, anaerobic conditions to, a dry 1-L reaction apparatus as described in the previous experiment, in which is already placed 1.66 g (0.01 mol) recrystallized fluorene. Then, enough fresh butyllithium solution (hexane or similar solvent) to provide 0.0097 mol active agent is transferred by dry syringe and the mixture stirred under nitrogen for an hour, when metallation of the fluorene should be complete. The mixture is cooled to −70°C in dry ice-acetone and 83 g previously chilled (−20°C at least), purified, dry (see prior preparation) methyl methacrylate is added by syringe. The polymerization is continued for 6 h to give a viscous solution, which, after addition of a few milliliters of methanol and warming to room temperature, is poured with stirring into about 5 L petroleum ether. The filtered polymer is redissolved in benzene, centrifuged to remove insoluble material, reprecipitated into petroleum ether, filtered, and washed well with methanol in a blender to give nearly 100% conversion of poly(methyl methacrylate) with a J value of about 117 (see previous preparation), indicating a syndiotactic character. The viscosity average molecular weight is about 16,000. Swelling with n-propyl ketone or annealing at 100 to 130°C may be necessary to effect crystallization.

Vinylidene cyanide, $CH_2{=}C(CN)_2$, has been studied extensively (164 to 166). It polymerizes readily by an anionic mechanism, but the polymer is easily depolymerized. It is more useful in copolymers.

235. Group Transfer Polymerization (167)

It has already been pointed out that anionic polymerization can provide living polymer, i.e., one in which the polymer chains grow without termination or chain transfer. While classical anionic polymerizations require low temperatures to minimize chain transfer and achieve high molecular weight, several modified anionic polymerization processes have been reported that give high molecular weight living polymers at temperatures up to 100°C. The most versatile of these is group transfer polymerization (GTP), which, however, is essentially limited to methacrylates (167, 167a). Organic chemists will recognize a strong relationship between GTP and the Michael addition reaction (168).

GTP is commonly initiated by trimethylsilyl ketene acetals and nucleophilic catalysts. The polymerization operates best with mild anionic catalysts, such as tetrabutylammonium carboxylates, tetrabutylammonium fluoride trihydrate, or *tris* (dimethylamino) sulfonium bifluoride but also proceeds with Lewis acid catalysts and sometimes with no catalyst under high pressure. The mechanism of group transfer polymerization is shown below. This mechanism proposes that the trialkylsilyl group remains close to the chain end while the monomer is adding (169). Other work with a different catalyst suggests that the silyl group is dissociated from the polymer chain end (170).

$$(CH_3)_3SiO(CH_3O)C{=}C(CH_3)_2 + n\ H_2C{=}C(CH_3)COOCH_3 \xrightarrow[\text{THF}]{(C_4H_9)_4N\ OAc.HOAc}$$

$$CH_3O{-}C({=}O){-}C(CH_3)_2{-}\left(CH_2{-}C(CH_3)(COOCH_3)\right)_{n-1}{-}CH_2{-}C(CH_3){=}C(OCH_3)OSi(CH_3)_3 \xrightarrow{CH_3OH}$$

$$CH_3O{-}C({=}O){-}C(CH_3)_2{-}\left(CH_2{-}C(CH_3)(COOCH_3)\right)_{n-1}{-}CH_2{-}CH(CH_3){-}C({=}O){-}OCH_3 + (CH_3)_3Si{-}OCH_3$$

The key to generating living polymers (anionic or cationic) appears to be close association of the growing end with a bulky counterion. Obviously, this is difficult to maintain in free-radical polymerizations. Evidence for the effect of a bulky counterion has been presented by Reetz (171), who showed that tetraalkylammonium salts with suitable nucleophilic counterions will initiate living polymerizations of acrylates at room temperature. Thus tetrabutylammonium thiolates gave poly(butyl acrylate) of >20,000 molecular weight and $M_w/M_n = 1.3$. Webster (172) reviewed a range of living polymer systems and identified a number of mechanistic similarities. His key points are listed below. Again, the main criterion appears to be close association of the growing end with a bulky counterion.

> As conditions continue to be worked out for living polymerization of more and more monomers, a number of mechanistic similarities are evident. The three characteristics listed below seem to apply to many of the new procedures and include anionic, cationic, and covalent living polymerizations.
>
> (a) The growing chain ends exist mainly as a stabilized latent form that is activated reversibly by a catalyst or by heat. The trace amounts of highly active ends that are produced add monomer and then revert to the latent form.
>
> (b) Narrower polydispersity of molecular weights and better molecular weight control are obtained when excess monomer is maintained during the polymerization run. This stabilizing effect can be explained by complex formation of reactive ends with monomer or by an increased rate of propagation vs chain termination.
>
> (c) Large counter ions or counter ion clusters seem to aid the progress of living cationic and anionic polymerizations.
>
> In agreement with feature (c) of living polymerizations, metal acetates do not catalyze GTP.

To obtain polymer of low dispersity the rate of initiation must be comparable to or faster than the rate of propagation, otherwise the first-formed chains will grow to higher molecular weight than the latter. To ensure that the initiation step is comparable in rate to propagation, the initiator chosen should be close in structure to the growing chain end. Thus for GTP of methyl methacrylate, trialkylsilyl ketene acetals (where the alkyl groups are preferably methyl) are ideal.

236. Preparation of Poly(methyl Methacrylate) by GTP (167)

All polymerization equipment should be dried at 150°C for 24 h and cooled to room temperature under a stream of dry oxygen-free argon. Solvent and reagents transfers are made by dry syringes. THF is distilled from sodium and benzophenone immediately before use. The methyl methacrylate (MMA) is obtained from Aldrich and purified by passing through a dried column of Woelm alumina, activity grade 1, under dry argon. 1-Methoxy-1-(trimethylsiloxy-2-methyl-1-propene) (MTS) initiator can be purchased from Huls America. This can be purified by fractional distillation through a column and selecting the fraction that boils at 59°C/28 torr). The tetra-*n*-butylammonium fluoride trihydrate catalyst is available from Fluka.

A 500-mL, four-necked round-bottom flask, fitted with a magnetic stirring bar, argon inlet, thermocouple well, serum cap, and 125-mL pressure-equalizing dropping funnel equipped with a serum cap is assembled under argon while hot. The entire apparatus is then heated with a heat gun while being flushed with argon. The argon flush is continued until the apparatus is cool. The assembly is then maintained under a slight positive pressure of argon. Then 225 mL THF is transferred to the flask via a cannula. Then 75 mL (0.69 mol) MMA is transferred to the dropping funnel via a cannula. Then 0.37 mL (1.16 mmol) of 3 and 58 μL of tetrabutylammonium fluoride trihydrate (0.02 M in THF) are introduced with syringes. MMA is then added dropwise over a period of 90 min. After stirring for 16 h, the reaction mixture is quenched with 10 mL of methanol, stirred for 1 h, and poured very slowly into a fivefold volume of hexane that is being agitated strongly in a blender. After filtration and washing with hexane, the polymer is air-dried at room temperature and then dried for 48 h to constant weight in a vacuum oven at 65°C, under a slight nitrogen purge. The polymethyl methacrylate (PMMA) weighed 71.2 g (quantitative conversion), which is slightly more than theory due to the presence of traces of solvent (observed only in thermogravimetric analysis): M_n(GPC), 62300; M_w, 71,500; M_w/M_n, 1.15; M_n(theory), 60,180.

9.9. LIVING FREE-RADICAL POLYMERS

Living anionic and cationic polymers are well documented and owe their existence, to a large extent, to control by closely held counterions. This mechanism obviously is not applicable to free-radical polymerization and considerable effort has been made to develop other routes. Apart from the challenge, the driving force to identify routes to living free-radical polymers is that this will surely broaden the range of polymer types that can be made and copolymerized versus the limited range from ionic polymerizations. The focus on routes to living free-radical polymerizations has been largely twofold: metal-mediated and radical-trap-mediated routes. The first route uses a broad variety of metal complexes of Ru(II), Cu(I), Fe(II), Ni(II), Rh(I), Pd(II) and Re(V). $RuCl_2(PPh_3)_3$ is one example of such a catalyst. These living radical polymerizations proceed via the metal-mediated homolytic and reversible activation of dormant carbon-halogen bonds at the polymer terminal originated from a halide as initiator (173) PMMA over 1 million number-average molecular weight and $M_w/M_n = 1.04$ to 1.22 has been made with this catalyst system.

$$\text{R—X} \underset{}{\overset{\text{Ru}^{\text{II}}}{\rightleftharpoons}} \text{R}\cdot\ \text{XRu}^{\text{III}} \underset{\text{Al(O}i\text{-Pr)}_3}{\overset{\text{MMA}}{\rightleftharpoons}} \text{R}\sim\text{CH}_2\text{—C}(\text{CH}_3)(\text{CO}_2\text{CH}_3)\cdot\ \text{XRu}^{\text{III}}$$

$$\underset{\text{Dormant}}{\sim\text{CH}_2\text{—C}(\text{CH}_3)(\text{CO}_2\text{CH}_3)\text{—X}} \underset{}{\overset{\text{Ru}^{\text{II}}/\text{Al(O}i\text{-Pr)}_3}{\rightleftharpoons}} \underset{\text{Active}}{\sim\text{CH}_2\text{—C}(\text{CH}_3)(\text{CO}_2\text{CH}_3)\cdot\ \text{XRu}^{\text{III}}}$$

$$\underset{\textbf{2b}}{\text{CH}_3\text{—C}(\text{CH}_3)(\text{CO}_2\text{CH}_3)\text{—CH}_2\text{—C}(\text{CH}_3)(\text{CO}_2\text{CH}_3)\text{—Br}}$$

A large amount of living free-radical research has focused on organic- or organic/metal-assisted catalysts. Major advances have occurred in use of nitroxide-mediated processes and atom transfer radical procedures (ATRP). The nitroxide-mediated processes appear to be simpler than the metal-mediated. but have been more limited in range of monomers polymerized. Indeed it has been suggested that the nitroxides are radical traps and are successful in providing a narrow molecular weight range because a limited number of chains are initiated and termination is low.

More recently, Hawker and co-workers used alkoxyamines/nitroxides (**9.1** and **9.29**) to prepare polystyrene, acrylate, acrylamide, acrylonitrile, and copolymers with $M_n = 40{,}000$ and polydispersities in the 1.05 to 1.15 range. Typical experimental conditions are given below as well as the structures of the preferred alkoxyamine and nitroxides. Syntheses of these have been described (174). 3904,

R
•O—N

9.1

N Ph
O

9.29

237. Metal-Mediated Living Free-Radical Polymerization

237.A. Living Polymerization

The polymerization is carried out under dry nitrogen in baked and sealed glass tubes. All reagents are used after ordinary purifications, and the toluene solvent is bubbled with dry nitrogen for more than 15 min immediately before use. A typical example with **2b** is given here. In a 50-mL round-bottom flask were placed 92.0 mg $RuCl_2(PPh_3)_3$, 3.86 mL toluene, 0.384 mL *n*-octane, 2.05 mL MMA, 125 mm $Al(O\text{-}i\text{-}Pr)_3$ in 3.07 mL toluene, and 800 mm 2b in 0.240 mL toluene, sequentially in this order at room temperature under dry nitrogen. The total volume of the reaction mixture is thus 9.60 mL. Immediately after mixing, aliquots (1.20 mL each) of the solution are injected into baked glass tubes, which are then sealed and placed in an oil bath kept at 80°C. In predetermined intervals, the polymerization is terminated by cooling the reaction mixtures to −78°C. Monomer conversion is determined from the concentration of residual monomer measured by gas chromatography with *n*-octane as an internal standard. The quenched reaction solutions are diluted with about 20 mL toluene and rigorously shaken with about 5 g of an absorbent [KYOWAAD-2000G-7 ($Mg_{0.7}Al_{0.3}O_{1.15}$); Kyowa Chemical Industry] to remove the metal-containing residues. After the absorbent is separated by filtration (Whatman 113V), the filtrate is washed with water and evaporated to dryness to give the products, which are subsequently vacuum dried overnight.

237.B. Polymer Characterization

The $\overline{M}_n$, $\overline{M}_w/\overline{M}_n$, and MWD of the polymers are determined by size-exclusion chromatography in chloroform at 40°C on three polystyrene gel columns (Shodex K-805L × 3) that are connected to a Jasco PU-980 precision pump and a Jasco RI-930 refractive index detector. The columns are calibrated against 11 standard PMMA samples (Polymer Laboratories; $\overline{M}_n = 630$ to 1,200,000; $\overline{M}_w/\overline{M}_n = 1.04$ to 1.22) as well as the monomer.

238. Alkoxyamine/nitroxide Free-Radical Polymerization

238.A. General Procedure for Styrene Polymerization; Preparation of Polystyrene, 9.45, of DP = 250 from 9.29

A mixture of 325 mg (1.0 mmol) of the desired alkoxyamine, **9.29**, 204 mg (2.0 mmol) acetic anhydride, and 26.0 g (250 mmol) styrene are degassed by three freeze-thaw cycles, sealed under argon, and heated at 125°C under nitrogen for 8 h. The solidified reaction mixture is then dissolved in 50 mL dichloromethane and precipitated twice into 2 L methanol. The precipitate is then collected by vacuum filtration and dried to give the desired polystyrene, 9.45, as a white solid (23.1 g, 87.8% yield), $M_n = 23{,}000$, PD = 1.08.

238.B. General Procedure for Acrylate Polymerization; Preparation of Poly(n-butyl acrylate), 9.47, of DP = 250 from 9.29

A mixture of 325 mg (1.0 mmol) of the desired alkoxyamine, 9.29, 11 mg (0.05 mmol) of the corresponding nitroxide, 9.1, and 32.0 g (250 mmol) *n*-butyl acrylate are degassed by three freeze-thaw cycles, sealed under argon, and heated at 125°C under nitrogen for 16 h. The viscous reaction mixture is then dissolved in 50 mL dichloromethane and precipitated twice into 2 L methanol. The gummy precipitate is then collected and dried to give the desired poly(n-butyl acrylate), 9.47, as a colorless gum (26.9 g, 83.2% yield), $M_n = 24{,}000$, PD = 1.09.

In conventional polymerization of a substituted olefin, a new asymmetric center is produced as each monomer unit is added, but there is no control of the configuration of each succeeding center.

$$\sim CH_2-\overset{R}{\overset{|}{C}}H\cdot + CH_2=\overset{R}{\overset{|}{C}}H \longrightarrow -CH_2-\overset{R}{\overset{|*}{C}}H-CH_2-\overset{R}{\overset{|*}{C}}H\sim$$

The result is a completely random configuration of the chain.

These configurations may be visualized by imagining the carbon-carbon polymer chain laid out on a plane in the extended zigzag conformation. If the substituents from the monosubstituted vinyl monomer are arranged at random above and below the plane of the carbon chain, the polymer lacks stereochemical regularity and is called *atactic*:

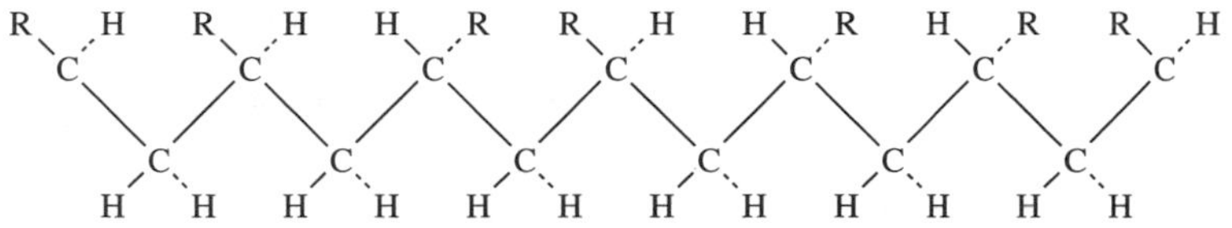

```
R  H   R  H   H  R   R  H   H  R   H  R   R  H
 \.'    \.'    \.'    \.'    \.'    \.'    \.'
  C      C      C      C      C      C      C
   \    / \    / \    / \    / \    / \    /
    C        C      C      C      C      C
   / \      / \    / \    / \    / \    / \
  H   H    H   H  H   H  H   H  H   H  H   H
```

The coordination catalysts give a stereoregular chain. If the substituents all fall on one side of the plane, the polymer is said to be *isotactic*:

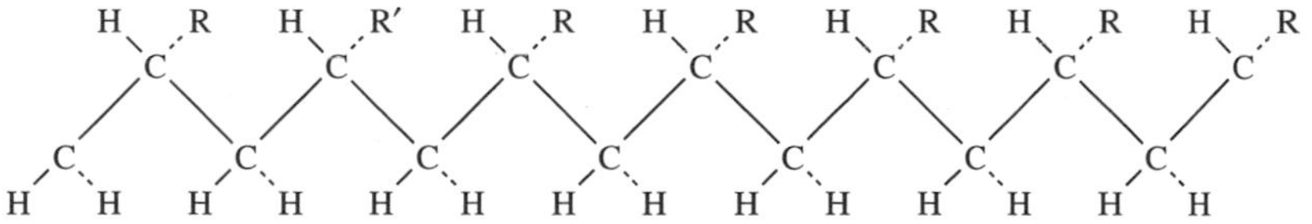

```
    H  R   H  R'  H  R   H  R   H  R   H  R   H  R
     \.'    \.'    \.'    \.'    \.'    \.'    \.'
      C      C      C      C      C      C      C
     / \    / \    / \    / \    / \    / \    /
    C    C      C      C      C      C      C
   / \  / \    / \    / \    / \    / \    / \
  H   H H  H  H   H  H   H  H   H  H   H  H   H
```

Finally, if the substituents fall alternately above and below the plane of the chain, the polymer is designated *syndiotactic*:

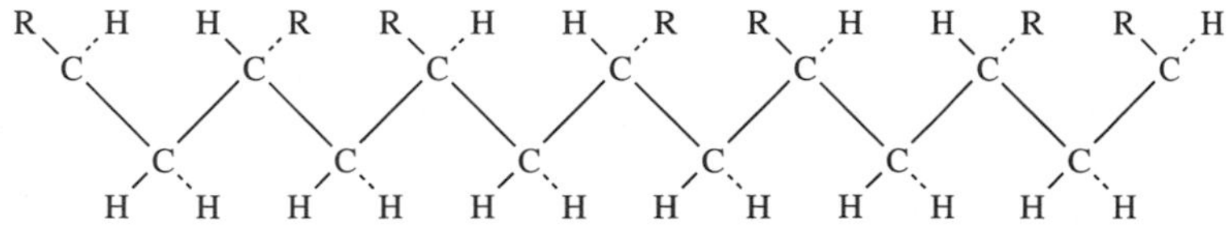

```
R  H   H  R   R  H   H  R   R  H   H  R   R  H
 \.'    \.'    \.'    \.'    \.'    \.'    \.'
  C      C      C      C      C      C      C
   \    / \    / \    / \    / \    / \    /
    C        C      C      C      C      C
   / \      / \    / \    / \    / \    / \
  H   H    H   H  H   H  H   H  H   H  H   H
```

9.10. STEREOREGULAR POLYMERIZATION OF OLEFINS

It was noted earlier that that among the olefins only ethylene and styrene can be readily polymerized by free radicals, the former only at high temperature and pressure. In one of the most important discoveries in polymer chemistry in the second half of the twentieth century, it was found in the 1950s that catalysts composed of transition metal compounds, especially the halides, and certain organometallic compounds will polymerize ethylene and most α-olefins (6, 175–179) to high molecular weight, usually crystalline, polymers. Most remarkably, the olefins from propylene and higher (180–182) could be made not only to be high molecular weight and linear (almost zero branching) but of stereoregular configuration.

These catalysts were called initially and variously coordination-, mixed metal-, bimetallic-, transition metal-, and Ziegler-Natta catalysts, the latter coming from the names of their primary discoverers, who shared a Nobel Prize for their work. At about the same time, the category of metal oxide catalysts were discovered that could also polymerize ethylene and propylene to linear and, in the case of the latter, stereoregular polymers (183). This category of catalysis spawned an immense patent and scientific literature, as researchers sought to define the catalysts, operative chemical limits and the properties of the possible new polymers and the mechanisms by which they acted—not to mention distinctions that could provide commercial advantages and patent protection under which to operate. The scramble for the latter lasted decades, and is a topic for chemical historians to describe.

The coordination catalysts were effective at room temperature and atmospheric pressure, for the most part, simplifying their use in investigative studies at small scale in readily accessible laboratory glassware. It became feasible to easily achieve high molecular weight products, sometimes too high for practical fabrication. As with any polymerization process, it became necessary to understand the factors affecting rate, stereoregulation, reproducibility, and molecular weight control, among others.

Stereoregularity permits the chains to crystallize in most cases; hence the properties of the polymers differ markedly from the random counterpart. Thus free-radical polystyrene is clear, noncrystalline, and low melting, whereas stereoregular polystyrene is hazy like nylon, crystallizable, and high melting. The nature of R also affects the melting point markedly; the bulkier it is, the higher melting the polymer (184, 185). The preparations of some selected polymers follow.

The stereoregular polymers have been discussed extensively by Natta (180–182) and others (6).

239. Preparation of Lithium Aluminum Tetradecyl (175–177)

$$LiAlH_4 + C_8H_{17}CH{=}CH_2 \longrightarrow LIAl(C_{10}H_{21})_4$$

A mixture of 700 ml tetrahydronaphthalene 150 ml 1-decene, (7.6 g) and lithium aluminum hydride is heated with stirring to 135°C. A mildly exothermic reaction occurs, and the temperature is allowed to rise to about 180°C. It is maintained at this temperature for about 2 h, then cooled. The originally clear liquid containing lumps of white hydride has become a solution in which is suspended a gray-black flocculent solid. The flask is transferred to a nitrogen-filled dry box and filtered while still warm through a Celite pad. The filtered solution is stored, and the pyrophoric residue on the funnels destroyed at once with isopropanol. For standardization, a 5.0-mL aliquot of the solution is removed, dissolved in some alcohol, then 200 mL water is added. The resulting mixture is titrated potentiometrically to pH 7 with standard acid. Under these conditions, only the lithium is titrated, and normality = molarity. This preparation should have concentrations of around 0.2*M*. It is necessary to prevent contact of oxygen and moisture with the lithium aluminum tetradecyl solution. Otherwise, no problems should be encountered during storage.

240. Preparation of Titanium Tetrachloride Solution

A commercial product is distilled at 136.5°C after removal of an appreciable forecut. A stock solution is made up in cyclohexane in 500-mL batches as needed. A convenient concentration that should be sought for is about 0.5*M*. The exact concentration is determined gravimetrically. Pure $TiCl_4$ may be available from specially suppliers.

241. Preparation of Catalyst and Polymerization: General Procedure (175–178, 184)

In general, measured amounts of the catalyst components (titanium tetrachloride and lithium aluminum tetradecyl) are mixed in some inert solvent, usually cyclohexane. Stirring is rapid during mixing and during polymerization. The major cause of trouble in polymerizations is oxygen, which is excluded during all phases of the polymerization and catalyst preparation. Also, water and other electron-rich substances must be excluded, because they also inactivate the catalyst.

In the actual experiments listed, the volumes and concentrations used are taken directly from the literature. It is obvious that other concentrations could be used, the volumes of the component solutions being adjusted accordingly.

If aluminum triisobutyl or triethyl are on hand, it may be advantageous to use these materials. However, it should be recognized that they are dangerous reagents. A catalyst is then prepared as follows.

A 1-L three-necked flask equipped with stirrer, a nitrogen inlet, a thermometer, a dropping funnel, and a condenser is flushed with nitrogen and maintained with an atmosphere of nitrogen throughout the entire preparation. Then 100 mL decahydronaphthalene is introduced into the reaction flask, and heat is applied through a heating mantle. To the decahydronaphthalene is added in succession 6 mL 1*M* titanium tetrachloride, and 2 mL 1*M* triisobutylaluminum, both solutions having previously been made up in decahydronaphthalene. Addition of the aluminum alkyl causes precipitation of a brownish black product. The temperature of the flask is then increased as rapidly as possible to 180 to 185°C and maintained at this temperature for 40 min. The color of the suspended complex changes to a deep violet, which is quite intense. The solution is then cooled, and 400 mL cyclohexane is added followed by 12 mL 1*M* aluminum triisobutyl solution. The violet color becomes at once a deep purplish black.

All of the polymers described below are prepared in essentially the same manner. A three-necked flask or a resin kettle is equipped with a stirrer, a nitrogen inlet, and a simple outlet. A catalyst suspension is prepared in the flask by mixing appropriate amounts of the catalyst components under nitrogen with or without additional solvent. Monomer is then added and the polymerization is allowed to proceed. The polymer is isolated by addition of alcohol followed by filtration. It is purified by washing in a high-speed home mixer with additional quantities of alcohol, then freed of organic solvents in an appropriate manner, usually either with steam or with dry nitrogen at 100 to 120°C.

242. Preparation of Linear Polyethylene (175–177, 186)

$$CH_2{=}CH_2 \longrightarrow [-CH_2-CH_2-]_n$$

A three-necked 4-L resin kettle is equipped with a stirrer, a gas inlet tube, and an outlet consisting of a simple glass tube 24 in long, protected with a drying tube. The kettle is flushed with nitrogen, and a catalyst suspension is prepared in the following manner: 2-L cyclohexane is added to the kettle followed by 200 mL a 0.2*M* solution of lithium aluminum tetradecyl as previously made. The mixture is cooled externally in a water-ice bath and 0.029 mol titanium tetrachloride is added as a cyclohexane solution. The mixture immediately becomes brownish black and is an active catalyst for the polymerization of ethylene. A cylinder of ethylene is attached to the gas inlet tube through a safety trap. Ethylene is bubbled through the vigorously stirred catalyst suspension, and polyethylene begins to separate with the evolution of some heat. The polymerization may be continued as long as desired until the solution becomes unstirrable. The slurry of polymer and solvent containing catalyst is poured into a large excess of isopropyl alcohol with vigorous stirring, and the finely divided, precipitated polymer is isolated by filtration. The product is usually obtained as a white powder, which can be molded to clear, tough films or extruded to tough fibers. The product has a crystalline mp in the range of 130°C.

As an alternative to the metal alkyl used above, some of the commercially available aluminum alkyls may be used, as triisobutyl aluminum or triethylaluminum and aluminum sesquichloride. For the above experiment, 8 mL (0.060 mol) triethylaluminum and 4 mL (0.036 mol) titanium tetrachloride may be used in about 2 L total solvent.

The ethylene gas stream can be purified if necessary by passing it through an aluminum alkyl solution before introducing it to the polymerization flask.

When $TiCl_4$ is reduced by aluminum alkyls as in preparation 242, the $TiCl_3$ produced is present in primarily its β form, which is brown in color. It is capable of polymerizing propylene but a large part of the resulting polymer will be the amorphous atactic variety. To maximize the generation of crystalline isotactic polypropylene, the reduction of $TiCl_4$ should be carried out at 185 to 200°C., where γ-$TiCl_3$ is formed. This variety of $TiCl_3$ is capable of a much higher degree of stereoregulation as is α-$TiCl_3$, which results from reduction of $TiCl_4$ with hydrogen or aluminum at high temperatures. The α- and γ-$TiCl_3$ are purple in color and normally contain aluminum chloride when produced, in their respective cases, from aluminum or aluminum alkyl reduction steps (175).

The activating aluminum alkyl used to form the bimetallic coordination catalyst system is also influential in the isotactic stereoregulation of the propagation step (187). Trialkylaluminum compounds decrease in effectiveness with increasing alkyl chain length. Dialkylaluminum monohalides are distinctly more stereoregulating than the corresponding trialkylaluminum, and the monohalides themselves induce higher isotacticity increasingly in the order -Cl, -Br, -I (233). A host of modified catalysts are known that are capable of polymerizing propylene with considerable stereoregulating effectiveness, and a veritable torrent of patent literature exists on the subject (188). The catalysts shown in the following examples are illustrative and are convenient for laboratory work. The second example, using an activated α-$TiCl_3$ from aluminum reduction in conjunction with Et_2AlCl, is highly effective for the polymerization of most α-olefins to highly isotactic polymers.

243. Preparation of Isotactic Polypropylene

$$CH_2{=}CHCH_3 \longrightarrow [-CH_2-\underset{}{\overset{CH_3}{\overset{|}{C}H}}-]_n$$

Polypropylene may be polymerized over either the catalyst prepared in the previous experiment or over a catalyst prepared as follows. In a 4-L three-necked resin kettle equipped with a stirrer, gas inlet, and an outlet consisting of a simple glass tube 24 in long protected with a drying tube is placed 500 mL decahydronaphthalene. The flask is swept with nitrogen and heated moderately with a heating mantle. To the decahydronaphthalene is added, in succession, 30 mL 1.0 titanium tetrachloride and 10 mL 1.0 M triisobutylaluminum, both the latter solutions in decahydronaphthalene.

Addition of the aluminum alkyl causes precipitation of a brownish black product. The temperature of the flask is then increased as rapidly as possible to about 185°C and maintained at this temperature for 40 min. The color of the suspended complex changes to a deep violet. This suspension is then cooled, and

2000 mL cyclohexane is added, followed by 60 mL 1.0 M aluminum triisobutyl solution. The purplish black suspension is then an effective catalyst for the polymerization of gaseous propylene. The propylene is bubbled through the catalyst suspension as in the previous experiment, and the slurry of polypropylene is precipitated into isopropyl alcohol, filtered, washed, and dried. Isotactic polypropylene can be pressed to clear, tough films and molded to other objects. The crystalline mp is about 165°C. The yield is determined by the length of time the propylene is bubbled into the catalyst.

244. Preparation of Isotactic Polypropylene over a Preformed $TiCl_3$ Catalyst Diethylaluminum chloride can be obtained commercially or it can be easily prepared by mixing in a dry box 33.5 g (0.29 mol) triethylaluminum and 19.0 g (0.14 mol) sublimed aluminum chloride in 60 mL dry n-heptane in a septum-capped beverage bottle. The reaction occurs with little evolution of heat as the aluminum chloride dissolves via reaction. The solution will contain about 3.8 mmol Et_2AlCl mL. α-Titanium trichloride can be obtained from AKZO & Crompton as the AA grade which contains 0.33 mol of $AlCl_3$/1mol $TiCl_3$. It is obtained from high-temperature reduction of $TiCl_4$ with aluminum.

To a 200-mL beverage bottle, baked dry at 115°C and cooled under nitrogen, is added 100 mL dry n-heptane; 0.24 g (2 mmol) diethylaluminum chloride, conveniently in n-heptane solution; and 0.198 g (1 mmol) titanium trichloride AA (see above). The latter can be added as a slurry in cyclohexane or n-heptane by ordinary syringe technique if the needle is not less than 18 gauge. Propylene is pressured into the bottle at 40 psi. The polymerization can then be carried out under continuous propylene pressure of 40 psi if the bottle is heated in an upright position in a shaker assembly. Alternatively, the monomer can be charged only to its limit of saturation, 6 to 7 g/100 mL at 40 psi. The bottle is then tumbled at 70°C for at least 4 h.

In either case, the bottle is cooled and carefully vented with a hypodermic; 20 mL isopropanol is added to deactivate the catalyst. The mixture is slurried in a blender with 600 mL isopropanol containing some HCl. The polymer is filtered and washed additionally with methanol, then dried at 60°C under nitrogen in a vacuum oven overnight. If propylene was charged only initially, about 6 g polymer should be obtained; if charged continuously, at least 25 g should be had. A sample of polymer is subjected to continuous extraction with n-heptane in a vapor-jacketed Soxhlet (Kumagawa) extractor for 10 h; a ceramic thimble is advantageous. At least 90% of the polymer should be insoluble in boiling n-heptane in this manner, and 94% is quite within reach. This is the isotactic index value. The soluble, amorphous polymer can be recovered from the heptane extract. The crystalline, isotactic polymer has a T_m of about 165°C, though values of 170°C are sometimes reported. The inherent viscosity should be 2.5 to 3.5 (0.1 g 100 mL in Decalin, 130°C).

Essentially the same product can be obtained by running the polymerization in a stirred flask or resin kettle; propylene is passed into the catalyst-solvent slurry as in the polyethylene case.

A useful intrinsic viscosity-molecular weight relationship for polypropylene is (189):

$$[\eta] = 1.10 \times 10^{-4} M_v^{0.80}$$

It should not be out of place at this point to remark that isotactic polypropylene in a way stands as an introduction to the enormous contributions of Natta to polymer science. The *Journal of Polymer Science*, in an issue dedicated entirely to him (190), called him "The Father of Stereoregular Polymers," which is no less than the truth, and the reader is commended to the brief biography to be found there. To no one's surprise, he and Ziegler shared the Nobel Prize in chemistry in 1963. Natta's Nobel award address was published in *Science* (191). While the phenomenon of polymer stereoregularity had early been predicted to be a matter of importance in controlling physical properties and while a few synthetic examples were recognized or believed to exist, it was not until Natta's work that the vast scientific richness of the field was realized in fact. The numerous examples of preparations of stereoregulated polymers in this book are only representative of the full harvest; many are derived directly from Natta's work, but those that are not are often not far removed in lineage.

Polypropylene with mainly a syndiotactic placement of monomers has been prepared at quite low temperatures with certain coordination catalyst components, e.g., VCl_4 and i-Bu_2AlCl with anisole added (192 to 194). The resulting polymer is of a relatively lower molecular weight than obtained customarily with isotactic polypropylene and is also less crystalline, due in all likelihood to a stereoblock structure with the syndio-structure dominant but not exclusive.

245. Preparation of Primarily Syndiotactic Polypropylene (192)

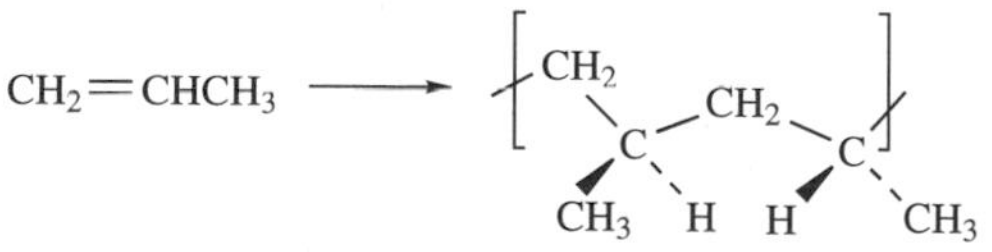

$CH_2{=}CHCH_3 \longrightarrow$ [CH_2–C(CH_3)(H)–CH_2–C(H)(CH_3)]

A 250-mL flask with stirrer, dry ice condenser, nitrogen inlet, and serum cap, and a means for adding propylene below the dry ice condenser is baked dry and assembled in a stream of nitrogen. In a stream of nitrogen, the flask is charged with 100 mL dry toluene and cooled to −78°C in a dry ice-methanol bath. Then the following are introduced in sequence by syringe: 0.108 g (10^{-3} mol) anisole in 5 mL toluene, 0.193 g (10^{-3} mol) VCl_4 as a solution in toluene, and 0.885 g (5×10^{-3} mol) i-Bu_2AlCl also as a solution in toluene. Then 80 high-purity propylene (bp −48°C) is added; and after initially mixing the components, the polymerization is allowed to proceed for 24 h without stirring. The polymerization can also be executed in a septum-capped beverage bottle immersed at −78°C and agitated occasionally by hand.

The slightly viscous, purple mixture is poured into excess methanol containing a small amount of HCl. The precipitated polymer is washed repeatedly with methanol and dried in vacuum at 60°C to give about 1 g of product with an intrinsic viscosity of 1.1 dL/g (tetralin, 135°C). The T_m of this syndiotactic polypropylene varies with steric purity is but about 131°C. It has been previously estimated (194) that the ultimate T_m for syndio- is 166°C. For the isotactic polypropylene, the ultimate T_m is estimated in the range 188 to 200°C, although the usual practically observed value is around 165°C. See Metallocene catalysts.

246. Preparation of Poly(4-methyl-1-pentene) (184)

$$CH_2{=}CH{-}CH_2CH(CH_3)_2 \longrightarrow [{-}CH_2{-}CH(CH_2{-}CH(CH_3)_2){-}]_n$$

A three-necked, 4-L resin kettle is equipped with a stirrer, nitrogen inlet, and an outlet consisting of a simple glass tube 24 in long. The kettle is flushed with nitrogen and a catalyst suspension is prepared in the following manner: 2L cyclohexane is added to the kettle followed by 200 mL of a 0.20-*M* solution of lithium aluminum tetradecyl as previously made. The mixture is cooled in a water bath, and 33 mL of a 0.87 M titanium tetrachloride solution in cyclohexane is added. The mixture immediately becomes brownish black and is then an active catalyst for the polymerization of an α-olefin. To the catalyst suspension is then added 450 g distilled, dry 4-methyl-1-pentene, which may be obtained commercially. The mixture warms up moderately and rapidly becomes viscous. Polymerization may be allowed to proceed for 12 h. It is essentially complete in a much shorter time, and the reaction may be worked up after 2 h if it is desired, although the yield may be diminished by a few percent. Because the polymerization is mildly exothermic, a certain amount of monomer may be lost by distillation and entrainment from the reaction vessel. This may be trapped in a condenser system if desired.

After a 12-h period the polymerization is essentially complete, and the reaction mixture has the appearance of a solid, blackish brown lump of rubber. This mass of polymer that is swollen by the cyclohexane present is cut up into convenient size lumps with scissors or a knife and then cut up very finely in a high-speed cutter, such as a household blender, with isopropyl alcohol. The color of the original polymer is discharged immediately on contact with the alcohol, and the resulting product is isolated by filtration as a pure, white rather granular product. The polymer may be washed repeatedly with alcohol and then dried either in a vacuum oven at pressures less than 1 mm and temperatures of 100°C for 8 h or by treatment with a rapid stream of live steam or nitrogen at 100°C, with vigorous agitation. The yield of dry polymer is about 250 g and the inherent viscosity is 4 to 6, measured at 130°C on 1 0.1% solution in decahydronaphthalene. The extremely high molecular weight product may be fabricated directly to a clear tough film by pressing at temperatures in excess of 250°C. It also may be extruded from a spinning cell in the form of a continuous filament that may be after drawn on a hot pin at 100 to 150°C. The polymer shows a crystalline mp of about 240°C as measured on a polarizing microscope.

4-Methyl-1-pentene may also be polymerized over a catalyst prepared as above from aluminum triisobutyl. To this catalyst solution is added 100 mL 4-methyl-1-pentene. The mixture is stirred for a period of 90 to 100 min, at which time the mixture is so thick that it wraps around the stirrer blade and breaks away from the walls of the vessel. The polymer is isolated, then dried in a vacuum oven at 50°C by methods previously described. Polymer prepared in this manner is essentially the same as prepared in the previous experiments. Yields will be of the order of 65 to 80%. The inherent viscosity is 2 to 4 (0.1% in decahydronaphthalene at 130°C).

The Et_2AlCl-$TiCl_3$ catalyst system (the later component preformed) used for polymerizing propylene also works well in this instance. In fact, such a catalyst gives a distinctly lower amount of boiling heptane extractables (amorphous polymer) than any of the others. Poly(4-methyl-1-pentane) with as high as 98% heptane-insoluble content can be realized. The polymerization rate is slower with Et_2AlCl-$TiCl_3$ than with the others, however.

247. Polymerization of 3-Methyl-1-butene (184)

$$CH_2{=}CH{-}CH(CH_3){-}CH_3 \longrightarrow [{-}CH_2{-}CH(CH(CH_3)_2){-}]_n$$

A catalyst suspension is prepared as in the preceding experiment by mixing in a 4-L resin kettle under nitrogen 250 mL 0.19 *M* lithium aluminum tetradecyl and 40 mL 1.08 *M* titanium tetrachloride in 1 L cyclohexane. Then 100 mL 3-methyl-1-butene (b p 20°C) is collected in an ice bath and added to the mixture. After 20 h, the polymer is isolated by filtration and washed with alcohol. Only 13.8 g, approximately 22%, is isolated in this manner. The inherent viscosity cannot be determined since the product is incompletely soluble in decahydronaphthalene. The crystalline melting point (310°C) is much higher than that of poly(4-methyl-1-pentene), and it is necessary to go to temperatures above 310°C to fabricate the polymer into a clear, transparent film. Strips of this film can be oriented by stretching at temperatures on the order of 250 to 300°C, or fibers may be drawn from the melt at 310°C. Oriented film strips or fibers show extremely high crystallinity and very high orientation. The crystalline melting point is on the order of 310°C.

Once again, the Et_2AlCl-$TiCl_3$ catalyst (see poly(4-methyl-1-pentene) preparation for comments) gives a higher content of stereoregular polymer.

The stereoregular polymerization of styrene and other vinylaromatics exhibits some peculiarities not seen in the polymerization of ethylene and vinylalkanes. For example, high conversions are best attained in solvents that dissolve or swell the polymer. Benzene is the preferred solvent; cyclohexane is suitable for styrene but not for 4-vinylbiphenyl. With vinylaromatics, Et_3Al-$TiCl_3$ is the preferred catalyst system. The Et_2AlCl-$TiCl_3$ system gives both lower conversion and poorer stereoregularity. Perhaps this is because the stereoregular polymerization with the latter catalyst is so slow that cationic polymerization leading to atactic polymers becomes important. Frequently, isotactic polystyrene contains a toluene-insoluble fraction (195). There are indications that the proportion of toluene-insoluble polymer depends on the particular batch of $TiCl_3$ that is used. The cause of the toluene insolubility is not definitely established. It has been found, however, that polymerization in the presence of vinyl chloride (a molecular weight regulator) decreases the toluene insolubility (too much vinyl chloride greatly lowers conversion and isotaxy), and that controlled pyrolysis of the polymer (300°C) also decreases toluene insolubility.

*248. Preparation of Isotactic Polystyrene (116)**

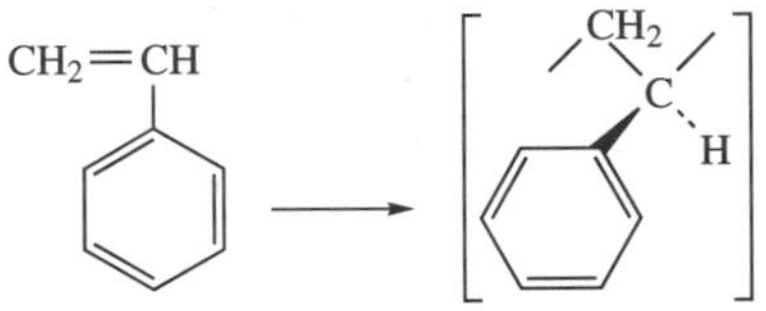

An argon-flushed beverage bottle containing 50 mL dry benzene and a Teflon-covered stirring bar is capped with a Buna N septum. Then 0.41 mL (3 mmol) triethylaluminum and 3 mL (~ 1.8 mmol) of a "1*M* suspension" of $TiCl_3$ AA in cyclohexane are transferred to the bottle with syringes (18-gauge needle, 5-ml syringe for the $TiCl_3$) in an inert atmosphere. Finally, 5 g styrene is introduced with a syringe, and the bottle is agitated at 60°C by placing it in a water bath on a hot plate with a magnetic stirrer. After 20 h, the thick black mass is transferred to a blender and washed twice with isopropanol. If not yet colorless, the polymer is washed additionally once or twice with 2-butanone and dried in vacuo at 80–100°C. Conversion is 95%.

Continuous extraction with 2-butanone removes the polymer (about 5%), which has low stereoregularity. Continuous extraction of the butanone-in-soluble (isotactic) polymer with toluene gives a toluene-insoluble fraction, A (about 10% of the total polymer) and a toluene-soluble fraction, B. The latter fraction is precipitated with methanol. Films of A and B may be pressed at 250°C; after annealing for 30 min at 180°C, both give the characteristic spectrum of crystalline polystyrene (196). The polymer melts at about 230°C. Although the isolated polymer B, is readily redissolved in toluene only on heating to near reflux, the polymer remains dissolved when the solution returns to room temperature. The inherent viscosity (0.1 g/100 mL in toluene at 30°C) is 1.6.

In $CDCl_3$, isotactic polystyrene displays broad NMR signals for the benzylic and methylenic protons at 7.96 and 8.54_{τ}, respectively, while the corresponding signals for the atactic polymer are at 8.17 and 8.59_{τ}.

Isotactic co-polyhydrocarbons can be made, as in the following example.

* Dr Donald B. Miller was kind enough to provide this example and much of the foregoing commentary.

249. Random Copolymer of 4-Methyl-1-pentene and 1-Hexene (197)

$$CH_3CH_2CH_2CH_3CH{=}CH_2 + (CH_3)_2CHCH_2CH{=}CH_2 \longrightarrow \text{Copolymer}$$

To illustrate the effects of copolymerization on the properties of poly-α-olefins, the following experiments can be performed. A mixture of 40 ml 4-methyl-1-pentene and 10 mL 1-hexene is polymerized for 3 h, as described above. The polymer is isolated by precipitation in alcohol followed by filtration. The inherent viscosity is 2.5 to 3.5 as measured in cyclohexane (0.5%) at room temperature. This polymer may be pressed to clear film at about 120°C. It may be drawn and oriented at 125°C. These strips show a crystalline melting point of approximately 195°C. The interesting thing about this copolymer is that it is readily soluble in cyclohexane to give a viscous but true solution in contrast to poly(4-methyl-1-pentene), which forms with cyclohexane a swollen, gel-like mass. The added 1-hexene has changed the solubility characteristics quite markedly.

The commercially most significant olefin copolymer is that from ethylene and propylene. When prepared to minimize long blocks of either monomer and by a catalyst that is not favorable to highly isotactic placement of consecutive propylene units when they do occur, the polymer is completely amorphous, has a low T_g, and has many useful elastomeric properties. It can be vulcanized with peroxides, but it is generally the custom to prepare a terpolymer with a diene so that unsaturation is introduced to allow sulfur vulcanization. Dicyclo-pentadiene is frequently used to make the terpolymer (EPT), because its double bonds are of unequal reactivity; one of them does not readily enter into copolymerization.

*250. Preparation of an Elastomeric Terpolymer of Ethylene, Propylene, and Dicyclopentadiene (198)**

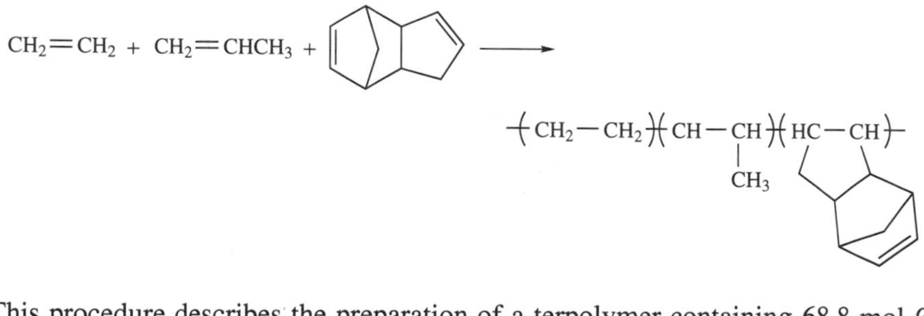

$$CH_2{=}CH_2 + CH_2{=}CHCH_3 + \text{(dicyclopentadiene)} \longrightarrow$$

$$-\!\!(CH_2-CH_2)\!\!-\!\!(CH-CH(CH_3))\!\!-\!\!(HC-CH)\!\!- \text{(with fused norbornene ring)}$$

This procedure describes the preparation of a terpolymer containing 68.8 mol % (56.5 wt %) ethylene (E), 29.3 mol % (36.1 wt %) propylene (P), and 1.9 mol % (7.4 wt %) dicyclopentadiene (D).

The procedure is analogous to that described elsewhere (5) for the preparation of EP copolymer containing 30 mol % P, except that D is also added initially and

* We are very grateful to Dr. Carl Lukach for making available this extensively detailed preparation.

in many increments (or continuously) during the polymerization. To prepare a terpolymer of uniform composition, all three monomer concentrations in solution must be kept constant at a precalculated ratio throughout the polymerization. This requirement is fulfilled by using monomer reactivity ratios to calculate the initial monomer concentrations, and then supplying each monomer during the polymerization at a rate corresponding to its rate of consumption, under conditions of sufficient agitation and constant temperature.

The procedure involves a relatively slow rate of catalyst addition, and therefore of polymerization, to facilitate equilibration of gas and liquid phases, and to minimize the temperature rise.

250.A. Apparatus

The apparatus used is a jacketed, 1-L resin kettle equipped with condenser, thermometer, gas inlet tube, and paddle stirrer (with glass shaft and Teflon half-moon blade) mounted from an air-driven motor (electric motors should not be used in the immediate vicinity of the reactor because of the possibility of a gas leak). One neck of the reactor is provided with a rubber stopule so that injections can be made with a hypodermic syringe.

The gas inlet tube, projecting into the reactor about 1 in above the stirrer blade, is connected to an exterior monomer inlet line containing two sets of flowmeters for ethylene and two sets for propylene. Each set consists of one 06-150/13 and one 08-150/13 Fischer-Porter flowmeter (steel ball) connected in parallel through top and bottom stopcocks to provide a total flow rate range of at least 1500 cm^3/min (0–300 cm^3/min with the 08-, and higher flow rates with the 06-flowmeter). Each set is connected to the appropriate monomer cylinder. The inlet line is also provided with a manometer, T-tube, and stopcock between the row of eight flowmeters and the reactor, for introducing nitrogen, or evacuating (aspirator) the reactor.

The excess gas exits from the reactor through the condenser. A two-way stopcock is connected to the top of the condenser for removing gas aliquots for analysis, if desired. The exit gas passes through two bottle traps, the last containing glass wool to trap any solvent mist carried by the gas, and then through an 06-150/13 flowmeter (to measure the exit gas flow rate) before being vented into the hood.

Hot water is circulated through the outer jacket of the resin kettle. A temperature-regulating device is used for adjusting the water temperature. A manually operated needle valve is also provided for introducing cold water into the constant-temperature stream to control any exothermic heat rise and maintain a constant temperature.

Both catalyst components (ethyl aluminum sesquichloride and VCl_4) are best added separately and continuously at a constant Al/V ratio. Either motor-driven syringe drivers, microbellows pumps, or Hershberg dropping funnels can be used for this purpose, with the catalyst components being pumped directly into the reactor slightly above the liquid level via stainless-steel, small-diameter tubing. It is convenient to provide the catalyst pump system with a timer (such as a

Flex-O-Pulse Timer, Eagle Signal Corp., Moline, Ill.) so that the motor on-off cycle can be adjusted to control the rate of delivery.

250.B. Procedure

The flowmeters are calibrated at several settings with their intended monomers, either ethylene or propylene, by timing the rise of a soap bubble in a graduated column attached to the flowmeter. A nonlinear plot of flow rate versus flowmeter setting is prepared for each flowmeter.

After shutting the appropriate stopcocks, the reactor (exclusive of monomer inlet line or off-gas line) is alternately evacuated and filled with nitrogen twice, while hot water (50°C) is circulated through the outer jacket of the resin kettle. Then 500 mL pure-grade Phillips *n*-heptane (previously dried by being passed through a column of Linde Type 4A molecular sieves) is added via stainless-steel tubing from a beverage bottle under nitrogen pressure (10 psig), or by syringe. The apparatus is again evacuated and then filled with nitrogen to atmospheric pressure. The temperature is adjusted so that the heptane is at 50°C. After the stopcocks are opened, the heptane is saturated with a mixture of E and P containing 68 mol % P fed at a total gas rate of 700 cm^3/ min from one set of E and one set of P flowmeters. When the exit gas flow rate becomes constant, this saturating stream is reduced to 300 cm^3/ min, and **kept at this flow rate throughout the polymerization** without changing its composition. Thus, before polymerization occurs, the inlet and exit gas flow rate will each be 300 cm^3/ min, and the gas composition will be 68 mol % P and 32 mol % E.

Then 0.137 g (1.04 mmol) dicyclopentadiene and 1.0 mL 0.5*N* ethyl-aluminum sesquichloride are added directly to the saturated heptane by syringe. The syringe driver is turned on to feed heptane solutions of both ethyl-aluminum sesquichloride (0.30*N* in Al) and VCl_4 (0.06*N* in V) at a rate of 0.09 mL/min. Shortly after the VCl_4 addition begins, the start of polymerization is heralded by a decrease in the exit gas flow rate to below 300cm^3/ min.

At this point, a second monomer stream containing 30 mol % P, controlled by the second set of E and P flowmeters, is also opened and fed at whatever rate is necessary to bring the total exit gas flow rate back to 300 cm^3/ min and keep it there. This second monomer input stream, therefore, supplies E and P of the composition desired in the product, at a rate corresponding to their rate of consumption. The rate of addition of this 30% C_3 monomer stream is adjusted throughout the run so that the exit gas flow rate remains constant. While this is being done, the first monomer input stream is also feeding a 68% P composition at a constant rate of 300 cm^3/ min.

Dicyclopentadiene must also be added during the run to keep its concentration in solution constant at the initial value. This objective is best accomplished by basing the D addition rate on the rate of consumption of either E, P, or both, as measured by the appropriate flowmeter of the second monomer input stream. In the present example, 0.105 g D is added, as 0.29 mL of a 1 : 2 D : heptane solution (v/v), for every 700 cm^3 of E (or alternatively every 300 cm^3 of P) consumed throughout the run. This amount corresponds to 0.08 mmol D/1.21 mmol of P consumed. The addition of D is easily programmed in this way with the help of the flowmeter calibration charts prepared previously.

Manipulation of the addition rate of E, P, and D is continued for 75 min, during which time the solution becomes viscous but remains clear and free from any gel or precipitate. During this period, the catalyst syringe driver adds 6.75 mL 0.3*M* ethylaluminum sesquichloride (2.03 mmol Al) and 6.75 mL 0.06*M* VCl_4 (0.41 mmol V), while 1.21 g (9.15 mmol) D is added in 12 equal increments (as the D heptane solution). The rate of reaction of E and P (as measured by the second input stream flowmeters) gradually increases for 20 min and then remains essentially constant at a total EP rate of 155 cm^3/min. Periodic gas chromatographic analysis of both the exit gas and the reaction mixture indicate that the exit gas composition remains at 66 to 68 mol % P, and the D concentration remains essentially constant throughout the terpolymerization.

The terpolymerization is stopped after 75 min by discontinuing catalyst addition and *slowly* adding 5 mL *n*-butanol to the reaction mixture. After the initial reaction of the butanol and aluminum alkyl subsides, the monomer inlet stream is replaced with nitrogen, which is bubbled through the reaction mixture for 30 min.

The reaction mixture is cooled and washed from the reactor with heptane. It is then washed with three 500-mL portions of water (preferably hot). The organic layer is then poured into a shallow Pyrex baking dish, which is placed in a nearly closed hood overnight. The product (17 g; 27.2 g/Lh; 41.4 g/mmol V) is obtained as a clear film easily peeled from the baking dish. Its reduced specific viscosity (Decalin, 0.1%, 135°C) is 1.8 and it contains 7–9 wt. % dicyclopentadiene linkages by analysis.

The terpolymerization may be run for a longer time but polymerization to a higher product concentration becomes progressively more difficult since equilibrium with the gas phase suffers severely at high solution viscosities.

250.C. Calculations

In the preparation of any terpolymer composition, the ideal copolymerization equation is assumed to apply for any pair of monomers, i.e.,

$$\frac{dm_1/dt}{dm_2/dt} = r_{12}\frac{M_1}{M_2}$$

where m refers to copolymer composition, M to monomer composition, and r_{12} is the reactivity ratio of monomer 1 with respect to monomer 2. The relationship between P and D is then given by:

$$\frac{dm_D/dt}{dm_P/dt} = r_{DP}\frac{M_D}{M_P}$$

where M_D is the concentration of D in solution, M_P is the concentration of P in solution, r_{DP} is the reactivity ratio of D with respect to P, dm_D/dt is the rate of consumption of D, and dm_P/dt is the rate of consumption of P.

The mol % of D and P desired in the product can be used for the left side of the equation. For example, in the present case, for 500 mL solvent and 50°C,

$$dm_D/dt = 1.93$$

$$dm_P/dt = 29.3$$

$$r_{DP} = 6.0$$

$$M_P = 94.4 \text{ mmol } (3.97 \text{ g}) \text{ at } 0.554 \text{ atm. of P}$$

where M_P is based on a P solubility value of 2.1 g/100 g heptane at 1 atm and 50°C, and a vapor pressure value of heptane at 50°C of 141 mm. With these values, M_D is calculated from the equation to be 1.04 mmol (0.137 g) for 500 mL solvent. This is the amount of D added initially, and determines the D concentration to be kept constant throughout the polymerization. The addition rate used for D is based on the rate of consumption of P; it is calculated so that the mole ratio of D added to P consumed is 1.93 to 29.3.

Similarly for E and P, the relationship is

$$\frac{dm_E/dt}{dm_P/dt} = r_{EP}\frac{M_E}{M_P}$$

and $r_{EP} = 17$ to 20 (1). Knowing two reactivity ratios, a third can be calculated from:

$$r_{ED} = \frac{r_{EP}}{r_{DP}}$$

thus enabling the use of any combination of the three monomers in heptane at 50°C with an ethylaluminum sesquichloride-VCl_4 catalyst system. The reactivity ratio values will be different if any drastic departures from these specific conditions are used, especially with a different catalyst system. The solubility of ethylene and propylene will also change with different solvents.

250.D. Possible Procedure Variations

1. A 1-L round-bottom flask immersed in a thermostat can be used in place of a resin kettle. The stirrer must be able to create a rapidly renewed suspension of gas bubbles. With ordinary laboratory stirrers, this is possible only with a relatively shallow depth of liquid. Adequate interior baffling or a creased flask is desirable.
2. The constant off-gas stream of 68 mol % P provides a convenient source of gas for analysis, serves to purge inert gases, and helps correct minor errors in the feed ratio of P to E. If the quality of feed gas is good and control is adequate, the off-gas stream can be dispensed with during polymerization. Even so, the system must be first presaturated with the appropriate monomer composition (68 mol % P), and a different monomer composition (30 mol % P) must be fed during the run.

3. The elaborate system of flowmeters used was set up so that terpolymer of any composition could be prepared. If only the composition of this example is desired, it would be more convenient to provide two sources of mixed gases, of compositions 68 mol % P for presaturation, and 30 mol % P for feeding during the run. The proper rate of addition of D could then be calculated from the pressure drop of the container holding the 30 mol % P mixture.
4. The molecular weight of the product decreases as its propylene content increases. The catalyst requirement is also higher at higher propylene content. Both effects can be counteracted by operation at a lower temperature. As the temperature is lowered, the same gas compositions may be used for presaturation and running, but higher initial concentrations of D will be needed.
5. Diethylaluminum chloride may be used instead of ethylaluminum sesquichloride. The VCl_4 may also be replaced with either $VOCl_3$ or a vanadium ester such as triethyl orthovanadate, but a slightly higher Al/V ratio should be used, and reactivity ratios may be slightly different.

250.E. Procedure at High Pressure

A modified Sutherland reactor (5, 199) can be used. The solvent is saturated with the proper E and P composition (in this case, 68 mol % P) at the higher pressure. Total pressure is held constant by automatic feed of an EP mixture of the composition desired (here, 30 mol % P). The diene is added initially and is kept constant by adding D at a rate dependent on the rate of consumption of E or P, measured by the pressure drop in the feed mixture. Both M_P and M_E will have different values at higher pressures in the calculations used for determining the proper M_D.

Polymerization of Olefins with Metallocene Catalysts

Catalysts of the above designation were noted in the introduction to the section on stereoregular polymerization of olefins. The metallocenes began to come into their own in 1980, and constitute an evolution from the early-era Ziegler-Natta systems. The metallocenes are based on so-called "sandwich" compounds, the prototypical case being bis(cyclopentadienyl)titanium dichloride, dating from around 1951. Since then, the synthesis and study of metallocenes has amounted to a tour de force in organometallic chemistry, with one result being more advantageous olefin polymerization catalysts. Major differences between the early-era catalysts and the metallocenes include the heterogenous, insoluble nature of the former vs. the homogenous, soluble character of the latter. More significant is the so-called single site nature of the metallocenes, while the old-era catalysts are considered to have more than one kind of chemical site, few or none of the same reactivity and stereoregulating capacity.

The battle in controlling the performance of early era catalysts often consisted of knowing precisely the polymerization conditions that would produce a particular desired result, then controlling the process to keep it in those empirically determined conditions. Many factors were involved—the purity and crystalline form of the transition metal halide component, its particle size distribution, the optimum aluminum alkyl to be used and the conditions of their being brought together in the process. There was often only modest understanding of the underlying mechanistic events. Still, something approaching fifty million metric tons of polyethylene and polypropylene were produced world-wide in 2000 by modern variants of the early-era catalysts, but a much smaller amount by metallocenes. But the latter is growing rapidly and is expected to be the dominant catalytic system in time.

While all the Ziegler-Natta and metallocene catalysts that polymerize olefins involve an active site made up of a cationic alkyl that is presumably stabilized by ligands, the metallocene complexes differ markedly in that they have well understood structures, being essentially pure, single compounds. They can be molecularly constructed to have a specific type of polymerization site having essentially one function, in terms of monomer acceptance and resulting stereo-regulating capacity. It is also easier to control molecular weight and distribution than with older catalysts, and turnover—grams polymer/gram catalyst—can be very high. Molecular weight distributions are remarkably narrow, in fact; e.g., values for polyethylene are typically 1.6–2.6, for isotactic polypropylenes, 1.9–2.6. Polyethylene with little long or short chain branching, or ethylene copolymers with longer olefins to impart uniform and controlled branching, cyclic olefin polymers, and both iso- and syndiotactic polypropylene—all these and more can be made with appropriate metallocenes. Regarding syndiotactic polypropylene (syndioPP), as noted earlier the Tm depends on the degree of such tacticity, as it does for the isotactic polymer. With an isopropylidene-bridged cyclopentadiene-fluorene-Zr catalyst with MAO, syndioPPs have been made at different temperatures, yielding varying degrees of syndio status and varying Tm values as a result: at 0°C and 1.7 bar in toluene, the Tm was 186°C; at 80°C and 6 bar, it was 166°C; and at 60°C and 4 bar in pentane, Tm was 135°C (249).

The most remarkable aspect of metallocene chemistry is their status as discrete organometallic compounds, capable of being structurally characterized, often by x-ray diffraction, and they can be tailor-made by synthetic methods. It is understandable that these factors have changed the ground of olefin polymerization to the point where comparison of catalyst development with drug development is not so far fetched. The metallocenes have been discussed extensively by Huang and Rempel (250).

Linear polyethylenes of varying densities, can be made using earlier Ziegler-Natta catalysts and olefins such as 1-butene or 1-octene. But these copolymers, termed linear low density polyethylene, or LLDPE, are more readily accomplished with metallocenes because of more uniform placement of the comonomer. The result is better impact strength, toughness clarity among other properties. Metallocenes based on titanium and zirconium (all Group IVB) in particular, combined with different cyclopentadienes, have been studied. A large number of substituted cyclopentadienes have been explored, also, from fully substituted cyclopentadiene to the bulky, flat indene and fluorene. Patent battles reminiscent of those from the early years of Ziegler-Natta era have erupted. Some of the systems studied have been open-faced sandwiches, i.e., with only one cyclopentadienediene ligand attached to the metal, but that unit bridged to the metal by such groupings as the dimethylsilyl. Some of the currently most-favored catalysts require a bridging unit connecting two cyclopentadienes in true sandwich style.

Methylaluminoxane, termed MAO, is necessary as a cocatalyst with many of the present metallocene systems. MAO latter is prepared by controlled hydrolysis of trimethylaluminum (251). The latter forms the most effective aluminoxane thus far. The structure of the aluminoxanes is not clearly understood, but it is thought to consist of linear and cyclic oligomers:

$$(CH_3)_3Al + nH_2O \longrightarrow (CH_3)_2(CH_3{-}Al{-}O)(CH_3)_2 + \{-[Al(CH_3){-}O{-}Al(CH_3)]_m{-}O-\} + 2nCH_4$$

Four-coordinate aluminum has also been suggested as a possible structure, giving two-dimensional network structure to the aluminoxane. An apparently preferred method of bringing water and trimethylaluminum together is to mix the organoaluminum with a hydrate of copper or aluminum sulfate in a hydrocarbon solvent in the desired ratio of water to aluminum-carbon bonds. The fact that MAO and most of the metallocenes themselves are sensitive to moisture and oxygen requires careful purification of solvents, monomers and sweep gases, plus scrupulous drying of equipment. Glove box handling of reagents is essential for lab experimentation. Polymerizations can usually be run at one atmosphere, but even on a research level, few are. Conditions simulating plant-scale operations would need small, highly instrumented and controlled pressure and temperature reactors.

The aluminoxane is thought to methylate the metal of the catalyst itself by reaction with the chlorides of the liganded metal dichloride. The underlying strategy appears to be to use a cyclopentadiene capable of having a bridge constructed between two such ligands attached to the central metal; e.g., zirconium IV dichloride sandwiched between two indenyl rings, themselves bridged to each other by an ethylene group. This bridged molecule has a chirality that permits racemic (rac-) and meso forms. These are shown in Figure YY, below.

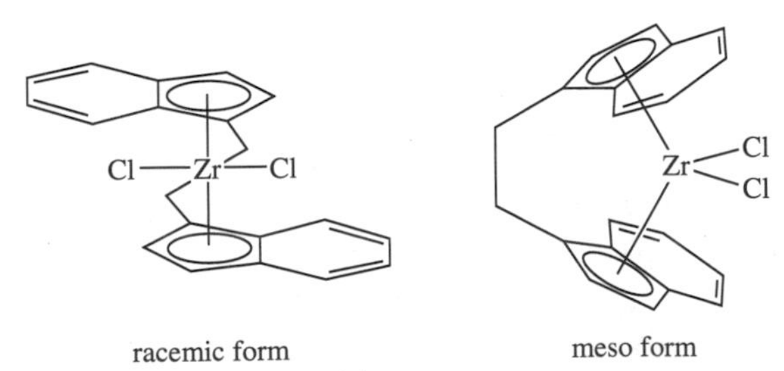

Figure YY. Racemic and meso forms of a bridged bis(indenyl)zirconium metallocene.

The racemic can polymerize propylene in a stereoregular manner, while the meso gives atactic polymer. (The racemic version is used in the first experiment, below). The electronic and steric factors presented by the cyclodiene also influence catalyst activity; e.g., Zr (IV)Cl_2 having cyclopentadiene ligands is not stereoregulating, but with ethylene-bridged bis(indenyl) as the ligand system, it is. In a monocyclopentadieny titanium catalyst having a covalent single atom bridge from the metal to the ring, a constrained geometry (CG, the term applied to this catalyst type) is imposed by the bridging unit. The molecule thus has a fixed tilt between the ring, the heteroatom (e.g., a substituted nitrogen) and the titanium.

Indeed, the structural variety accessible to metallocene chemistry can be very great, and so also are the results from catalysts so-derived. Cyclopentene and norbornene are among the cyclic dienes that have been polymerized through the double bond without ring opening, to isotactic homopolymers, no less. Cyclopolymerization of 1,5-hexadiene occurs to give optically active polymethylene-1,3-cyclopentane, when (R,R)-ethylenebis(hexahydroindenyl) zirconium-1,1′-bi-2-naphtholate (252) is used.

Outcome distinctions between the earlier heterogenous Ziegler-Natta catalysts and homogenous (soluble) metallocenes come, among other things, from the fact that all the metal sites of the metallocenes can be active polymerization sites. And as single-site catalysts, the polymers from them can have narrow Mw/Mn ratios, usually 2 or less, and therefore have much narrower melting points. But all catalyst types have their idiosyncrasies. Increasing the polymerization temperature of metallocenes can result in a marked, abrupt drop off in stereoregulating capacity. Molecular weights of polymers from metallocenes tend to be lower than from old-era Ziegler-Natta catalysts, which usually required a chain transfer agent (e.g. hydrogen). The MW problem with metallocenes can be coped with by reducing polymerization temperature, using hafnocene instead of zircocenes or titanocenes or rendering the metallocene heterogenous, tactics that may not be practical. Reduced activity is a consequence in each case. The cost of MAO is burdensome to process economics, a matter that may be ameliorated in time by less costly activators or by metallocenes needing no activator. Nevertheless, the potential for new polymeric materials ranging from superior elastomers to those of high strength and stiffness is sufficient to drive the field.

Licensing and corporate mergers in the late 1990s have probably distilled the catalyst chemistry toward a few commercially feasible, sufficiently "optimum" systems, but the field is still too new, too much in ferment, to allow its contenders to remain fixed on only a few catalyst types indefinitely (353). New metallocenes, their nearest relatives and a plethora of modifications are certain to be seen (254).

The following preparative example uses a bis(indenyl)-based catalyst wherein the two indenyl rings of the sandwich are bridged by a $-CH_2CH_2-$ grouping. Such bridged systems are considered to more tightly define the catalyst site and thus add to the control of the monomer placement in each addition step.

*9-m. Preparation of a Polyethylene with a rac-Ethylenebis (indenyl)zirconium/MAO system**

A 250 mL Schlenk flask is equipped with a magnetic stirring bar, a drying tube and a dip tube positioned to instill ethylene above the surface of the stirred reaction mixture. Admitting ethylene below the surface of the solvent results in evaporative cooling to a degree that overly cools the reaction. Then, 125 mL of toluene dried over, and distilled from, sodium is added and the system purged with dry nitrogen (passed through fresh silica gel) during subsequent manipulations. The toluene is deemed dry when a few drops added to a dry benzophenone solution in toluene turns deep purple from formation of the sodium benzophenone radical anion.

The catalyst is rac-ethylenebis(indenyl)zirconium(IV) dichloride, which can be purchased (neat, in amber screw cap bottles) from Aldrich Chemical (catalog no. 456–748) or from Boulder Scientific (catalog no.BSC-365). Methylaluminoxane can also be obtained from Aldrich (catalog no. 40,459-4) as 10 weight percent in toluene (Sure/Seal bottles) and from Crompton Corp, Marshall, TX. (Trimethylaluminum is available from Akzo Chemicals, Polymer chemicals Group, Dobbs Ferry, NY, and Crompton, but purchase of MAO is urged over hydrolysis). These are added in the amounts of 1000uM aluminum and 300uM zirconium. The ratio can be maintained if larger scale runs are to be made, but in no instance less than the Al/Zr ratio used here. Glove box manipulation and loading is recommended. If the solvent is properly dried, a striking color change is noted on mixing. After addition of catalyst and cocatalyst, ethylene is introduced over the stirred mixture. Polymer starts to precipitate fairly quickly. The gas flow rate should be increased if polymerization is not evident. Once started, gas should be continued for 30–60 minutes, at which point the gas should be stopped, the flask opened to allow escape of residual ethylene, and the suspension, containing a small amount of polymer in solution, is poured into about three times its volume of stirred methanol containing phenylnaphthylamine, filtered and the solid polymer air dried followed by 30 minutes in a vacuum oven at 50°C. The polymer is highly crystalline and of high molecular weight. The nontypical conditions of atmospheric pressure and absence of tight temperature control don't allow an approximate estimate of molecular weight or polydispersity.

Welborn and Ewen (255), have shown how closely related metallocenes can control the molecular weight of homopolyethylene and the amount of propylene incorporated in ethylene-propylene copolymers. The following procedures use an autoclave equipped for temperature control. Only those laboratories equipped for safely handling combustible gases under high pressures should undertake these or similar experiments. The catalysts are unbridged sandwich types in which a cyclopentadiene is the ligand to zirconium.

* We are indebted to Dr. Eric Wasserman, Union Carbide Corporation, Piscataway, NJ, for outlining the experimental procedure for this polymerization.

9-n. Polymerization of Ethylene with Bis(cyclopentadienyl) zirconium (225)
A 1 liter stainless steel autoclave equipped with stirrer, a jacket for external cooling, a septum inlet and vent line and connected to sources of dry ethylene and nitrogen, is thoroughly swept with nitrogen to dry it and remove oxygen. It is charged with 400 mL dry, degassed toluene, followed by methylaluminoxane (0.785 mmolar in aluminum) in toluene through the septum via a gas tight syringe. (See the prior experimental procedure for cocatalyst availability. The metallocene used here is available from Boulder Scientific.) After the mixture is stirred for 5 minutes at 80°C, 0.210 mg. bis(methylcyclopentadienyl)zirconim(IV) in 2 mL dry toluene is injected through the septum to give an Al/Zr mole ratio of about 1200. After one minute ethylene at 60 psig is admitted for at least 30 minutes into the stirred autoclave maintained at 80°C. After venting, about 28 g. polyethylene is filtered, washed with methanol containing some phenyl-β-naphthylamine, then dried for an hour at 50° in a vacuum oven. The Mw is around 212,000, the Mn, 55,000, for polydispersity index of 3.8.

9nn. An Ethylene-Propylene Copolymer Made with a Metallocene Catalyst (225)
Into the same kind of reactor, prepared and equipped as in the preceding example, are charged 0.8 mmol methyaluminoxane, followed by 200 mL liquid propylene, resulting in a pressue of 126.2 psig at 25°C. Then 0.113 mg of bis(cyclopentadienyl)zirconium(IV) dimethyl in 10 mL toluene is added to give an Al/Zr mole ratio of 2293. Ethylene at 152.1 psig is admitted and the stirred reactor held at 50°C. After at least 30 minutes the reactor is vented, the mixture poured into a considerable excess of methanol containing the previously used antioxidant, the copolymer filtered, washed and dried as in the preceding example to give about 66 g. of product. The latter has a density of 0.854 and a propylene content of 31 mol percent. The inherent viscosity in hot toluene at 0.1% is around 0.74, indicating a reasonable molecular weight for many applications.

Grubbs and coworkers (256) have reported a nickel-based catalyst class bearing a salicylaldimine monoligand that is highly active and steroregulating, but unlike the Ziegler-Natta and metallocene types, is stable in the presence of functional polar species, e.g., ethers, ketones and esters. Water was also deemed to have little effect, e.g., 1500 equivalents of water, based on molar catalyst amount still allowed high catalyst activity in grams polyethylene/mol Ni/hr.

9-p. Preparation of a Polyethylene Using a Nickel (II) Having a Salicylaldimine Catalyst
The catalyst can be prepared according to the method of Wang (257), et al, who synthesized a variety of nickel (II) complexes of the type

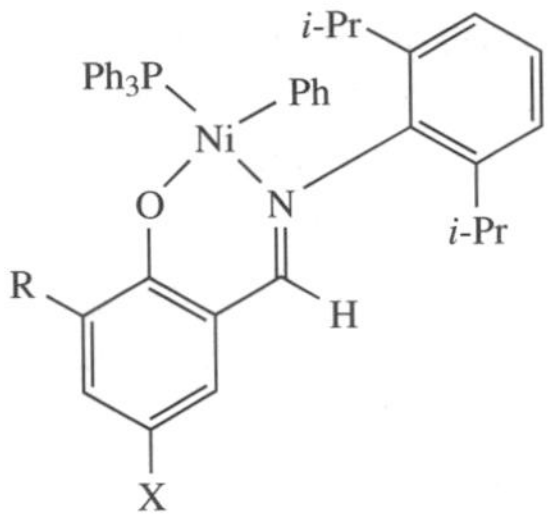

The catalyst to be used in this polymerization has R = phenyl and X=H. It's synthesis requires four steps from o-phenylphenol: formylation to give the substituted salicylaldehyde; condensation of the latter with 2,6-diisopropylaniline; conversion of the phenol to its sodium salt with NaH; reaction of the latter with trans-[$NiCl(Ph)(PPh_3)_2$] (258). Thus, the catalyst is an experience in synthetic organometallic chemistry; experimenters should consult up to date catalogs, since catalysts of the category may be available over time, or at least some of the intermediates.

Polymerization is, as for the preceding example, normally conducted at high pressure (e.g., 250 psig), but 1 atm. is considered operable, though the amount of polymer per unit of catalyst and time will be much lower than at the higher pressure. A vessel similar to the one used in the preceding example can be used at 1 atm., but the outcome is harder to predict. In the following, the results are representative of those from a stirred autoclave. To an autoclave of about 200 mL capacity and immersed in a 10°C cooling bath is charged 90 mL toluene followed by 65 umol of the catalyst in toluene. Ethylene at 100psig is introduced continuously under the surface of the liquid for one to two hours. The polyethylene thus formed can be filtered and washed with methanol containing some phenyl-β-napthylamine. The Mw by gel permeation chromatography should be around 200,000, the Mw/Mn around 2.5, and C_1-C_4 branching about 5/1000 carbon atoms.

251. Alternating Copolymerization of Ethylene and Carbon Monoxide Using Palladium Catalysts

Much of the credit for establishing the basis of metal complex polymerizations should be given to Reppe, who discovered nickel-catalyzed copolymerization of carbon monoxide and ethylene (200). Newer developments in transition metal chemistry have led to new catalyst systems that produce high molecular weight, perfectly alternating polyketones by processes that are economically and industrially attractive. This breakthrough is based on highly efficient homogeneous palladium catalyst systems. These new systems are formed by combination of an equimolar quantity of a suitable bidentate phosphine ligand with a palladium(II) species, in which the counter ions are weakly coordinating (201). For a series of diphenylphosphinoalkanes of the general formula $Ph_2P(CH_3)mPPh_2$, the most efficient catalyst system for high molecular weight polyketones is with $m = 3$. High rates with conversions of more than 1 million molecules of carbon monoxide and ethylene per palladium center are obtained (201). Other palladium complex catalysts, based on reaction of (2,2′-bipyridine) Pd (CH_3) with $Ar'_4B^{\ominus}Et_2OH^{\oplus}$ in acetonitrile are effective in copolymerizing *t*-butylstyrene and carbon monoxide (202). ($Ar' = 3,5 - (CF_3)_2C_6H_3$).

251.A. Intermediates
Copper and palladium compounds are obtained from Merck, phosphines from Strem or Aldrich, and carbon monoxide from Matheson.

251.B. Polymerization
Polymerizations are carried out in a 250-mL magnetically driven Hasteloy autoclave. The reactor is purged with ethylene/carbon monoxide, and 175 mL degassed is methanol added. The reactor is then sealed and raised to 65°C over 15 min. The system is pressurized to 4.5 mPa with a 1/1 mixture of ethylene and carbon monoxide, and a solution of 0.012 mmol 1,3-bis(diphenylphosphino) propane, 0.01 mmol palladium acetate, and 0.02 mmol cupric toluenesulfonate in 25 mL degassed methanol is injected all at once. Toluenesulfonic acid can be used instead of the copper salt, with a slight reduction in rate. Pressure is maintained by a continuous supply of the E/CO mixture, and the gas uptake is monitored by a mass flowmeter in the feed line. At the end of a chosen time period (e.g., 30 to 90 min) the reactive gasses are replaced with nitrogen. The autoclave is then cooled rapidly to room temperature, and any residual E/CO is vented. The autoclave is then opened, and the product slurry filtered, washed with fresh methanol, and dried in a vacuum oven at 60°C. The polymer is a white powder with a sharp melt temperature of 257°C and is highly crystalline. Tough films and moldings can be made by melt pressing. Molecular weights run as high as 20 to 35 × 10^4, depending on duration of the polymerization. The system is essentially a living polymerization, and molecular weight depends on amount of monomers used. Other monomers, such as propylene, can be terpolymerized with E/CO and this catalyst system.

9.11. ACETYLENE POLYMERS

Acetylene and simple monofunctional derivatives are polymerized to black, highly conjugated, and intractable polymers. However, when a sufficiently bulky side group is introduced, complete conjugation is prevented, and soluble products of yellow color are obtained. Thus polymerization of 4-methyl-1-hexyne gives a high polymer with maxima at 230 and 325 μm (203–205).

252. Preparation of Iron-tris-Acetylacetonate (206)

$$FeCl_3 + CH_3COCH_2COCH_3 \longrightarrow Fe\left[\begin{array}{c} CH_3 \\ | \\ O{=}C \\ \swarrow \quad \diagdown \\ \quad CH \\ \diagdown \quad /\!/ \\ O{-}C \\ | \\ CH_3 \end{array}\right]_3$$

To a solution of 67.6 g (0.25 mol) $FeCl_3 \cdot 6H_2O$ in 250 mL water is added a solution of 75 g acetylacetone (0.75 mol) in 150 mL methanol, with stirring. To the mixture is added 0.75 mol sodium acetate in 200 mL water, and the product is heated briefly on a hot plate, cooled to room temperature, and then refrigerated several hours. The product is filtered, recrystallized from methanol-water mixture, and dried thoroughly (mp 178 to 183°C).

253. Polymerization of Racemic 4-Methyl-1-hexyne (203)

$$CH_2{=}C{\equiv}CH\text{-}CH_2\text{-}CH(CH_3)(CH_2CH_3) \quad \longrightarrow \quad \left[-\underset{\underset{\underset{CH(CH_3)(CH_2CH_3)}{|}}{CH_2}}{\overset{|}{C}}=CH-\right]_n$$

4-Methyl-1-hexyne is made from the corresponding olefin by adding bromine and treating with sodium amide in mineral oil at 160°C (204, 205). $\eta_D^{25} = 1.4055$, bp 91 to 92°C.

Polymerization is carried out in a 100-mL pear-shaped flask equipped with nitrogen inlet, stirrer, condenser, and thermometer. Into the reaction vessel under nitrogen is introduced 0.735 g iron acetylacetonate, 3 mL isooctane, and 1.238 g aluminum triisobutyl. The mixture is heated and stirred at 70°C for 15 min. At this point, 42 mL isooctane and 5.00 g 4-methyl-1-hexyne are added at room temperature, and the mixture is warmed 3 h at 30 to 35°C then left to stand 45 h at room temperature. The polymerization is interrupted by the addition of diethyl ether to dissolve the polymer and dilute hydrochloric acid to kill the catalyst. The organic layer is treated with $NaHCO_3$ solution, washed with water, and evaporated to yield 4.65 g polymer. This product is extracted with boiling acetone to give 4.50 g acetone-insoluble product, intrinsic viscosity 2.7 to 3.5 dL/g at 30°C in toluene. The polymer has a uv spectrum with $\lambda_{max} = 230(\log E = 3.43)$ and $\lambda_{max} = 325(\log E = 3.45)$ and is amorphous at room temperature.

9.12. DIENE POLYMERS AND COPOLYMERS

The preparation of diene polymers and copolymers is an important technology that has occupied the attention of many scientists since before World War 1. These polymers are, in general, elastomers and have gone far toward supplementing and, perhaps, ultimately supplanting natural rubber.

The volume of published work in this field is large and has been well summarized in many excellent books (14, 15, 31, 35, 207, 208) and papers. Briefly, dienes will polymerize by a variety of techniques and with a variety of catalysts. With butadiene, the resulting polymerization may give the following structural units:

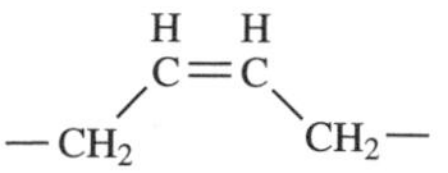

$$\underset{cis\ 1,4\text{-}}{-CH_2-CH=CH-CH_2-} \qquad \underset{trans\ 1,4\text{-}}{-CH_2-CH=CH-CH_2-}$$

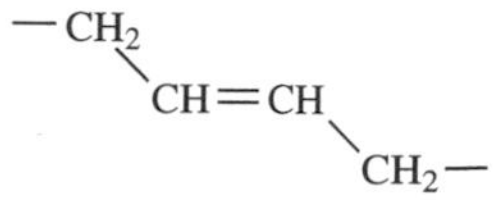

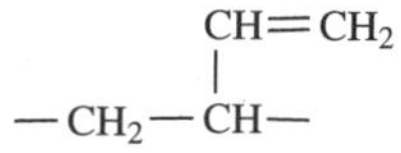

$$\underset{1,2}{-CH_2-\underset{|}{CH}(CH=CH_2)-}$$

Ordinarily, the free radical homopolymer will contain varying amounts of these structural units making it actually a copolymer. The proportion of units will vary with the technique or catalyst used.

Isoprene, 2-methylbutadiene, gives both 1,2- and 3,4-polymers because of its unsymmetrical structure, as well as cis- and trans-1,4. On the other hand, chloroprene, which is basically similar, appears to give only trans-1,4 polymerization with free radicals. 2,3-Dichlorobutadiene polymerizes similarly. Interestingly, the product from this monomer is a hard plastic, not an elastomer.

Polybutadienes prepared by the emulsion and stereospecific techniques differ in average isomer content. The percentages also vary within a given type, depending on temperature of polymerization and the other factors. For example, emulsion polybutadiene runs about 80% total cis and trans 1,4 addition and polybutadiene from stereoregulating catalysts can be almost 100% cis- or trans-1,4, or iso- or syndiotactic-1,2. An example of emulsion polybutadiene was given earlier (prep. 189) as part of an ABS synthesis.

An effective catalyst for polymerization of butadiene to a highly cis-1,4 structure involves a cobalt salt with an alkylaluminum halide. Both diethylaluminum chloride and ethylaluminum sesquichloride are effective with cobalt octoate, but the former requires water for activation. No polymerization occurs in an anhydrous system; 10 to 15 mol% water based on aluminum gives optimum cis-1,4 levels. The sesquihalide, in contrast, requires no water.

254. Polybutadiene with a High cis-*1,4 Content with an Aluminum Alkyl-cobalt Octoate Catalyst (209)*

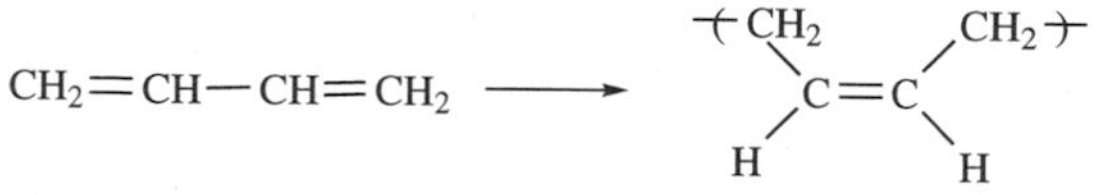

$$CH_2=CH-CH=CH_2 \longrightarrow \left[CH_2-CH=CH-CH_2\right] \ (cis)$$

To a previously oven-dried and nitrogen-flushed beverage bottle of about 1 qt. capacity is charged 450 mL thiophene-free benzene, which has been purified as described below. Slightly in excess of 50 g polymerization-grade butadiene is condensed into the bottle and enough allowed to boil out to expel residual air and to give a final weight of about 50 g monomer. The bottle is closed with a neoprene gasket and a metal crown cap with a hole for hypodermic insertion.

Then, 10 mmol ethylaluminum sesquichloride (calculated as $Et_{1.5}AlCl_{1.5}$) and 0.02 mmol cobalt octoate are added through the septum by syringe. The aluminum alkyl can be conveniently added as 10 mL of a 1*M* solution in benzene and the cobalt salt as 1 mL of a 0.02*M* solution in benzene.

The small quantities of catalyst used requires that a highly purified system be used. The benzene should be distilled initially and the first 20% discarded, the subsequent distillate passed through a silica gel column, and finally stored over calcium hydride. Higher amounts of catalyst may be required if sufficient purity is not maintained.

The polymerization is conducted by rotating the bottle in a 5°C bath for 2 h, in which time conversion should exceed 90%. The polymer is coagulated by pouring the solution into excess isopropanol containing phenyl-β-naphthylamine. The solid is leached for 24 h in fresh isopropanol-antioxidant, blotted dry, and covered with a pentane solution of antioxidant. The swelled polymer is dried in vacuum at 50°C, under nitrogen. The final products should have a cis-1,4 content of 95–98% by (210). Other methods have been used, but the film technique of this reference has been commonly used for comparison with polybutadiene prepared and reported in early work. The inherent viscosity is 3 to 4 in benzene (0.5 g/100 mL) at 25°C.

High cis-1,4-polybutadiene can also be prepared using a catalyst based on an aluminum alkyl and titanium tetraiodide. The titanium chlorides do not have a cis-1,4 stereodirecting capacity equal to that of the iodides. The following example uses TiI_4 in conjunction with tetrabutyltitanate and triisobutylaluminum. The vast differences in coordination catalyst action as a function of relatively minor changes in composition is illustrated by the fact that the latter two components alone give a polybutadiene high in 1,2-units.

255. Polybutadiene with a High cis-1,4 Content Using an Aluminum/Titanium Catalyst (211)

$$CH_2{=}CH{-}CH{=}CH_2 \longrightarrow \left[\begin{array}{c} CH_2 \\ \diagdown \\ C \\ \diagup \\ H \end{array} \!\!=\!\! \begin{array}{c} CH_2 \\ \diagup \\ C \\ \diagdown \\ H \end{array} \right]$$

A solution is made up in a small beverage bottle consisting of 100 mL toluene (dried over and distilled from calcium hydride), and 2.93 mmol of both 1.00 g tetra-*n*-butyl titanate and 1.63 g titanium tetraiodide. The bottle is capped with a neoprene septum and shaken to achieve mixing. Titanium tetraiodide is slow to dissolve. It should be ground to a fine particle size to speed its solution, but grinding should be done with care (e.g., in a good, nitrogen-flushed dry box).

A dry beverage bottle of about 1 qt. capacity is charged with 500 mL dry toluene, then flushed thoroughly with nitrogen, capped as above, and pressured with nitrogen to about 10 lb. Then 2.2 mmol triisobutylaluminum is added, conveniently as 10 mL of a 0.22*M* solution in toluene. (The bottle is prepressured with nitrogen to preclude entry of air during subsequent charging. This step is not essential but is a useful precaution against contamination. If it is employed, great care must be exercised in adding catalysts from a syringe to prevent the plunger and contents from being forced out of the syringe.) Then, 0.26 mmol titanium tetraiodide and tetra-*n*-butyl titanate is added as 9.3 mL of the mixture made as described, after which 50 g polymerization-grade butadiene is carefully pressured into the cooled bottle through the septum.

The bottle is tumbled for 17 h at 5°C, after which 0.5 mL of a solution of 2,2′-methylene bis(4-methyl-6-*t*-butylphenol) (52 g in 4 L toluene and 100 mL isopropanol) is added and thoroughly mixed in. The contents of the bottle is then poured into 1-L isopropanol and stirred vigorously. The polymer is treated with a phenyl-β-naphthyl amine solution in pentane as in the preceding example and dried under nitrogen in vacuum at 50°C. The conversion is 85-90%, and the inherent viscosity is about 3.5 (benzene, 25°C, 0.5 g/100 mL). The cis-1,4 content is about 96%, trans-1,4 is 0.7%, and the remaining 3.3% is 1,2. These values are determined by examination of a 25% solution of the polymer in carbon disulfide (212).

A polybutadiene with only a somewhat lower cis-1,4 content than the above (e.g., 87%) can be prepared by using titanium tetrachloride in place of the tetrabutyl titanate. A mole ratio of components such as 15 : 1 : 1 for triisobutylaluminum : titanium tetraiodide : titanium tetrachloride (213).

Polybutadiene with a high degree of trans-1,4 enchainment is a crystalline polymer that can be made with trialkylaluminum-titanium (214) or vanadium (215) halide catalysts, or a mixture of the two transition metal halides (216). At least 95% trans-1,4 structure can be made to prevail. A catalyst derived from lithium aluminum hydride and titanium tetraiodide (mole ratio 0.86/1.31) in benzene produces a polymer with only 78 to 86% trans-1,4-structure (217); this polymer displays useful elastomeric properties. The trans-1,4 polymer can also be made in a water emulsion by means of rhodium chloride and related salts and their complexes (218). Molecular weights are modest, as indicated by intrinsic viscosity (e.g., 0.4).

256. Preparation of trans-*1,4-Polybutadiene (216)**

$$CH_2{=}CH{-}CH{=}CH_2 \longrightarrow \left[\diagdown CH_2{-}\overset{H}{C}{=}\underset{H}{C}{-}CH_2 \diagdown \right]$$

In this example, the monomer is bubbled at atmospheric pressure through the catalyst-solvent mixture rather than being added all at once to either a bottle or an autoclave that is then sealed. If the catalyst is sufficiently active, loss of butadiene is not significant in the present case. Solvent and monomer are conveniently dried over Linde molecular sieves no. 4, which may also be used to dry the nitrogen used to protect the system during the operation. The system must also be kept oxygen free.

* We wish to thank Dr. Jack Laskey for making the details of this preparation available to us.

A 500-mL three-necked flask is equipped with a stirrer, gas inlet reaching below the liquid level in the flask, and a gas outlet through a condenser and drying tube. One inlet is glass stoppered. The flask assembly is flamed out in a stream of nitrogen and cooled while being purged; the glass stopper is removed and replaced by a serum cap for catalyst addition. To the flask is charged 300 mL dry benzene, 0.227 g (1.2 mmol) titanium tetrachloride, 0.058 g (0.3 mmol) vanadium tetrachloride, and 0.063 g (0.55 mol) aluminum triethyl. The mixture is stirred and heated to reflux for 1 h. The transition metal halides are both reduced to primarily the trivalent state and are coprecipitated, a fact that appears to enhance their activity. If this reduction of the halides with aluminum triethyl is carried out at 170°C in a higher boiling solvent or in an autoclave, an even more highly active catalyst results, though its stereoregulating capacity is not very changed. The aluminum chloride formed in the reduction in either case is believed to be present in the crystal lattice of the transition metal trihalides and is not washed out during polymerization. Reduction at 80°C forms β-$TiCl_3$ while the γ-isomer forms at 170°C (219).

When the mixture has cooled, 0.513 g (4.5 mmol) additional aluminum triethyl is added to activate the trihalides. The advantage of the two-step catalyst preparation (i.e., reduction and activation) is that reduction of titanium and vanadium to valence states lower than three is avoided, thus minimizing the number of differing catalyst sites. An Al/(Ti + V) mole ratio of slightly more than 1/3 is maintained in the reduction, and 3/1 in the activation.

Butadiene (high purity polymerization grade) is then passed through the stirred solvent-catalyst with the temperature maintained at 60°C by means of a heating bath. The initial rate of polymerization should be upward of 18 g/h. After 1 h, the mixture is poured into about 1.5 L methanol containing 1% phenyl-β-naphthylamine. The polymer is washed at least three times in a blender with methanol containing the antioxidant and dried under a nitrogen stream in a vacuum oven at 50°C. The polymer is judged to be greater than 95% of the trans-1,4 type, by (band at 10.36 μ, a solution of 10 g/L in $CS_2 : CCl_4$, 1 : 1). It has a crystalline melting point of around 146°C using a polarizing microscope. This represents a second crystalline modification; the first is stable up to 75°C, going over to the second, higher melting, form at that temperature.

Butadiene can be polymerized in an essentially complete 1,2-enchainment and syndiotactic placement of the pendant vinyl group by using a cobalt salt complexed with pyridine in conjunction with aluminum triethyl (220). Other catalysts are known, e.g., aluminum triethyl-titanium tetrabutoxide and vanadium or molybdenum acetylacetonates with metal alkyls (221), but in general do not give quite as stereospecific a synthesis.

It is interesting to note that diethylaluminum chloride in place of aluminum triethyl will produce, in the system of the following example, a polybutadiene of 96% cis-1,4 content. Mixtures of the two aluminum alkyls will generate highly syndiotactic 1,2-polymer with an Et_3Al/Et_2AlCl mole ratio as low as 0.84; at an intermediate ratio of 1.0, the crystallinity of the polymer is higher than with Et_3Al alone. At ratios lower than 0.84, cis-1,4 content increases abruptly.

*257. Preparation of Syndiotactic 1,2-Polybutadiene (220)**

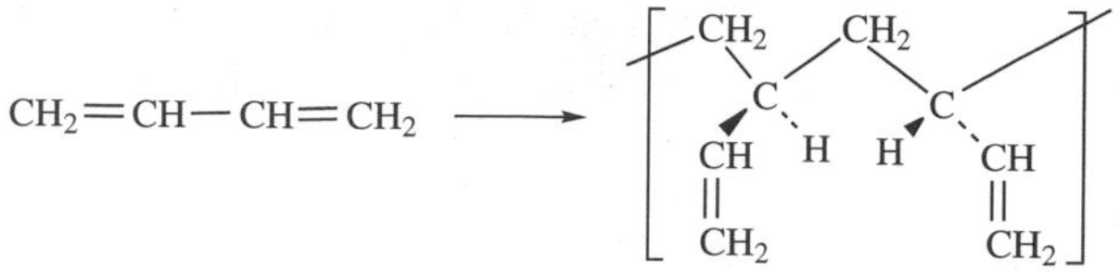

$$CH_2{=}CH{-}CH{=}CH_2 \longrightarrow \left[\begin{array}{c} CH_2 \quad CH_2 \\ C \qquad C \\ CH \quad H \quad H \quad CH \\ \| \qquad\qquad \| \\ CH_2 \qquad\quad CH_2 \end{array} \right]$$

A 1-qt., dry, nitrogen-flushed beverage bottle sealed with a rubber septum and cap arrangement for addition of components by syringe provides a convenient reaction vessel. A stainless-steel stirred autoclave can be used to advantage, if available.

To the vessel is added, in order, 400 mL benzene (distilled from calcium hydride or, better, K-Na alloy), 9.72 mg (7.48×10^{-5} mol) cobalt(II) chloride, and 0.0118 g (1.49×10^{-4} mol) dry, distilled pyridine. The objective is to form the soluble $CoCl_2{\cdot}2C_5H_5N$ complex, and the vessel should be closed and agitated for 1 to 2 h. Then, somewhat more than 40 g polymerization-grade product is subsequently allowed to distill out. The vessel is closed and 1.57 g (1.33×10^{-2} mol) aluminum triethyl is added. An Al/Co mole ratio anywhere between 0.3 and 600 will produce highly syndiotactic polymer. The polymerization is conducted at 16°C, and the mixture should be at this temperature when the final component is added. If catalyst components are added in benzene solution, the final amount of benzene should be 400 mL. As an alternative to the above, the $CoCl_2{\cdot}2C_5H_5N$ complex can be prepared by mixing 0.2 g anhydrous $CoCl_2$ and 1.5 mL dry pyridine in 180 mL dry benzene under nitrogen and with exclusion of moisture and stirring for 3 to 4 h. A solution containing 0.17 to 0.18 g/L of $CoCl_2$ is obtained, the requisite amount of which is used in preparing the polymerization catalyst solution.

After 15 h, about 50 mL of methanol is added to terminate the polymerization. The finely divided, white powder, which separates during reaction, is filtered and washed thoroughly in a blender with methanol, then dried at room temperature in vacuum to give about 9 g polymer. The polymer should be protected by a nitrogen blanket throughout these operations. If the polymer is successively extracted in a jacketed Soxhlet with boiling acetone, ether, and benzene, about 3% of the crude polymer is removed. The residue is judged to be about 85% crystalline and to be 98% or more of a syndiotactic configuration, as judged by x-ray diffraction and, respectively, the latter by a band at 660 cm^1. The polymer should be protected by addition of an antioxidant in the manner shown for previous polybutadiene if it is desired to handle the polymer in air for any operations. T_m for the syndiotactic-1,2 polymer is about 156°C. The insolubility of the polymer makes intrinsic viscosity determination difficult.

Copolymers of butadiene have shown much usefulness in technical applications. Thus important copolymers with styrene and acrylonitrile have been developed. Copolymers of styrene with butadiene were the basis of the synthetic

* We are grateful to Dr. Ermanno Susa for his helpful comments on this experiment.

rubbers made in large amounts during World War II. They were prepared using the so-called Mutual, or GR-S recipe, which was chosen for its simplicity and reproducibility (Table 9.1). The butadiene units in GR-S made at 50°C are about 60% trans-1,4; 20% cis-1,4; and 20% 1,2. Trans-1,4 content increases at a lower temperature.

258. Preparation of GR-S Rubber (35)

$$CH_2{=}CH{-}CH{=}CH_2 + \phi CH{=}CH_2 \longrightarrow \text{Copolymer}$$

The mixture given in Table 9.1 is polymerized at 50°C. The conversion is approximately 6% h, and the polymerization is ordinarily short-stopped after about 75% conversion, i.e., after about 12 h of polymerization.

Polymerization may be carried out in 4-oz. screw-cap bottles mechanically tumbled end over end in a constant temperature bath. Bottles larger than 4 oz. should not be used because the reaction is exothermic and might get out of control if the heat cannot be dissipated rapidly. It should be borne in mind that all traces of air must be kept out of the system. That is, the water used should be boiled to deaerate it, while the reaction vessel should be flushed by allowing a small amount of the butadiene to boil out before capping.

After approximately 12 h, the polymerization is short stopped by adding about 0.2 g hydroquinone. The polymer latex so obtained is poured into a beaker of suitable size, and steam is passed through it to remove unreacted butadiene and styrene. To this stripped latex is then added an antioxidant such as phenyl-β-naphthylamine, and the latex is coagulated. This may be done by first adding sodium chloride solution, which causes partial coagulation ("creaming") of the mixture. Coagulation is then completed by the addition of dilute sulfuric acid, which converts the dispersing salts to the free acids. The product is obtained in the form of crumbs, which are filtered, washed well with water, and dried.

Like butadiene, isoprene may be polymerized by a variety of techniques. High molecular weight polyisoprene has been obtained in a solution polymerization at 50°C, with metallic sodium, and with organometallic catalysts similar to those described for polymerization of butadiene. Because natural rubber is a polyisoprene, it is not surprising that chemists have directed their attention

Table 9.1. Mutual (GR-3) Rubber in Its Simplest Form

Constituent	Quantity, parts
Butadiene	75
Styrene	25
Commercial dodecyl mercaptan	0.5
Potassium persulfate	0.3
Soap flakes	5
Water (freshly boiled)	180

to the synthesis of polyisoprene of the same chemical configuration as the natural product. The two principal types of naturally occurring polyisoprene are hevea — which is about 97.8% cis-1,4 and 2.2% of 3,4-polyisoprene — and balata — which is about 98.7% trans-1,4 and about 1.3% 3,4-polyisoprene. The synthetic polyisoprenes made in an emulsion system have 12 to 14% of 1,2 product, while those made with sodium have between 50 and 55% 1,2 addition; the remainder is mainly trans-1,4 addition product, with some cis. It has been found that polymerization of isoprene with organotitanium compounds, with lithium alkyls, or with lithium aluminum alkyls produces a polyisoprene that is essentially identical to the cis-1,4-polyisoprene of hevea rubber. Furthermore, although metallic sodium in a finely divided condition gives a product containing large percentages of 1,2 addition, it has been found that finely divided lithium metal will give essentially the same results as the organometallic derivatives. The following preparations are typical of the polymerization of isoprene with this type of catalyst. In all these polymerizations it must be recognized that success depends on a number of factors, the most important of which is purity of the monomer and absence of contaminants such as moisture or air in the system.

259. Preparation of the Catalysts

The organotitanium type catalyst is prepared as described earlier. The butyllithium may be purchased as a commercial product dissolved in pentane, or synthesized. Finely divided metallic lithium may also be obtained commercially and is much more satisfactory than any product that may be made in the laboratory with ordinary equipment.

260. Polymerization of Isoprene with n-Butyllithium (223, 224)

$$CH_2{=}CH{-}\underset{}{\overset{CH_3}{\overset{|}{C}}}{=}CH_2 \longrightarrow (C_5H_8)_n$$

The butyllithium solution is standardized by hydrolysis and titration of lithium hydroxide with standard acid. The concentration is adjusted with n-pentane to 1.0 molar. The polymerization is carried out simply by adding 84 mL pure isoprene, 180 mL petroleum ether or n-pentane, and 3.0 mL of the butyllithium solution to a round-bottom flask, which should be dried thoroughly and flushed with an inert gas, preferably helium or argon. All transfers should be made under absolutely anaerobic and anhydrous conditions. The flask is stoppered and placed in a water bath at 30°C. Polymerization time is approximately 18 h. At the end of this period, the flask is removed from the water bath, and the contents are poured into methanol, which contains about 3% based on weight of polymer of an antioxidant such as phenyl-β-naphthylamine. The polymer is coagulated, filtered, washed thoroughly with methyl alcohol, and dried in a vacuum at 50°C. Polymer made in this way is approximately 77.4% cis-1,4, 13.0% trans-1,4, and about 9.5% 3,4.

261. Polymerization of Isoprene over Finely Divided Lithium (225)

Polymerization of isoprene over finely divided lithium is most conveniently carried out in the laboratory according to the following technique. To purify isoprene, it is refluxed over metallic sodium for 4 h, then distilled, passed through a silica column, and used immediately. It is kept out of contact with air or moisture. The silica column should be freed of air before use by passing a stream of oxygen-free helium through the column. Absolutely dry glass bottles sealed with aluminum-lined crown caps are charged with 100 mL isoprene. To the isoprene contained in the crown-capped bottles is added 0.1 g lithium as a 35% dispersion in vaseline or petroleum oil. The cap is placed on the bottle loosely and the bottle and contents are brought to a vigorous boil. Approximately 10% of the total isoprene is allowed to boil out to completely free the reaction vessel of traces of oxygen and moisture. The bottle is then rapidly sealed and placed in a constant temperature bath at 30 to 40°C. It is allowed to remain at this temperature either agitated or unagitated until the content have been converted to a solid chunk of polymer. This may require 30 min to 3 days, depending on the purity of the ingredients. The unagitated polymerization is somewhat slower. However, it is also safer, since polymerization is exothermic and may become dangerous if not properly controlled.

When polymerization is deemed to be complete, the cooled bottle is broken, and the solid chunk of polymer is removed and soaked in isopropanol containing a trace of acetic acid to remove the catalyst and a small amount of a suitable antioxidant such as phenyl-β-naphthylamine. Infrared examination of the polymer should indicate that it is of the order of 98% cis-1,4 structure, the remainder being 3,4. The product may be milled and compounded exactly as is natural rubber.

262. Polymerization of Isoprene over a Titanium-Based Catalyst (226–228)

Petroleum ether is purified by treating with concentrated sulfuric acid until no further discoloration is observed. It is then washed with water, dried by passing through an alumina column, and distilled from metallic sodium. The isoprene is distilled, then refluxed immediately before use with sodium, and passed through a silica column as described in the previous preparation. The catalyst is prepared by mixing equimolar quantities of triisobutyl aluminum and titanium tetrachloride. In view of the hazardous nature of triisobutyl aluminum, all possible care must be taken in handling this material and the experimenter should be completely familiar with all the safety hazards inherent in triisobutyl aluminum. If desired, lithium aluminum tetraalkyls may be used advantageously in place of the triiosobutyl aluminum (227). It is convenient to prepare a 1-molar solution of either alkyl compound in anhydrous, olefin-free heptane to be used as needed. All monomers, solvents, and catalyst components are kept absolutely free of moisture and air and are stored in an inert atmosphere. The bottles are conveniently capped with a self-sealing stopper of the type used on serum bottles (see chapter 2). A measured amount may be removed by use of a hypodermic syringe.

A mixture of 75 mL petroleum ether and 25 mL isoprene is placed in a beverage bottle and heated to vigorous boiling in a hot water bath. A quantity of petroleum ether is distilled from the bottle sufficiently to ensure the absence of all moisture and air from the polymerization vessel. Then 2 mL 1 *M* triisobutyl aluminum is added under inert atmosphere, and the bottles are stoppered with a serum-type rubber stopper through which can be inserted a hypodermic syringe. Bottles containing the monomer plus alkyl aluminum are cooled to the desired polymerization temperature and an amount of titanium tetrachloride equivalent in moles to the aluminum alkyl used is added from a hypodermic syringe. The titanium tetrachloride may be added either as undiluted catalyst or as a premixed 1-molar solution in heptane. Polymerization is quite rapid, and the mixture of catalyst and monomer rapidly becomes sufficiently viscous to disperse the catalyst particles making stirring unnecessary. However, stirring or shaking during the first part of the polymerization cycle is recommended.

Polymerization is allowed to proceed for approximately 24 h when the polymer is isolated and soaked in alcohol for 24 h to destroy the catalyst components. It is then mixed with an antioxidant, milled, or otherwise treated as in previous experiments.

To obtain polymer with a low gel content in high yield with a high inherent viscosity, it is necessary to have the molar ratio of trialkylaluminum to titanium tetrachloride at about 1 : 1. It may be desirable to have a very slight excess of alkylaluminium because this appears to minimize the production of gel.

The temperature at which the polymerization is run is also important in determining the molecular weight of the resulting polymer and, to a lesser extent, the microstructure of the polymer. Temperatures between 5 and 25°C appear to be most suitable for this purpose. Above about 25°C, the molecular weights obtained are too low to be useful. Polymer prepared at 25°C is approximately 96% cis-1,4 and 4% 3,4 polyisoprene. No trans-1,4 structure is observed if the polymer is made according to the instructions previously given.

2,3-Dimethylbutadiene gives a polymer by free radical initiation having about 13% 1,2-addition and the remainder an undetermined mixture of cis-and trans-1,4 (222). However, with an Al/Ti catalyst similar to that in the previous example, an essentially cis-1,4-polymer is produced as a crystalline powder with T_m of about 189°C (229). Conditions are i-$Bu_3Al/TiCl_4$ mole ratio 1 : 1; 0.06 mol total catalyst/mol of monomer; monomer/solvent (benzene) of 1/6; temperature, 25°C; time, 72 h; conversion, about 25°C; inherent viscosity, about 0.7 (0.2 g/100 mL, tetralin at 100°C, with antioxidant).

Aluminum trialkyls and vandium trichlorides and tetrachlorides are capable of converting isoprene and butadiene to trans-1,4 polymers. It has been reported that VCl_4 is more efficient than VCl_3 in converting butadiene to the trans polymer (230); but the reverse appears to be true for the preparation of trans-1,4-polyisoprene (231). This is another example of monomer-catalyst variation that affects the overall effectiveness of catalysis (efficiency defined as grams polymer/grams catalyst component) rather than altering the structural features of the polymer.

In the following example, the efficiency of VCl_3 for isoprene polymerization is enhanced further by supporting it on an inert carrier such as titanium dioxide, kaolite, or magnesium oxide. Mixed Ti/V halides have also been used to prepare the trans-1,4 polymer (216).

trans-1,4-Polyisoprene is a natural polymer bearing the common name of balata, a tough, crystalline thermoplastic finding use in golf ball covers (232). Its laboratory synthesis is as remarkable, though not so commercially significant, as that of its cis counterpart, remarked previously.

263. Preparation of trans-1,4-Polyisoprene (231)

$$CH_2{=}C(CH_3){-}CH{=}CH_2 \longrightarrow \left(CH_2{-}C(CH_3){=}C(H){-}CH_2 \right)$$

All operations should be carried out with equipment that has been baked dry (120°C) and cooled under nitrogen. Handling and transfers of catalyst components should be done in a dry and oxygen-free atmosphere.

A sample of kaolin (Continental Clay, R. T. Vanderbilt Co., about 10 m^2/g surface area) is thoroughly dried by heating at 125°C for 15 h. After cooling under nitrogen, 8.0 g is added to 70 mL benzene, which has been dried over sodium or Linde 4A molecular sieves, in a 250-mL three-necked flask with stirrer, condenser, and inert gas inlet. Then, 1.5 mL (2.72 g) vanadium tetrachloride is added, conveniently as a solution in 10 to 15 mL dry benzene. The mixture is refluxed for 3 h with stirring, care being taken to blanket the reaction with nitrogen throughout, including mixing, cooling, and workup. The tetrachloride decomposes smoothly to the trichloride and chlorine, the former depositing on the surface of the kaolin. Vanadium trichloride is formed in 80 to 90% yield, and amounts to 18 to 19% of the carrier. The solid is filtered in a dry box, washed well with dry benzene, and dried at 1 mm in a vacuum oven under nitrogen for 24 h.

To a 1-qt. beverage bottle flushed with nitrogen is quickly added 300 mL dry benzene, 0.78 g of the VCl_3-on-kaolin (containing about 0.14 g VCl_3, 0.98 mmol), 0.2 mL (0.44 mmol) tetra-2-ethylbutyl titanate in 10 mL dry benzene, 2.3 mL (8.9 mmol) aluminum triisobutyl in 10 mL dry benzene, and 200 g (300 mL) polymerization-grade isoprene that has been distilled from sodium and stored under nitrogen. The mixture is heated at 50°C for 6 h with agitation. An equivalent amount of tetraisopropyl titanate may be used, it should be mentioned, in place of the 2-ethylbutyl titanate ester described.

The mixture is cooled and poured into 2 L methanol containing 3 g of an antioxidant such as 2,6-di-*t*-butyl-4-methylphenol. The polymer is filtered, washed repeatedly with methanol and antioxidant and dried at 40°C in a vacuum oven under a nitrogen stream. About 90 g of trans-1,4-polyisoprene with a dilatometric melting point of 56 and 62°C., representing the phase changes of two crystalline modifications.

The resulting polymer has some gel content, preventing full solubility and thus solution viscosity determination. If it is milled for only a few minutes at 240°C, it is rendered completely soluble in benzene and will have an intrinsic viscosity of three or more at 30°C. In this respect, it is much higher in molecular weight than natural balata (intrinsic viscosity, 1.0 or less).

264. Preparation of 3,4-Polyisoprene (221)

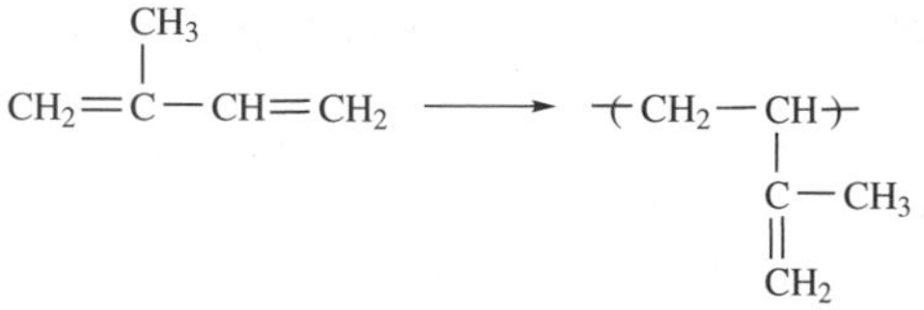

$$CH_2{=}C(CH_3){-}CH{=}CH_2 \longrightarrow {-}\!\!\left(CH_2{-}CH(C(CH_3){=}CH_2)\right)\!\!{-}$$

In a suitable stirred flask or beverage bottle, taking customary precautions to exclude air and moisture, is placed 100 mL toluene (distilled from calcium hydride or, better, sodium hydride), 1.25 g (10.92 mmol) aluminum triethyl, and 0.50 mL (0.517 g, 1.82 mmol) tetra-*n*-propyl titanate (distilled before use). The catalyst is aged 12 min at 18°C. An Al/Ti mole ratio of 6 gives a maximum conversion. Then, 30 mL polymerization-grade isoprene is added. Polymerization is continued with stirring or agitation for 8 h at 24°C The polymer is poured into excess methanol and washed thoroughly with methanol in a blender, then dried at room temperature under nitrogen. An antioxidant such as phenyl-β-naphthylamine is added to the polymer via the last methanol wash. About 10 g of added polymer is obtained that cannot be crystallized. Infrared indicates that the polymer is around 95% 3,4-; the remainder is cis-1,4. If the polymerization is carried out at −10°C., the product is about 99%, 3,4- and is completely amorphous. The failure to crystallize despite the high 3,4- content is believed to mean that the polymer is nonstereoregular (i.e., is atactic at the unsymmetrical carbon of each chemical repeat unit). The intrinsic viscosity is about 2.5 dL/g in toluene at 25°C.

All the uses for polyisobutylene itself are somewhat limited by the fact that it is, even at high molecular weight, a relatively soft, plastic material. It was found that copolymerization of a small amount of a diene with isobutylene gives a product that can be cross-linked (vulcanized) to a useful product known as butyl rubber, e.g., tires and inner tubes.

265. Preparation of an Isobutylene-Isoprene Copolymer by Cationic Catalysis (233, 234)

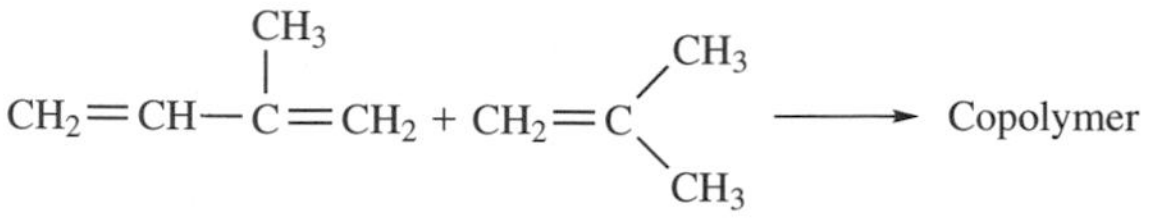

$$CH_2{=}CH{-}C(CH_3){=}CH_2 + CH_2{=}C(CH_3)_2 \longrightarrow \text{Copolymer}$$

Pure isobutylene (100 mL) and 1.5 mL pure isoprene are placed in a 1-L three-necked flask equipped with a mechanical stirrer, a gas inlet tube, and a dry ice condenser. Approximately 300 mL methyl chloride is then added from a cylinder, and the mixture is cooled to about −100°C by means of a sludge of alcohol in liquid nitrogen. The solution of isobutylene and isoprene is stirred gently, and successive 1-mL portions of a solution of 0.2 g anhydrous aluminum chloride in 40 mL methyl chloride are added. As each portion is poured into the solution, an insoluble mass of copolymer is produced that floats in the cold solution. Catalyst may be added until conversion of the isobutylene-isoprene to copolymer is about 50%. Higher conversions give improper monomer balance to the copolymer because of the different reactivity of the monomers. The polymerization mixture is quenched by the addition of a small amount of isopropyl alcohol previously cooled to −100°C, and the mixture is then allowed to warm to room temperature. The lumps of rubbery copolymer are removed, washed with alcohol, dried in a vacuum oven, and converted to sheet products by conventional techniques.

Certain highly substituted butadienes, e.g., 2,5-dimethyl-2,4-hexadie (1,1,4,4-tetramethylbutadiene) polymerize like isobutylene, i.e., at low temperature with Lewis acids. The product is completely 1,4-trans (235).

Chloroprene ($CH_2{=}C(Cl){-}CH_2{-}CH_2$) is a monomer that duplicates in many ways the over-all geometry of isoprene; however, the polymer differs in many respects. This monomer is the basis for the neoprenes that have been successfully commercialized in a variety of forms.

Chloroprene is synthesized by addition of hydrogen chloride to monovinyl acetylene. The vinyl acetylene is prepared by dimerization of acetylene, **an operation that is not recommended for anyone not familiar with the hazards of acetylene chemistry**, or by the dehydrohalogenation of 1,4-dichloro-2-butene (236). Chloroprene itself is relatively stable, and may be obtained commercially as a 50% solution in toluene. When pure, it will polymerize spontaneously in about 10 days to a high molecular weight, clear, tough product, the so-called μ polymer described by Carothers and co-workers (237). Polychloroprene shows a structural dependence on polymerization temperature. The typical product (238) is composed mainly of trans- and cis-1,4 units, with the latter not in excess of 20%. There are 10 to 15% of the 1,4-units present in head-to-head, tail-to-tail enchainment along with the dominant head-to-tail structure. Essentially pure trans-1,4 polymer results from free radical polymerization at −30°C. The cis-1,4 polymer has been prepared by polymerization of 2-(tributyltin)-1,3-butadiene followed by chlorinolysis of the substituent (239).

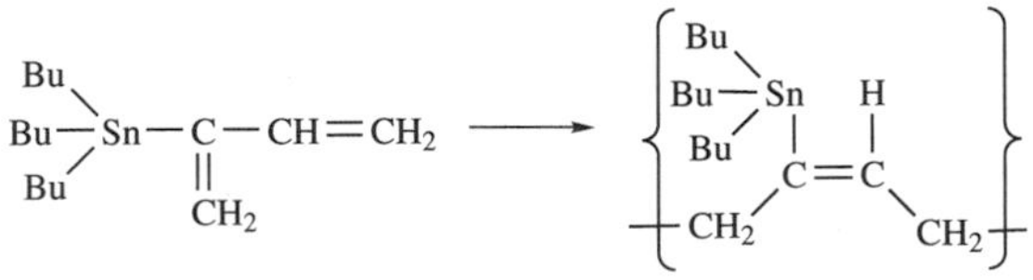

$$Bu_3Sn{-}C({=}CH_2){-}CH{=}CH_2 \longrightarrow \left[{-}CH_2{-}C(SnBu_3){=}CH{-}CH_2{-} \right]$$

I

Never try to polymerize chloroprene with an organotitanium catalyst; a violent explosion may result.

266. Polymerization of Chloroprene in an Emulsion (240)

$$CH_2{=}CH{-}\overset{\displaystyle Cl}{\overset{|}{C}}{=}CH_2 \longrightarrow \left[-CH_2-CH{=}\overset{\displaystyle Cl}{\overset{|}{C}}-CH_2-\right]_n$$

Emulsion polymerization of chloroprene may be carried out satisfactorily according to the following procedure. A total of 100 g freshly distilled chloroprene is emulsified in 150 mL water containing 4 g wood resin, 0.6 g sulfur, 0.8 g sodium hydroxide, 0.5 g potassium persulfate, and 0.7 g the sodium salts of naphthalene sulfonic acid-formaldehyde condensation product, or other emulsifier. The progress of the polymerization is followed by means of specific gravity changes (241). The density of the emulsion increases with time and polymerization may be considered complete when the specific gravity is between 1.068 and 1.070. The polymer then may be precipitated by acidification with 2 mL dilute acetic acid or the emulsion may be allowed to age in the presence of tetraethylthiuram disulfide. During this aging period, the polymer properties improve to a more desirable level as a result of the action of the disulfide or sulfur linkages in the polymer (242) yielding a more soluble, plastic product. The coagulated polymer prepared by either method is filtered, washed thoroughly with water, and air dried at 120°C. The polymer is then compounded for the specific purpose for which it is desired.

An interesting halogenated diene polymer that is a hard, tough, material and not an elastomer, is prepared by persulfate initiated polymerization of 2,3-dichlorobutadiene. This polymer may be made by the methods described by Kuhn who has defined specialized conditions claimed to be necessary to produce a polymer that can be molded and shaped (243).

Allene is the simplest of possible dienes in structure, but several structural variants are possible when it is polymerized. With bimetallic coordination catalysts, the polymer contains varying amounts of vinylidene-, vinyl-, and *cis*-unsaturation (244). The propagating species can be considered as an anion (in an admitted oversimplification) and the mechanism leading to the structural units visualized as follows:

$$C{=}C{=}C + M^{\oplus}R^{\ominus} \longrightarrow R-C-\overset{\displaystyle C}{\overset{\|}{C}}{\ominus}M^{\oplus} \xrightarrow{A} R-C-\overset{\displaystyle C}{\overset{\|}{C}}-C-\overset{\displaystyle C}{\overset{\|}{C}}{\ominus}M^{\oplus}\ \ (\mathrm{I})$$

$$\mathrm{I} \longrightarrow \left(C-\overset{\displaystyle C}{\overset{\|}{C}}\right)_n \quad \text{vinylidene}$$

$$\mathrm{I} \longrightarrow R-C-\overset{\displaystyle C}{\overset{\|}{C}}-\underset{\displaystyle M^{\oplus}}{\underset{\ominus}{C}}-\overset{\displaystyle C}{\overset{\|}{C}} \longleftrightarrow \text{II (allylic anion ring C–C–C with } M^{\oplus}\text{)}-C-C-R$$

$$\mathrm{II} \xrightarrow{A} R-\overset{\displaystyle C}{\overset{\|}{C}}-\overset{\displaystyle C{=}C}{\overset{|}{C}}-C-\overset{\displaystyle C}{\overset{\|}{C}}{\ominus}M^{\oplus} \quad \text{vinyl}$$

$$\mathrm{II} \xrightarrow{A} R-C-\overset{\displaystyle C}{\overset{\|}{C}}-C{=}C-C-C-\overset{\displaystyle C}{\overset{\|}{C}}{\ominus}M^{\oplus} \quad \text{cis}$$

The relative amounts of each vary with polymerization conditions: increasing monomer concentration will result in an increase of vinylidene units because addition of allene to the carbanion I will tend to occur before hydride shift to carbanion II. Polymerization at low pressure reduces vinyl content markedly, Heterogeneous polymerization and low polymer solubility or swelling will favor vinyl formation, and vigorous agitation will decrease it, because of their effect on monomer concentration at the growing sites. Polymers as prepared can range from amorphous to crystalline, depending on reaction conditions, but both types are soluble in halogenated hydrocarbons in particular. High crystallinity can evidently be induced by film casting or pressing, to give rather brittle films. T_m is about 122°C (244).

267. Preparation of Polyallene (244)

$$CH_2{=}C{=}CH_2 \longrightarrow -\!\!\left(CH_2-\underset{\displaystyle \overset{\|}{C}}{}\!\!\!\!\overset{CH_2}{\overset{\|}{C}}\right)\!\!\left(CH{=}CHCH_2\right)\!\!\left(\overset{CH=CH_2}{\overset{|}{CH}}\right)\!\!-$$

To a l-L three-necked flask, previously baked 24 h at 140°C and cooled under nitrogen, and equipped with a stirrer, thermometer, and a gas inlet and outlet, is added 500 mL benzene (distilled from calcium hydride) under nitrogen. The solvent is then saturated with allene that has preferably been distilled and passed through a silica gel column. Pure allene is essential to successful polymerization. Then, 0.05 mL $VOCl_3$ is added, causing a dark red color. Addition of 2.5 mL 2.2*M* triisobutylaluminum in cyclohexane changes the color to a cherry red. Allene is then bubbled through the solution and the polymerization is conducted at 30°C until the gelled or precipitated polymer prevents easy stirring and monomer uptake is difficult. Cooling may be required. Then 50 mL ethanol is added, and the mass slurried into excess ethanol. The filtered polymer is washed repeatedly in a home blender with 10% ethanolic HCl, ethanol, water, and acetone; the last wash should contain an antioxidant such as 2,2′-methylenebis(4-methyl-6-*t*-butylphenol). The polymer is dried under nitrogen in vacuum at 50°C. It should have an inherent viscosity in bromobenzene (0.5 g/100 mL, with an antioxidant present) of about 2. The yield is about 15 g. There are three crystalline forms of polyallene, two of which are metastable. The T_m value of 122°C mentioned earlier is from an x-ray melting point on a stable crystalline form.

Allene can also be polymerized to a crystalline polyallene, having the majority of the vinylidene units $CH_2{-}C({=}CH_2)$ in its structure as a 2_1 helix, using a pi-allylnickel bromide catalyst (245). The bimetallic coordination catalyst apparently can give a spectrum of polymers, of which the polymer from the pi-allylnickel bromide catalyst is probably one (246).

It has been possible to polymerize conjugated trienes with bimetallic coordination catalysts to give polymers with primarily 1,6-enchainment and, consequently, a conjugated diene in the chemical repeat unit of the polymer (247). Other mixed enchainments (e.g., 1,4- and 1,6-) can also be induced with substantial resultant changes in the character of the polymer. Examples are 1,3,5-hexatriene, 1,3,5-heptatriene and 2,4,6-octatriene. Generally, a Lewis base is used to restrain the activity of the catalyst toward secondary reactions on the polymeric diene (247).

The polytrienes can be modified by addition of a dienophile to the conjugated double bond units in a typical Diels-Alder reaction (248). Maleic anhydride, sulfur dioxide, tetracyanoethylene, 1,2-dicyano-1,2-di(trifluoromethyl)ethylene, and dimethyl acetylenedicarboxylate are active dienophiles, although the degree of acceptance by the diene ranges from only about 10% of theory for sulfur dioxide to greater than 75% for the acetylene dicarboxylate. Models suggest that the buildup of successive cyclohexene rings along the chain imposes a great deal of steric strain, so that the reaction of more than every other diene unit would appear difficult. It was also clear that trans-trans diene units in the polymer chain underwent reaction much more readily than cis-trans dienes, in keeping with observations on conventional diene compounds.

REFERENCES

1. C. H. Bamford, W. G. Barb, A. D. Jenkins, and P. F. Onyon, *The Kinetics of Vinyl Polymerization by Radical Mechanisms*, Academic Press, New York, 1958.
2. P. A. Plesch, Ed., *Chemistry of Cationic Polymerization*, Macmillan, New York, 1963.
3. F. W. Billmeyer, Jr., *Textbook of Polymer Science*, Interscience, New York, 1962.
4. N. G. Gaylord and H. F. Mark, *Linear and Stereospecific Addition Polymers*, Interscience, New York, 1959.
5. G. Ham, Ed., *Copolymerization*, Interscience, New York, 1964.
6. R. A. V. Raff and K. W. Doak, Eds., *Crystalline Olefin Polymers*, Parts 1 and 2, Interscience, New York, 1964; H. V. Boenig, *Polyolefins: Structure and Properties*, American Elsevier, New York, 1966.
7. P. J. Flory, *Principles of Polymer Chemistry*, Cornell University Press, Ithaca, New York, 1953.
8. W. E. Goode, F. H. Owens, and W. L. Myers, *J. Polymer Sci.*, **47**, 75 (1960).
9. H. Brody, M. Ladacki, R. Milkovich, and M. Szwarc, *J. Polymer Sci.*, **25**, 221 (1957).
10. M. Szwarc, *Nature*, **178**, 1168 (1956).
11. M. Szwarc, M. Levy, and R. Milkovich, *J. Am. Chem. Soc.*, **78**, 2656 (1956).
12. M. P. Dreyfuss and P. Dreyfuss, *J. Polymer Sci. A-1*, **4**, 2179 (1966).
13. J. K. Stille, *Introduction to Polymer Chemistry*, Wiley, New York, 1962.
14. F. Marchionna, *Butalistic Polymers*, Reinhold, New York, 1946.
15. G. S. Whitby, Ed., *Synthetic Rubber*, Wiley, New York, 1954.
16. H. Mark, Ed., *Collected Papers of Wallace H. Carothers on High Polymeric Substances*, Interscience, New York, 1940.
17. Y. Yamashita, T. Tsuda, M. Okada, and S. Iwatsuki, *J. Polymer Sci. A-1*, **4**, 2121 (1966).
18. C. E. Schildknecht, *Vinyl and Related Polymers*, Wiley, New York, 1952.

19. A. H. Frazer, private communication.
20. F. J. Glavis, L. L. Ryden, and C. S. Marvel, *J. Am. Chem. Soc.*, **59**, 707 (1937).
21. M. Hunt and C. S. Marvel, *J. Am. Chem. Soc.*, **57**, 1691 (1935).
22. L. L. Ryden and C. S. Marvel, *J. Am. Chem. Soc.*, **57**, 2311 (1935).
23. N. L. Zutty, C. W. Wilson III, G. H. Potter, D. C. Priest, and C. J. Whitworth, *J. Polymer Sci. -A*, **3**, 2781 (1965).
24. W. G. Barb, *J. Am. Chem. Soc.*, **75**, 224 (1953).
25. F. S. Dainton and K. J. Ivin, *Proc. Roy. Soc. (London)* A212, **66** (1952).
26. W. J. Burlant and A. S. Hoffman, *Block and Graft Copolymers*, Reinhold, New York, 1960.
27. R. J. Ceresa, *Block and Graft Copolymers*, Butterworths, London, 1962.
28. P. W. Allen, Ed., *Techniques of Polymer Characterization*, Academic Press, New York, 1959.
29. G. Smets and R. Hart, *Advan. Polymer Sci.*, **2**, 173 (1960).
30. B. D. Gesner, *J. Polymer Sci. A*, **3**, 3825 (1965).
31. G. Bier, *Angew. Chem.*, **73**, (6), 186 (1961).
32. M. A. Luftglass, W. R. Hendricks, G. Holden, and J. T. Bailey, *Tech. Papers*, Vol. 12, *Soc. Plastics Engrs. 22nd ANTEC, Montreal*, March 7, 1966, p. XIV-6.
33. British Pat. 1,000,090, (to Shell Intern. Res.) Aug. 4, (1965).
34. C. E. Schildknecht, *Polymer Processes*, Interscience, New York, 1956.
35. F. A. Bovey, I. M. Kolthoff, A. I. Medalia, and E. J. Meehan, *Emulsion Polymerization*, Interscience, New York, 1955.
36. P. A. Lovell, Ed., *Emulsion Polymerization and Emulsion Polymers*, Wiley, New York, 1997.
37. A. Renfrew and P. Morgan, Eds., *Polythene: The Technology and Uses of Ethylene Polymers*, Interscience, New York, 1960.
38. R. H. Boundy and R. F. Boyer, Eds., *Styrene, Its Polymers, Copolymers and Derivatives*, Reinhold, New York, 1952.
39. Dow Chemical Company, Product Bulletin, *The Polymerization of Styrene*.
40. I. M. Kolthoff and W. J. Dale, *J. Am. Chem. Soc.*, **69**, 442 (1947).
41. J. R. Hiltner and W. F. Bartoe, U.S. Pat. 2,264,376 (Dec. 2, 1941).
42. L. E. Daly, U.S. Pat. 2,439,202 (to U.S. Rubber) (April 6, 1948).
43. J. L. Amos, J. L. McCurdy, and O. R. McIntire, U.S. Pat. 2,694,692 (to Dow Chemical) (November 16, 1954).
44. W. C. Calvert, U.S. Pat. 2,908,661 (to Borg-Warner) (October 13, 1959).
45. G. H. Fremon and W. N. Stoops, U.S. Pat. 3,168,593, (Jan 15, 1953) (to Union Carbide)
46. H. Kainer, *Polyvinylchlorid and Vinyl Chlorid-Mischpolymerisate*, Springer-Verlag, Berlin/New York, 1965; W. S. Penn, *PVC Technology*, MacLaren, London, 1962; F. Chevassus and R. Broutelles (C. J. R. Eichhorn and E. E Sarmiento, Transl.), *Stabilization of Polyvinyl Chloride*, Edward Arnold, London, 1963.
47. D. D. Coffman and T. A. Ford, U.S. Pat. 2,419,008 to DuPont (April 15, 1947).
48. D. D. Coffman and T. A. Ford, U.S. Pat. 2,419,009 to DuPont (April 15, 1947).
49. D. D. Coffman and T. A. Ford, U.S. Pat. 2,419,010 to DuPont (April 15, 1947).

50. P. J. Manno, *Nucleonics*, **22** (2), 49; **22** (6), 64; **22** (9), 72 (1964).
51. M. L. Dannis and F. L. Ramp, U.S. Pat. 2,996,489, (Aug. 15, 1961) to B. F. Goodrich Co.
52. V. E. Shashoua, private communication.
52a. W. Sweeny. Private Communication (1991).
53. R. G. Beaman, private communication.
54. W. G. Vosburgh, private communication.
55. N. Grassie and I. C. McNeill, *J. Polymer Sci.*, **39**, 211 (1959).
56. R. C. Houtz, *Textile Res. J.*, **20**, 796 (1950).
57. British Pat. 786,960 (Nov. 27, 1957).
58. *Emulsion Polymerization I*, Alcolac Chemical Corp.
59. M. Hunt, U.S. Pat. 2,471,959 (May 31, 1949).
60. W. O. Herrmann and W. Hachnel, *Ber.*, **60**, 1658 (1927).
61. W. O. Herrmann and W. Hachnel, U.S. Pat. 1,672,156 (June 5, 1928).
62. W. O. Herrmann and W. Hachnel, U.S. Pat. 2,109,883 (May 1, 1938).
63. L. M. Minsk, W. J. Priest, and W. O. Kenyon, *J. Am. Chem. Soc.*, **63**, 2715 (1941).
64. R. C. Houtz, U.S. Pat. 2,388,325 (Nov. 6, 1945), to DuPont.
65. D. L. Wilson, U.S. Pat. 2,399,970 (May 11, 1946), to DuPont.
66. E. A. Land and C. D. West, *Colloid Chemistry*, Vol. 6, Reinhold, New York, 1946, p. 160.
67. G. Kranzlein and H. Reis, Ger. Pat. 765,265 (Jan. 1, 1954).
68. G. S. Stamatoff, U.S. Pat. 2,400,957 (May 18, 1946).
69. R. C. Houtz, U.S. Pat. 2,341,553 (Feb. 15, 1944).
70. H. N. Friedlander, H. E. Harris, and J. G. Pritchard, *J. Polymer Sci. A-1*, **4**, 649 (1966).
71. H. E. Harris, J. F. Kenney, G. W. Willcockson, R. Chiang, and H. N. Friedlander, *J. Polymer Sci. A-1*, **4**, 665 (1966).
72. W. C. Tincher, *Makromol. Chem.*, **85**, 46 (1965).
73. S. Okamura, T. Kodama, and T. Higashimura, *Makromol. Chem.*, **53**, 180 (1962).
74. S. Murahashi, H. Yuki, T. Sano, U. Yonemura, H. Tadokoro, and V. Chatani, *J. Polymer Sci.*, **62**, 877 (1962).
75. S. Murahashi, S. Nozakura, and M. Sumi, *J. Polymer Sci. B*, **3**, 245 (1965); *ibid.*, **4**, 65 (1966); *ibid.*, **4**, 59 (1966).
76. J. F. Kenney and G. W. Willcockson, *J. Polymer Sci. A-1*, **4**, 679 (1966).
77. F. E. Kung, U.S. Pat. 2,377,085 (May 29, 1945).
78. L. A. R. Hall, W. J. Belanger, W. Kirk, Jr., and Y. Sundstrom, *J. Appl. Polymer Soc.*, **2**, 246 (1959).
79. *American Cyanamid Co. New Prod. Bull.*, Coll. Vol. III.
80. W. C. Mast and C. H. Fisher, *Ind. Eng. Chem.*, **41**, 790 (1949).
81. H. Fikentscher and K. Herrle, *Modern Plastics*, **23**, 157 (1945).
82. W. Reppe and C. Schuster, U.S. Pat. 2,265,450 (Dec. 9, 1941).
83. C. Schuster, R. Sauerbier, and H. Fikentscher, U.S. Pat. 2,335,454 (Nov. 30, 1943).
84. J. H. Werntz, U.S. Pat. 2,497,705 (Feb. 14, 1950).

85. R. F. Conaway, U.S. Pat. 2,008,577 (Aug. 3, 1937).
86. R. C. Schulz, C. H. Cherdron, and W. Kern, *Makromol. Chem.*, **24**, 141 (1957).
87. E. E. Ryder and P. Pezzaglia, *J. Polymer Sci. A*, **3**, 3459 (1965).
88. F. L. Ramp, E. J. DeWitt, and L. E. Trepasso, *J. Polymer Sci. A-1*, **4**, 2267 (1966).
89. J. W. C. Crawford, *J. Soc. Chem. Ind.* (Trans.), **68**, 201 (1949).
90. D. E. Strain, R. G. Kennelly, and H. R. Dittmar, *Ind. Eng. Chem.*, **31**, 382 (1939).
91. See References 89,90.
92. Shell Development Company, Report S-9976, June 25, 1947.
93. M. A. Pollach, U.S. Pat. 2,870,193 (Jan. 20, 1959).
94. British Pat. 741,236 (Nov. 30, 1955); N. Grassie and E. M. Grant, *J. Polymer Sci. A-1*, **4**, 1821 (1966).
95. R. C. Reinhardt, *Chem. Eng. News*, **25**, 2136 (1947).
96. Dow Chemical Company, Product Bulletin, *Handling Precautions for Vinylidene Chloride Monomer*.
97. British Pat. 570,711 (July 19, 1945), to ICI.
98. M. Hauptscheien, U.S. Pat. 3,193,539 (to Pennsalt) (July 6, 1965).
99. M. S. Newman and R. W. Addor, *J. Am. Chem. Soc.*, **75**, 1263 (1953).
100. M. S. Newman and R. W. Addor, *J. Am. Chem. Soc.*, **77**, 3789 (1955).
101. N. D. Field and J. R. Schaefgen, *J. Polymer Sci.*, **58**, 533 (1962).
102. R. M. Thomas, U.S. Pat. 2,873,230 (Feb. 10, 1959), to DuPont.
103. W. E. Hanford, U.S. Pat. 2,396,785 (March 19, 1946), to DuPont.
104. H. W. Arnold, M. M. Brubaker, and G. L. Dorough, U.S. Pat. 2,301,356 (Nov. 10, 1942).
105. British Pat. 505,120 (May 5, 1939), to Standard Oil De.
106. J. L. Lang, W. A. Pavelich, and H. D. Clarey, *Polymer Preprints*, **2** (2), 36 (1961), Polymer Div., Am. Chem. Soc., September 1961; G. H. Potter and N. L. Zutty, U.S. pat. 3,280,080 (to Union Carbide) (October 18, 1966).
107. J. P. Roth, P. Remp, and J. Parrod, *J. Polymer Sci. C*, **4**, 1347 (1964).
108. G. D. Jones, *J. Org. Chem.*, **9**, 500 (1944).
109. French Pat. 922,429 (July 9, 1947).
110. F. Scholoffer and O. Scherer, Ger. Pat. 677,071 (June 17, 1939).
111. R. M. Joyce, U.S. Pat. 2,394,243 (Feb. 5, 1946), to DuPont.
112. R. J. Plunkett, U.S. Pat. 2,230,654 to DuPont. (Feb. 4, 1941).
113. F. P. Reding, *J. Polymer Sci.*, **21**, 547 (1956).
114. T. Higashimura, Macromolecules, **17**, 2228 (1984).
115. J. P. Kennedy and Marechal, "Carbocationic Polymerization", J. Wiley and Sons, N.Y., 1982; R. Faust and J. P. Kennedy, J. Poly. Sci. (a), 25, 1847 (1987); Mishra and J. P. Kennedy, J. Macromol. Sci. Chem A, 24 (8), 933 (1987).
116. D. B. Miller, private communication.
117. J. P. Kennedy and R. M. Thomas, *J. Polymer Sci.*, **55**, 311 (1961).
118. A. J. Morway and F. L. Miller, U.S. Pat. 2,243,470 (May 27, 1941).
119. M. Müller-Cunradi and M. Otto, U.S. Pat. 2,203,873 (June 11, 1940) to Jasco.
120. W. Repp, E. Keysner, and E. Dorrer, U.S. Pat. 2,072,465 (March 2, 1937) to IG.

121. D. E. Sargeant, U.S. Pat. 2,560,251 (July 10, 1951) to DuPont.
122. H. F. Miller and R. G. Flowers, U.S. Pat. 2,445,181 (July 13, 1948) to General Electric.
123. M. Imoto and K. Takemoto, *J. Polymer Sci.*, **15**, 271 (1955).
124. J. Springer, K. Ueberreiter, and R. Wenzel, *Makromol. Chem.*, **96**, 122 (1966).
125. C. E. Schildknecht, A. O. Zoss, and C. McKinley, *Ind. Eng. Chem.* **39**, 180 (1947).
126. S. Okamura, T. Higashimura, and H. Yamamoto, *J. Polymer Sci.*, **33**, 510 (1958).
127. S. Okamura, T. Higashimura, and I. Sakurada, *J. Polymer Sci.*, **39**, 507 (1959).
128. S. Okamura, T. Higashimura, T. Yonezawa, and K. Fukui, *J. Polymer Sci.*, **39**, 487 (1959).
129. D. J. Cram and K. R. Kopecky, *J. Am. Chem. Soc.*, **81**, 2748 (1959).
130. A. D. Ketley, *J. Polymer Sci.*, **62**, 581 (1962).
131. G. J. Blake and A. M. Carlson, *J. Polymer Sci. A-1*, **4**, 1813 (1966).
132. G. Natta, G. Dall 'Asta, G. Mazzanti, U. Giannini, and S. Cesca, *Angew. Chem.*, **71**, 205 (1959).
133. J. Furukawa, T. Saegusa, S. Yasui, and S. Akutsu, *Makromol. Chem.*, **94**, 74 (1965).
134. R. C. Schulz and E. Kaiser, *Advan. Polymer Sci.*, **4**, 236 (1965).
135. P. Pino, F. Ciardelli, and G. P. Lorenzi, *Makromol. Chem.*, **70**, 182 (1964); P. Pino, F. Ciardelli, and G. Montagnoli, *Symp. Macromol. Chem.*, Prague, 1965, Preprints, p. 428.
136. G. Natta, P. Pino, and S. Valenti, *Makromol. Chem.*, **67**, 225 (1963).
137. P. Pino, *Adv. Polymer Sci.*, **4**, 393 (1966).
138. M. Farina and G. Bresson, *Makromol. Chem.*, **61**, 79 (1963); Y. Takeda, V. Hayakawa, T. Fueno, and J. Furukawa, *ibid.*, **83**, 234 (1965).
139. G. Natta and M. Farina, *Tetrahedron Letters*, 1963, 603.
140. G. Bresson, M. Farina, and G. Natta, *Makromol. Chem.*, **93**, 283 (1966).
141. J. E. Mulvaney, C. G. Overberger, and A. M. Schiller, *Advan. Polymer Sci.*, **3**, 106 (1961).
142. S. Bywater, *ibid.*, **4**, 66 (1965).
143. M. Roha, *ibid.*, **4**, 353 (1965).
144. M. Szwarc, *ibid.*, **2**, 173 (1960).
145. T. E. Werkema, U.S. Pat. 2,658,058 (Nov. 3, 1953) to Dow Chemical.
146. R. G. Beaman, *J. Am. Chem. Soc.*, **70**, 3115 (1948).
147. Y. Joh, T. Yoshihara, Y. Kotake, F. Ide, and K. Nakatsuka, *J. Polymer Sci. B*, **3**, 933 (1965); **4**, 673 (1966).
148. A. L. Serge, F. Ciampelli, and G. Dall 'Asta, *J. Polymer Sci. B*, **4**, 633 (1966).
149. K. Ishigure, Y. Tabata, and K. Oshima, *J. Polymer Sci. B*, **4**, 669 (1966).
150. R. Waack, A. Rembaum, J. D. Coombes, and M. Szwarc, *J. Am. Chem. Soc.*, **79**, 2026 (1957).
151. P. J. Flory, *J. Am. Chem. Soc.*, **62**, 1561 (1940).
152. M. Morton, R. Milkovich, D. B. McIntyre, and L. J. Bradley, *J. Polymer Sci. A*, **1**, 443 (1963).
153. F. Wenger and Shiao-Ping S. Yen, *Makromol. Chem.*, **43**, 1 (1961).

154. D. P. Wyman and T. G. Fox, *Tech. Doc. Rept. No. ASD-TDR-62-1110* (January, 1963).
155. N. D. Scott, U.S. Pat. 2,181,771 (Nov. 28, 1939) to DuPont.
156. N. D. Scott, J. F. Walker, and V. L. Hansley, *J. Am. Chem. Soc.*, **58**, 2442 (1936).
157. S. L. Jung, private communication.
158. A. A. Morton, B. Darling, and J. Davidson, *Ind. Eng. Chem. Anal. Ed.*, **14**, 734 (1942).
159. E. F. Evans, A. Goodman, and L. D. Grandine, private communication.
160. L. S. Luskin and R. J. Meyers, *Encyclopedia of Polymer Science and Technology*, Vol. 1, Interscience, New York, 1964, p. 290 ff.
161. W. E. Goode, R. P. Fellmann, and F. H. Owens, *Macromolecular Syntheses*, Vol. 1, C. G. Overberger, Ed., Wiley, New York, 1963, p. 25.
162. C. F. Ryan and J. J. Gormley, *ibid.*, p. 30.
163. W. E. Goode, F. H. Owens, R. P. Fellmann, W. H. Snyder, and J. E. Moore, *J. Polymer Sci.*, **46**, 317 (1960); D. L. Glusker, E. Stiles, and B. Yoncoskie, *ibid.*, **49**, 297 (1961).
164. A. E. Ardis, U.S. Pat. 2,615,865 (Oct. 28, 1952) to B. F. Goodrich.
165. A. E. Ardis, S. T. Averill, H. Gilbert, F. F. Miller, R. F. Schmidt, F. D. Stewart, and H. L. Trumball, *J. Am. Chem. Soc.*, **72**, 1305 (1950).
166. H. Gilbert and F. F. Miller, U.S. Pat. 2,615,867 (Oct. 28, 1952) to B. F. Goodrich.
167. Private Communication, G. M. Cohen & O. W. Webster.
167a. O. W. Webster & B. C. Anderson, New Methods for Polymer Synthesis, W. J. Mijs (Ed.), Plenum, New York (1988); D. Y. Sogah, W. R. Hertler, O. W. Webster and B. M. Cohen, *Macromolecules*, **20**, 1473 (1988).
168. K. Narasaka, K. Soai, Y. Aykawa and T. Mukaiyama, *Bull. Soc. Chem. Japan*, **49** 779 (1976); K. Saigo, M. Osaki and T. Mukaiyama, Chem. Lett., 1976, 163.
169. D. Y. Sogah and W. B. Farnham, *Organosilicon and Bioörganic Silicon*, H. Sakurai (Ed.), J. Wiley & Sons, New York (1985).
170. R. P. Quirk and G. P. Baldinger, *Polymer Bull.*, **22**, 63 (1989).
171. M. T. Reetz, T. Knouf, U. Minet and C. Bingel, *Angew. Chem. Int. Ed.*, **27**, 1373 (1988).
172. O. W. Webster, *Makromol. Chem., Macromol. Sympos.*, **60**, 287 (1992).
173. T. Ando, M. Kamigaito and M. Sawamoto, *Macromolecules*, **33**, 2819 (2000).
174. D. Benoit, V. Chaplinski, R. Breslau and C. J. Hawker, *J. Am. Chem. Soc.*, **121**, 3904 (1999); see also C&EN, Apr. 17, p. 43, 2000.
175. A. W. Anderson et al., U. S. Pat. 2,721,189 (Oct. 18, 1955) to DuPont.
176. A. W. Anderson, N. G. Merckling, and P. H. Settlage, U. S. Pat. 2,799,668 (July 16, 1957).
177. British Pat. 777,538 (June 26, 1957).
178. K. Ziegler and H. Martin, *Makromol. Chem.*, **18/19**, 186 (1956).
179. K. Ziegler, E. Holzlcamp, H. Breil, and H. Martin, *Angew. Chem.*, **67**, 426, 541 (1955).
180. G. Natta, *J. Polymer Sci.*, **16**, 143 (1955).
181. G. Natta, *Angew. Chem.*, **68**, 393 (1956).

182. G. Natta, P. Pino, P. Corradini, F. Danusso, E. Manteca, G. Mazzanti, and G. Moragli, *J. Am. Chem. Soc.*, **77**, 1708 (1955).
183. J. P. Hogan & R. L. Banks, U.S. Patent 2,825,721 (to Phillips Petroleum), Mar. 4, 1958; Y. V. Kissin, *Encyc. Chem. Technol.*, **17**, 773 (1996), Wiley, New York.
184. T. W. Campbell and A. C. Haven, Jr., *J. Appli. Polymer Sci.*, **1**, 79 (1959).
185. F. P. Reding, *J. Polymer. Sci.*, **21**, 547 (1956).
186. A. W. Larchar and D. C. Pease, U.S. Pat. 2,816,883 (Dec. 17, 1957) to DuPont.
187. A. K. Ingberman, I. J. Levine, and R. J. Turbett, *J. Polymer Sci. A-1*, **4**, 2781 (1966).
188. F. W. Breuer, L. E. Geipel, and A. B. Loebel, in *Crystalline Olefin Polymers*, Part I, R. A. V. Raff and K. W. Doak, Eds., Interscience, New York, 1965.
189. J. L. Jezl, T. L. Kelley, H. M. Khelghatian, and E. M. Honeycutt, in *Manufacture of Plastics*, Vol. 1, W. M. Smith, Ed., Reinhold, 1, New York, 1964.
190. Issue dedicated to G. Natta, *Science*, **51**, 383 (1961).
191. G. Natta, *Science*, **147**, 261 (1965).
192. A. Zambelli, G. Natta, and I. Pasquon, *J. Polymer Sci. C*, **4**, 411 (1963).
193. G. Natta, A. Zambelli, G. Lanzi, I. Pasquon, and E. R. Magnaschi, *Makromol. Chem.*, **81**, 161 (1965).
194. J. Boor, Jr. and E. A. Youngman, *J. Polymer Sci. A-1*, **4**, 1861 (1966).
195. F. Ang, *J. Polymer Sci.*, **25**, 126 (1957).
196. H. Tadokoro, N. Nishiyama, S. Nozakura, and S. Murahashi, *J. Polymer Sci.*, **36**, 553 (1959).
197. T. W. Campbell, Italian Pat. 579,572 (July 16, 1958) to DuPont.
198. C. A. Lukach, private communication.
199. J. D. Sutherland and J. P. McKenzie, *Ind. Eng. Chem.*, **48**, 17 (1956).
200. W. Reppe and A. Magin, US Patent 2,577,208 (to I. G. Farben), 1951.
201. E. Drent, J. Van Broekhoven and M. Doyle, *J. Organometallic Chem.*, **417**, 235 (1991).
202. M. Brookhart, *J. Am. Chem. Soc.*, **114**, 5894 (1992).
203. F. Ciardelli, E. Bennedetti, and O. Pieroni, *Makromol. Chem.*, in press.
204. M. Bourgel, *Ann. Chim. France*, **3**, 191, 325 (1925).
205. L. Lardicci, C. Botteghi, and E. Bennedetti, *J. Org. Chem.*, **31**, 1534 (1966).
206. R. G. Charles and M. A. Paulikowski, *J. Phys. Chem.*, **62**, 440 (1958).
207. M. Morton, Ed., *Introduction to Rubber Technology*, Reinhold, New York, 1959.
208. J. P. Kennedy and E. Tornqvist, Eds., *The Polymer Chemistry of Synthetic Elastomers*, Interscience, New York, 1982.
209. M. Gippin, *Ind. Eng. Chem. Prod. Res. Develop.*, **4**, 160 (1965).
210. Br. Pat. 592, 477 (to Montecatini).
211. R. C. Farrar and F. E. Naylor, U.S. Pat. 3,223,692 (to Phillips Petroleum) (December 14, 1965).
212. R. S. Silas, J. Yates, and V. Thornton, *Anal. Chem.*, **31**, 529 (1959).
213. F. E. Naylor and J. R. Hooten, U.S. Pat. 3,205,212 (to Phillips Petroleum) (September 7, 1965).
214. N. G. Gaylord, T. K. Kwei, and H. F. Mark, *J. Polymer Sci.*, **42**, 417 (1960).

215. G. Natta, L. Porri, and A. Carbonaro, *Rend. Accad. Nazl. Lincei*, **31**, 189 (1961).
216. G. J. van Amerongen, in *Adv. in Chem.*, **52**, B. L. Johnson and M. Goodman, Eds., Am. Chem. Soc., 1966, p. 11.
217. R. P. Zelinski and D. R. Smith, U.S. Pat. 3,050,513 (to Phillips Petroleum Company) (August 21, 1962).
218. R. E. Rinehart, *ACS Polymer Div. Preprints*, **7**(2), 556 (1966).
219. G. Natta, I. Pasquon, A. Zambelli, and G. Gatti, *J. Polymer Sci.*, **51**, 383 (1961).
220. E. Susa, *J. Polymer Sci. C*, **4**, 399 (1964).
221. G. Natta, L. Porri, and A. Carbonaro, *Makromol. Chem.*, **77**, 126 (1964).
222. M. Morton and W. E. Gibbs, *J. Polymer Sci. -A*, **1**, 2679 (1963).
223. German Pat. 1,040, 795 (Oct. 9, 1958).
224. German Pat. 1,040,796 (Oct. 9, 1958).
225. F. W. Stavely et al., *Ind. Eng. Chem.*, **48**, 778 (1956).
226. H. E. Adams, R. S. Stearns, W. A. Smith, and J. L. Binder, *Ind. Eng. Chem.*, **50**, 1507 (1958).
227. British Pat. 776,326 (June 5, 1957).
228. S. E. Horne, J. P. Kiehl, J. J. Shipman, V. L. Folt, C. F. Gibbs, E. A. Wilson, E. B. Newton, and M. A. Reinhart, *Ind. Eng. Chem.*, **48**, 784 (1956).
229. T. F. Yen, *J. Polymer Sci.*, **35**, 533 (1959).
230. G. Natta, L. Porri, and A. Mazzei, *Chim. Ind.*, **41**, 116 (1959).
231. J. S. Lasky, H. K. Garner, and R. H. Ewart, *Ind. Eng. Chem. Prod. Res. Develop.*, **1**, 82 (1962).
232. E. G. Kent and F. B. Swinney, *Ind. Eng. Chem. Prod. Res. Develop.*, **5**, 134 (1966).
233. J. D. Calfie et al., U.S. Pat. 2,431,461 (Nov. 25, 1947).
234. R. M. Thomas and W. J. Sparks, U.S. Pat. 2,356,128 (Aug. 22, 1944).
235. F. B. Moody, *ACS Polymer Div. Preprints*, **2**, 285 (1961).
236. G. F. Hennion, C. C. Price, and T. F. McKeon, Jr., *Org. Syn.*, 1958, 70.
237. W. H. Carothers, I. Williams, A. M. Collins, and J. E. Kirby, *J. Am. Chem. Soc.*, **53**, 4203 (1931).
238. R. C. Ferguson, *Anal. Chem.*, **36**, 2204 (1964); *J. Polymer Sci. A.*, **2**, 4735 (1964).
239. C. A. Aufdermarsh, Jr. and R. Pariser, *J. Polymer Sci. A.*, **2**, 4727 (1964).
240. H. W. Walker and W. E. Mochel, *Proc. Rubber Technol. Conf. 2nd*, 69–78 (1948).
241. R. S. Barrows and G. W. Scott, *Ind. Eng. Chem.*, **40**, 2193 (1948).
242. W. E. Mochel and J. H. Peterson, *J. Am. Chem. Soc.*, **71**, 1426 (1949).
243. L. B. Hunt, Canadian Pat. 525,592 (May 29, 1956).
244. W. P. Baker, Jr., *J. Poly. Sci.-A*, **1**, 655 (1963).
245. S. Otsuka, *J. Am. Chem. Soc.*, **87**, 3017 (1965).
246. W. P. Baker, Jr., private communication.
247. V. L. Bell, *J. Polymer Sci. A*, **2**, 5291 (1964).
248. V. L. Bell, *J. Polymer Sci. A*, **2**, 5305 (1964).
249. G. Balbontin, D. Dainelli, M. Galimberti and G. Paganetto, *Makromol. Chem.*, **193**, 693 (1992).

250. J. Huang and G. L Rempel, Ziegler-Natta Catalysts, in *Progress in Polymer Science*, **20**, 459–526 (1995).

251. H. Sinn and W. Kaminsky, in *Advances in Organometallic Chemistry*, **18**, 99–148 (1980).

252. W. Kaminsky, A. Bark and R. Steiger, *J. Mol. Catal.* **74**, 109 (1992).

253. A. M. Thayer, C&EN, Sept. 11, 1995, p. 15; P. M. Morse, C&EN, Dec. 7, 1998, p. 25; Sci./Technol Note, C&EN, May 4, 1998; U.S. Pat. 5,324,800, to H. C. Welborn and J. A. Ewen, *Process and Catalyst for Polyolefin Density and Molecular Weight Control* (to Exxon Chemical Patnets, Inc.), June 28, 1994.

254. G. J. P. Britovsek, V. C. Gibson and Duncan F. Wass, The Search for New generation Olefin Polymerization Catalysts: Life Beyond Metallocenes; *Angew. Chem. Int. Ed.*, **38**, 428–447 (1999).

255. H. C. Welborn, Jr., and J. A. Ewen, US. Patent 5,324,800 (to Exxon Chemical Patents), June 28, 1994.

256. T. R. Younkin, E. F. Connor, J. I. Henderson, S. K. Friederich, R. H. Grubbs and D. A. Bansleben, *Science*, **287**, 460–462 (2000).

257. C. Wang, S. Friedrich, T. R. Younkin, R. T. Li, R. H. Grubbs, D. A. Bansleben and M. W. Day, Neutral Nickel (II)-Based Catalysts for Ethylene Polymerization; *Organometallics*, **17**, 3149 (1998).

258. M. Hiodai, T. Kashiwagi, T. Ikeuchi, Y. Uchida, *J. Organomet. Chem.* **30**, 279–282 (1971).

10

ELASTOMERS FOR FABRICS

Natural rubber is the best known elastomer and is made by vulcanizing natural latex with sulfur. The latex is largely composed of cis-polyisoprene, which is partially cross-linked in vulcanizing. Between 1 and 8% sulfur is used; larger amounts result in too much cross-linking and reduce elasticity. Deficiencies of rubber in garment use (discoloration, allergies related to the sulfur compounds and latex proteins, relatively low retractive force requiring thick threads) led to the development of synthetic elastomers (called spandex), the best known of which is the DuPont Company's Lycra. These synthetic elastomers were developed from copolymer formulations, based on 1000 to 3000 molecular weight aliphatic polyesters or polyethers having biterminal hydroxyl, amine, or carboxyl groups, that are capable of further reaction with other difunctional compounds such as diisocyanates. Their reaction can be used to couple low molecular weight polyesters or polyethers via urethane links (1) or the diisocyanate may be used in excess so that it becomes a terminal group (2, 3). In the latter case, these macrodiisocyanates may be further coupled by means of still another reagent, such as water, diols, amino-alcohols or diamines (2), with subsequent formation of high polymer. Such elastomeric products are complex block copolymers. The keys to preparing a good polyurethane/urea elastomer are as follows:

1. Bifunctional polyglycol of 2000 to 3000 $\overline{M}_n$ with $\overline{M}_w/\overline{M}_n$ of 1.8 to 2.0.
2. A capping ratio of about 1.60 to 1.7. This is the molar ratio of added diisocyanate groups to hydroxyl groups in the starting polydiol and controls the amount of diamine/dialkylamine finally added to couple the isocyanate-capped macrodiol and provide the finished polymer.

3. The macrodiol must contain <100 ppm water to avoid excessive branching and gelation. Temperature control in capping should be 80 to 100°C. Temperature on adding the diamine extender/dialkylamine-terminating agent must be carefully controlled at about 80°C.
4. Above all, good mixing and filtration are required at all aspects of the polymerization (**10.1**) if fiber spinning is planned.

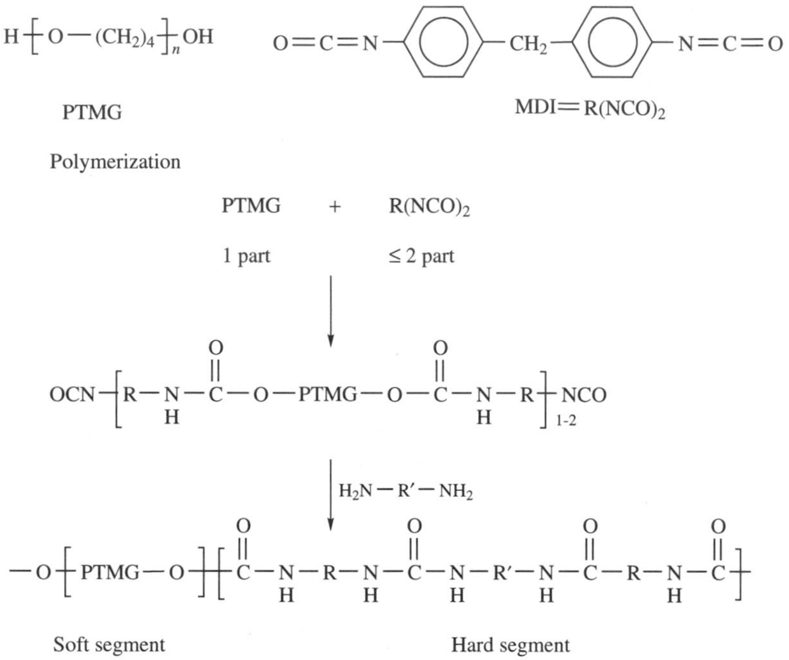

$H{+}O{-}(CH_2)_4{+}_nOH$ $\quad$ $O{=}C{=}N{-}C_6H_4{-}CH_2{-}C_6H_4{-}N{=}C{=}O$

PTMG $\quad$ MDI$=$R(NCO)$_2$

Polymerization

PTMG + R(NCO)$_2$

1 part $\quad$ ≤ 2 part

↓

$OCN{-}[R{-}NH{-}C(=O){-}O{-}PTMG{-}O{-}C(=O){-}NH{-}R]_{1\text{-}2}{-}NCO$

↓ $H_2N{-}R'{-}NH_2$

$-O{-}[PTMG{-}O]{-}[C(=O){-}NH{-}R{-}NH{-}C(=O){-}NH{-}R'{-}NH{-}C(=O){-}R{-}NH{-}C(=O)]{-}$

Soft segment $\quad$ Hard segment

Schematic 10.1

Commercially, polyester-diol used is about 2000 $\overline{M}_n$ and polyester about 2800 $\overline{M}_n$. The coupling diamine is often ethylenediamine or ethylenediamine plus another diamine or triamine. The terminating component is often diethylamine or in some cases dipropylamine, which provides better storage stability because it is less likely to disassociate from the polymer by evaporation. In all cases the calculated total combined molar amount of coupling agent plus terminating component is determined by the molar amount of diisocyanate capping used minus 1. Remember that 1 mol of the macropolyether is 2000 and that the diol, diisocyanate, and the extender are all bifunctional. For example, if 1 mol of diisocyanate is added to 1 mol of diol, the capping ratio is 1 (thus there are no free isocyanate ends). Hence the molar amount of required extender + terminator is $1 - 1 = 0$. If 2 mol of the diisocyanate are added to 1 mol of the macrodiol, the capping ratio is 2 (there are isocyanate groups on both ends of the macrodiol). Then the molar amount of required extender + terminator is $2 - 1 = 1$. Closer to commercial practice, a diisocyanate capping agent in a molar amount of 1.7 requires 0.7 mol of diamine extender + terminator. Thus the amount of the extender used would be 0.6 to 0.63 mol and the terminator, because it is monofunctional, would be 0.2 to 0.14 mol. An empirical relationship between capped glycol molecular weight and capping ratio is shown in Figure 10.1.

The theoretical requirements for elastomers are quite simple, although reduction to practice to obtain a high-performance product is more complex. Most elastomers are composed of "soft" and "hard" segments, as shown in Figure 10.2. In the case of rubber, the hard segment is the site of cross-linking. The function of the hard segment is to maintain structural integrity of the elastomer under stress, i.e., to act as tie points from which the soft segment can extend and retract to. The hard segment prevents plastic flow of the polymer under stress. The hard segment thus should be high melting and crystalline or could be a site of cross-linking, either real or by virtue of hydrogen bonding. The soft segment should be of high enough molecular weight to provide significant extension under stress. It should

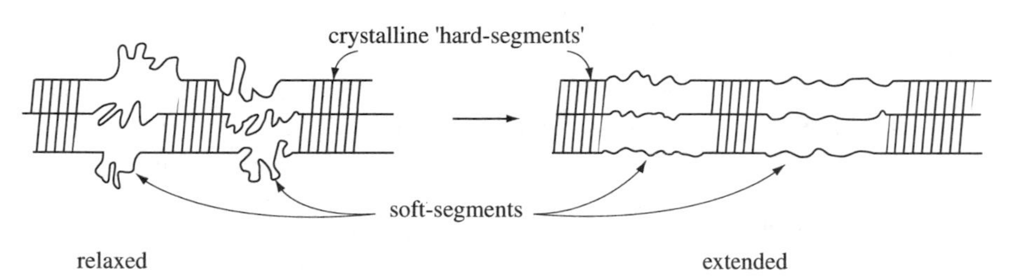

Figure 10.1

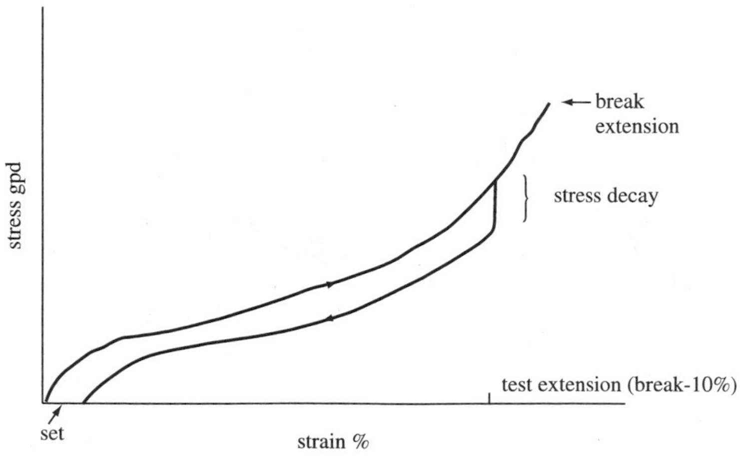

Figure 10.2

also be low melting (lower than room temperature or any use temperature) and not crystallize even under high stress, unless the crystals then melt rapidly on stress reduction.

The elastic properties derive from the retractive forces developed on stretching the soft segment and are entropic. Maximizing retractive forces requires a low-melting, essentially linear, soft segment that does not crystallize under stress, (or that remelts rapidly when stress is removed) and a high-melting hard segment. Melt temperature of the soft segment under stress should be at or below room temperature. The soft segment preferably should be unbranched and contain few, if any, large lateral substituents, e.g., ethyl or greater, that could hinder the rate of randomization of the soft segment during recovery from stretching and so result in a sluggish elastomer. Some methyl groups may be tolerated and indeed may be desirable, because they can break up the regularity of the soft segment, reduce crystallinity, and lower the melt temperature T_m. Preferred soft segments for commercial fiber use are poly(tetramethylene glycol) of 1800 to 2000 molecular weight, and poly(ethylene/tetramethylene glycol adipate) of about 3000 molecular weight and poly(ethylene/2,2-dimethylpropylene glycol adipate) and poly(ethylene/methylethylene glycol adipate) of similar molecular weights. Shorter chain length polydiols are less desirable because they are higher melting and provide reduced extensibility. Elastomers of lower extensibility, although not desirable for fibers, are useful in plastics.

Most of the soft segment macro-intermediates have hydroxyl end groups because they are easy to make with good bifunctionality. Amine-ended macro segments are sometimes used, e.g., for melt processing products but are more difficult to make with good bifunctionality. The parent soft segment should have minimal hydrogen bonding or polar interchain forces that could reduce the rate of retraction after extension.

The hard segment should crystallize readily and form a separate phase from the soft segment. In other words, the hard- and soft-segment phases should be distinct to maximize their separate functions. Obviously, this will occur most readily in a solvent process (as distinct from a melt process) in which solvent removal is slow and polymer mobility is high, e.g., in dry spinning and to a lesser extent in wet spinning. Segmental aggregation is favored in such processes, in contrast to melt processing, in which the elastomer is cooled rapidly and the rapid increase in polymer viscosity inhibits good phase separation.

A possible route to a high-strength, high-modulus, high-extension melt- or solution-spun elastomer would be to use a higher molecular weight diol (e.g., 4000 to 5000 $\overline{M}_n$), to use a capping ratio of 1.8 to 2.0, and to preform part of the hard segment in the diamine/terminator solution. For example, for a total capping ratio of 1.9, the amount of diisocyanate to give a capping ratio of 1.8 would be added, as in the common practice, to the diol. The residual 0.1 mol of the diisocyanate would then be added to the 0.9 mol of extender plus terminator to form a low molecular weight amide + diamine (incipient hard segment). This hard segment would then be added to the capped diol and be incorporated in the new hard segment system together with the residual diamine extender and terminator. Obviously, some work would be required to optimize this proposed system.

Poly(urethane/urea) elastomers are usually prepared as shown in schematic **10.1**: Capping the macro soft segment with an aromatic diisocyanate (often bis-[isocyanatophenyl] methane) in a solvent such as dimethylacetamide. As discussed above, less than a 2/1 molar ratio of diisocyanate/polyglycol is commonly used to couple the macro-diol and enlarge the soft segment, but this varies with the molecular weight of the macrodiol. Higher molecular weight polydiol usually require a higher capping ratio to minimize extension of the soft segment unit and to maximize the hard segment units. Conversely, low molecular weight polydiols require lower capping ratios so that the size of the soft segment units will be increased. As a rule of thumb, the higher the molecular weight of the macro diol the more the ratio of the diisocyanate/macrodiol approaches 2/1. Capping ratios of 1.6 to 1.8 are commonly used with macrodiols of 2000 to 3000 $\overline{M}_n$. In special cases, the capping ratios in excess of 2 may be used. After capping, the macro diisocyanate is reacted with 90 to 95% of the calculated molar amount of diamine(s) for complete reaction with the residual isocyanate groups to form the urea hard segment, together with the 10 to 20% of monofunctional capping agent. This gives the almost completed elastomer. The capping groups are thermally labile secondary diamines used to inhibit premature cross-linking, excessive molecular weight growth, and gelation of the elastomer in solution. The capping stabilizes solution viscosity and allows easier processing.

In the final step, the solution is dry spun into a hot nitrogen atmosphere (some use hot air, but this chances possible explosion of the solvent/air mixture). The secondary amine end capper cleaves in hot gas atmosphere and further chain extension and build up of molecular weight occurs. Most polyurethane/urea elastomers also contain antioxidants and light stabilizers to reduce discoloration and degradation.

In wet spinning, the capped glycol is reacted in solution with less than the calculated amount of diamine for complete reaction with the residual isocyanate groups and is spun into an aqueous/solvent bath containing diamine (sometimes plus triamine). The final fiber is completed by reaction with diamine, water (and triamine) and is often cross-linked. Commercially, the dry-spinning process is most favored because finer elastomeric threads can be produced. An outline of the processes discussed is shown in Table 10.1.

Stress decay is usually caused by crystallization of the soft segment or by chain slippage due to some rearrangement of the hard segment. The stress decay is important, because if high, it implies that a garment made with the spandex will lose its holding power and will feel looser in wear vs the initial wearing. This factor, called set, results from slippage of the hard segment and/or sluggish return of the soft segment to its original random state. The slopes and stress levels of the outward and return curves of the stress-strain profile and the area between the curves (work loss) are important because

Table 10.1. Outline of Spandex Process

- $\sim 2000\overline{M}_n$ glycol ($\overline{M}_w/\overline{M}_n \sim 1.8$); 98% bifunctional; <1000 ppm water.
- Cap with aromatic diisocyanate at 80 to 100°C. Capping ratio 1.7 to 1.8. Use high-shear mixing.
- React with diamine(s)/terminator (9/1 molar) at <80°C and high-shear mixing in dimethylacctamide solvent.
- Add TiO_2, antioxidants, and other additives and dry-spin.
- Wind up after coating with finish.

they can be used to define the stretch range over which the stress level and retractive force are reasonably constant and define a fit range over which the garment will be comfortable in wear. A large work loss (area between the outward and return curves) implies heat generation and perhaps discomfort in wear.

10.1. PREPARATIONS OF BLOCK CONDENSATION ELASTOMER

268. Preparation of Poly(ethylene-co-propylene adipate)

$$\mathrm{HOCH_2CH_2OH + HOCH_2\overset{\overset{\displaystyle CH_3}{|}}{C}HOH + HO{-}\overset{\overset{\displaystyle O}{\|}}{C}{-}(CH_3)_4{-}\overset{\overset{\displaystyle O}{\|}}{C}{-}OH \longrightarrow}$$

$$\mathrm{H{+}O{\frown}O{-}CH_2{-}\overset{\overset{\displaystyle O}{\|}}{C}{-}(CH_2)_4{-}\overset{\overset{\displaystyle O}{\|}}{C}{-}OCH_2\overset{\overset{\displaystyle CH_3}{|}}{C}HO{-}\Big]_n{-}H{-}}$$

In a 500-mL three-necked flask equipped with stirrer, nitrogen inlet tube below the level of the reaction mixture, and a straight distilling head and condenser is placed 351.5 g (2.40 mol) adipic acid, 106.4 g (1.71 mol) distilled ethylene glycol, and 86.9 g (1.14 mol) distilled 1,2-propylene glycol. The reaction mixture is heated to 140°C by means of an oil bath, with stirring and nitrogen is slowly passed through the mixture until distillation of the water ceases. This may require 10 to 15 h. The temperature is then raised to 200°C, and the pressure gradually reduced to about 20 mm by means of a water aspirator. The reaction mixture is stirred, and the slow passage of nitrogen continued. Part of the excess glycols is removed during this step, and the molecular weight is determined by this heating and vacuum cycle. After a period of 23 h total, the mixture is cooled under nitrogen to give a white, waxy solid. A viscous syrup may result if insufficient glycol is removed, and the resulting polyester is of very low molecular weight.

A sample of the polyester is removed and analyzed for hydroxyl and carboxyl ends in the following manner (Makay, personal communication, 4,5). About 2 g polyester is weighed to the nearest milligram into a 100-mL round-bottom flask. To dissolve the sample, 25.0 mL of a mixture of 12 mL acetic anhydride and

500 mL dry pyridine is added by pipet. The solution is refluxed for 1 h. The condenser is then washed out by the addition of about 5 mL water through the top, and the heating is continued for 5 min. The heat is then removed and the condenser tube, and the tip is washed with 25 mL methanol. When the mixture has cooled to room temperature, it is titrated with approximately 0.5 N standard potassium hydroxide solution to a phenolphthalein end point. This is value A, (mg KOH/g polymer). Value A should never be $<65\%$ of value B. If it is, insufficient acetic anhydride was added for complete acylation.

A blank is then run on a mixture of the same volumes used above of acetic anhydride-pyridine reagent and water, which has been allowed to stand for 15 min. Methanol is added in the same amount, as in the preceding case, just before titrating. This is blank value B (mg KOH/g polymer).

Another titration is carried out on a sample of polymer of about the same size as the first dissolved in 25 mL pyridine, again with 0.5 N KOH, to a phenolphthalein end point. Here, value C (mg KOH/g polymer), gives the acid number of the polyester. The hydroxyl number of the polyester is obtained from value C + value B − value A. The method for hydroxyl number depends on a quantitative acetylation of −OH groups followed by titration of the acetic acid from the hydrolyzed, unreacted acetic anhydride plus any carboxyls in the polymer. The latter are determined separately by titration as described. Other active hydrogen compound, such as amines, interfere with −OH determination. The sum of acid and hydroxyl numbers permits a calculation of number average molecular weight:

$$(\text{acid number} + \text{hydroxyl number})/(2 \times 56.1 \times 1000) = \text{mol polymer/g}$$

$$\text{MW} = 1/\text{mol of polymer/g}$$

Because there arc two ends per polymer molecule, the total of hydroxyl and carboxyl ends equals twice the number of polymer molecules. In the polyester prepared above, the hydroxyl number should be 58 to 59 and the carboxyl number 3 to 4.

269. Preparation of Polycaprolactone Trimerized with 2,4-Toluenediisocyanate (TDI)

In a dry 1-L three-necked flask is placed 62 g (1 mol) freshly distilled ethylene glycol and 0.02 g tetraisopropyl titanate catalyst. The mixture is heated under nitrogen to 140 to 150°C, at which temperature 570 g (5 mol) epsilon-caprolactone containing 0.04 g tetraisopropyl titanate catalyst is added over a period of 60 to 90 min. The temperature gradually increases to 180 to 185°C during the lactone addition. The mixture is stirred an additional 2 h at 165 to 170°C. The polycaprolactone glycol is then cooled to 50 to 60°C and placed under vacuum at 0.05 mm Hg. A slow stream of nitrogen is allowed to pass intermittently through the glycol as its temperature is raised gradually to 95 to 100°C at 0.05 mm Hg. The distillate obtained (about 3 g) is primarily ethylene glycol.

The colorless polycaprolactone glycol solidifies to a white wax on cooling to room temperature. The glycol molecular weight calculated is 670 and determined by acetylation is 690. A mixture of 103.5 g (0.15 mol) of catalyst-free polycaprolactone glycol (690 M_n) and 17.4 g (0.10 mol) freshly distilled 2,4-tolylene diisocyanate is stirred under dry nitrogen at 90 to 100°C for 2 h. An ir spectrum of the product shows that all of the TDI has reacted (no NCO band at 4.3 μ).

270. Preparation of Polycaprolactone Glycol

In a 500-mL three-necked round bottom flask equipped with a stirrer, thermometer, and a pressure-equalizing dropping funnel with nitrogen inlet in the top is placed 8.02 g anhydrous ethylene glycol, and 16 mg dibutyltin dilaurate. The mixture is warmed to 150°C, and a mixture of 250 g caprolactone and 64 mg dibutyltin dilaurate is added dropwise. Addition is begun at a temperature of 150°C and allowed to rise to 180°C and held there until all the lactone is added, about 1.5 h. After addition is complete, the reaction is continued 2 h to ensure completion. Molecular weight should be about 2000 as determined by the acetylation method. The next step in preparing the block elastomer is to react the polyether or polyester glycol with diisocyanate. If 1 mol diisocyanate is used per mole of glycol, chain extension to high polymer is complete in one step. The next preparation illustrates this.

271. Chain Extension of Poly(ethylene-co-propylene Adipate) with Methylene Bis(4-phenylisocyanate) (1)

$$H\!-\!\left[O\!-\!CH_2CH_2\!-\!O\!-\!CH_2\!-\!\overset{O}{\overset{\|}{C}}\!-\!(CH_2)_4\!-\!\overset{O}{\overset{\|}{C}}\!-\!OCH_2\overset{CH_3}{\overset{|}{C}}HO\!-\right]_n\!-\!H$$

$$\xrightarrow{OCN-C_6H_4-CH_2-C_6H_4-NH-\overset{O}{\overset{\|}{C}}-}$$

$$\left\{-\overset{O}{\overset{\|}{C}}-NH-C_6H_4-CH_2-C_6H_4-NH-\overset{O}{\overset{\|}{C}}-\left[-OCH_2CH_2O-\overset{O}{\overset{\|}{C}}-(CH_2)_4-\overset{O}{\overset{\|}{C}}-OCH_2\overset{CH_3}{\overset{|}{C}}HO-\right]_n\right\}_m$$

In a 500-mL flask equipped with a stirrer and a nitrogen inlet tube reaching to the bottom of the flask is placed about 200 g of the polyester prepared above. The polymer should have a molecular weight of 2000 to 3000 and a ratio of hydroxyl to carboxyl numbers of at least 12. The polyester may be washed into the flask with a small quantity of methylene chloride. The temperature is raised to 120°C by means of an oil bath, with stirring and passage of a slow stream of nitrogen through the liquid ester. A quantity of solid 4-methylene bis(4-phenyl) isocyanate is added corresponding to 96% of the theoretical amount based on the number of moles of polyester used, as determined by the end group analysis described above. The mixture is stirred vigorously at 120°C for 40 min. After cooling under nitrogen, the polymer obtained is an off-white solid. The inherent viscosity of the polymer at 0.5% concentration is 1.0 to 1.5 in dimethylformamide (DMF). Films dry cast from DMF are elastomeric.

A curable polyester-urethane can be prepared by thorough mixing of with additional isocyanate on a rubber mill. When the polyester used has a molecular weight of 2000 and 96% of the theoretical diisocyanate is used in the chain extension reaction, 100 g of the final polymer can be mixed with 5.5 g diisocyanate, molded or pressed into a desired shape, and cured at 150°C for 70 min to give a strong and useful elastomer.

As an alternative to chain extending a hydroxy-ended polyester segment with a diisocyanate, the initial low polymer may be terminated with a diisocyanate at each end of the molecule (2, 3). A second difunctional molecule is used in the chain-extending step. When water is used, carbon dioxide is evolved and a urea is the connecting linkage of the final product.

272. Chain Extension of Isocyanate-Terminated Poly(ethylene-co-propylene Adipate) with Water (2)

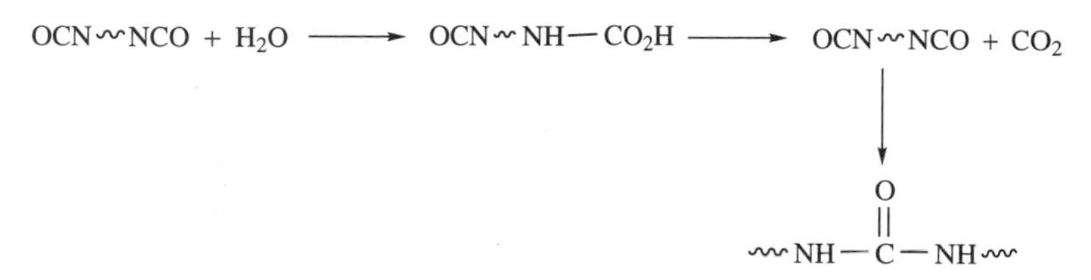

A sample of poly(ethylene-propylene adipate) prepared as described is analyzed for the number of hydroxyl end groups. The following directions are for use with a copolyester with an average molecular weight of 2000 to 3000 and a hydroxyl number of 55 to 60. A total of 200 g of the polyester is placed in a 1-L three-necked flask equipped with stirrer, condenser with drying tube, and a nitrogen inlet. To the flask, flushed with nitrogen, is added 2 molar equivalents of methylene bis(4-phenylisocyanate) based on the number of moles of polyester used, derived from its observed molecular weight. The mixture is stirred vigorously at 80°C under nitrogen for 3 h. It is then cooled to room temperature and 500 mL dry DMF is added. The mixture is brought into solution as rapidly as possible by stirring at room temperature. Another 100 mL DMF, containing a quantity of water equivalent in moles to one half the number of moles of diisocyanate used is added. The over-all stoichiometry is: 1 mol polyester + 2 mol diisocyanate + 1 mol water. The resulting viscous solutions may be cast and dried in a vacuum oven at 50°C to give elastomeric films that may be somewhat tacky. The final film may become insoluble after heating.

A low molecular weight poly(propylene oxide) terminated with hydroxyls may be used to prepare foamed polymers. The polyester glycol is reacted with excess diisocyanate and then treated with water. The foaming agent is the carbon dioxide liberated when the water reacts with free isocyanate groups in the prepolymer mixture (4). The foam effect results from trapped gas in the polymeric mass. Reactions of this variety are used to make some of the widely employed urethane foam products.

273. Preparation of a Resilient Polyurethane Foam from Poly(propylene oxide) Glycol and 2,4-Toluenediisocyanate (5)

In this preparation, amounts specified correspond to the use of a poly(propylene oxide) glycol of molecular weight 2000, a hydroxyl number of 56.1, and an equivalent weight of 1000. If the poly(propylene oxide) glycol actually used has a different hydroxyl number of molecular weight, suitable corrections must be made in the amounts of reactants that follow.

A 500-mL resin kettle equipped with a stirrer, condenser with drying tube, thermometer, and gas inlet with nitrogen flush, and in it is placed 200 g poly(propylene oxide) glycol. Water is added, if necessary, to bring the total water content to 0.8 g. If the original water content of the polyether glycol is not known, and a water analysis cannot be conveniently carried out, the polyether glycol can be dried by heating to 160°C under nitrogen at a pressure of 1 mm or less for a period of 3 to 4 h. To the dried macroglycol is then added 0.8 g water. The mixture is then stirred and heated for 30 min at 30 to 35°C with a heating mantle. This and all subsequent reactions must be carried out under nitrogen.

With continued stirring, but with removal of the heat source, 29.4 g 2,4-toluenediisocyanate is added. This is an amount calculated to give an NCO/OH ratio of 1.0. The reaction is slightly exothermic. After 30 min, the temperature is raised at the rate of about 2°C/min to 117 to 123°C, at which it is maintained for 90 min. The mixture is then cooled to 80°C. If the molecular weight and hydroxyl number of the macro-glycol used are not as stated for this preparation, the calculations for the amount of diisocyanate (DI) used are given by:

$$\begin{aligned} \text{A} &= \text{weight of DI for OH reaction} \\ &= (\text{Weight of diol/Equivalent weight of diol}) \\ &\quad \times \text{Equivolent weight of DI} \times 1.25 \\ \text{B} &= \text{Weight of DI for reaction with water present} \\ &= (\text{Weight of water/Equivalent weight of water}) \\ &\quad \times \text{Equivalent weight of DI} \end{aligned}$$

The sum of A and B is the required amount of diisocyanate. For the macrodiol stipulated above, this is 29.4 g. (The equivalent weight of the diisocyanate is 87; of water, 9.) At this point, a second addition of tolylenediisocyanate is necessary to bring the total isocyanate content of the mixture to the desired level of 9.5%.

to do this, the isocyanate content of the first reaction product must be determined as follows.

Weigh about 0.5 g polymer from the resin kettle to the nearest milligram and transfer to a dry 250-mL. Ehrlenmeyer flask. Add 25 mL dry toluene and stir magnetically for 5 min to dissolve the polymer. Add 25.0 mL of 0.1 N butylamine in toluene (prepared as a primary standard from distilled butylamine and dry toluene) and continue stirring for 15 min. Add 100 mL isopropyl alcohol and four to six drops of bromophenol blue indicator solution. Titrate with 0.1 N HCl to a yellow-green end point. Run a blank, including all reagents but omitting the sample.

$$\%\text{NCO} = [(\text{mL HCl for blank} - \text{mL HCl for sample}) \times (\text{Normality of HCl}) \times 4.202]/\text{Weight of sample}$$

In this example, the NCO content of the polymer at the present stage should be about 1.5 to 1.8%. The amount of diisocyanate needed to bring the total isocyanate content to 9.5% would be about 45 g and is obtained from the expression:

$$\text{Weight of DI to be added} = [(z - y)/(48.3 - z)] \times (\text{Weight of polymer})$$

where $z = \%$ NCO desired (9.5 in this experiment) and $y = \%$ NCO in polymer, determined as above. Weight of polymer in this experiment, at the end of the reaction after the first amount of DI added, is about 229.4 g. The amount of DI so determined is added to the reaction mixture at 80°C. Stirring is continued for 30 min while the temperature is allowed to fall to 40°C. The product at this point, usually termed the prepolymer, is ready for the subsequent foaming step. The prepolymer may be safely stored under nitrogen at room temperature for 6 months or more.

Foaming can be carried out in a large beaker or other suitable container. It is suggested that a small amount be tried initially to estimate the volume of the vessel needed to contain the foamed product. A blend of the following relative amounts is made at room temperature under nitrogen: 100 parts prepolymer with 9.5 % NCO + 0.5 part silicone oil, 50 centistokes (as Dow Corning DC-200) + 1.0 part *N*-methylmorpholine + 0.3 part triethylamine. To this mixture is added 2.25 parts of water (110% of theory). A foam results with a density of about 2.5 lb/ft^3. The following two preparations illustrate the use of a diamine in chain extension.

274. Preparation of Elastomer from Poly(caprolactam Trimer) Glycol Capped with Methylene Bis(4-phenyl Isocyanate) and Extended with Ethylenediamine To 120.9 g (0.05 mol) of the trimer glycol descibed above (cooled to 40 to 45°C) is added 25.0 g (0.1 mol) methylene bis(4-phenylisocyanate). The mixture is stirred under nitrogen at 90 to 95°C for 1 h, then cooled to 75 to 80°C and held for 2 to 3 h. The theoretical isocyanate (NCO) analysis for the methylene bis(4-phenyl-isocyanate) capped trimer is 2.88%. The capped trimer is viscous at room temperature but flows readily at 75 to 80°C.

To 30 g (0.0089 mol) of the above capped trimer is added to 130 mL dry dimethylacetamide at room temperature. To this dilute prepolymer solution is added gradually with stirring a 1 *M* ethylenediamine solution in dimethylacetamide (DMAc). A viscous solution (about 2000 p) is obtained after adding 8 to 9 mL of the extender solution; further addition may cause gelation. The viscous polymer solution obtained is degassed at 75 to 80°C for 2 h in a steam-heated oven. Clear, tough films may be cast in the usual way.

275. Preparation of Elastomer from Polycaprolactone (2000) Capped with Methylene Bis(4-phenylisocyanate) and Extended with Hydrazine

To cap the polycaprolactone glycol, 37 g is mixed with 9 g of methylene bis(4-phenylisocyanate), and the mixture is stirred under nitrogen at 80 to 90°C for 1 h. The resulting viscous material is diluted with 170 mL DMAc. To this solution, at room temperature, is slowly added a stock solution of hydrazine in DMAc. This solution is prepared by adding 3.3 g hydrazine (**Cacinogen!**) to 100 mL DMAc. The exact amount cannot be specified, but will be in the range of 14 to 15 mL. A good guide is solution viscosity. Addition should stop when the solution viscosity reaches about 2000 p; if too much hydrazine is added, a gel will form. At 2000 p, clear tough films can be cast in the usual way.

The majority of commercial activity in polyurethanes is in flexible and rigid foams. However, another important phase of polyurethane technology is the solid, cast elastomer. The chemistry is essentially the same as that for other urethane products. Of course, no blowing agent is added or generated. The following example is based on a polyester soft segment from ε-caprolactone and methyl-ε-caprolactone. Use of mixed lactones reduces the tendency of the product to crystallize.

276. Preparation of a Solid, Cast Polyurethane Elastomer from a Polyester Based on Caprolactone (6)

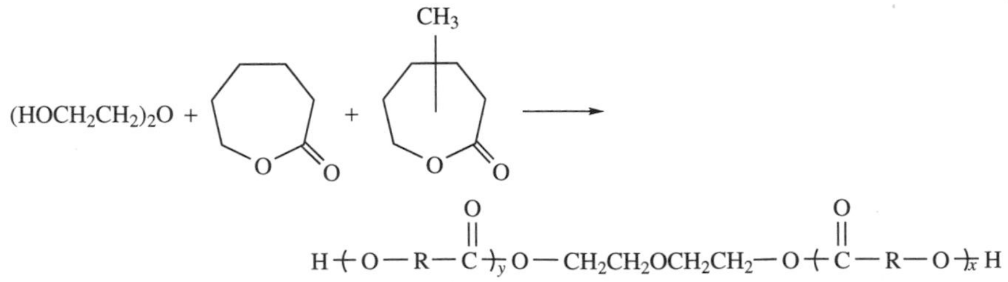

$$(HOCH_2CH_2)_2O + \text{ε-caprolactone} + \text{methyl-ε-caprolactone} \longrightarrow$$

$$H\!-\!\!\left[O-R-\overset{\displaystyle O}{\overset{\|}{C}}\right]_y\!\!O-CH_2CH_2OCH_2CH_2-O\!\left[\overset{\displaystyle O}{\overset{\|}{C}}-R-O\right]_x\!\!H$$

I

$$R = \left[CH_2\right]_5 \text{ and } \left[CH_2\right]_m\overset{\displaystyle CH_3}{\overset{|}{C}H}\left[CH_2\right]_n \; ; \; m + n = 4$$

$$\mathrm{I} + \mathrm{OCN{-}C_6H_4{-}NCO} \longrightarrow \mathrm{OCN{-}C_6H_4{-}NH\overset{O}{\overset{\|}{C}}{-}\!\left(O{-}R{-}\overset{O}{\overset{\|}{C}}\right)_{y}}$$

$$\mathrm{{-}OCH_2CH_2OCH_2CH_2O{-}\!\left(\overset{O}{\overset{\|}{C}}{-}R{-}O\right)_{x}\overset{O}{\overset{\|}{C}}NH{-}C_6H_4{-}NCO}$$

II

$$\mathrm{II} + \mathrm{HO(CH_2)_4OH} \longrightarrow \left\{\left[\mathrm{{-}O\overset{O}{\overset{\|}{C}}NH{-}C_6H_4{-}NH\overset{O}{\overset{\|}{C}}{-}\!\left(O{-}R{-}\overset{O}{\overset{\|}{C}}\right)_{y}}\right.\right.$$

$$\left.\left.\mathrm{{-}OCH_2CH_2OCH_2CH_2O{-}\!\left(\overset{O}{\overset{\|}{C}}{-}R{-}O\right)_{x}\overset{O}{\overset{\|}{C}}NH{-}C_6H_4{-}NH\overset{O}{\overset{\|}{C}}}\right]\mathrm{{-}O{-}\!\left(CH_2\right)_{4}{-}}\right\}$$

A 4-L resin kettle is charged with 191 g (1.80 mol) diethylene glycol, 1110 g (8.67 mol) methyl-ε-caprolactone, 2300 g (20.18 mol) ε-caprolactone, and 200 mL benzene. The flask is fitted with an air stirrer, jacketed thermocouple, condenser, Dean-Stark water trap, and nitrogen inlet tube and is heated by a mantle. The mixture is stirred and heated to reflux under a nitrogen stream sufficient to produce an inert blanket. Residual water is removed by azeotropic distillation, thus reducing the possibility that it will cleave the lactone ring and form hydroxycaproic acid. After removal of the water, the benzene is distilled off until a reaction temperature of 190°C is reached. This allows the polymerization reaction to proceed under anhydrous conditions. Benzoyl chloride (2 g) is then added and the heating continued at 190°C for 8 to 10 h until an acid number <1.0 is reached and the refractive index of the batch becomes constant. (Acid number determinations can be made only near the end of the polymerization reaction, as the alcoholic potassium hydroxide solution normally used tends to open the lactone ring giving an unstable end point). The residual benzene and unreacted lactone monomers are then stripped out by vacuum distillation. The finished ester is cooled. The hydroxyl number is 54.5 to 57.5, the acid number is 1.0, and the molecular weight is about 2000, the viscosity is about 3000 cp at 25°C.

A 4-L resin flask is charged with 600 g (3.75 mol) *p*-phenylene diisocyanate and 2 g benzoyl chloride. The flask is fitted with a stirrer, nitrogen inlet tube, reflux condenser, and heating mantle. The reactants are blanketed with a nitrogen stream and the diisocyanate is melted and held at 135°C. The reaction mixture is agitated, and 3380 g of the above polyester, heated to 80 to 90°C, is added rapidly. The heating mantle is dropped and the temperature is kept below 135°C. Rapid agitation is continued until a reaction temperature of 85 to 92°C is reached. The finished prepolymer is then transferred to a wide-mouth jar, and the top of the container flushed with dry nitrogen and sealed. This prepolymer is now ready for chain extension with a suitable curing agent. The viscosity of the prepolymer is about 28,000 cp at 25°C.

With both components at room temperature, sufficient dry 1,4-butanediol is added to a weighted amount of the above prepolymer so that the OH/NCO is 0.85. The NCO content of the prepolymer can be determined as previously described. The mixture is blended by hand mixing with a spatula and then thoroughly stirred with an air-driven stirrer. Where a casting formulation of extremely low viscosity is required, the prepolymer may be heated to 50 to 65°C to facilitate rapid mixing with little loss of working life noted. The mixture of the prepolymer and diol is degassed in a vacuum of 1 to 2 mm, then poured into open-faced slab and button molds, which are treated with a mold-release agent and preheated to 80 to 90°C. The liquid resin is again degassed by vacuum, then cured in a forced draft oven at 57°C for 8 h.

The cured elastomer is cross-linked through reaction of the NCO in excess of that required for the curing reaction with diol; the former react with in-chain urethane NH to form allophanate links. Typical properties are tensile strength, 3700 psi; elongation 600%; compression set, 46%; and tear strength, 269 lb/in.

Elastomers can be produced from macroglycols capped with 2 mol diisocyanate by methods not involving the addition of difunctional materials such as water or diamine. For example, the isocyanates alone can be coupled as carbodiimides using phospholine oxide as catalyst (7).

277. Preparation of Carbodiimide Elastomer Based on Poly(tetramethylene Ether) Glycol (7)

$$OCN-C_6H_4-CH_2-C_6H_4-NHCO_2(CH_2CH_2CH_2CH_2O)_n-\overset{O}{\overset{\|}{C}}NH-C_6H_4-CH_2-C_6H_4-NCO \longrightarrow$$

$$\left[-C_6H_4-CH_2-C_6H_4-NH\overset{O}{\overset{\|}{C}}O(CH_2CH_2CH_2CH_2O)_n-\overset{O}{\overset{\|}{C}}NH-C_6H_4-CH_2-C_6H_4-N{=}C{=}N-\right]_m$$

A mixture of 33.7 g poly(tetramethylene ether) glycol with an average molecular weight of about 2000 and 8.7 g methylene bis(p-phenyl isocyanate) is heated on a steam bath for 90 min. The capped glycol is diluted with 200 mL xylene containing 0.1 g of the preferred phospholine oxide catalyst. The mixture is refluxed and stirred for 1 h to give a viscous solution. A clear, tough, snappy elastomeric film results from casting and drying on a glass plate. The material can be elongated 600% and has good recovery from extension despite the fact it has no tie points from hydrogen bonding. It is probable that the tie points are areas where the carbodiimide linkages have cross-linked.

Other non-hydrogen-bonded elastomers have been described. One example is when the system is held together (i.e., chain slippage is reduced) by a very bulky hard segment (see prep. 278).

278. Preparation of Polycarbonate Elastomer (8)

$$y\,HO-C_6H_2X_2-R-C_6H_2X_2-OH + HO(CH_2CH_2CH_2CH_2O)_nH \xrightarrow[\text{Pyridine}]{COCl_2} \left[\left(O-C_6H_2X_2-R-C_6H_2X_2-O-\overset{O}{\overset{\|}{C}}-O\right)_y(-CH_2CH_2CH_2CH_2O)_n-\overset{O}{\overset{\|}{C}}-\right] + \text{Pyridine}\cdot HCl$$

In a well-vented hood, a three-necked flask is fitted with a stirrer, thermometer, and glass dropping funnel. The stirrer control box is wired with a milliammeter that indicates the amount of torque on the stirrer. This gives a relative measure of the viscosity of the elastomer. To the flask are added 16.8 g (0.004 mol) poly(tetramethylene ether) glycol of 4200 MW, 9.1 g (0.0218 mol) 4,4′-(2-norbornylidene) bis(2,6-dichlorophenol), 105 mL methylene chloride, and 25 mL pyridine. A 10% phosgene (**Hazard!**) stock solution in methylene chloride is added to the rapidly stirred reaction mixture from a graduated dropping funnel. The temperature is maintained at 20 to 30°C by means of a water bath. Thickening of the elastomeric solution is indicated by the milliammeter. Slow addition of the phosgene solution is continued until the meter shows no further buildup; usually the mixture has wrapped around the stirrer. The volume of phosgene used is about 29 mL (10 to 30 molar excess of phosgene is usually required). Water is added and the elastomer is diluted with methylene chloride. The solution is stirred with dilute hydrochloric acid and then washed with a large quantity of water to remove pyridine hydrochloride. After the aqueous layer is separated, 1.0 g antioxident (an alkylated phenol) is added to the solution. The elastomer is precipitated in 50/50 acetone/methanol solution and dried overnight in a vacuum oven. The inherent viscosity of the elastomer in chloroform is about 3 (0.5%).

279. Preparation of Norbornylidene Bis(2,6-dichlorophenol) Used in Preparation 278

OH $)_2$ $\xrightarrow{4Cl_2}$ Cl OH Cl $)_2$

A 5-L three-necked flask is fitted with a nitrogen sweep, stirrer, thermometer, and a water-cooled condenser. The top of the condenser is connected to a water trap to remove any HCl fumes that come off. To the flask are added 385 g (3.5 mol) distilled 2-norbornanone (bp 169 to 171°C; mp 80 to 90°C) 1320 g (14 mol) phenol, 2220 mL concentrated HCl, and 18 mL 3-mercaptopropionic acid. The mixture is stirred at 50°C for 7 h. During the reaction, the product separates as pink balls about the size of small peas. The mixture is allowed to stand overnight at room temperature, and the aqueous phase is decanted. The product is washed with water several times on a filter; then, while still moist, it is dissolved in 2500 mL hot acetic acid, decolorized with carbon and filtered with the aid of filter aid. The filtrate is then heated to 100°C, and hot water (about 90°C) is added until crystallization begins. This requires about 2700 mL water. After crystallization is complete and the mixture has cooled, the bisphenol is collected and washed with 50% aqueous acetic acid, and then with water, and finally dried. Yield is about 83 calculated as the hydrate. If the bisphenol has a pink cast it can be recrystallized from hot acetic acid by addition of an equal amount of boiled water. The bisphenol hydrate as small colorless crystals melts at 177 to 179°C. The unsolvated bisphenol is obtained by dissolving in hot xylene-acetone, removing the water as an azeotrope, and allowing the bisphenol to crystallize (mp 199 to 200°C).

A 2-L three-necked flask is fitted with a stirrer, thermometer, gas inlet tube, and condenser. A glass tube extending 3 to 4 in below the surface of the water in a small flask is connected to a trap, which is connected to the top of the condenser. A lecture bottle (small cylinder) of chlorine on a balance is connected to a bubbler containing sulfuric acid, which is attached to a trap connected to the gas inlet tube.

Then 149 g (0.50 mol 4,4′-(2-norbornylidene)diphenol hydrate and 800 mL acetic acid are placed in the flask. While the mixture is stirred, it is heated to 40°C for about 10 min. The heat is then removed, and chlorine is passed through the inlet at such a rate that no bubbles are emitted under the water at the other end of the system. Because the reaction is slightly exothermic, the mixture is cooled occasionally with a water bath keeping the temperature close to 45°C. During the chlorine addition, all of the bisphenol dissolves. In the later stages, the product starts to crystallize. The total amount of chlorine added is 149 g (2.1 mol). Stirring is continued while the mixture cools to room temperature. The product is collected and washed thoroughly once with cool acetic acid. After drying, the chlorinated bisphenol weighs about 171 g (82%). It consists as small white crystals (mp 182 to 184°C) and can be recrystallized from 1,2-dichloroethane (**Toxic!**) (mp 183 to 184°C).

Although melt processible elastomers are unusual from isocyanate-based polymers they can also be prepared from preformed hard-segments via ester or amide linkages, because of the thermal instability of the urea and or urethane linkages. The former are preferred because macrodiols are easier to prepare with good bifunctionality vs. macrodiamines.

280. Preparation of Melt-Processible Polyester Elastomer from Alicyclic Hard-Segment Components (9)

In a 350-mL glass resin kettle equipped with a mechanical stirrer, a nitrogen inlet and a distillation column are placed 10.18 g (0.051 mol) dimethyl trans-cyclohexane-1,4-dicarboxylate, 4.23 g (0.036 mol), 50% trans cyclohexane-1,4-diol, and 28.84 g (0.0144 mol) poly(tetramethylene glycol) (2000 $\overline{M}_n$). Then 150 mg AO-330 antioxidant and 1.5 mL of a solution of tetrabutyl titanate catalyst in ethylene glycol are added. The kettle is purged with nitrogen and placed in a Woods metal bath heated to and maintained at 240°C. The ingredients are stirred rapidly for 45 min; then, over the course of 90 min, the pressure inside the kettle is reduced to 0.5 to 0.25 mm Hg. Stirring is continued for 1 to 4 h until a viscous melt is obtained. The vacuum is released with nitrogen, and the molten polymer is removed from the kettle and processed into strip and film. The resulting elastomer can be extruded at 220°C. The polymer has an inherent viscosity number of 1.64 dL/g at 0.5% concentration in *m*-cresol. Elongation at break is 645%.

Using the same procedure, another polymer is made from 9.33 g of the trans ester, 4.33 g cyclohexane diol, and 29.42 g poly(tetramethylene glycol) (2900 M_n). The resulting polymer has an inherent viscosity of 1.65 dL/g (0.5%) and an elongation to break of 840%.

281. *Preparation of a Polyester-Based Elastomer from a Performed Hard Segment (10)*

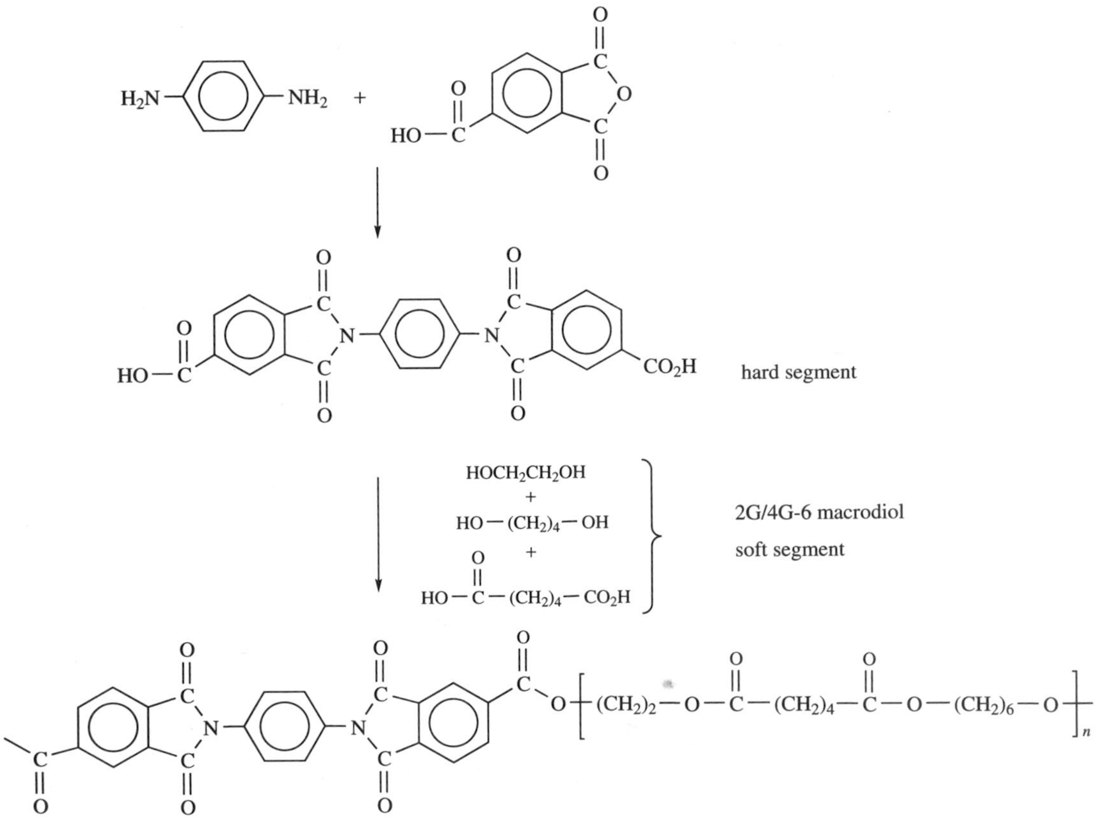

hard segment

$HOCH_2CH_2OH$ + $HO-(CH_2)_4-OH$ + $HO-C(=O)-(CH_2)_4-CO_2H$

2G/4G-6 macrodiol

soft segment

$$-\!\!\left[(CH_2)_2-O-\overset{O}{\overset{\|}{C}}-(CH_2)_4-\overset{O}{\overset{\|}{C}}-O-(CH_2)_6-O\right]_n\!\!-$$

The hard segment, *p*-phenylene *N,N′* bis-(trimellitic acid imide) is prepared by adding 54 g *p*-phenylenediamine in 200 mL anhydrous DMF to a refluxing solution of 192 g trimellitic anhydride in 500 mL DMF. After refluxing for 2 h and cooling, the dispersion is filtered and dried in a vacuum oven at 110°C. The yield is 191 g. A portion is crystalized from DMF and has a DSC melt temperature of 455°C. Several esters are made from the dibasic acid for further characterization. The dibutyl ester melts at 275°C and has an nuclear magnetic resonance (NMR) spectrum consistent with the attributed structure.

Preparation of elastomers is as follows: The polymerization is run in a 350-mL resin-kettle equipped with a mechanical stirrer, a nitrogen inlet, a Woods metal heating bath, and a vacuum distillation column To the resin kettle is added 18.44 g (0.0404 mol) *p*-phenylene *N,N′*-bis-trimellitimide, 2.72 g (0.0238 mol) *trans*-diaminocyclohexane, 47.48 g (0.325 mol) adipic acid, 20.0 g (0.3226 mol) ethylene glycol, and 10.89 g (0.1210 mol) of 1,4-butanediol. Also added are 0.3 g Ethyl Antioxidant 330 and 4 mL of a 1% solution of antimony trioxide in ethylene glycol. The kettle is purged with nitrogen and while maintaining the flow of nitrogen, the kettle is placed in the Woods metal bath at 220°C. The ingredients are stirred rapidly for 45 min, then the bath temperature is raised to and maintained at 280 to 285°C with rapid stirring for 1 to 2 h until a homogeneous viscous melt is obtained. The temperature is then lowered to 270°C, and the pressure is reduced to 0.5 to 0.25 mm Hg. over the course of 1.5 h. Stirring is continued for another 30 min. The vacuum is then released with nitrogen, and the viscous melt is removed from the kettle. Samples of the polymer are extruded at 200 to 210°C, giving elastomeric strips. The yield of polymer is 62.8 g, and the inherent viscosity is 0.9 in *m*-cresol at 0.5% concentration at 30°C. The strips can be elongated about 600%, and the break strength is 0.23 gpd based on initial denier.

REFERENCES

1. U.S. Pat. 2625535 (Jan. 13, 1953), T. G. Mastin and N. V. Seeger to Wingfoot Corp.
2. U.S. Pat. 2755266 (July 1956), W. Brenschede to Bayer AG.
3. U.S. Pat. 2621166 (Dec. 9, 1952), F. W. Schmidt to Bayer AG.
4. C. L. Ogg, W. L. Porter, and C. O. Willits, *Ind. Eng. Chem. Anal. Educ.*, **17**, 394.
5. C. M. Barringer, *Elastomer Chemistry Department Bulletin HR-26*, DuPont, *Wilmington*, DC, 1958. T. E. MacKay, personal communication.
6. C. H. Smith, *Ind. Chem. Prod. Res. Develop.*, **4**, 9 (1965).
7. T. W. Campbell and K. C. Smeltz, *J. Org. Chem.*, **28**, 2069 (1963).
8. K. H. Perry, W. J. Jackson Jr., and J. R. Caldwell, *J. Appl. Poly. Sci.*, **9**, 3451 (1963).
9. U.S. Pat. 4810770 (March 7, 1989), G. Figuly to DuPont.
10. U.S. Pat. 4731435 (March 15, 1988), G. Figuly and R. Greene to DuPont.

11

SYNTHETIC RESINS AND COMPOSITES

11.1. RESINS

The term *resin* had its original use in connection with certain naturally occurring materials, obtained in most cases from coniferous trees, which found use as hard, protective coatings when solutions of these materials were allowed to dry in air. Today the term *resin* covers a multitude of polymer types, including the classical phenol-formaldehyde condensates, epoxy resins, polystyrene, methyl methacrylate, and a host of others. The term is widely applied to meltable linear polymers, but mostly to thermoset (cross-linkable) polymers that are used in molding, casting, and extrusion operations.

In composites, thermosettable resins (mostly epoxy types) have a distinct advantage over linear polymers (such as the aromatic polyether-type, nylons, and polyesters) in that, as used initially, they are low molecular weight resins with low viscosity that can efficiently coat the support fiber (glass, carbon) free of voids, whereas the linear resins have high viscosity and are inefficient in providing void-free structures.

11.1.1. Cross-Linked Polyesters

Polyesters that have been rendered insoluble and infusible by cross-linking are commercially important types of polymeric materials. Cross-linking may be accomplished in several ways. One of these is to start with an unsaturated polyester made by conventional esterification of a glycol with an unsaturate such as maleic acid. The double bonds in the polyester can then be used as sites for copolymerization with vinyl monomers, such as styrene. The resulting polymer then is a hybrid of a condensation and vinyl type polymer. The final product is complex in structure.

Usually, the linear unsaturated polyester is prepared and mixed with the vinyl monomer to give a viscous solution, which is treated with free-radical catalyst to initiate the cross-linking polymerization step. This latter step is generally carried out after the polymer solution has been placed in the casting, laminating, or other operational process from which the desired product has to be obtained. The possible combinations of unsaturated polyester and vinyl monomer are great. The most widely used unsaturated acids are maleic and fumaric. Ethylene, propylene, and diethylene glycols are widely used in the ester polymerization. The vinyl component is often styrene, dially phthalate, triallyl cyanurate, and diallyl diglycol carbonate.

282. Preparation of a Polyester Resin Based on Poly(oxydiethylene Maleate) and Styrene (1)

$$HOCH_2CH_2OCH_2CH_2OH + \text{(maleic anhydride)} \longrightarrow \left[OCH_2CH_2OCH_2CH_2O-\overset{O}{\overset{\|}{C}}-CH{=}CH-\overset{O}{\overset{\|}{C}} \right] \quad \text{(I)}$$

$$\text{(I)} + CH_2{=}CH-Ph \longrightarrow \left[OCH_2CH_2OCH_2CH_2O-\overset{O}{\overset{\|}{C}}-\overset{|}{C}H-\underset{CH_2\underset{|}{C}H-Ph}{\underset{|}{C}H}-\overset{O}{\overset{\|}{C}} \right]$$

A 1-L four-necked flask is equipped with a stirrer, siphon, nitrogen inlet, and thermometer, all reaching below the surface of the solution and a side arm with a condenser set for distillation. In the flask is placed 233.4 g (2.2 mol) diethylene glycol, which is heated to 80°C by means of a Glas-Col heater while nitrogen is passed through in a slow stream, and stirring is begun. Then, 196.1 g (2.0 mol) maleic anhydride is added. The temperature is raised to 150°C over 1-h, then to 190°C over 4 h. An exothermic reaction occurs about 100°C, and the heat should be removed until the reaction subsides. The temperature is kept at 190°C for 1 h, and a vacuum of 100 to 200 mm is applied. The temperature is then lowered to, and kept at, 170°C until the acid number of a sample of the polyester removed through the siphon is 50 or less. This requires about 1 h. The acid number is determined as described in chapter 10 using 75 mL acetone as solvent. The vacuum is removed and the reaction product allowed to cool to 100°C under nitrogen. About 0.02 g hydroquinone or *t*-butylcatechol is added as an inhibitor. At 100°C the liquid, slightly yellow polyester, is poured with good stirring into a sufficient quantity of styrene at 25°C to give a 70% solution of the polyester. The solution is relatively stable at room temperature when stored in nitrogen in a brown bottle.

Copolymerization of the solution can be effected by adding 4.0 g benzoyl peroxide to 200 g the polyester-styrene mixture. The mixture forms a hard, tough solid in about 2 h at room temperature. Castings can be obtained if the mixture is poured into a suitable mold or container and allowed to set up.

283. Preparation of a Polyester Resin Based on Poly(ethylene-oxydiethylene Maleate-phthalate-adipate) and Styrene (1)

In the same apparatus described in the preceding preparation, but using a 2-L flask, are placed 170 g (2.75 mol) ethylene glycol and 292 g (2.75 mol) diethylene glycol. The temperature is raised to 80°C and 343 g (3.5 mol) maleic anhydride, 111 g (0.75 mol) phthalic anhydride, and 109 g (0.75 mol) adipic acid are added. The temperature is raised to 150°C over 1 h and then increased at a rate of 10°C/h to 210°C. It is maintained at this level until the acid number of a sample of the polyester is 60 or less. A vacuum of about 100 mm is applied, and the temperature is dropped to 180°C. When a sample of the polyester has an acid number of 20 to 30, the product is allowed to cool to room temperature under nitrogen and a trace of *t*-butyl catechol is added as an inhibitor.

A solution of the polyester is made in styrene, containing a trace of inhibitor, in the proportion of 100 g polyester to 43 g styrene. A casting can be made in the following way: 0.5 g of a 6% commercial solution solution of cobalt naphthenate is added to 100 g of the polyester-styrene varnish and carefully mixed to avoid the formation of a large number of air bubbles. Then, 1.5 g of a 60% solution of methyl ethyl ketone hydroperoxide in dimethyl phthalate (Lupersol DDM) is dispersed carefully and thoroughly into the solution. The solution will cure in a mold at room temperature in about 2 h to a hard, tough casting.

11.1.2. Alkyd Resins

The term *alkyd*, a blend of the first part of the word alcohol and the last of acid, is generally applied to the polyesters from the reaction of alcohols and acids where the total functionality is capable of causing cross-linking directly. Such a combination would be glycerol with phthalic anhydride, a reaction product sometimes termed a glyptal resin. Fundamental studies (2) have shown that gelation occurs at 75% total esterification.

284. Preparation of Poly(glyceryl phthalate) (2)

O + $HOCH_2CHOHCH_2OH \longrightarrow$

C—OCH_2CHCH_2O—C C—OCH_2CHCH_2O

C=O

In a 600-mL beaker immersed in a silicon oil or Woods metal bath and equipped with a thermometer and a stirrer are placed 148.1 g (1.0 mol) phthalic anhydride and 61.4 g (0.67 mol) glycerol. The mixture is stirred and the temperature raised to and maintained at 200°C for 1.5 h. The acid value should be 127 to 132, determined in acetone. At this point, the product is still soluble in acetic acid, acetone, and others. Continued heating at 200°C for about 15 min causes the mixture to set to an immoble gel.

The possible combinations of glycerol with other dibasic acids is evident. Maleic acid is often substituted in whole or in part for phthalic acid and pentaerythritol for glycerol. A widely used modification of the basic polymerization technique is the addition of an air drying unsaturated fatty acid or oil to the glycerol-phthalic anhydride such that the addend becomes incorporated in the polyester. Such polymers are soluble in the hydrocarbon or ester solvents used in the paint and varnish industries. These drying oil-modified alkyds are extremely important in surface coating uses.

The following is an example of an oil-modified alkyd resin. Free fatty acids are used as the modifying agent. The unsaturated acids are esterified by the glycerol and become part of the poly(glyceryl phthalate) molecule.

285. Preparation of a Drying Oil-Modified Poly(glyceryl Phthalate) Alkyd by the Fatty Acid Process (3)

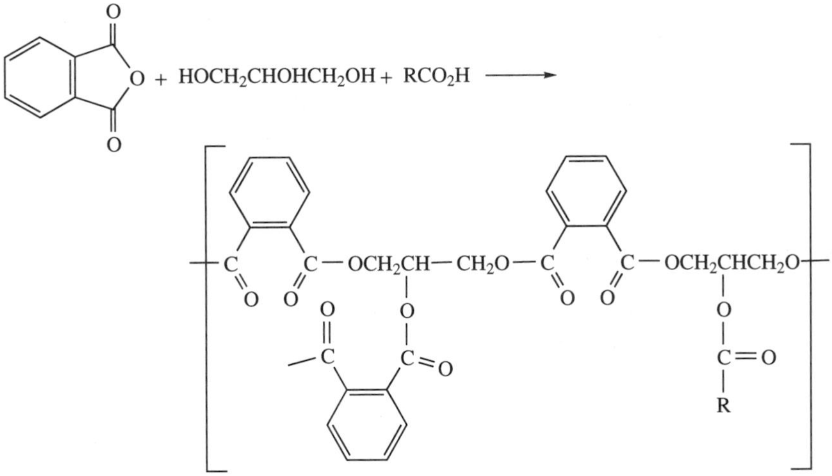

$$\text{(phthalic anhydride)} + HOCH_2CHOHCH_2OH + RCO_2H \longrightarrow$$

where $R = CH_3(CH_2)_4CH{=}CHCH_2CH{=}CH(CH_2)_7$-, and related unsaturates.

In a 1-L. beaker immersed in a silicon oil bath and equipped with a stirrer is placed 206 g phthalic anhydride An excess of the anhydride is used to compensate for an anticipated 15% loss by sublimation. The anhydride is melted by raising the bath temperature to 130 to 135°C and 200 g linseed oil fatty acid is added. The mixture is stirred and heated at 135 to 140°C until miscible. Then 92 g glycerol is added to the solution, and the reaction temperature raised at 1°C/min until 240°C is reached. This temperature is maintained for 10 to 15 min. Water is evolved during heating and some phthalic anhydride may sublime; for this reason, the reaction should be run in a hood. The product, when cooled, is a clear solid that is soluble in butyl acetate/toluene (75/25 v/v). A 50% solution of the resin in this solvent mixture may be cast onto a glass plate, smoothed out with a doctors knife or glass rod, and heated at 150°C for 2 h. A hard, tough, cross-linked coating results. If cobalt naphthenate drier (0.2 g/100 mL resin solution) is added, a film can be dried at room temperature.

11.1.3. Allyl Resins

Allyl resins are formed by the polymerization of allyl esters of di- and higher carboxylic acids; diallyl phthalate is the most widely employed ester in this technology, but others are diallyl isophthalate, triallyl cyanurate, and diethylene glycol bis(allyl carbonate) (4) The monomer is usually polymerized to a limited degree of conversion to give a relatively low molecular weight thermoplastic solid containing a large amount of residual unsaturation. On being mixed with a free-radical catalyst and raised to an elevated temperature, polymerization occurs through the remaining allyl groups to give a cross-linked resin. The polymers are noted for good heat and chemical stability and excellent electrical properties. The diethylene glycol bis(allyl carbonate) resin is noted for outstanding optical clarity, akin to poly(methyl methacrylate) but with greater surface hardness.

286. Preparation of an Allyl Resin from Diallyl Phthalate (5)

$C_6H_4(CO_2CH_2CH{=}CH_3)_2 \longrightarrow \text{[}CH_2{-}CH(CH_2{-}O{-}C({=}O){-}C_6H_4{-}C({=}O){-}O{-}CH_2CH{=}CH_2){-}\text{]}$

To a 2-L three-necked flask equipped with stirrer, thermometer, and condenser is added 886 g diallyl phthalate, 65 g isopropanol, and 7.5 g 50% hydrogen peroxide. The mixture is stirred and heated to reflux, which should correspond to a pot temperature of 104 to 108°C. The reaction should remain essentially homogeneous. After 10 h, the solution is cooled to 25°C and poured with stirring into 6-L isopropanol cooled to 0°C to precipitate the solid polymer. The mixture is held at 0°C for 1 h before filtering the polymer, washing with cold isopropanol on the filter, sucking dry, then drying in a vacuum oven at 45°C. The conversion of monomer to polymer is about 25%. The polymer softens at 80 to 105°C. The degree of unsaturation is indicated by an iodine number around 55. The polymer is soluble in methyl ethyl ketone, ethyl acetate, and benzene. If the dry, powdered polymer is mixed with 2% by weight of *t*-butyl perbenzoate and heated in a mold at 175°C for

15 min at 6000 psi pressure, a clear thermoset resin of high hardness and stiffness is formed, having a heat deflection temperature of 155°C (at 264 psi fiber stress).

Diethylene glycol bis(allyl carbonate) (6) can be polymerized to clear sheets by dissolving 3% by weight of dibenzoyl peroxide in the monomer and heating in a mold at a cycle determined by the thickness of the sheet to be cast, i.e., the thicker the sheet, the slower the temperature rise used and the longer the time cycle required. A 0.125-in sheet can be cast in an open mold under nitrogen by using the following cycle: from room temperature to 75°C in 5 h, to 80°C in 3 h, to 87°C in 4 h, and a postcure at 115°C for 2 h. The objective is to bring about heat evolution at a uniform rate throughout the polymerization.

11.1.4. Resins from the Reaction of Formaldehyde with Phenols

Formaldehyde condenses readily with phenols (7–11) primarily in ortho and para positions, to give, eventually, cross-linked polymers having aromatic rings linked together by methylene or oxydimethylene bridges. The reaction is carried out only to such a point that a soluble, meltable intermediate condensate, which can be formed readily, results. It is converted to an insoluble, infusible final polymeric product by a later treatment, usually heat plus additional catalyst or formaldehyde. The intermediate condensation products are of low or moderate molecular weight and may be one of two types. The first is often called a resol, and is formed by the reaction of excess formaldehyde with phenol, in about 1.5 : 1 mole ratio, in the presence of base. It contains hydroxymethyl groups that can condense further on heating.

The second type of intermediate is called a novolak. It arises from the reaction of less than equivalent of formaldehyde with phenol, about 1 : 0.8 mole ratio, in an acid catalyzed reaction. There are essentially no hydroxymethyl groups present for further condensation. The novolaks may have molecular weights up to 1200 to 1500. The resols are lower molecular weight, 300 to 700. Novolaks do not condense further without the addition of catalyst and more formaldehyde. Hexamethylenetetramine is frequently used as catalyst; it may also take part in the condensation, by providing formaldehyde by hydrolysis, or in formation of dibenzylamine bridges.

287. Preparation of Resol from Formaldehyde and Phenol (8)

OH
+ CH_2O ⟶
OH OH
CH_2 CH_2OH
CH_2OH CH_2
$-H_2C$ CH_2OH
OH

A 500-mL resin kettle is equipped with a reflux condenser, stirrer, thermometer, and a siphoning tube leading to a collecting trap for the removal of samples for testing. To the reaction vessel is added 94 g (1 mol) distilled phenol, 123 g (1.5 mol) aqueous formaldehyde (37% by weight), and 4.7 g barium hydroxide octahydrate. The reaction is stirred and heated in an oil bath at 70°C for 2 h. Two layers form if stirring is stopped. Sufficient 10% sulfuric acid is added to bring the pH to 6 to 7. Vacuum is then applied by water aspirator (pressure regulated to 30 to 50 mm), and water is removed through the condenser, which is now set up for distillation. The temperature *must not* exceed 70°C. Every 15 min 1- to 2-mL samples are withdrawn through the vacuum siphon takeoff and tested for gel time; by working with a spatula on a hot plate at 160°C. Gel time is taken as the time required for the resin to set up to a rubbery infusible solid. A portion of each sample should be cooled to room temperature and its brittleness noted. The dehydration should be stopped when the gel time is less than 10 s, or the resin is brittle and nontacky at room temperature.

This product has been termed *A-stage* resin. Further heating forms a resin that softens with heat but does not melt and is no longer soluble. This is referred to as *B-stage* resin. The final product from continued heating, *C-stage* resin, is hard, insoluble, and infusible. Resin at the first stage can be mixed with wood flour, $CaCO_3$, and pigments and used as a molding powder for conversion to the final stage C by means of heat. Much industrial use is made of the A-stage resin in laminates, adhesives, and varnishes. To observe the eventual hardening to the C-stage-resin, the resin prepared above should be removed at the A stage and heated in test tubes or small beakers at 100°C.

288. Preparation of a Novolak from Formaldehyde and Phenol (8)

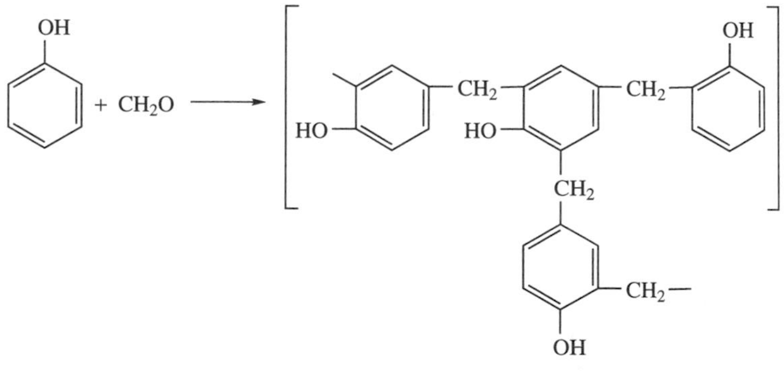

A 500-mL resin kettle is equipped as described in the preceding preparation and charged with 130 g (1.38 mol), phenol, 13-mL water, 92.4 g (1.14 mol) 37% aqueous formaldehyde, and 1 g oxalic acid dihydrate. The mixture is stirred and refluxed for 30 min. An additional 1 g oxalic acid is then added, and refluxing is continued for another 1 h. At this point, 400 mL water is added and the mixture cooled. The resin is permitted to settle for 30 min and the upper layer of water is decanted or withdrawn through the siphon. Heating is then begun with the condenser set for vacuum distillation. Water is distilled at 50 to 100 mm pressure until the pot temperature reaches 120°C or until a sample of the resin is brittle at room temperature. The resulting novolak is soluble in alcohol. About 140 g resin is obtained.

289. Preparation of a Molding Powder from a Novolak Resin (12)

A mixture of 46 g of the above finely ground novolak, 44.6 g dry wood flour (80 to 100 mesh), 6.7 g hexamethylenetetramine, 2 g magnesium oxide, and 1 g magnesium or calcium stearate are blended by tumbling in a jar or ball mill. The blended material can then be placed in a mold and heated under 2000 psi at 160°C for 5 min. A hard cured solid results.

290. Preparation of a Cast Phenolic Resin (12)

In a 1-L resin kettle equipped with a stirrer, condenser, thermometer, and vacuum siphon for sampling is placed 100 g (1.06 mol) phenol, 203 g (2.5 mol), 37% aqueous formaldehyde, and 3 g 20% aqueous sodium hydroxide. The reaction is stirred and heated to 80°C by means of an oil bath for 3 h. The reaction mixture is then concentrated at 30 mm pressure until a pot temperature of 65°C is reached. Then 6.5 g lactic acid is added, followed by 15 g glycerol. The removal of water is then continued at 30 mm, until a sample of the resin withdrawn through the vacuum siphon forms a ball that will just barely yield to pressure between the fingers when a drop is placed in 11 to 13°C water. This may require that a pot temperature of about 85°C be reached. Samples of this finished resin, a viscous liquid, may then be poured while hot into test tubes or beakers and heated at 80°C for 4 to 8 days to give hard castings that are clear if sufficient water was removed from the resin during preparation.

Although formaldehyde condenses preferentially at the ortho and para positions of phenol, it is not posssible to prepare only linear, soluble high polymers from formaldehyde and a phenol with either an ortho or para position blocked. Apparently some condensation takes place at open meta positions, since *o*- and *p*-cresol will eventually give an infusible, thermoset material, though times involved are long (13).

11.1.5. Resins from Reaction of Formaldehyde with Urea and Melamine

Formaldehyde and urea react under alkaline conditions to give monomethyl and dimethylol ureas (13–16). Such compounds are precursors of cross-linked urea-formaldehyde resins. The mechanism of polymerization encompasses both cyclic and noncyclic branched structures involving reaction of the methalol groups with residual amine groups (17, 18). The ratio of urea : formaldehyde used commercially is about 1 : 5. The commercial preparation of urea-formaldehyde resins usually involves the formation of soluble methalol-urea derivatives with basic catalysts. The intermediate condensate is then compounded with various fillers, pigments, and an accelerator. The latter is either an acidic material or one capable of functioning as an acid at high temperatures. The product can then be placed in a mold and heated to effect the final thermosetting, cross-linking reaction (13). Urea-formaldehyde condensates can also be used as additives for wet strength in papers and as finishes for fibers to impart crease resistance, and as adhesives.

$$3\,H_2N-\overset{O}{\overset{\|}{C}}-NHCH_2OH \longrightarrow \text{(triazine: } H_2NC(=O)-N,\ N-C(=O)-NH_2,\ N-C(=O)-NH_2\text{)} \xrightarrow{CH_2O}$$

$$\left[\ -CH_2-HN-\overset{O}{\overset{\|}{C}}-N\ (\text{ring, } N-C(=O)-NH-)\ N-\overset{O}{\overset{\|}{C}}-NH-CH_2-NH-\overset{O}{\overset{\|}{C}}-N\ (\text{ring, } N-C(=)-NH-)\ N-\overset{O}{\overset{\|}{C}}-NH-\ \right]$$

291. Preparation of Urea-Formaldehyde Resin (19)

To a 500-mL three-necked flask equipped with a stirrer and a reflux condenser is charged 130 g (1.6 mol) 37% aqueous formaldehyde that is brought to a pH of about 7.5 by addition of 10% sodium hydroxide solution. Then 60 g (1.0 mol) urea is added, and the mixture is gently refluxed and stirred for 2 h. The mixture is then concentrated to about 70% solids by distillation of 40 mL water under water aspirator pressure. The resulting syrup, after acidifying with acetic acid, can be heated further at 100°C for several hours to effect gelation.

292. Preparation of Urea-Formaldehyde Adhesive (19)

To prepare a plywood adhesive, 100 g of the unacidified concentrated syrup prepared above is mixed with 28 g furfural alcohol, 16 g wood flour (80 to 100 mesh), 1 g calcium phosphate, and 0.35 g triethanolamine by stirring, while raising the temperature to 90°C over 30 min. The temperature is held for 15 min, then the mixture is cooled slowly to room temperature. The mixture will set to a solid at room temperature if mixed with 2 g ammonium chloride and 3 mL water in a beaker. The ammonium chloride and water function as hardening catalysts, providing a working life for the adhesive of about 6 h. Before catalyst addition the resin is stable for weeks.

The methanolurea groupings in the soluble first stage of the urea-formaldehyde reaction reduce the compatibility of the product with many nonpolar organic solvents and oils. Solubility in such solvents can be achieved by carrying out the initial reaction in the presence of an alcohol. The methanol groups are partially etherified and the solubility of the product depends on the chain length of the alcohol used. Butanol will produce a toluene-soluble product. Etherification of the methalol group also reduces the likelihood of gelation at moderate temperatures.

293. Preparation of Urea-Formaldehyde Resin Modified with Butanol (20)

$$NH_2CONH_2 + CH_2O \longrightarrow -\overset{|}{N}CH_2OH \xrightarrow{BuOH} -\underset{|}{N}-CH_2OBu(n)$$

$$-NCH_2OBu(n) \longrightarrow \text{crosslinked polymer} + n-BuOH + H_2O$$

In a 1-L three-necked flask equipped with reflux condenser, thermometer, and stirrer are placed 243 g (3.0 mol) 37% aqueous formaldehyde and 4 to 6 g concentrated ammonium hydroxide to bring the pH to 7.5 to 8.5. Then 60 g (1.0 mol) urea is added with stirring, and the mixture is heated to 100°C over 1 h. This temperature is maintained for 30 min. Then 148 g (2 mol) *n*-butanol is added, followed by enough phosphoric acid to bring the pH to 5.5. The reaction is heated and stirred for 30 min at 100°C. The resin is freed of water by heating at 60 to 75°C under water aspirator pressure of 100 to 200 mm. The hot resin is pourable but becomes tacky at room temperature. The resin can be dissolved in butanol or toluene to give a 50 to 60% solution. When this solution is flowed onto glass or metal plates that are then heated for 30 min in an oven at 150°C, a hard clear coating results.

Melamine reacts with formaldehyde in much the same way as urea, forming compounds of varying degrees of *N*-methalol substitution, depending on the mole ratio of the reactants. The trimethalol and hexamethalol compounds are readily prepared and isolated and may be polymerized to cross-linked products. The cross-linking appears to result from ring formation and branching. In practice, the initial reaction is carried out to a soluble syrup, which can be mixed with fillers or used as such in molding or casting under conditions of heating, to give hard insoluble, infusible products.

294. Preparation and Polymerization of Hexamethalol Melamine (21)

$$C_3N_3(NH_2)_3 + 6CH_2O \longrightarrow C_3N_3[N(CH_2OH)_2]_3$$

In a 500-mL three-necked flask equipped with stirrer and condenser are placed 37.8 g (0.3 mol) melamine and 195 g (2.4 mol) 37% aqueous formaldehyde that has been made slightly basic (pH 7.50) with dilute aqueous sodium hydroxide. The mixture is stirred and heated on a steam-bath until a solution results. Heating is continued for 10 min more, and the reaction is cooled. A solid separates that is filtered, washed well with ethanol, and dried at 50°C. The melt temperature T_m is about 150°C. The clear melt resolidifies on further heating to a clear, hard, insoluble product.

295. Preparation of a Melamine-Formaldehyde Molding Powder

In a 1-L. resin kettle equipped with a condenser and stirrer are placed 126 g (1 mol) melamine and 365 g (4.5 mol) 37% neutralized aqueous formaldehyde. The mixture is stirred and heated at reflux for 40 min. Dilution of a sample of the solution with an equal volume of water precipitates the resin. The undiluted solution is cooled to room temperature and 235 g of the reaction mixture is kneaded with 50 g α-flock and 0.5 g zinc stearate in a dough mixer or by hand in a metal beaker, making sure that all lumps are broken up. The mass is then dried in a circulating air oven for 2 to 4 h at 70 to 80°C. It is ground to a uniform powder in a mechanical moll or by hand in a mortar to give a solid, which, if pressed in a mold or press at 145°C for 2 to 3 min at 2000 psi, gives a hard, lustrous, water-insensitive material.

At the end of the reflux period, thc solution prepared above, is concentrated to about 70% solids by distilling about 60 mL water at water aspirator pressure. As a softening agent, 20 g glycerol is stirred into the resin. The resulting syrup can be solidified in a mold or in a beaker by gradually raising the temperature to 150°C to give a clear hard product.

Melamine-formaldehyde resin can also be modified with butanol to give more stable and soluble solutions. A total of 50 g of the hexamethalolmelamine prepared previously is added to 80 g n-butanol and 0.5 g concentrated hydrochloric acid in a 250-mL flask equipped with a condenser. The reaction mixture is heated to reflux for 15 min. A clear solution is obtained, which forms a hard clear coating when a portion of it is evaporated to dryness on a surface, then heated at 150°C for 30 min.

Ion exchange resins can be prepared by condensing formaldehyde with a phenol, urea, or melamine in such a way to introduce ionic sites in the final resin (22–25). This can be achieved by using an ionic coreactant that enters into the polymer structure, by using a phenol to carry an ionic substituent, or by after treating the resin (e.g., sulfonation). Ion exchange resins are of two types: with basic groups capable of exchanging anions and with acidic groups capable of exchanging cations.

11.1.6. Epoxy Resins

Epoxy resins (26–28) are most commonly prepared by base-induced condensation of a polyhydroxy compound (usually a bisphenol) with, in most cases, epichlorohydrin to give a low molecular weight, essentially linear polymer, as an intermediate with terminal epoxide groups and pendant hydroxyls. An excess of epichlorohydrin in the reaction accounts for the termination of chains with epoxy groups.

$$\mathrm{HO{-}R{-}OH} + \overset{\overset{O}{\diagup\diagdown}}{\mathrm{CH_2{-}CH}}\mathrm{{-}CH_2Cl} \xrightarrow{\mathrm{NaOH}}$$

$$\mathrm{HO{-}R{-}O{-}CH_2{-}\underset{}{\overset{\overset{OH}{|}}{CH}}{-}CH_2Cl} \xrightarrow{\mathrm{NaOH}} \mathrm{HO{-}R{-}O{-}CH_2{-}}\overset{\overset{O}{\diagup\diagdown}}{\mathrm{CH{-}CH_2}} \longrightarrow$$

$$\mathrm{HO}\left[\mathrm{R{-}OCH_1{-}\overset{\overset{OH}{|}}{CH}{-}CH_2O}\right]_n\mathrm{R{-}OH} \xrightarrow[\mathrm{NaOH}]{\overset{\overset{O}{\diagup\diagdown}}{\mathrm{CH_2{-}CH}}\mathrm{{-}CH_2Cl}}$$

$$\overset{\overset{O}{\diagup\diagdown}}{\mathrm{CH_2{-}CH_2}}\mathrm{{-}CH_2{-}O}\left[\mathrm{R{-}OCH_2{-}\overset{\overset{OH}{|}}{CH}{-}CH_2{-}O}\right]_n\mathrm{R{-}OCH_2{-}}\overset{\overset{O}{\diagup\diagdown}}{\mathrm{CH{-}CH_2}}$$

The final cross-linking, to give an infusible, insoluble, hard product, can be carried out in a variety of ways, which usually involve either ring opening of the terminal epoxides and/or esterification of the chain hydroxyls. Among the more widely used cross-linking, or curing, agents are amines and dicarboxylic acids or their anhydrides (29). With primary amines, the reaction is presumed to be of the addition type; with tertiary amines, the reaction is thought to be a catalytic ring-opening polymerization of the epoxy groups, rather than a simple addition reaction. Secondary amines function in the same way after an initial single step addition reaction.

$$\mathrm{RNH_2} + \overset{\overset{O}{\diagup\diagdown}}{\mathrm{CH_2{-}CH}}\mathrm{\sim}\overset{\overset{O}{\diagup\diagdown}}{\mathrm{CH{-}CH_2}} \longrightarrow \mathrm{R{-}N}\begin{array}{l} \diagup \mathrm{CH_2{-}\overset{\overset{OH}{|}}{CH}\sim\overset{\overset{OH}{|}}{CH}{-}CH_2{-}\overset{\overset{\overset{\overset{\wr}{CH{-}OH}}{|}}{\overset{CH_2}{|}}}{N}{-}R} \\ \diagdown \mathrm{CH_2{-}\underset{\underset{OH}{|}}{CH}\sim\underset{\underset{OH}{|}}{CH}{-}CH_2{-}\overset{\overset{R}{|}}{N}{-}CH_2{-}\underset{\underset{OH}{|}}{CH}\sim} \end{array}$$

[where R = $—C_6H_4C(CH_3)_2C_6H_4—$]

Tertiary amines are more generally used in conjunction with anhydrides, and catalyze reaction of the latter with the epoxide function. Anhydrides of dibasic acids react first with a chain -OH. The free carboxyl can then esterify an –OH of another chain or open a terminal epoxide. The epoxy resins are stable to heat and have little tendency to cross-link before a curing agent has been added. After compounding with a suitable curing agent, the epoxy resin is then used in its soluble, fusible state for a variety of applications (adhesives, coatings, laminates, foams, etc.). Curing to the final hard resin is effected either by application of heat or, more slowly, at ambient temperature.

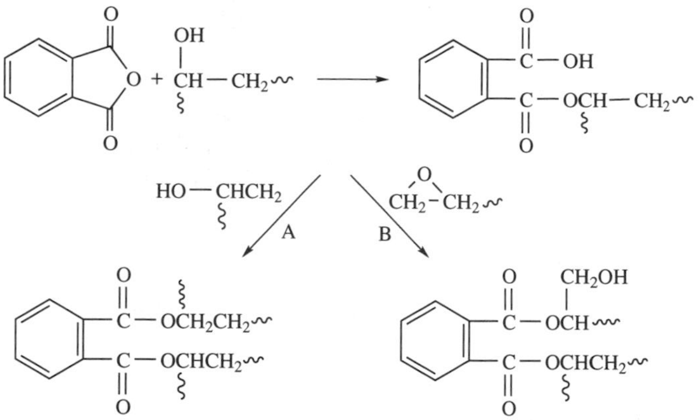

$$\text{Phthalic anhydride} + \text{HO–CH(}\sim\text{)–CH}_2\sim \longrightarrow C_6H_4(\text{COOH})(\text{COOCH(}\sim\text{)CH}_2\sim)$$

A: $+ \text{HO–CHCH}_2(\sim) \longrightarrow C_6H_4(\text{COOCH}_2\text{CH}_2\sim)(\text{COOCHCH}_2\sim)$

B: $+ \text{CH}_2\text{–CH(O)–CH}_2\sim \longrightarrow C_6H_4(\text{COOCH(CH}_2\text{OH)}\sim)(\text{COOCHCH}_2\sim)$

296. Preparation of Epoxy Resins from 2,2-Bis(4-hydroxyphenyl)propane (Bisphenol A) and Epichlorohydrin (30, 31)

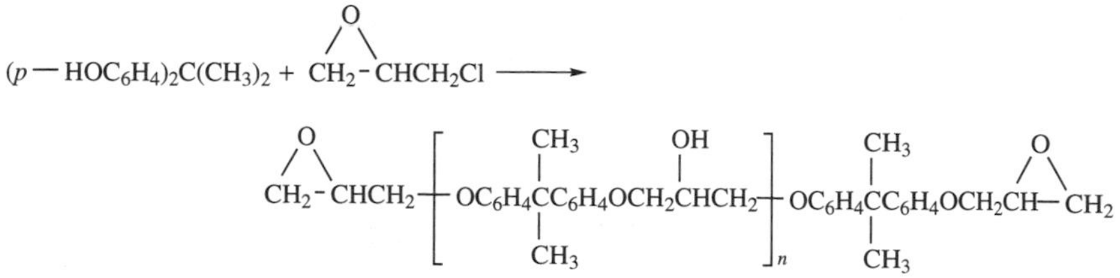

$$(p\text{–HOC}_6\text{H}_4)_2\text{C(CH}_3)_2 + \overset{\text{O}}{\text{CH}_2\text{–CH}}\text{CH}_2\text{Cl} \longrightarrow$$

$$\overset{\text{O}}{\text{CH}_2\text{–CH}}\text{CH}_2\left[\text{OC}_6\text{H}_4\text{C(CH}_3)_2\text{C}_6\text{H}_4\text{OCH}_2\text{CH(OH)CH}_2\right]_n\text{OC}_6\text{H}_4\text{C(CH}_3)_2\text{C}_6\text{H}_4\text{OCH}_2\overset{\text{O}}{\text{CH–CH}_2}$$

I

296.A. Resin A: MW 370

In a 2-L resin kettle equipped with stirrer, thermometer, condenser, and dropping funnel is placed a mixture of 228 g (1 mol) bisphenol A, 925 g (10 mol) epichlorohydrin, and 5 mL water. A total of 82 g (2.05 mol) solid sodium hydroxide is added in portions. First 13 g of the base is added and the mixture heated with stirring. Heating is stopped at 80°C and the Glass-Col heating mantle is replaced by an ice water bath so that the temperature does not exceed 100°C. When the reaction falls to 95°C. another 13 g base is added. Temperature control is exercised as before. The remainder of the sodium hydroxide is added in 13- to 14-g. increments. After the final addition of base, no cooling is applied. When the exothermic reaction subsides, the excess epichlorohydrin is distilled at a pressure of about 50 mm with a pot temperature not exceeding 150°C. The residue is cooled to 70°C and 50 mL benzene (**Carcinogen**!) is added to precipitate the salt present. The salt is removed by vacuum filtration and washed with 50 mL benzene. The benzene solutions are combined and the benzene distilled. When the pot temperature reaches 125°C, a vacuum of about 25 mm is applied, and the distillation continued until a pot temperature of 170°C is reached. The resulting clear, highly viscous liquid epoxy resin has an average MW of 370, as determined ebullioscopically in ethylene dichloride. The epoxide content is determined by heating 1 g of the resin at reflux for 20 min with 25 mL of a standardized solution prepared from 16 mL concentrated HCl diluted with pyridine to 1 L. After cooling, the excess hydrochloric acid is back titrated with 0.1 *N* sodium hydroxide in methanol to a phenolphthalein end point. One HCl is considered equivalent to one epoxide group. For the resin prepared above, the epoxy content per 100 g is about 0.5, which corresponds to 1.85 epoxy groups per molecule for a MW of 370.

296.B. Resin B: MW 900

In a 1-L. resin kettle equipped as above, with the added provision for a siphon, are placed 228 g (1 mol) bisphenol A and 75 g (1.88 mol) sodium hydroxide as a 10% aqueous solution; the mixture is heated to 45°C. Then, 145 g (1.57 mol) epichlorohydrin is added rapidly with stirring. The mixture is then heated to and maintained at 95°C for 80 min. The mixture separates into two phases. The aqueous layer can be siphoned off from the taffy-like product. The latter is washed with hot water with stirring while molten until the wash water is neutral to litmus. The resin is then removed while hot and is dried by heating in an air oven at 130°C. The molecular weight is about 900.

296.C. Resin C: MW 1400

The procedure for resin B is followed except that 54.8 g (1.37 mol) sodium hydroxide solution (10%) and 113 g (1.22 mol) epichlorohydrin are used.

296.D. Resin D: MW 2900

A total of 100 g (0.071 mol) resin C (containing 0.103 mol epoxy groups) is placed in a 250-mL beaker and heated with stirring to 150°C by means of an oil bath. Then 5 g (0.022 mol) bisphenol A is added and the mixture heated to 200°C over a 2-h period. The resulting resin has a softening point about 130°C. There are approximately 0.05 epoxy group 100 g and it has a molecular weight of about 2900.

297. The Curing of Epoxy Resins

The method of choice in curing an epoxy resin is determined by a combination of factors such as end use, hardness, and stability to heat or light. The following examples demonstrate only a few of the types of curing agents used.

297.A. With a Tertiary Amine

A surface coating can be made from a solution of 10 g resin C and 1 g benzyldimethylamine in 5 mL *each* of xylene and methyl cellosolve acetate. The solution is spread uniformly on a glass plate. The solvent is evaporated in air and the plate is heated at 100°C for 30 min. A hard film coating results.

297.B. With a Secondary Amine

To 142 g diethylamine in a 1-L resin kettle equipped with stirrer, condenser, and thermometer is added 125 g resin A in 125 g dioxane, with stirring. A slightly exothermic reaction occurs. The mixture is heated to reflux (55 to 60°C) for 3 h. The resulting mixture is poured into 750 mL water in a 2-L beaker, and the sticky product is washed repeatedly with water by stirring and decantation to remove excess amine and dioxane. The resin is then dissolved in 500 mL diethyl ether, and the solution extracted with 500-mL portions of water until the washings are neutral to litmus. The ether solution is dried over Drierite and the ether is removed by distillation on a steam bath. The product (about 92 g) is a viscous liquid at room temperature, but fluid at 60°C. The resin is still soluble in methyl ethyl ketone and chloroform. The large amount of diethylamine used provides for a one-to-one reaction of amine with epoxy groups, and no cross-linking is thus possible. If, however, 50 g resin A is mixed with 2.5 g diethylamine and kept at room temperature for 60 h followed by 65°C for 24 h, a solid, insoluble product is obtained. In this case, each amine group, present in low quantity, effects the catalytic polymerization of several epoxides, with cross-linking as the result. If 50 g resin A is then mixed with 7.5 g of the soluble resin prepared above from resin A and excess of diethylamine, the mixture will cure to a hard, clear solid in 1 to 2 h at 60°C. The combined amine groups in the amine-modified resin A serve to cross-link the mixture through the epoxides of the added resin A.

297.C. With an Acid (32)

A solution is prepared in a 50-mL Ehrlenmeyer flask by dissolving 10 g resin C in 12 mL methyl ethyl ketone, to which is then added 1 g oxalic acid dihydrate. The mixture is gently warmed on the steam bath to assist in dissolving the acid. The solution is cast on a glass plate with aid of a doctor knife and the solvent allowed to evaporate at room temperature. The glass plate is then heated in an oven at 150°C for 30 to 60 min to give a cured coating of considerable hardness.

297.D. With an Anhydride (33, 34)

A total of 50 g resin B is placed in a 200-mL tall-form beaker in an oil bath. The resin is heated to 120°C. Then 15 g molten phthalic anhydride is added and stirred into the resin. Part of the phthalic anhydride will precipitate if the resin is cooled at this point to 60°C or below, but reheating will redissolve it. The beaker is covered with a glass plate, and the mixture is held at 120°C for 1 h, at which point it is still soluble in acetone and chloroform. Heating to 170 to 180°C will effect the final cure and produce a clear, hard, insoluble resin in 1 to 2 h.

If 0.5 g *N,N′*-dimethylaniline is added to the reaction immediately after or along with the phthalic anhydride, the resin will cure in about 1 h at 120°C. Tertiary amines function as accelerators for anhydride curing of epoxy resins.

The condensation of bisphenol A with epichlorohydrin can be carried out with equimolar amounts of each reactant to give a thermoplastic solid polymer of sufficient molecular weight to have good mechanical properties and thus practical utility in its linear state. The chemical repeat unit of the polymer has secondary hydroxyl, and the polymer can be modified chemically through reaction at that point (e.g., acetylation) and thus modified in its physical properties (35).

298. Preparation of a Poly(hydroxy ether) from Bisphenol A and Epichlorohydrin (36)

$$HO-C_6H_4-C(CH_3)_2-C_6H_4-OH + \overset{O}{CH_2-CH}-CH_2Cl \longrightarrow \left[C_6H_4-C(CH_3)_2-C_6H_4-O-CH_2-\underset{OH}{CH}-CH_2-O \right]$$

To a 2-L three-necked flask equipped with a stirrer, thermometer, and condenser is added 228.3 g (1.0 mol) 2,2-bis(*p*-hydroxyphenyl)propane (recrystallized from toluene), 92.5 g (1.0 mol) epichlorohydrin (distilled, heart cut), 256 g ethanol, 40.0 g (1.0 mol) sodium hydroxide, and 160 mL water. The caustic should be added first as a standard solution containing 1 mol of base. The bisphenol A is added second, and the alcohol and the epichlorohydrin in that order after solution of the bisphenol A. The mixture is stirred at room temperature for 6 h, when 6 g (0.15 mol) sodium hydroxide in 24 mL water is added. The mixture is then heated to reflux, which should occur at a pot temperature of about 80°C. Distilled chlorobenzene is added in portions: 60 mL after 30 min of reflux, 30 mL after 45 min, and another 30 mL after a total of 60 min at reflux. Refluxing is continued for 4 h, when 9.4 g (0.1 mol) phenol in 60 mL chlorobenzene is added. Refluxing is continued for 2 h more. The mixture is cooled, the aqueous phase is separated by decantation, and the chlorobenzene-polyether phase is washed three times with 400-mL portions of water. Then 400-mL chloroform is added to dilute the chlorobenzene-polyether, and this solution is acidified by adding with stirring a solution of 20 mL 87% phosphoric acid in 100 mL water. The organic phase is washed eight times with 400-mL portions of water. The polymer solution is precipitated by pouring it into 2-L isopropanol with stirring. The precipitated polyhydroxyester is filtered, washed well with isopropanol, and dried in a vacuum at 65°C for 24 h or more. The inherent viscosity is about 0.6 (0.2 g/100 mL in tetrahydrofuran (THF) at 25°C). The polymer is amorphous and is soluble in chlorinated solvents. Tough films can be cast from methylene chloride or melt pressed above 160°C. The glass-transition temperature T_g is about 100°C.

The cross-linking reaction of diepoxides with polycarboxylic acids can be used to make surface coatings valuable as enamels for appliances and other things. A typical vinyl copolymer is made in solution wherein one of the comonomers is acrylic acid. Bisphenol A diglycidyl ether is added to the copolymer solution followed by application of the solution to a surface and a heat curing step. The result is a cross-linked coating whose properties (metal adhesion, hardness, etc.) reflect the original vinyl copolymer constituency and the epoxy cross-linking agent. Such a resin is derived, therefore, from vinyl and epoxy technology.

299. Preparation of Poly(styrene-co-methyl Acrylate-co-acrylic acid [72/20/8 wt%]) Cross-linked by Reaction with Bisphenol A Diglycidyl Ether (37)

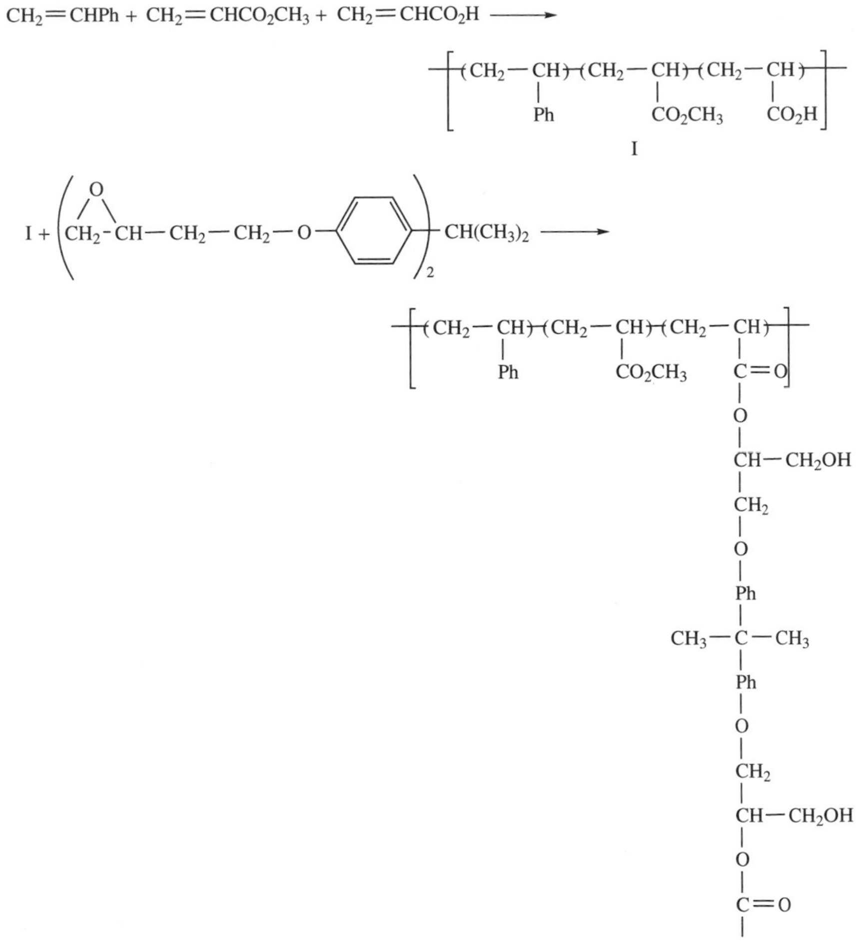

$$CH_2{=}CHPh + CH_2{=}CHCO_2CH_3 + CH_2{=}CHCO_2H \longrightarrow \left[(CH_2-\underset{Ph}{CH})(CH_2-\underset{CO_2CH_3}{CH})(CH_2-\underset{CO_2H}{CH}) \right] \quad \text{I}$$

$$\text{I} + \left(\overset{O}{CH_2-CH}-CH_2-CH_2-O-C_6H_4 \right)_2 CH(CH_3)_2 \longrightarrow$$

$$\left[(CH_2-\underset{Ph}{CH})(CH_2-\underset{CO_2CH_3}{CH})(CH_2-\underset{C=O}{CH}) \right]$$

$$C(=O)-O-CH(CH_2OH)-CH_2-O-Ph-C(CH_3)_2-Ph-O-CH_2-CH(CH_2OH)-O-C(=O)-$$

A 500-mL three-necked flask is equipped with condenser, nitrogen inlet, and stirrer. It is flushed with nitrogen, and 250 mL xylene is added. Four separate monomer-initiator batches are prepared: each one contains 18 g styrene, 5 g methyl acrylate, 0.5 g azoisobutryonitrile, and amounts of acrylic acid in successive batches of 1.8 g, 1.76 g, 2.24 g, and 2.72 g, respectively. The xylene is brought to reflux under a light stream of nitrogen so that essentially no volatiles escape the condenser, and the four monomer-initiator batches are added in succession at a constant rate over a period of 20 min each. This technique is to provide a homogeneous distribution of acrylic acid in the copolymer; it is essential for satisfactory curing that each copolymer molecule has as nearly as possible the same average number of carboxyl groups.

When addition of the last batch of monomers is completed, refluxing is continued for another hour; the solution is then allowed to cool. A portion of the copolymer can be isolated by precipitation into methanol followed by solution and reprecipitation and with methanol washing in the usual way. The weight average molecular weight can be fairly well estimated from the intrinsic viscosity in methyl ethyl ketone by the following relationship, determined for the copolymer esterified by reaction with diazomethane:

$$[\eta] = 1.59/1000\ M_w^{0.42}$$

To the cooled xylene solution of the copolymer is stirred in 18.7 g of a bisphenol A diglycidyl ether (see resin A). This should give approximately equimolar amounts of carboxyl and epoxy, although not all of the former needs to, or appears to, react with an epoxy group to give good final film properties. As catalyst to the cross-linking reaction, 0.75 g benzyldimethylamine or triethylenediamine is added. The solution can be cast onto metal plates, the xylene evaporated in a forced-air oven at 100°C, and the film cured for about 30 min at 150°C to give a hard chemical- and heat resistant coating. If desired, the polymer solution can be pigmented by dispersing titanium dioxide or other materials in it.

11.1.7. Thermoplastic Resins

Many of the thermoplastic resins used in composites have been described in earlier chapters. These include nylon, polyethylene terephthalate; and for aerospace uses, a variety of high-melting, high-T_g polymers, primarily polyether, ketones, and polyimides. We refer readers to the appropriate chapters for the preparation of these polymers.

11.2. COMPOSITES

11.2.1. Fiber and Resin Requirements

A major application of thermoset, and to a lesser extent thermoplastic, resins is to combine them with strong, stiff fibers to form composites. The fiber component is often glass, carbon, or synthetic fibers such as poly (ethyleneterephthalate) (PET). The fiber can be used in short chopped lengths (0.5 in), long lengths, or as continuous windings, disposed randomly or uniformly (uniaxially) within the resin. Figures 11.1 and 11.2 illustrate the various types of reinforcements.

The combinations of fiber and resin provide products with high strength, stiffness, and low density compared to metals. When the fiber is continuous and the matrix is weak and of low stiffness, the fiber bears almost all the load. The purposes of the resin are (1) to combine the fibers together to support a load, (2) to coat the fibers to reduce abrasion, (3) to provide lateral support to prevent buckling when the fibers are loaded in compression, (4) to act as a shear transfer medium, and (5) to provide a solid material by filling the voids between the fibers.

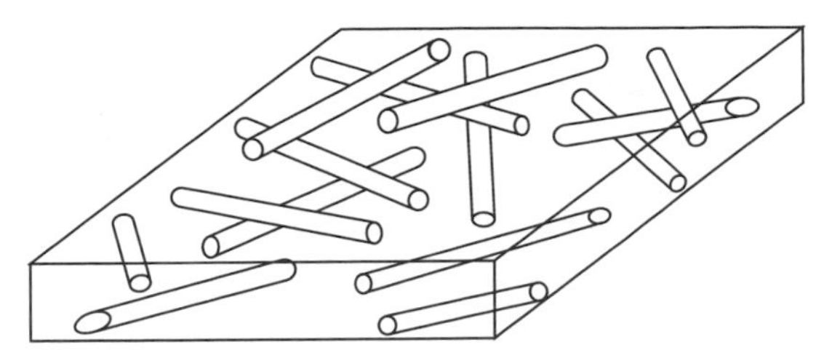

Two-dimensional (mat)

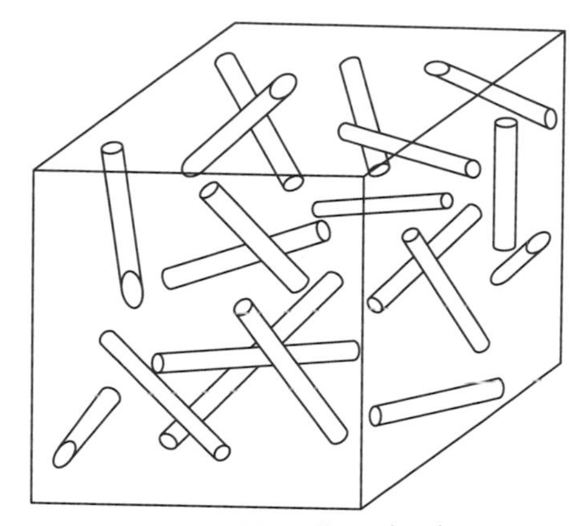

Three-dimensional

Figure 11.1. Chopped-fiber reinforcement.

Continuous-fiber reinforcement

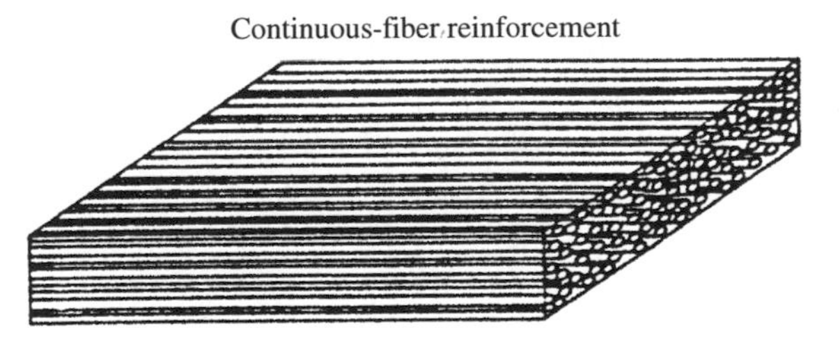

Colimated tape

Fabric

Figure 11.2. Continuous-fiber reinforcement.

Listings of some fibers and thermoplastic resins commonly used in composites, especially for high-temperature use, are provided in Tables 11.1 and 11.2. The advantages of composites over metals are (1) high specific strength and stiffness (specific properties are properties relative to density, e.g., stiffness/density), (2) easy formability to complex shapes, (3) nonconductive, (4) corrosion resistant, (5) controlled low thermal expansion, and (6) good fatigue resistance.

Several factors affect composite strength and stiffness properties. **For fiber**, the critical factors are modulus, strength characteristics, strain to failure, compressive properties, fiber length ($L/d \gg 1$), diameter, curvature/crimp (fiber should be straight). Volume fraction of fiber in the composite should be high, and fiber well dispersed. **For matrix**, the critical factors are modulus, elongation to failure, stress/strain behavior (brittle, elastic, plastic, etc.), amount of voids, and coefficient of thermal expansion. Other factors are fiber-to-matrix bond strength and method of fabrication. Critical to the formation of high-strength composites is uniform, and essentially complete, wetting of the fibers with the matrix resin during composite formation. This is one of the major advantages of thermoset matrices over thermoplastic, simply because the thermoset monomer or prepolymer, because of its much lower viscosity, wets the fiber well and results in fewer voids.

Thermoset resins also require lower processing temperatures and, once crosslinked, have a lower propensity to creep at high temperatures. The advantages

Table 11.1. Thermoplastic Matrices

Resin	Density, g/mt	Tensile Strengths psi $\times 10^3$	Modulus psi $\times 10^6$	Maximum Use Temperature, °C.
Nylon 6,6	1.14	12	0.5	80
Polysulfone	1.24	10	0.4	150
Polyether sulfone	1.37	12	0.4	150
Polyaryl sulfone	1.36	13	0.4	260
Polyphenylene sulfide	1.34	10	0.5	260
Polyamide/imide	1.40	27	0.7	340

Table 11.2. Fibers for Composites

Fiber	Density g/mL	Tensile, psi $\times 10^3$	Modulus psi $\times 10^6$
E glass	2.54	372	10
S glass	2.48	550	12
Nomex	1.38	90	2
Kevlar-49	1.44	400	18
Alumina	3.90	225	55
Graphite HT	1.72	360	32
Graphite HM	1.97	270	75
Boron	2.68	500	58

of thermoplastic matrix resins are higher toughness and better thermal stability. Improved thermosets, made by altering the chemical structure units between cross-links, are closing the gap in toughness relative to thermoplastics.

11.2.2. Fabrication Processes

A variety of methods are used for making composites. The simplest is the hand lay-up process (Fig. 11.3). This is a process for molding a part by manual addition of catalyzed resin and reinforcement to an open mold, followed by room temperature curing. Most common resins used are polyesters and epoxys. Reinforcement includes chopped glass mat, E-glass, and polyethylene terephthalate rovings and fabrics. Fiber volume is usually low (20 to 40%), and void content is high. The process is inexpensive and there is almost no shape or size limitation. One side of the part is rough and the other smooth. The quality is operator dependent and labor intensive, and the output is low.

An alternative route is a spray process whereby catalyzed resin and chopped fiber (usually glass) are simultaneously deposited into the mold with a spray gun. This is usually used in conjunction with hand lay-up and is likewise operator dependent.

Another widely used route is compression molding in which premixed resin and reinforcement (usually chopped fiber (0.25 to 0.5 in) is molded by heat and pressure in a die. The resin and fiber are often used in the form of a preformed sheet (Figs. 11.4 and 11.5). Properties of molded parts increase directly with increase in fiber content and fiber length.

Three other methods should be described because they are most frequently used in high-performance composites. The first is pultrusion, which is a process for producing continuous lengths or shapes with constant cross-section, e.g., rods, tubes, and channels. The process uses a high content of fiber in unidirectional reinforcement (Fig. 11.6). Matrix resins commonly are polyesters or epoxies, combined with high-modulus glass, carbon fiber, Kevlar, and high-modulus inorganic fibers. The process involves transport of fiber roving, resin

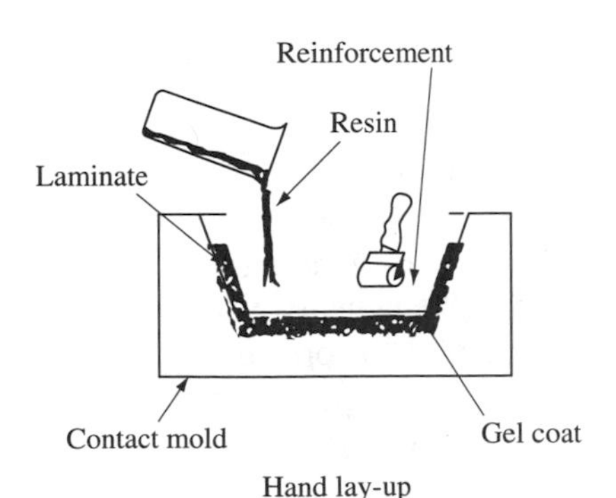

Figure 11.3. Hand lay-up process.

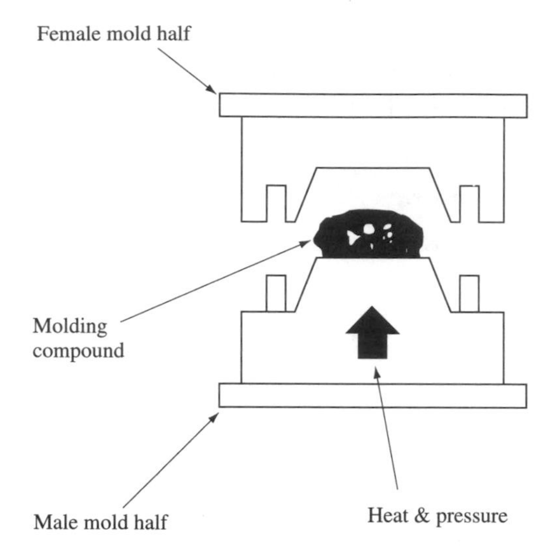

Figure 11.4. Compression molding technique.

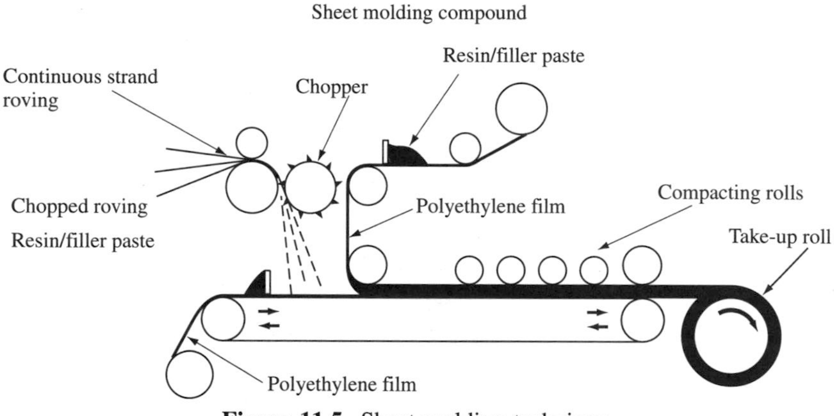

Figure 11.5. Sheet molding technique.

impregnation, shaping, preheating, and traverse through a die and cure. Speeds are usually 0.5 to 10 ft/min, depending on part size.

The second process is filament winding (Fig. 11.7). This involves winding a high-strength, high-quality, filamentary yarn of low twist, impregnated with a quality resin, around a mandril to form pipes, tanks, pressure vessels, rocket cases, etc. Fibers used are carbon, E-glass, Kevlar, etc., and the resins are usually polyesters or epoxies. The filament winding pattern is usually helical (mandril rotates and fiber traverses) or planar (mandril is stationary and the fiber feed rotates around the longitudinal axis).

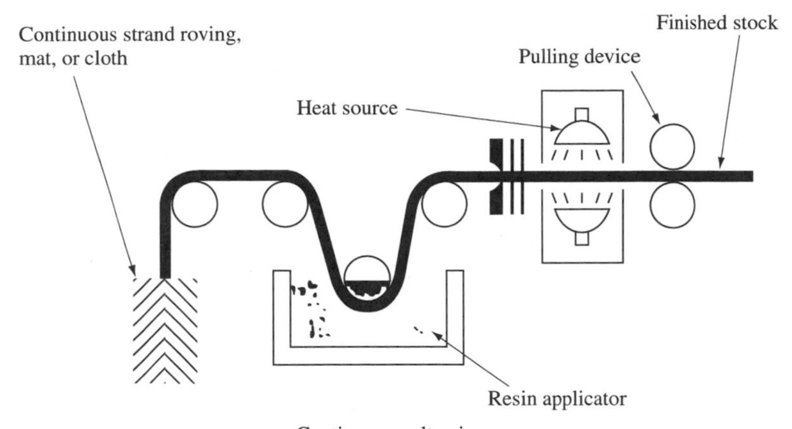

Figure 11.6. Pultrusion technique.

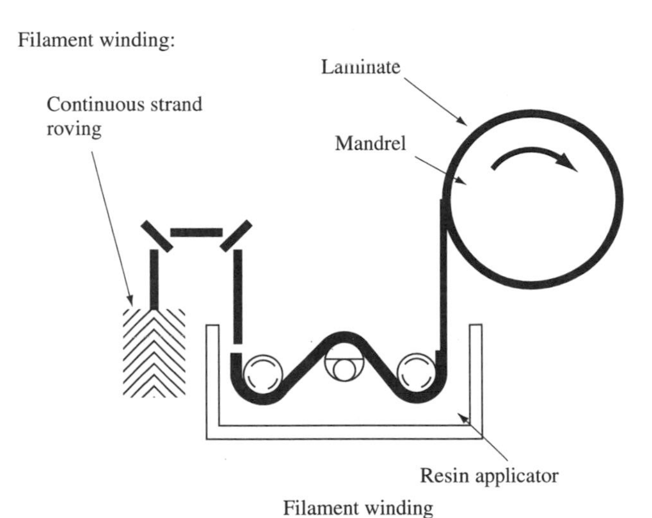

Figure 11.7. Filament winding.

The last and most widely used process for composites is preimpregnated reinforcements (prepregs). Prepregs are ready-to-mold combinations of resins and reinforcement that have been partially cured. They are commonly maintained in this condition for extended periods at low temperatures. A variety of resins and fibers are used. The resins are often epoxies (best property balance), polyesters (low cost), phenolic resins (high strength, heat resistant), and polyimide thermoplastics (high temperature applications). The process usually involves dipping the fiber reinforcement in a solution of the resin/curing agent, passing the wet reinforcement through a drying tower to remove solvent and partially cure the resin, applying a liner or separating film while winding up on a roll, and storing at low temperature to maximize shelf life. The prepreg sheet can then be molded by a variety of processes, using a vacuum bag or pressure bag to apply the molding pressure for the part to conform to the mold.

11.2.3. Testing

Testing composite properties in the laboratory is usually done with a unibar (usually $6 \times 0.5 \times 0.125$ in). This is made by impregnating yarn with resin, winding around a flat aluminum plate (larger than the desired sample size to allow for trimming to the proper dimensions), polymerizing, and then cutting the cured sample from the plate for trimming to size and testing. The fiber can be impregnated by a variety of methods:

- Using a solution of the polymer matrix, followed by drying (multitreatments may be necessary to get the desired fiber to resin ratio).
- Impregnating the yarn with molten resin.
- Winding the yarn onto the plate and stacking a thermoplastic matrix resin between layers. Thermal treatment and consolidation are used to distribute the resin throughout the fiber. High viscosity of thermoplastic resins, when used, however makes uniform wetting of the fiber difficult.
- Using comingling, which requires that the matrix resin is in fiber form and that the resin and reinforcing fiber are thoroughly intermingled. Wetting of the fiber with resin is achieved by heat and pressure consolidation.

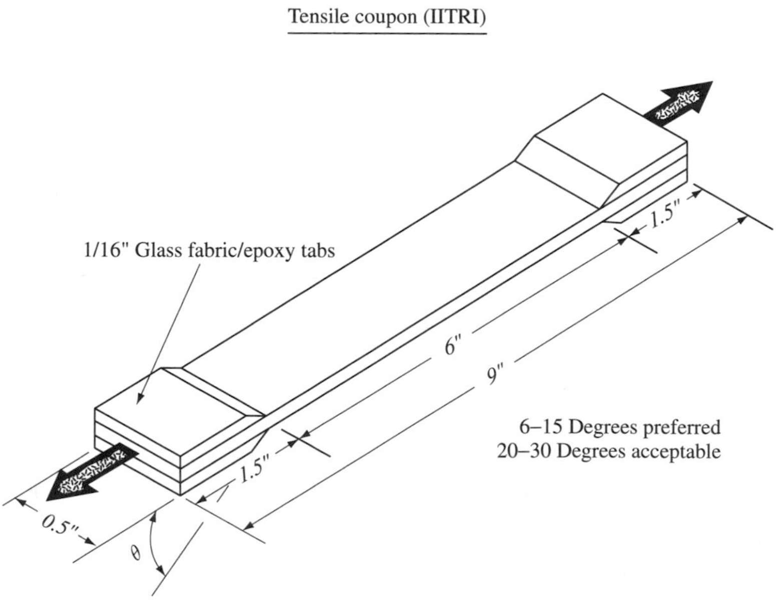

Figure 11.8. Molded piece (coupon) for testing.

- Using dry powder impregnation. Dry thermoplastic powder is introduced into the fiber tow and sticks by electrostatic attraction, heating is then used to sinter the powder particles. The tow is heated and passed through a die to produce an impregnated tow.

The unibar allows easy testing of extensional stress and strain to failure and also flex-life to failure. Composites are typically tested for mechanical properties in tension, compression, in-plane shear; interlaminar shear, and flexure; fatigue; impact; and fracture.

Most test samples for tensile testing use a $6 \times 0.5 \times 0.125$-in test bar with glass epoxy tabs for clamping to prevent damage to the test sample (Fig. 11.8). Clamp damage and sample misalignment are common causes of inaccurate results. Test methods for flexural modulus, flexural strength and short beam shear are described in ASTM D2344 and D790 testing procedures. Other test methods can be found in textbooks dealing specifically with composites (38).

REFERENCES

1. National Aniline Division, Composition and Utilization of Polyesters, Allied Chemical and Dye Corp., Morristown, New Jersey, 1954.
2. R. H. Kienle et al., *J. Am. Chem. Soc.*, **61**, 2258 (1939).
3. U.S. Pat. 1,888,849 (November 22 1932), E. S. Dawson to GE Corp.
4. H. Raech Jr., *Allylic Resins and Monomers*, Reinhold, New York, 1965.
5. U.S. Pat. 3,096,310 (July 2, 1963), (to FMC Corp.) C. A. Heiberger.
6. W. R. Dial et al., *Ind. Eng. Chem.*, **47**, 2447 (1955).
7. T. S. Carswell, *Phenoplasts: Their Structure, Properties and Chemical Technology*, Wiley-Interscience, New York, 1947.
8. Office of Technology. Service, P.B. Report 25,642, U.S. Department. of Commerce, Washington, D.C., 1945.
9. D. F. Gould, *Phenolic Resins*, Reinhold, New York, 1959.
10. R. W. Martin, *The Chemistry of Phenolic Resins*, John Wiley & Sons, New York, 1956.
11. N. J. L. Megson, *Phenolic Resin Chemistry*, Academic Press, New York, 1958.
12. U.S. Pat. 1,909,786 (May 16 1933), O. Pantke.
13. R. S. Morell and H. M. Langton, eds., *Synthetic Resins and Allied Plastics*, 3rd ed., Oxford University Press, London, 1951.
14. C. Ellis, *The Chemistry of Synthetic Resins*, Reinhold, New York, 1935.
15. C. E. Schildkneck, ed., *Polymer Processes*, Wiley-Interscience, New York, 1956.
16. J. F. Blais, *Amino Resins*, Reinhold, New York, 1959.
17. C. S. Marvel et al., *J. Am. Chem. Soc.*, **68**, 1681 (1946)
18. J. I. de Jong and J. de Jonge, *Rec. Trav. Chim.*, **72**, 1027 (1953).
19. U.S. Pat. 2,518,388 (August 8 1950), W. G. Simons to American Cyanamide Corp.
20. U.S. Pat. 2,226,518 (December 24 1940) T. S. Hodgins and A. G. Hovey, to Reichhold Corp.

21. U.S. Pat. 2,328,592 (September 7 1943), G. Widmer and W. Fisch.
22. R. Kunin, *Ion Exchange Resins*, 2nd ed., John Wiley & Sons, New York, 1958.
23. U.S. Pat. 2,477,328 (July 26 1949), H. M. Day.
24. U.S. Pat. 2,204,539 (June 11 1940), H. Wasseneger and K. Jaeger.
25. U.S. Pat. 2,285,750 (June 9 1942), R. C. Swain.
26. *Technical Bulletin on Epoxy Resins*, Shell Chemical Corp.
27. H. Lee and K. Neville, *Epoxy Resins: Their Applications and Technology*, McGraw-Hill, New York, 1957.
28. *Reports on the Progress of Applied Chemistry*, Vol. 62, Society of the Chemistry. Industry, London, 1957.
29. S. H. Christie, III, *Modern Plastics*, **42**, 134 (1965). Y. Tanoka and H. Kakiuchi, *J. Poly. Sci.*, **2**, 3405 (1964) and *J. Appl. Poly. Sci.*, **7**, 1063 (1963).
30. U.S. Pat. 2,643,329 (June 23 1953), E. C. Shokal, H. A. Newey, and T. F. Bradley.
31. U.S. Pat. 2,681,901 (June 22 1954), Q. T. Wiles and D. Elam.
32. U.S. Pat. 2,500,449 (March 14 1950), T. F. Bradley.
33. U.S. Pat. 2,324,483 (July 20 1943), P. Castan.
34. E. C. Dearborn et al., *Ind. Eng. Chem.*, **45**, 2715 (1953).
35. N. H. Reinking et al., *J. Appl. Poly. Sci.*, **7**, 2135 (1963).
36. Fr. Pat. 1,309,401 (October 8, 1962) to Union Carbide, J. Wynstra et al.
37. J. D. Murdock and G. H. Segall, *Official. Digest. Federation Soc. Paint Technol.*, **33**, 709 (1961).
38. F. P. Gerstle Jr., ed., Composites, In *Encyclopedia of Polymer Science and Technology*, vol. 3, Wiley-Interscience, New York, 1985.

12

NONCLASSICAL ROUTES TO POLYMERS

This chapter includes polymers prepared by via distinctive chemistry, such as oxidative coupling, and coupling via precious metal catalysis. Obviously, not every approach has been covered, but we have tried to provide a selection that would be of interest for most polymer chemists.

12.1. CYCLOPOLYMERIZATION

It is possible for certain structurally favorable unconjugated dienes to copolymerize according to the equation

```
R· + C      C            R—C     C
     \\     \\              \  ↗ \\
       C      C  ——→         C·    C   ——→
       |      |              |     |
       C      C              C     C
        \    /                \   /
          X                     X

                         C                       C
                       /   \                   /   \
            R—C—C         C·   ——→   R[ —C—C        C— ]
                |          |               |         |
                C          C               C         C
                  \      /                   \     /
                     X                          X        n
```

Polymerizations of this type were first reported by Butler annd Angelo (1), Marvel and Vest (2) and Jones (3). The following example is from Jones.

300. Preparation of Linear Poly(acrylic Anhydride) by Cyclopolymerization(3)

$$CH_2{=}CH{-}C({=}O){-}O{-}C({=}O){-}C{=}CH_2 \longrightarrow \left[-CH_2- \text{(six-membered anhydride ring)} - \right]_n$$

A solution of 20 g acrylic anhydride of at least 98% purity in 200 mL anhydrous benzene is mixed with 0.4 g benzoyl peroxide in a polymer tube and sealed under an atmosphere of nitrogen. The tube is agitated at 50°C for 25 h, then the resulting thick polymer slurry is cooled to room temperature and the polymer is isolated by suction filtration under nitrogen. It is dried at 50°C in a high vacuum for 24 h. The yield is 20 g, with an inherent viscosity of about 2, measured in dimethylacetamide (DMF) at 0.5% concentration. This corresponds to a molecular weight of about 200,000. Although the monomer from which the polymer is made is difunctional, the polymer is linear, as evidenced by the fact that it is completely soluble in the solvents DMF, *N*-methylpyrrolidinone (NMP), etc. The polymer can be pressed to bars and thin films at 180°C. The films of poly(acrylic anhydride) hydrolyze quite readily, and the product is identical in every respect to polyacrylic acid.

12.2. POLYMERIZATION OF THE CARBONYL GROUP

12.2.1. Polymerization of Aldehydes

With few exceptions, the only unsaturated compounds polymerizable to high molecular weight are those containing the C=C linkage. One notable exception is the carbonyl group. The polymerization of monomeric formaldehyde has been known for many years (4). Early studies (5, 6) produced films and fibrous material, but it was left to Schneider (7) to be first to produce tough, high molecular weight polymer by polymerizing trioxane with a variety of Friedel-Craft type catalysts. MacDonald (8) prepared high molecular weight linear polymer, with a high degree of thermal stability, by pyrolyzing paraformaldehyde to gaseous formaldehyde and passing the gas into a variety of different catalysts.

Two polyformaldehyde resins became commercially available: Delrin (DuPont) and Celcon (Celanese). The former is made thermally stable by acetate end capping and the latter by copolymerizing with ethylene oxide. Depolymerization occurs by chain unzippering from the chain ends and this is minimized by either of these protective capping processes.

301. Polymerization of Formaldehyde Vapor to High Molecular Weight Polyoxymethylene (8)

$$CH_2O \longrightarrow [-OCH_2-]_n$$

A total of 75 g of anhydrous paraformaldehyde is pyrolyzed over a 2-h period to produce monomeric gaseous formaldehyde, which is passed through two traps at −15°C and then into a reaction vessel containing 800 mL carbon tetrachloride, 0.05 g diphenylamine, and 0.078 g tributylamine. The reaction mixture is stirred vigorously and maintained at about 25°C. Polymerization proceeds throughout the addition of the monomeric formaldehyde to give a slurry of polymer. This is filtered, washed with ether, and air dried to give about a 35% yield of snow white, high molecular weight polyformaldehyde. The polymer should have an inherent viscosity of about 1.8 in *p*-chlorophenol at 0.5% concentration. Tough translucent films can be compression molded at 190°C and can be stretched and oriented.

302. Polymerization of Trioxane with Molybdenum Dioxydiacetylacetonate (9)
The catalyst, MoO_2 $(CH_2COCH{=}C{-}(CH_3){-}O{-})_2$, is prepared from molybdenum trioxide and acetylacetone by the method of Berardinelli et al. (10) For this polymerization, it should be washed with additional dry *n*-heptane and dried in vacuum at 50°C overnight.

To a dry, nitrogen-flushed, 8-oz beverage bottle is charged 60 g trioxane recrystallized from methylene chloride (4 g/mL) and 30 mL dry cyclohexane. The mixture is heated in a hot water bath and swirled by hand to bring the trioxane into solution. The bottle is again well flushed with nitrogen, and 0.1 g of the molybdenum catalyst is added. The bottle is closed with a septum and crown cap, shaken to disperse the catalyst, and immersed in a 100°C oil bath (a proper laboratory shield should be used at all times). The speed of the reaction, once it begins, is interesting to observe, however. Within a few minutes after the contents reach 85°C, polymer begins to separate with the appearance of falling snow. The mass forms quickly, giving nearly a solid clump. (If no polymer forms in 30 min, the bottle can be reopened and additional 0.01 g catalyst added.) Rather impure trioxane can deactivate the catalyst. The mixture is heated another 10 min and removed with caution from the bath. (**Face shield, heavy gloves, and tongs should be used when removing the hot bottle.**) After the bottle is allowed to cool, it is opened and the contents scraped into a blender containing 40 mL concentrated ammonium hydroxide and washed thoroughly. The mixture is filtered and the polymer allowed to stand 1 h in ammonium hydroxide solution. This treatment serves to remove most of the catalyst and the blue color. The polymer is filtered, washed with water in a blender three times, filtered, and dried in vacuum at 50°C. The inherent viscosity is about 1.4 (0.5 g in 100 mL *p*-chlorophenol/α-pinene (98/2) at 60°C). The overall yield is is about 30 g (65%).

303. Stabilization of Polyformaldehyde (10)

$$\sim\!\sim OCH_2OH + (CH_3CO)_2O \longrightarrow \sim\!\sim OCH_2O\overset{\overset{\displaystyle O}{\|}}{C}{-}CH_3$$

To a nitrogen-flushed 1-L three-necked flask, equipped with stirrer, condenser, and drying tube is charged 50 g polyformaldehyde (from either monomer or trioxane), 40 mL acetic anhydride, and 0.2 g sodium acetate. The mixture is refluxed for 45 min, and cooled; the precipitated polymer is freed from acetic anhydride by repeated washing with acetone in a blender and air dried at 60°C. The recovery should be about 80%, with some decrease in inherent viscosity from the original. The thermal stability exceeds that of the original polymer.

304. Copolymerization of Trioxane with Ethylene Oxide (10)

An 8-oz. beverage bottle is thoroughly baked dry and cooled in a nitrogen stream. It is charged with 50 g purified trioxane, flushed with nitrogen, and capped by means of a septum and crown cap with a hole. The trioxane is melted at 65°C and 1.5 g liquid ethylene oxide is injected below the surface of the trioxane using a cold syringe with a long needle. Then, 0.007 mL boron trifluoride dibutyl etherate in 5 mL cyclohexane is added. The bottle is then tumbled in a bath at 60°C. (The customary precautions should be taken, when handling the closed bottle.) After 40 min, the contents of the bottle, after cooling and venting cautiously, are washed into 1-L acetone containing 3 mL tributylamine, filtered, washed well with acetone, and air dried at 60°C. The yield is about 45 g (90%) and the inherent viscosity is about 1.0 (0.1 g in 100 mL *p*-chlorophenol/α-pinene (98/2) at 60°C). The crystallinity is somewhat less than homopolymeric polyformaldehyde, and the melt temperature T_m is about 173°C compared to 180°C for the homopolymer.

An interesting property of trioxane is its ability to polymerize in the solid state. The polymerization may be effected with cationic catalysts (11, 12) or ionizing radiation (13). The polymerization is an example of a topochemical reaction (see chapter 7) in that the crystal structure of the product is directly related to the crystal structure of the monomer.

305. Solid-State Polymerization of Trioxane

$$\text{trioxane} \longrightarrow (-CH_2O-)_n$$

An argon- or nitrogen-flushed beverage-bottle containing 20 g granular trioxane is capped with a rubber and partially evacuated. (Commercial trioxane is usually satisfactory; recrystallized material is better.) Then 100 mL silicon tetrafluoride gas is introduced into the bottle with a syringe. After 30 h at 50°C, 50 mL 5% sodium hydroxide is added, and the polymer is washed successively in a blender with 5% sodium hydroxide, water, and methanol. Conversion is 50 to 90%. Polyoxymethylene from solid-state polymerization consists of characteristic microscopic fibers that give pulp-like slurries and felt-like mats after filtration. The polymer melts about 185°C and gives tough, films on melt pressing. Inherent viscosities in *p*-chlorophenol/α-pinene (98/2) at 60°C (0.5% concentration) are usually greater than 1.0. Capping is necessary.

Trithiane, the sulfur analog of trioxane, has also been polymerized by BF_3 gas in the melt (14) and by γ-radiation in the solid state (15). The T_m of the polymer from ring opening is 245 to 260°C. Other interesting polyacetals have been prepared by the polymerization of the carbonyl group of acrolein, acetaldehyde, trichloroacetaldehyde, and others and of the thiocarbonyl group of thiocarbonyl fluoride.

306. Preparation of Polyacrolein by Anionic Polymerization (16)*

$$\underset{\substack{|\\ HC{=}CH_2}}{CH{=}O} \xrightarrow{NaCN} \left[-\underset{\substack{|\\ HC{=}CH_2}}{CH}-O- \right]_n$$

A 250 m-L round-bottom flask with stirrer, dropping funnel, thermometer, nitrogen inlet, and an opening plugged with a serum cap is flushed with nitrogen and heated with a bare flame to remove traces of water. Alternatively, the equipment (less serum cap) can be baked in an oven and allowed to cool under dry nitrogen. After cooling under dry nitrogen, a mixture of 20 mL pure acrolein and 80 mL dry tetrahydrofuran (THF) is added to the flask through the dropping funnel under nitrogen. (The acrolein should be at least 99% pure and should be distilled before using and dried by refluxing over calcium hydride. Likewise the THF should be dried over sodium metal or over calcium hydride). The reaction mixture is cooled to −50°C with vigorous stirring and 2 mL of a 0.07 *N* solution of anhydrous sodium cyanide in dry dimethylformamide is added from a hypodermic through the septum cap. The temperature during the course of the polymerization must not rise above −50°C. After 100 min, 2 mL methanol is added to stop the polymerization. The viscous solution is poured with stirring into 1-L petroleum ether to precipitate the polymer. It may be further purified by dissolving in THF and precipitating a second time with petroleum ether. The yield is about 9 g (54%).

307. Preparation of Polyacetaldehyde (17)

$$\underset{\substack{|\\ H}}{\overset{\substack{CH_3\\ |}}{C}}{=}O \xrightarrow[\text{ethylene}]{BF_3} \left[-O-\underset{\substack{|\\ H}}{\overset{\substack{CH_3\\ |}}{C}}-O-\underset{\substack{|\\ H}}{\overset{\substack{CH_3\\ |}}{C}}- \right]_n$$

Using a liquid nitrogen bath, 250 mL polymerization-grade ethylene (boiling point, bp, −104°C) is condensed in a 1-L four-necked flask equipped with a stirrer, thermometer (reading to −200°C), gas inlet tube, and a short condenser. The condenser terminates in a gooseneck adapter that is connected with a glass T to a dry nitrogen line and a double bubbler filled with mineral oil. After the ethylene has condensed, the gas inlet tube is replaced by an adapter with a serum stopper cap. Moisture must be rigorously excluded throughout the operation. Stirring is started, and 39 g (50 mL, 0.88 mol) pure acetaldehyde is injected

* D. B. Miller, personal communication.

slowly from a hypodermic to avoid excessive boiling of the ethylene (the syringe may be precooled in a dry polyethylene bag in a refrigerator at 0°C). The internal temperature of the reaction mixture is kept between −100 and −120°C, by lowering and raising the nitrogen bath. **Care should be taken that no freezing of the reaction mixture occurs.** Then 3 mL (13 mmol) gaseous BF_3 is injected into the mixture with a hypodermic syringe. In 10 to 20 min, the stirrer stops due to the formation of polymer. The reaction is allowed to stand for an additional 30 min while the ethylene is allowed to evaporate slowly; excess ethylene may be decanted with nitrogen blanketing. Then 100 mL anhydrous pyridine is added; on stirring at room temperature the polymer dissolves completely. The polymer may be isolated at this stage by pouring the pyridine solution into cold water, but it is relatively unstable. To increase the stability, the polymer is end capped by adding 300 mL acetic anhydride to the pyridine solution of the polymer, and stirring the mixture under nitrogen for 1 to 2 h. The brownish green viscous solution is poured into 1 kg ice and 1 L water, thereby precipitating the polymer. The polymer is then kneaded by hand (**Rubber gloves!**) in several changes of water and ice to destroy the residual acetic anhydride completely. The washing is complete when the wash water is colorless. The polymer is then dried under vacuum at 40°C. Yield is 34 to 37 g (87 to 94%). The inherent viscosity in butanone at 0.5% concentration is 1 to 2. An even more stable product is obtained if the polymer is dissolved in ether, washed with 1% acetic acid to remove pyridine, then 1% sodium carbonate, and finally several times with distilled water. The ether is dried over anhydrous sodium sulfate, filtered, and evaporated to give dry polymer.

Dichcloroacetaldehyde and trichloroacetaldehyde (chloral) have been polymerized to high molecular weight (18, 19). Both are rather unstable, with decomposition starting at fairly low temperatures; however, they can both be stabilized reasonably well by end capping via esterification. Polychloral is intractable but polydichloroacetaldehyde is soluble in convenient solvents, from which films can be cast.

308. Preparation of Polydichloroacetaldehyde (18)

$$Cl_2CHCHO \longrightarrow \left[\underset{\substack{|\\ CHCl_2}}{CH} - O \right]_n$$

Dichloroacetaldehyde can be purified by distilling over phosphorus pentoxide and fractionally distilling with retention of the heart cut for polymerization (bp 88 to 90°C/760 mm Hg). The aldehyde thus obtained is about 95% pure, with chloral as the major contaminant. Little or no chloral is incorporated in the polymer derived from this material.

In a 250-L resin kettle — dried at 120°C, cooled under dry nitrogen, and equipped with a nitrogen inlet and exit, drying tube, and serum cap — is placed 150 g dichloroacetaldehyde, purified as above. The aldehyde is cooled to 0°C, and 0.20 mmol boron trifluoride diethyl etherate is injected with a dry syringe.

The mixture is stirred at 0°C for 5 h. The polymer is transferred to a blender and washed three times with methanol, then air dried 3 to 4 h at room temperature in circulating air by spreading the polymer out in a thin layer in a flat dish or pan. The yield is about 105 g (70%).

The polymer should be stabilized without delay by refluxing 100 g polymer in 600 mL acetic anhydride containing 0.5 mL pyridine for 15 min. After cooling and filtering, the polymer is washed repeatedly with methanol in a blender and air dried as above. Uncapped polymer is destroyed by heating the polymer in DMF (1 g/100 mL) at 130°C for 30 min. The polymer is precipitated into a large excess of methanol, washing repeatedly in methanol, then drying in a vacuum at 50°C for 12 h. The reduced viscosity is about 3.5 (0.1 g in 100 mL THF at 25°C). The overall yield of capped material is about 80%. The polymer cannot be melted without decomposition. However, films can be cast from THF. Capped polymer is relatively stable at 190°C (weight loss 0.26%/min), but degradation becomes rapid at 220°C (3.45%/min).

12.2.2. Polymerization of Thiocarbonyl Fluoride

Thiocarbonyl fluoride is an unusual monomer, made by pyrolysis of the dimer. It polymerizes at low temperature, catalyzed by weak bases such as DMF. The polymer is tough and crystalline and melts about 35°C, above which temperature it is an elastomer. It can be copolymerized with olefins, using as catalyst a mixture of triethylboron and oxygen.

309. Preparation of Thiocarbonyl Fluoride (20)

$$CF_2\langle S_2 \rangle CF_2 \longrightarrow CF_2S$$

A total of 108 g thiophosgene dimer (21), 179 g antimony trifluoride, and 250 mL tetramethylene sulfone are heated with stirring at 90 to 100°C for 2 h. During this time, volatile products are formed, which are collected in an ice-cooled trap. By distillation of the condensate, there is obtained about 45 g 2,2,4,4-tetrafluoro-1,3-dithietane (bp 47 to 48°C). This compound is purified by shaking first with 25 mL 10% sodium hydroxide and then with 5 mL 30% hydrogen peroxide. After drying over silica gel, the colorless dithietane is distilled to give pure material, (bp 48°C; melting point; mp, −6°C).

The above dithietane is pyrolyzed by dropwise addition over a 2 h period to 0.5-in-dia., 25-in-long platinum tube that is heated to 475 to 500°C over a length of 12-in. During the pyrolysis, about 20 mL helium/min is passed through the tube. Gases emerging from the tube are condensed by passage through a trap externally cooled with solid carbon dioxide and acetone and then through a trap cooled with liquid nitrogen. Contents of both traps are combined and distilled through a column packed with Hastelloy helices to obtain thiocarbonyl fluoride as a colorless liquid that boils at −54°C.

310. Anionic Polymerization of Thiocarbonyl Fluoride (20)

$$CF_2S \longrightarrow (-CF_2S-)_n$$

A glass flask containing 75 mL (about 100 g) thiocarbonyl fluoride, is connected by means of glass tubing to a dry polymerization flask that has a small neck covered with a rubber septum cap and is fitted with a stirrer. Then 100 mL dry diethyl ether is added to the polymerization flask. Both flasks are frozen by cooling with liquid nitrogen, the system is evacuated, and the thiocarbonyl fluoride is distilled into the polymerization flask. The liquid nitrogen bath surrounding the polymerization flask is replaced with a solid carbon dioxide/acetone bath and dry nitrogen is added until the pressure inside the flask is at atmospheric pressure. When the contents of the flask are completely melted, the stirrer is started, and 5 drops of DMF are added from a no. 22 hypodermic needle. Polymerization begins almost at once and is essentially complete after 2 h, during which time the polymer separates as a white, spongy mass. It is removed from the flask and boiled in water containing 5 mL 50% nitric acid to destroy the DMF and dispel the ether. Inherent viscosity in chloroform at 0.5% concentration is 4 to 6. Films obtained by pressing at 150°C and 10,000 lb ram pressure are elastomeric as removed from the press. At room temperature, they become slowly opaque as the polymer crystallizes. The crystalline mp of the polymer is 35°C.

311. Copolymerization of Thiocarbonyl Fluoride and Propylene (22)

$$CF_2S + \underset{}{\overset{CH_3}{\overset{|}{C}}}H{=}CH_2 \longrightarrow \text{Copolymer}$$

A polymerization flask, as in the previous preparation, with a small neck covered with a serum cap and fitted with a stirrer is cooled with a solid carbon dioxide/acetone bath and charged with 300 mL dichlorodifluoromethane and 70 mL (43 g) propylene. A glass flask containing 35 mL (47 g) thiocarbonyl fluoride as measured at −78°C is connected to the polymerization flask and then frozen by cooling in liquid nitrogen. The system is evacuated, and the thiocarbonyl fluoride is distilled into the polymerization flask. The liquid nitrogen bath is replaced with solid a carbon dioxide–acetone bath, and the solid contents of the polymerization flask are allowed to melt under vacuum to remove dissolved oxygen. As soon as the melting is complete, nitrogen is added until atmospheric pressure is established. The solution is stirred and 0.75 mL 10% triethylboron in heptane is injected through the serum cap followed by injection of 9 mL oxygen. After 2.5 h, methanol is added to the polymerization mixture, which is then allowed to rise to room temperature. Dichlorodifluoromethane (**Re-cover!**) and excess propylene are distilled, and the remaining polymer is dissolved in 300 mL chloroform and precipitated by pouring into methanol. About 52 g polymer is obtained, which has a sulfur content of 32.02%, corresponding to a mole ratio of CF_2S/propylene of 2.34/1.

12.2.3. Polymerization of Aldehydes by the Tischenko Reaction

The Tischenko reaction does not normally go in high enough yield to be a polymer forming reaction. Sweeny found that with the proper selection of catalyst, aromatic dialdehydes such as terephthalaldehyde can be converted to polyesters with sufficiently high enough molecular weight to give tough films. The melt temperature, however, is relatively low because the polymer formed is a random copolymer of terephthalic acid, hydroxymethylbenzoic acid, and 1,4-bis(hydroxymethyl) benzene.

$$-OCH_2-C_6H_4-C(=O)-, \quad -C(=O)-C_6H_4-C(=O)-$$

$$\text{and } -OCH_2-C_6H_4-CH_2-$$

312. Tischenko Polymerization of Terephthaldehyde (23)

$$OHC-C_6H_4-CHO \xrightarrow{Et_3Al} \left[-C(=O)-C_6H_4-CH_2-O-\right]_n$$

A total of 2.5 g terephthaldehyde is suspended in 50 mL dry cyclohexane under nitrogen. The suspension is warmed to 80°C, and 0.5 mL (0.8 *M*) of a solution of triethylaluminum in cyclohexane is added. After stirring overnight at 80°C under nitrogen, a light yellow suspension remains in the flask. Then 100 mL methanol is added to the reaction mixture, and the polymer is filtered, washed with 200 mL methanol, and air dried. The polymer is purified by dissolving in 20 mL trifluoroacetic acid containing 1 mL tetrachloroethane. The solution is filtered, and the polymer reprecipitated by pouring into a large excess of methanol. The polymer is filtered, washed with methanol, and dried at 60°C under vacuum to give 1.9 g of white polymer. The inherent viscosity is 0.31 in tetrachloroethane/phenol (40/60 w/w) at 0.5% concentration. The polyester can be melt pressed at 120°C. to give tough, flexible films.

12.3. POLYMERIZATION OF MONOISOCYANATES

For many years isocyanates have been known to trimerize in the presence of certain basic catalysts. However, polymerization to linear polymer without cyclization was not observed until Shashoua, et al. (24) showed that basic catalysts, such as sodium cyanide in DMF, polymerize the –CN group to linear high polymer, provided the reaction is carried out at low temperatures. Structurally, the polymers are 1-nylons, i.e., polyamides of the hypothetical *N*-substituted carbamic acids; and many are liquid crystalline. The polymers can be obtained with molecular weights approaching 1 million. If the nitrogen substituent is sufficiently large, e.g., *n*-butyl, the polymers are soluble in benzene and tough films can be cast. The *n*-butyl polymer melts over 200°C and is crystalline. 1-Nylons tend to depolymerize in the melt or in solution at room temperature in the presence of catalyst. High molecular weight products may have inherent viscosities as high as 15 in benzene or DMF at 0.5% concentration. Very dilute solutions (2 to 3%) are viscous enough to cast very thin films.

313. Polymerization of n-Butyl Isocyanate (24)

$$CH_3CH_2CH_2CH_2N{=}C{=}O \longrightarrow [-\underset{\displaystyle C_4H_9}{\overset{|}{N}}-\overset{\displaystyle O}{\overset{\|}{C}}-]_n$$

Butyl isocyanate is prepared by the Curtius reaction of sodium azide with butyryl chloride. The isocyanate is distilled immediately before use (bp 113 to 115°C) and is stored under nitrogen. Catalyst for the polymerization is prepared by saturating dry dimethylformamide with anhydrous sodium cyanide under nitrogen for about 1 h. The resulting solution is a powerful anionic catalyst. A 250-mL three-necked flask is now equipped with a stirrer and two side-arm adapters. To one of these is attached a calcium chloride tube and a low-temperature immersion thermometer to determine the temperature of the reaction medium. The other is fitted with a nitrogen T inlet tube, the vertical arm of which is sealed with a rubber bulb. The empty flask is flamed out under nitrogen and allowed to cool in an inert atmosphere. Then 30 mL DMF is added and the flask and contents cooled to –58°C, which is close to the melting point of pure DMF. Then 10 mL freshly distilled *n*-butyl isocyanate is added, and the mixture is stirred and allowed to cool again to –58°C. The rubber bulb is pierced with a hypodermic needle and 1 mL catalyst solution is added dropwise over 3 min with vigorous stirring. After stirring 15 min at –58°C, 50 mL methanol is added to inactivate the catalyst and precipitate the polymer. The polymer is filtered and washed repeatedly with methanol, then dried at 40°C in a vacuum. The molecular weight is extremely high, and inherent viscosities on the order of 15 in benzene are routinely obtained. The yield is about 75%. The polymer is soluble in benzene but only to the extent of 2 to 3%. The viscosity is so high that solutions of this concentration can be cast successfully into film. These are clear and tough and similar in appearance to polyethylene. Poly(*n*-butyl-1-nylon) has a softening temperature of about 180°C and a melt temperature of 209°C. It gradually reverts to monomeric products when maintained at this temperature.

314. Polymerization of p-Methoxyphenyl Isocyanate (24)

$$CH_3O-C_6H_4-NCO \longrightarrow \left[-N(C_6H_4OCH_3)-C(=O)- \right]_n$$

Polymerization of *p*-methoxyphenyl isocyanate is carried out essentially in the same manner as in the previous example. A total of 35 g p-methoxyphenyl isocyanate (bp 82°C/2 mm) is mixed with 100 g dry DMF at −58°C and treated with 12 mL of a saturated solution of sodium cyanide in anhydrous DMF. After polymerization is completed at −58°C, the polymer is isolated as in the previous example in a yield of about 35%. The molecular weight is somewhat lower, the viscosity being in the range of 0.7 in DMF. In the case of the *p*-methoxyphenyl isocyanate, the polymer tends to revert to dimer in the presence of the polymerization catalyst, so the polymerization must be quenched as soon as high molecular weight is reached. Poly(*p*-methoxyphenyl-1-nylon) is soluble in DMF at fairly high concentrations, and films can be cast from this solvent and viscosity determinations carried out. The polymer melt temperature (PMT) is about 212°C.

This type of polymerization can be applied to 1,2-diisocyanates. In these cases, cyclopolymerization occurs.

$$O{=}C{=}N-CH(CH_3)-CH_2-N{=}C{=}O \longrightarrow \left[\text{cyclic } -N(-C(=O)-)-CH(CH_3)-CH_2-N-C(=O)- \right]_n$$

315. Preparation of 1,2-Propylene Diisocyanate (25)

(1) $N_2H_4 + CH_3O_2CCH(CH_3)CH_2CO_2CH_3 \longrightarrow NH_2NHCOCH(CH_3)CH_2CONHNH_2$

(2) $NH_2NHCOCH(CH_3)CH_2CONHNH_2 \longrightarrow OCN-CH(CH_3)CH_2NCO$

In this preparation, the diisocyanate is prepared from the diazide via the dihydrazide.

315.A. 2-Methylsuccinic Dihydrazide

A total of 545 g (10.9 mol) hydrazine hydrate dissolved in 7 L 95% ethanol and 1-L water is mixed with 429 g (2.7 mol) dimethyl 2-methyl succinate. This ester (bp 85 to 86°C/16 mm) is obtained by catalytic hydrogenation of dimethyl itaconate. The mixture is refluxed overnight and then concentrated; 367 g (84%) 2-methylsuccinic dihydrazide (mp 160 to 162°C) crystallizes.

315.B. 1,2-Propylene Diisocyanate

A 2-L beaker equipped with a thermometer and a mechanical stirrer and externally cooled in an ice bath is charged with 600 g ice, 100 mL benzene, 70 mL concentrated HCl, and 57 g (0.36 mol) 2-methyl-succinic dihydrazide. A solution of 50 g (0.73 mol) sodium nitrite in 100 mL water precooled to 0°C is added dropwise to the stirred mixture over 20 min. The temperature of the reaction mixture is maintained below 8°C by addition of pieces of ice, and stirring is continued for 30 min after all the nitrite is added. The layers are separated, and the benzene solution of the azide is combined with two 150-mL benzene extracts of the aqueous layer. The benzene solution is dried overnight over anhydrous calcium chloride, filtered, added to a dry 1-L flask equipped with a reflux condenser equipped with a drying tube. The solution is refluxed for 4 h to decompose the azide. The refluxing solution develops a blue color that disappears when the decomposition is essentially complete. The condenser is replaced with a distillation head, and the benzene removed by distillation. Under nitrogen, the residue is transferred to a dry 100-mL round-bottom flask, containing a Teflon-coated magnet, and equipped with a 4-in. Vigreux column, receiver condenser, and a three-flask collection adapter. All equipment is previously oven dried. The residue is then distilled under reduced pressure, and after a short forerun, about 19.6 g (42%) of pure 1,2-propylene diiscyanate is collected as a lachrymatory liquid (bp 83.5°C/25 mm).

316. Polymerization of 1,2-Propylene Diisocyanate (25)

$$CH_3-CH(N{=}C{=}O)-CH_2-N{=}C{=}O \longrightarrow \left[-N(C{=}O)N(C{=}O)- \text{ (4-methylimidazolidin-2-one ring)} \right]_n$$

A 100-mL three-necked flask equipped with a mechanical stirrer, calcium chloride tube a low-temperature thermometer, and T fitted with a rubber bulb, is charged with 50 mL dry DMF and then cooled to −40°C in a dry ice–acetone bath. Then 5.2 g 1,2-propylene diisocyanate, precooled at 0°C, is added with stirring to the flask through the rubber bulb, using a dry hypodermic syringe. When the temperature of the contents are again at −40°C, 1 mL of a saturated solution of dry sodium cyanide is added dropwise over 3 min through the rubber bulb, using a dry hypodermic syringe. The temperature rises to about −14°C within 6 min after the initial addition of the initiator and the polymerization mixture thickens to a smooth, viscous dope. The cold bath is removed, and the mixture is stirred for 30 min. Polymer is isolated by precipitation into methanol and stirred vigorously in a blender. (The blender motor should be blanketed with nitrogen to avoid risk of fire.) The polymer is filtered, washed with excess methanol, and dried at 40°C under vacuum. The dried polymer weighs about 4.6 g (89%). The inherent viscosity should be in the range of 1.3 to 1.4 at 0.5% concentration in DMF. Clear, flexible films are obtained when a 10% solution in formic acid or nitromethane is spread on a glass plate and the solvent evaporated.

12.4. POLYCARBODIIMIDES

Phospholene oxides convert isocyanates to carbodiimides and hence diisocyanates to polycarbodiimides. This reaction has been studied in detail (26–28). The yields are high, and traces of catalyst convert diisocyanates to high molecular weight polymer, capable of being melt pressed to tough film. Melt viscosities are high and solubility low, possibly from reversible bridging of the $-N{=}C{=}N-$ groups. This is supported by the fact that a low-melting, easily soluble polymer results from the hindered mesitylene diisocyanate.

Metastable solutions of polycarbodiimides can be made by adding small amounts of highly polar substances such as dimethyl sulfoxide to the reaction medium. In the solid state, polycarbodiimides are quite inert to water, amines, etc. but are quite reactive in solution. For example, polyguanidines are easily made by reaction of amines with solutions of polycarbodiimides (29).

317. Preparation of 1-Ethyl-3-methyl-3-phospholene-1-oxide (26, 30)

$$CH_2{=}C(CH_3){-}CH{=}CH_2 + C_2H_5PCl_2 \longrightarrow \text{1-ethyl-3-methyl-3-phospholene-}P,P\text{-dichloride } (C_2H_5PCl_2 \text{ ring}) \longrightarrow \text{1-ethyl-3-methyl-3-phospholene-1-oxide } (C_2H_5P{\rightarrow}O \text{ ring})$$

A 2-L four-necked flask is fitted with a spiral condenser topped with a dry ice condenser, thermometer, 1-L dropping funnel with pressure equalizing side arm, and magnetic stirrer. To this flask is added 1 g copper stearate, 780 g (5.96 mol) dichloroethylphosphene, and from the dropping funnel 447 g (6.56 mol) freshly distilled isoprene. The reaction mixture is stirred and refluxed under nitrogen for 42 h, cooled, allowed to stand for 2 days, and then refluxed without stirring for 5 days. Excess isoprene is then distilled from the mixture, and 850 mL water is added dropwise with stirring to the reaction flask, which is cooled in an ice bath. The dark brown aqueous solution is transferred to a 5-L flask, and 1250 mL 30% aqueous sodium hydroxide is added gradually to make the solution slightly alkaine (pH 8). The mixture is filtered, and the aqueous solution is extracted continuously with chloroform for 12 days. The chloroform is removed, and the residue vacuum distilled through a 25-cm Vigreux column to give 435 g (51%) of water white liquid with a slight odor of phosphine. The product is further purified by oxidation at 50°C with excess 3% hydrogen peroxide for 6 h. The aqueous mixture is extracted continuously with benzene, and the oxide is recovered by distillation (bp 115 to 119°C 1 to 2 mm). This is the preferred catalyst for forming polycarbodiimides.

318. Polymerization of Methylene Bis(4-phenyl Isocyanate) (28)

$$OCN-C_6H_4-CH_2-C_6H_4-NCO \longrightarrow \left[-N{=}C{=}N-C_6H_4-CH_2-C_6H_4-\right]_n + CO_2$$

In a three-necked 500 m-L flask equipped with stirrer, condenser, and nitrogen inlet is placed 150 mL xylene, 20 g methylene bis(4-phenyl isocyanate), and 0.03 g 1-ethyl-3-methyl-3-phospholene-1-oxide. The mixture is heated to reflux; first the solution becomes milky, then a second liquid phase begins to separate. The liquid phase becomes more and more viscous and eventually yields high molecular weight fibrous material. The fibrous nature of the product results from the shearing action of the stirrer blade on the rapidly thickening prepolymer that separates initially. After the polymer is separated and air dried, these short filaments can be separated manually from the bulk of the polymer and can be cold drawn. The drawn filaments are cream colored and quite tough. X-ray examination of the fibers showed approximately 30% lateral crystallinity and 5% longitudinal crystallinity, coupled with a high degree of orientation.

319. Preparation of Poly(3,3′-dimethoxy-4,4′-biphenylene Carbodiimide) (28)

$$OCN-C_6H_3(OCH_3)-C_6H_3(OCH_3)-NCO \longrightarrow \left[-C_6H_3(OCH_3)-C_6H_3(OCH_3)-N{=}C{=}N-\right]_n$$

A total of 10 g 3,3′-dimethoxy-4,4′-biphenylene diisocyanate is dissolved in 100 mL hot (about 100°C) xylene. The solution is filtered free of foreign matter and polymerized with 0.04 g 1-ethyl-3-methyl-3-phospholene oxide in a three-necked flask with refluxing and stirring. Polymerization is rapid; however, for high molecular weight, the mixture should reflux for 4 to 6 h. The white, finely divided polymer is filtered, washed with benzene, and dried. The yield is 8 to 9 g. The polymer has high x-ray crystallinity. It can be melt pressed at 250°C to a clear film with a slight yellow color. Cut strips of the film can be stretched three to four times at 160°C. The film exhibits typical necking phenomena on stretching, and shows strong birefringence that disappears at about 190°C. X-ray examination of the drawn polymeric material shows that the product is crystalline with good longitudinal order and a high degree of orientation.

320. Preparation of Polycarbodiimide from 2,4-Tolylene Diisocyanate (29)

$$CH_3C_6H_3(NCO)_2 \longrightarrow \left[-C_6H_3(CH_3)-N{=}C{=}N- \right]_n$$

A mixture of 11 g 2,4-tolylene diisocyanate, 50 mL toluene, 4.0 mL dimethylsulfoxide, and 0.03 g of 3-methyl-1-ethyl-3-phospholene-1-oxide is refluxed for 1 h. The extremely viscous material that results can be cast to a tough film. Strips of the film can be hot drawn, but develop no crystallinity.

321. Preparation of Polyguanidine (29)

$$\left[-C_6H_3(CH_3)-N{=}C{=}N- \right]_n + RNH_2 \longrightarrow \left[-C_6H_3(CH_3)-NH-C(NHR){=}N- \right]_n$$

If the solution prepared in the previous experiment is poured into a rapidly agitated solution of 15 g aniline and 100 mL toluene, a polyguanidine precipitates, which on filtering and drying can be pressed into stiff clear film at 275°C. Other amines may be used; *t*-butylamine gives a soluble product. Polyguanidines are soluble in dilute acids and have surfactant-like properties.

12.5. POLYMERIZATION OF DIAZO COMPOUNDS

An interesting type of polymerization is the conversion of diazoalkanes to hydrocarbons in the presence of catalysts such as boron trifluoride.

$$RCHN_2 \longrightarrow N_2 + -[-CH-(R)-]_n-$$

Several different mechanisms may be involved, depending on the catalyst. For example, with boron fluoride or boron esters, an intermediate such as $BF_3 \cdot CH_2^{\oplus}$ may be involved; whereas with copper powder or colloidal gold, a carbene structure may be important (31–33).

High molecular weight polymethylene can be prepared from diazomethane, but laterally substituted diazoalkanes yield lower molecular weight hydrocarbons. Apparently, steric factors inhibit the formation of higher molecular weight products. Diazo polymers seem to be living polymers, and block polymers can be produced (34, 35). Isotactic polyethylidine has been prepared using finely divided gold as catalyst, and the polymer has a crystalline mp of 195°C (33, 36). Because diazoalkanes are toxic and potentially explosive, due safety precautions should be taken when handling.

322. Polymerization of Diazomethane (32)

$$CH_2N_2 \longrightarrow -(CH_2)-$$

A solution of 13.7 g diazomethane in about 700 mL ether is treated with 0.3 g freshly distilled trimethyl borate at 0°C. Nitrogen evolves slowly, and a precipitate begins to form in the solution. After 24 h, the diazomethane is completely decomposed, as shown by lack of color in the solution. A rubbery solid is obtained by filtration and weighs 3 to 5 g. It may be pressed into thin films in a Carver press at 180 to 200°C.

323. Polymerization of Diazododecane (32)

A mixture of 160 g dodecylurea and 640 mL glacial acetic acid is heated on a steam bath until a clear solution is obtained. The solution is cooled to 0°C with vigorous stirring and a solution of 200 g sodium nitrite in 350 mL water is added with stirring over 15 min. After 15 min at 0°C, an equal volume of ice water is added and the precipitated nitrosodecylurea is collected on a filter at 0°C. Then 250 mL 40% aqueous potassium hydroxide, 500 mL ethanol, and 500 mL ligroin are mixed and cooled to 0°C. The crude nitroso decylurea is then added over 30 min with vigorous agitation. After a further 15 min stirring at 0°C, the upper layer is removed and filtered. This solution of 1-diazododecane is used without purification in subsequent polymerizations.

For estimating the concentration of diazododecane, an aliquot of about 30 mL is added to 0.5 g benzoic acid in benzene. Nitrogen is evolved, and the color of the diazo compound is discharged. The excess benzoic acid is then titrated with standard sodium hydroxide solution and the quantity of diazododecane present in the original aliquot is calculated. One molecule of benzoic acid is equivalent to one of the diazo compound. The yield of diazo compound is low, about 5.2 g.

324. Preparation of a Copolymer of Diazomethane and Diazododecane (32)

To a solution of 2.2 g diazododecane and 6 g diazomethane in a total of about 700 mL ether-ligroin is added 0.5 g freshly distilled trimethylborate. The mixture is allowed to stand for 24 h, and the precipitated polymer is collected, washed, and dried. The yield is about 2 g of a translucent rubbery solid, soluble in chloroform, benzene, and other such solvents. It becomes soft at about 250°C but does not decompose completely until 370°C. It is of very high molecular weight with an inherent viscosity in the range of 6 to 7 in chloroform.

12.6. PREPARATION OF POLYXYLYLENE

Before 1960, two methods were available for the preparation of poly(*p*-xylylene); the first is the pyrolysis of *p*-xylene to form quinodimethane, which polymerizes on quenching (37, 38); the second involves the decomposition of a *p*-xylylene trimethylammonium halide in strong alkali, a method claimed to give less branching and cross-linking (39–41). In both cases, the cyclic dimer di-*p*-xylylene (paracyclophane) is an interesting by-product and, as it turns out, a useful one in its own right for the synthesis of poly(*p*-xylylene).

A number of newer synthetic methods have been developed, and aliphatic- and aromatic-substituted poly(*p*-xylylenes) can be prepared with acceptable molecular weight. These include *(1)* pyrolysis of the cyclic dimer and its ring-substituted derivatives (42), *(2)* treatment of α-halogenated *p*-xylenes from (α-halo to $\alpha,\alpha,\alpha',\alpha'$-pentahalo-) with base (43), *(3)* pyrolysis of bis(trichloromethyl)benzene over copper above 300°C (44), and *(4)* electrolysis of α,α'-dihalobenzenes (45).

Poly(*p*-xylylene) is stable at elevated temperature (300°C) and is crystalline. It is hard to fabricate from melt or solution; and only weak, brittle fibers have been obtained. Tough, clear, drawable films can be obtained directly from quenching of the monomer on the interior surface of the collection flask.

325. Preparation of Poly(p-xylylene) by Pyrolysis of Xylene (46, 47)

$$CH_3-C_6H_4-CH_3 \xrightarrow{H_2} CH_2{=}C_6H_4{=}CH_2 \longrightarrow \left[-CH_2-C_6H_4-CH_2-\right]_n$$

The successful dehydrogenation of *p*-xylene requires high temperatures (700 to 1100°C) and reduced pressures (1 to 5 mm) under nitrogen. The most suitable technique to convert *p*-xylene to the transient quinodimethide, which polymerizes on cooling, is to pass vapors of *p*-xylene through a quartz tube heated by an external heater to temperatures of 900 to 1000°C. The xylene is vaporized by boiling at reduced pressure, and the vapors are first passed through a capillary tube to give a constant flow of gas and then into the quartz pyrolysis tube, 19 in long by 2 in wide, held at the required temperature (900 to 1000°C), preferably by use of a multiple unit furnace. One such as used in combustion analysis is satisfactory. The exit gases are passed from the hot tube directly into a series of four traps. The first consists of a 1-L round-bottom standard taper flask held at room temperature, followed by three cold traps maintained at dry ice temperature. The entire system is kept at a pressure of 1 to 5 mm by the use of an oil pump.

The *p*-xylene vapors are passed through the quartz tube, where they are dehydrogenated. On exiting the flask maintained at room temperature, the monomer polymerizes on the surface of the flask wall as a continuous tough, clear film. In addition, smaller amounts of polymer are collected as a solid fluff in the succeeding traps maintained at −80°C. Residual xylene also collects in the traps as well as gaseous by-products, exclusive of hydrogen. The polymerization is surprisingly efficient under the conditions described. Conversion to polymer ranges from 12 to 20%, the remaining xylene is recovered as such.

The film of polymer laid down on the inside walls of the collection flask can usually be removed in one piece. The film can be stretched at very high temperatures to give an oriented product. Inherent viscosities may be determined at very high temperatures (305°C) in benzyl benzoate. The viscosities obtained are often anomalously low because polymers prepared in this manner are not strictly linear and are partially cross-linked.

326. Purification of Trimethyl(p-methylbenzyl)ammonium Bromide (42)

$$CH_3-C_6H_4-CH_2Br + N(CH_3)_3 \longrightarrow CH_3-C_6H_4-CH_2\overset{+}{N}(CH_3)_3Br$$

Trimethy (*p*-methylbenzyl)ammonium bromide is prepared in about 90% yield by quaternization of *p*-methylbenzyl bromide with trimethylamine. The salt is purified as follows: 300 g of the salt is dissolved in 200 mL boiling absolute ethanol. The hot solution is filtered and cooled to about 1°C in an ice bath. The product is filtered, air dried, and oven dried at 110°C in a vacuum. About 65% of the crude salt is recovered as a white, pure product (mp 201 to 203°C).

327. Preparation of Poly(p-xylylene) from Trimethyl(p-methylbenzyl)ammonium Bromide (42)

$$CH_3-C_6H_4-CH_2-N^+(CH_3)_3 \; {}^-Br \xrightarrow{NaOH} \left[-CH_2-C_6H_4-CH_2-\right]_n$$

A 3-L three-necked flask is equipped with a paddle-type stirrer, a reflux condenser, and a nitrogen inlet with a stoppered opening for adding chemicals rapidly to the flask. In the flask is placed a solution of 800 g sodium hydroxide in 1200 g distilled water. The solution is heated to the boiling point with stirring. To the boiling alkali, blanketed with nitrogen, is added all at once a solution of 234 g trimethyl(*p*-methylbenzyl)ammonium bromide dissolved in 250 mL water. The mixture is refluxed for 4 h, during which time the solid poly-*p*-xylylene is formed as a suspended white solid. The reaction mixture is poured into 10-L water, and the suspension is filtered. The solid polymer is washed on the filter with several portions of hot water, and then extracted with 250-mL portions of boiling ethanol. The extracted polymer is then washed with ether and dried in a vacuum. The polymer yield is about 60 g. Poly-*p*-xylylene is soluble in benzyl benzoate at 305°C. Poly-*p*-xylylene made this way is very high melting and crystalline and difficult to fabricate. However, it can be pressed to films in a laboratory press at temperatures in excess of 305°C. Films are highly crystalline and brittle.

Di-*p*-xylylene and a variety of its substituted derivatives undergo pyrolytic polymerization (42), as does *p*-xylene; but yields are higher, a linear polymer is formed, pyrolysis temperatures are lower, low molecular weight by-products are absent, and a much greater variety of substituted polymers can be made because of the lower temperatures used. The pyrolysis vapors are condensed on a cool surface, as with *p*-xylene. A threshold surface temperature is observed for each di-*p*-xylylene above which condensation and polymerization will not occur at the reduced pressure in question. Thus if monoacetyl-di-*p*-xylylene is pyrolyzed, and the vapors are passed first over a surface at 90°C and then a second surface at 25°C, the polymer recovered from the first surface is poly(acetyl-*p*-xylylene) and from the second poly(*p*-xylylene).

This indicates that the active species of both halves of the starting acetyl-*p*-xylylene (i.e., one with acetyl attached and one without) are present in the pyrolysate and polymerize only below their respective threshold temperatures. Whether the active units are present in the quinoid or diradical form is not known with certainty.

328. Preparation of Poly(p-xylylene) from Di-p-xylylene by Pyrolytic Polymerization (42)

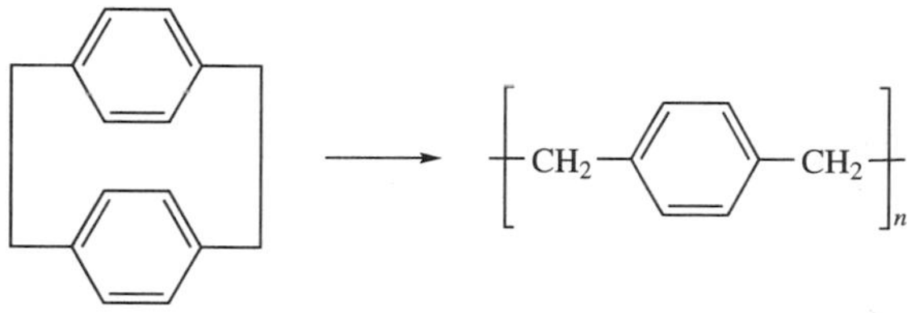

$$\longrightarrow \left[-CH_2-C_6H_4-CH_2- \right]_n$$

The isolation of di-*p*-xylylene (paracyclophane) has been described from a Hoffman elimination reaction on *p*-methylbenzyl trimethylammonium hydroxide (48). Poly(*p*-xylylene) is formed as the main product. The reaction is similar to that described in preparation 327. The cyclic dimer is obtained in 10 to 12% yield.

The apparatus consists of a 1-in inner diameter (id). Vycor tube about 24-in in length. It is mounted horizontally in a suitable furnace, with the first 6 in of the tube extending from the furnace. This is the zone from which distillation or sublimation of the sample is effected and is brought to the necessary temperature by heating tapes; a thermocouple is fixed against the glass to estimate the temperature. As an alternative, the distillation zone can be a separate piece of glass consisting of a sample chamber fixed for heating by means of refluxing solvent vapor in the manner of an Abderhalden drying apparatus, and connected to the pyrolysis tube by a standard Vycor joint. The temperature of the pyrolysis section is measured by a thermocouple in the middle of the furnace between the tube and the furnace wall. Glass wool is packed around the tube at the ends of the furnace to reduce air currents and temperature fluctuations in the furnace chamber. The pyrolysis tube leads into another 24-in-long by 1-in-id piece of glass tubing, serving as the deposition area, the first 15 in of which are heated by ir lamps to 90°C, and the remaining 9 in are kept at room temperature. The threshold temperature of *p*-xylylene is 30°C. The end of the deposition tube is connected by vacuum tubing to a dry ice trap and then a vacuum pump.

Then 1.5 g di-*p*-xylylene is placed in the distillation zone in a porcelain boat; the pressure is reduced to about 100 μ, and the distillation zone is brought to 175°C, after the pyrolysis section has first been heated to 500 to 600°C. Distillation occurs over a period of about 20 min to give essentially complete conversion to a film of polymer that can be stripped from the wall of the deposition tube, extracted with boiling carbon tetrachloride (**Carcinogen!**), and dried. It is soluble only, but completely, in benzyl benzoate and chlorinated biphenyls above 200°C. The films have good mechanical strength (6800 psi tensile strength; 10 to 15% elongation; 350,000 psi tensile modulus). The crystalline melt temperature (by x-ray) is 400°C and the glass-transition temperature T_g is 80°C.

It is of interest to note that *p*-bis(trichloromethyl)benzene can be pyrolyzed at 300 to 600°C over copper gauze to give the isolable (at −78°C) monomer α,α,α′,α′,-tetrachloro-*p*-xylylene. This can be polymerized directly by allowing the pyrolysis vapors to condense on a surface below the threshold of 140°C. The resulting polymer softens at 280 to 290°C but does not melt up to 350°C (44).

$$Cl_3C-C_6H_4-CCl_3 \xrightarrow{Cu} Cl_2C{=}C_6H_4{=}CCl_2 + CuCl \quad (1)$$

$$(1) \longrightarrow \left[C_6H_4-CCl_2-CCl_2 \right]$$

$$ClCH_2-C_6H_4-CH_2Cl \xrightarrow{^{-}B} ClCH_2-C_6H_4-{}^{-}CHCl + BH \quad (1)$$

$$(1) \longrightarrow H_2C{=}C_6H_4{=}CHCl + HCl \quad (2)$$

$$(2) \longrightarrow \left[CH_2-C_6H_4-CHCl \right]_n \quad (3)$$

The action of strong bases on α-halo-*p*-xylenes affords an approach to a variety of poly(*p*-xylylenes), and at the same time gives some insight into the mechanism of how these compounds interact with bases (43). The reaction evidently proceeds through a 1,6-elimination of hydrogen halide to give a quinodimethane that undergoes self-addition to high molecular weight polymer. When a dilute solution of potassium *t*-butoxide is added to an excess of α,α'-dichloro-*p*-xylene, HCl is eliminated and poly(α-chloro-*p*-xylylene) is formed. With excess base, another mole of HCl is eliminated to yield poly(1,4-vinylenebenzene). The more α-halogens present in a *p*-xylene, the weaker the base necessary to effect polymerization. In the tetrachloro compound, sodium hydroxide is sufficient. Higher molecular weights result when the reaction is carried out in *t*-butanol/methanol mixture, but some HCl elimination and methoxyl substitution also occur, although to a minor degree.

329. Preparation of Poly(α,α,α′-trichloro-p-xylylene) (43)

$$Cl_2CH-C_6H_4-CHCl_2 \longrightarrow \left[-C_6H_4-CHCl-CCl_2- \right]_n$$

A total of 200 g terephthaldehyde is refluxed and stirred for 1 h with a mixture of 600 g phosphorus pentachloride in 1500 mL carbon tetrachloride, using a condenser with a drying tube to exclude moisture. The major part of the excess phosphorus pentachloride and solvent is distilled, and the residue poured into about 500 g cracked ice. The resulting solid is filtered and recrystallized several times from heptane to give about 260 g (70%) $\alpha,\alpha,\alpha',\alpha'$-tetrachloro-*p*-xylene (mp 92 to 93°C). To a solution of 0.80 g (0.02 mol) sodium hydroxide in 30 mL 95/5 (v/v) *t*-butanol–methanol is added 4.88 g (0.02 mol) $\alpha,\alpha,\alpha',\alpha'$-tetrachloro-*p*-xylene. The mixture is stirred and refluxed for 1 h then poured into 300 mL methanol, filtered, washed in a blender with methanol and water (successively), and finally dried in a vacuum at 40°C. About 2.1 g (50%) of white polymer is obtained, which is soluble in THF. The reduced viscosity of the polymer in THF is about 3.0 (0.2 g/100 mL at 20°C). Strong films can be obtained by casting from THF solution or by compression molding at 190°C. The T_g is around 160°C.

In the electrolytic methods for synthesizing poly(*p*-xylylenes), a mercury or lead cathode is used in conjunction with a carbon rod anode and the Antrol constant potential power supply, with the monomer dissolved in acidified dioxane-water solution. The reaction probably goes through the intermediate quinodimethane, common to all poly(*p*-xylylene) syntheses. A host of substituted polymers can be prepared by this method.

12.7. POLYMERIZATION OF NORBORNYLENE

The polymerization of olefins by organometallic titanium catalysts usually proceeds without any rearrangement of the carbon skeleton. However, in the case of norbornylene it is possible to get two types of polymer, the structures of which depend on the ratio of the catalyst components. With a molar ratio of titanium tetrachloride to lithium tetraheptyl >1, a low yield of a stiff brittle polymer A is produced; with an excess of the lithium component, polymer B is formed by a unique ring-opening polymerization (49–51). Polymerization of norbornylene is discussed in chapter 8.

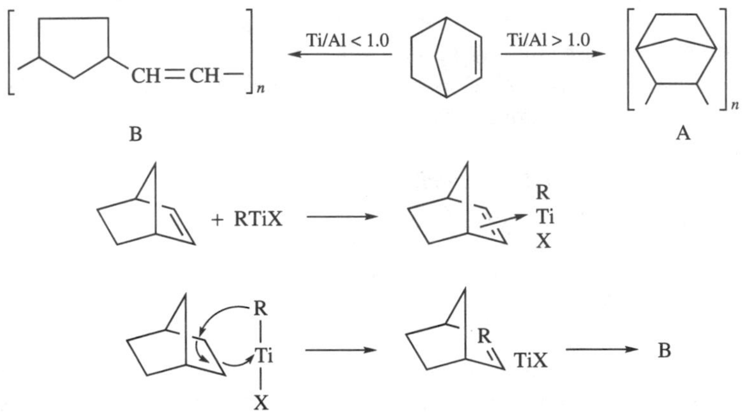

330. Preparation of Lithiumaluminum Tetraoctyl (51)

$$LiAlH_4 + CH_3(CH_2)_5CH{=}CH_2 \longrightarrow LiAl(C_8H_{16})_4$$

A mixture of 13 g (0.34 mol) lithium aluminumhydride, 285 mL (2.0 mol) 1-octene, and 300 mL decahydronaphthalene is heated to reflux under nitrogen in a 1-L three-necked flask, fitted with a stirrer, reflux condenser, and Glascol heater. The temperature gradually rises from 115 to 135°C over 5 h, at which time the reaction is complete. The mixture is filtered hot with suction through paper under nitrogen. The insoluble residue on the paper and in the flask is pyrophoric, and should be quenched quickly in isopropanol. The filtrate on cooling, deposits crystals that are conveniently freed of solvent under nitrogen by using a filter stick. The last trace of solvent is removed by drying under vacuum at room temperature. The white crystalline solid is dissolved in about 1-L xylene, and standardized by titration of an aliquot with standard acid using a pH meter. The concentration of the solution usually ranges from 0.35 to 0.40 molar in lithium aluminum tetraoctyl. Exposure of the solid or solution to air leads to loss of activity.

331. Polymerization of Norbornylene (51)

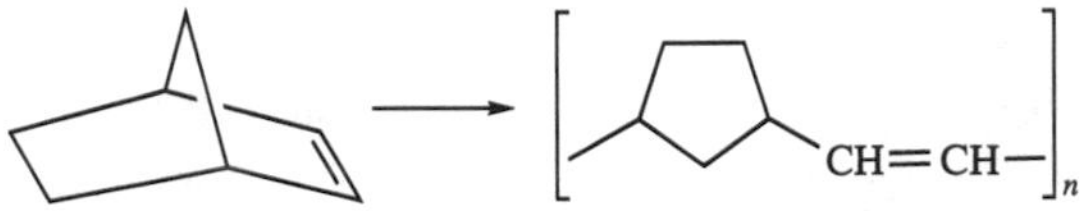

This polymerization is carried out in an inert atmosphere in any convenient equipment. A solution of 0.02 mol lithium aluminumtetraoctyl in xylene is added to 1.1 mL (0.01 mol) titanium tetrachloride in 50 mL dry decahydronaphthalene. The mixture is allowed to stand for 10 min and then 47 g (0.5 mol) norbornylene in 94 mL benzene is added. After 24 h, the polymer is worked up to give 14.5 g (31%) of white powder. The powder can be pressed to a clear, stiff, tough film at 225°C. Low molecular weight benzene-soluble polynorbornylene can be separated by extraction of 34 g of the polymer using 300 mL benzene in a Soxhlet extractor. The extraction is performed under nitrogen.

12.8. 1-*n*-POLYAMIDES

Polyamides based on methylenediamine may be prepared by the reaction of formaldehyde with dinitriles in strongly acidic media (52–55). The products are 1-*n* nylons, where *n* is the number of carbon atoms in the nitrile. The reactions involved are reversible, and the reaction product must be washed free of acid catalyst; otherwise, rapid degradation will result. Although the polymer structure is mostly that of linear polymethyleneamide, there is evidence of chain branching and presence of considerable –CN groups. The PMT for 1-6 nylon is about 300°C (35°C higher than for 6-6). The difference is due primarily to the closer arrangement of the amide groups in the 1-6 nylon.

332. Preparation of Poly(methyleneadipamide) (1,6-Nylon)

$$NC-(CH_2)_4-CN + CH_2O + H_2O \longrightarrow \left[-\overset{O}{\overset{\|}{C}}-(CH_2)_4-\overset{O}{\overset{\|}{C}}-NH-CH_2-NH- \right]_n$$

In a 5-L three-necked flask equipped with a stirrer and dropping funnel are put 78.0 g (0.50 mol) adiponitrile, 15.4 g (0.171 mol) trioxane, and 600 mL 98% formic acid. The solution is cooled to 10°C, and 200 g (2.0 mol) concentrated sulfuric acid is added with stirring and cooling over 10 min. The solution is then stirred for 1 h at 26 to 28°C. At this point, a gel forms, to which is then added, with vigorous stirring, 4 L water. The white powder formed is filtered and washed successively with dilute sodium carbonate solution, water, and ethanol. After drying in a vacuum at 70°C, the yield of polymer is 31 to 47 g (40 to 60%). The inherent viscosity in cresol is 0.6 to 1.2 at 0.5% concentration. The polymer melts at 290 to 300°C with decomposition.

12.9. POLY(PHENYLENE ETHERS) BY OXIDATIVE POLYMERIZATION

The preparation of high molecular weight polyphenylene ethers has been of interest to polymer chemists for some time. Early work (56, 57) prepared polymers of moderate molecular weight by the ferricyanide oxidation of 2,6-dialkyl-4-halophenols. This was presumed to involve the displacement of halogen by an aryloxy radical. The most thorough and enlightening work in this area was reported by Hay's group (58–60). These workers found that passing air through a vigorously agitated solution of 2,6-dialkyl phenol in an organic solvent containing an amine (e.g., pyridine) and copper(I) catalyst gave high molecular weight polymer.

333. Preparation of Poly(2,6-dimethyl-p-phenylene Ether) (58–60)

$$n\,\text{HO}{-}C_6H_3(CH_3)_2 + \tfrac{1}{2}n\,O_2 \longrightarrow \left[{-}C_6H_2(CH_3)_2{-}O{-} \right]_n + \tfrac{1}{2}n\,H_2O$$

A mixture of 9 mL pure pyridine and 30 mL pure nitrobenzene and 0.04 g cuprous chloride is shaken in an atmosphere of oxygen until the Cu(I) is converted to Cu(II). This occurs fairly rapidly; if desired, this reaction and the subsequent polymerization may be followed quantitatively by using a closed system and a gas burette. After the catalyst is prepared, 0.977 g (0.008 mol) pure 2,6-dimethylphenol is added, and vigorous agitation is continued. The absorption of oxygen is complete in about 30 min. The polymer is precipitated by pouring into 150 mL 1% concentrated aqueous HCl in methanol. The solid is filtered, slurried with 5% concentrated HCl in methanol, filtered, dissolved in chloroform, filtered, and reprecipitated into methanol. The yield is about 0.8 g polymer with an inherent viscosity of about 1.0 measured in chloroform at 0.5% concentration. An osmotic molecular weight of 28,000 was determined.

12.10. DIMETHYLKETENE POLYMERS

Ketenes show the structural possibilities of polymerization through the opening of a carbon-carbon double bond and through the carbonyl in the fashion of an aldehyde. Both types of polymer are known as well as a mixed species (61, 62). Polymerization of the double bond using a strong Friedel-Crafts catalyst at low temperatures gives a polyketone, and polymerization of the carbonyl to a polyacetal occurs with an anionic initiator also at low temperatures. A third dialkylketene polymer has also been made that has a polyester structure resulting from the alternation of the first two modes of polymerization (63). In this case, a trialkylaluminum catalyst at −25°C is needed. Dimethylketene and acetone have also been copolymerized to a polyester using a lithium alkoxide catalyst (64). Dimethylketene boils at 34°C at atmospheric pressure and may be prepared by the pyrolysis of the commercially available dimer 1,1,3,3-tetramethylcyclobutanedione (from Eastman, Organic Chemicals). The method has been detailed (65).

334. Preparation of a Polyketone from Dimethylketene (60–63)

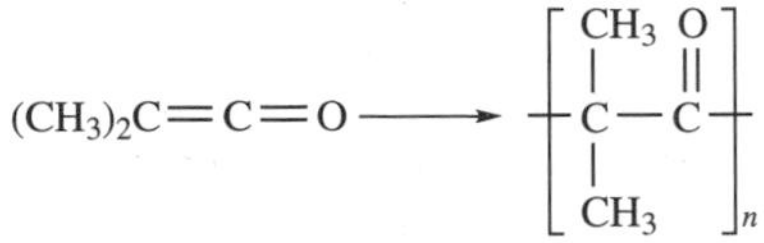

$$(CH_3)_2C{=}C{=}O \longrightarrow \left[\begin{array}{c} CH_3\ \ O \\ |\quad\ \ \| \\ -C-C- \\ | \\ CH_3 \end{array} \right]_n$$

Dimethylketene forms an explosive peroxide when exposed to air at low temperatures, so it must be thoroughly protected by an inert atmosphere in all phases of its preparation and use (66). Dimethylketene is purified by cooling a sample to −80°C in an inert atmosphere and adding about 1% (w/v) aluminum triethyl. At this temperature, polymerization is too slow to be a problem. After about 3 h, the pressure is reduced and 15 mL dimethylketene is allowed to distill into a small reaction vessel (e.g., a 100-mL test tube) kept at −78°C, with a side arm to receive the distillate. An atmosphere of dry, oxygen-free nitrogen, is maintained throughout the operation. When distillation is complete, the reaction vessel is brought to atmospheric pressure with nitrogen, and 15 mL dry toluene, previously cooled to −78°C, is added followed by 1 mL dry heptane containing 0.1 g distilled aluminum bromide. The bath temperature is raised to and held at −50°C for 20 h, during which time the viscosity becomes very great. Methanol is then added slowly to destroy the residual dimethylketene, and finally the polymer is coagulated completely with excess methanol. The polymer is washed in a blender with cold water repeatedly and then with warm 1% HCl, and again with methanol. About 3 g polymer is obtained after drying at 60°C in vacuum. If the polymer is successively extracted with boiling solvents, about 6% is removed with acetone, about 1% with ether, about 8% with benzene, and about 4% with toluene. The remainder is crystalline and has an inherent viscosity of about 0.7 in nitrobenzene at 135°C. The T_m is about 250°C.

By using aluminum bromide diethyl etherate, a higher conversion is possible. A lower temperature and higher monomer concentration increase the molecular weight.

335. Preparation of a Polyester from Dimethylketene by Alternating Double Bond/Carbonyl Polymerization (64)

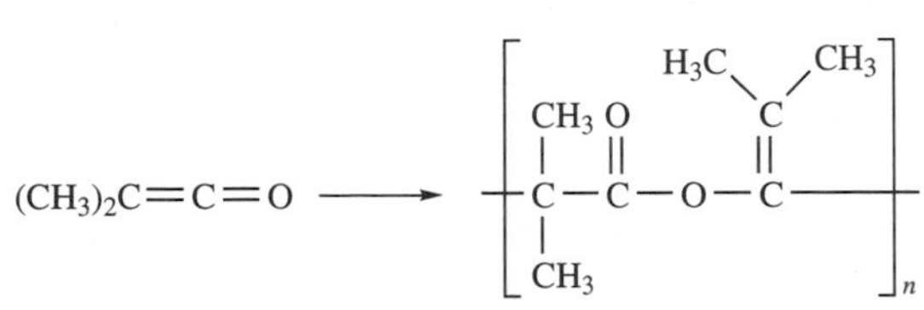

$$(CH_3)_2C{=}C{=}O \longrightarrow \left[\begin{array}{c} \qquad\qquad\quad H_3C\ \ \ CH_3 \\ \qquad\qquad\qquad \diagdown\ \diagup \\ CH_3\ \ O \qquad\quad C \\ |\quad\ \ \| \qquad\quad\ \ \| \\ -C-C-O-C- \\ | \qquad\qquad\qquad\quad \\ CH_3 \qquad\qquad\qquad \end{array} \right]_n$$

This polymerization is run as in the preceding reaction, using 0.1 g triethylaluminum to 20 mL dimethylketene, without solvent. The temperature is maintained at −25°C for 12 h, when the mixture has solidified. Again, unreacted monomer can be distilled at reduced pressure. Excess methanol is added, the polymer is washed thoroughly in 1% HCl and again in methanol, and dried in a vacuum at 60°C to give about 8 g of white powder. About 22% of the polymer is removed by boiling ether extraction. Most of the remainder is soluble in boiling benzene but is insoluble in boiling acetone. It is this benzene extractable fraction that is assigned the regular polyester structure. It is crystalline, with a T_m of 160 to 170°C; the intrinsic viscosity in tetralin at 135°C is about 0.4.

336. Preparation of a Polyacetal from Dimethylketene (62)

$$(CH_3)_2C{=}C{=}O \longrightarrow \left[\begin{array}{c} H_3C\diagdown\ \diagup CH_3 \\ C \\ \| \\ -C-O- \end{array} \right]_n$$

This polymerization is run in about the same way as the preceding two, except that a 250-mL three-necked flask with stirrer can be conveniently used, if desired. Then 30 mL dry ether is used as solvent; after cooling to −78°C, 9 g dimethylketene is added, followed by 1 mmol butyl lithium in pentane or similar hydrocarbon solvent. After 7 min, the mixture is poured into excess methanol to give, after methanol washing and vacuum drying, about 7 g white polymer, soluble in cold benzene and carbon tetrachloride and, for the most part, in refluxing acetone and ether. Its intrinsic viscosity in chloroform is 0.21. It is thermally stable up to 170°C and softens from 180 to 200°C. It is more thermally stable than polyisobutyraldehyde, which it structurally resembles. There is some evidence, both ir and degradative, for the presence of isopropylketone units, although the acetal structure predominates.

12.11. ELECTRICALLY CONDUCTIVE POLYMERS

It has been known for some time that certain polymers can exhibit fairly high electrical conductivity after having been treated with a doping agent. Figure 12.1 shows the chemical structures of some important conductive polymers (67). The conductivity range of these polymers is compared to that of conventional materials in Figure 12.2. All the "conducting" polymers, (e.g., polyacetylene) have a coupled (or potentially coupled) series of double bonds but are not strongly conductive unless they are doped. The dopant intercalates between the polymer chain layers and functions either by supplying electrons to *p*-type or removing electrons from *n*-type polymer energy band. Typical p dopants are iodine, bromine, and ferric chloride, and n dopants are alkali metals. At high doping levels, the energy band gap is suppressed, and high conductivity is achieved. One interesting difference between metals, such as copper and doped polymeric conductors, is that the conductivity of the latter decreases as temperature is decreased (semiconductive), whereas with metals the reverse is true.

Polyacetylenes are usually not very oxidatively stable and conductivity of doped samples invariably decrease with time. The conductive polymer that has evoked much interest is polyaniline, because it is easy to make, is reasonably stable, is readily formed into fibers and films, and provides a high level of conductivity when doped (68).

12.11. Polyanilines (69)

Polyanilines constitute a large family of polymers that are formed by the chemical or electrochemical oxidative polymerization of aniline and its derivatives. The polymers are usually electrically insulating but can be doped either chemically or electrochemically to increase conductivity about 10 orders of magnitude to give powders, fibers, or films with conductivity in the metallic conductive regime (**12.1**). Polyaniline (the emeraldine base form) in *N*-methyl-2-pyrrolidinone (NMP) at fiber-spinnable concentrations (e.g., 20%) gels so readily at room temperature that solutions cannot be spun through a spinning orifice to form fiber. Addition

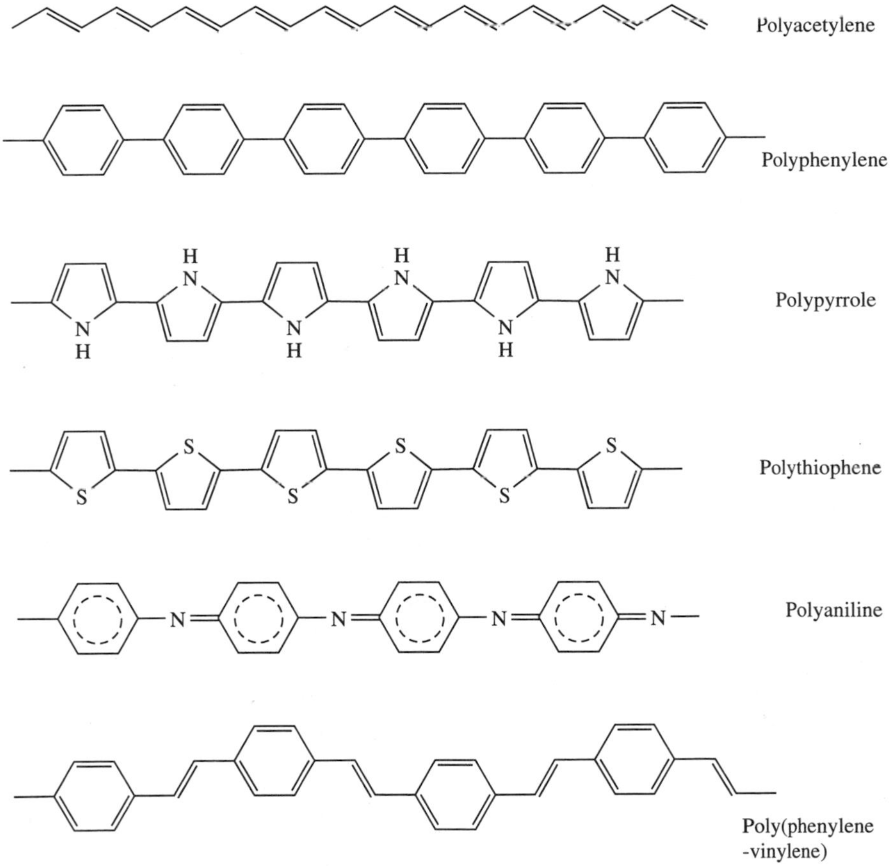

Polyacetylene

Polyphenylene

Polypyrrole

Polythiophene

Polyaniline

Poly(phenylene -vinylene)

Figure 12.1. Structures of some electrically conductive polymers (67).

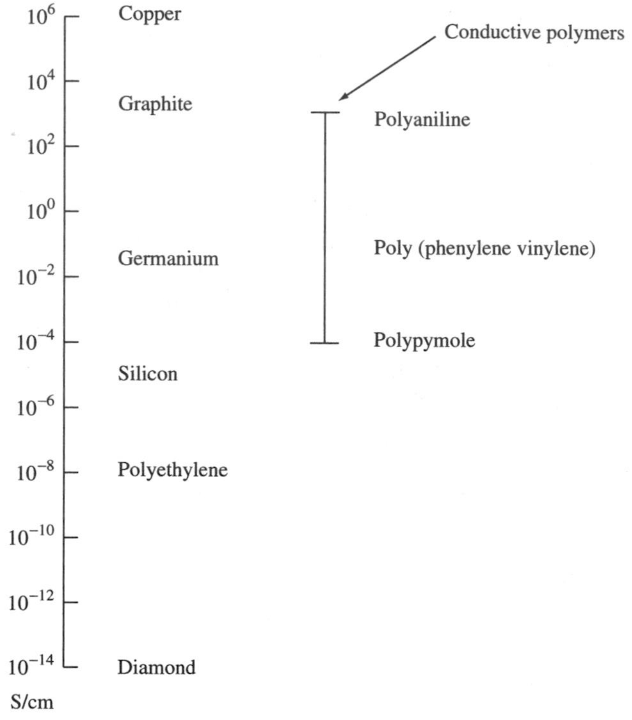

Figure 12.2. Conductivity range of selected polymers (67).

of certain amines (e.g., diaminocyclohexane) to NMP or using certain amines as solvent provides stable spinning solutions. Continuous dry-jet wet spinning can be operated successfully with a 1,4-diaminocyclohexane/polyaniline solution. The as-spun fiber is crystalline and can be oriented at about 215°C. The drawn fiber has high strength of about 4 gpd (400 MPa). Highest electrical conductivity is with drawn fiber doped with aqueous sulfuric acid (320.5 S/cm) and with HCl (157.8 S/cm). The tensile strength of the HCl-doped fiber is about 1.4 gpd.*

*337. Preparation of Polyaniline (Emeraldine Base)**

In a 2-L glass-reaction vessel, fitted with a cooling jacket, a 3-in twin-bladed stirrer, a nitrogen inlet and outlet, and a stoppered inlet, is added 134.24 g aniline, 194.34 g 37% aqueous HCl, and 1350.4 g deaerated deionized water. The system is blanketed with nitrogen, and medium speed stirring is started. The flask and contents are cooled to −8°C by circulating coolant through the reactor jacket. The stopper is replaced with a 500-mL graduated dropping funnel, and a solution of 154.85 g ammonium persulfate in 269.70 mL deaerated deionized water is added

* Thanks are due to C–H, Hsu (DuPont Company) for experimental details on preparing polyaniline.

Free amine form

Ammonium salt form

$C_6H_5{-}NH_2 \xrightarrow{\text{Oxidation}}$ $[{-}C_6H_4{-}NH{-}C_6H_4{-}NH{-}]_x$ $\underset{\text{aqKOH}}{\overset{\text{aqHA}}{\rightleftharpoons}}$ $[({-}C_6H_4{-}NH{-}C_6H_4{-}NH{-})^{+}A^{-}]_x$

1A
Insulating

1S
Insulating

oxidize ⇅ reduce

oxidize ⇅ reduce

$[{=}C_6H_4{=}N{-}C_6H_4{-}N{=}]_x$ $\underset{\text{aqKOH}}{\overset{\text{aqHA}}{\rightleftharpoons}}$ $[({=}C_6H_4{=}N{-}C_6H_4{-}NH{=})^{+}A^{-}]_x$

2A
Insulating

2S
Metallic

$[{=}C_6H_4{=}N{-}C_6H_4{-}N{=}]_x \longleftrightarrow [{=}C_6H_4{=}N{=}C_6H_4{=}N{=}]_x \longleftrightarrow [{=}C_6H_4{-}N{=}C_6H_4{=}N^{-}]_x$

Schematic 12.1

through it at a rate of 1.94 mL/min. Blanketing is continued with nitrogen after completion of the addition, and stirring and cooling are continued for 2.5 days. The solids content of the reactor are filtered using a sintered glass funnel and washed with about 25-L deionized water. The polymer, which is obtained in the doped, salt form, is converted to the base form by stirring the powder in 1 to 1.5-L 0.15% aqueous ammonium hydroxide solution for 24 h, repeating the process, and then filtering. The filtered polymer is dried to constant weight in a vacuum oven at 40°C. The polymer has an inherent viscosity of 1.31 at 0.5% concentration in 98% sulfuric acid at 30°C.

338. Preparation of Polymer Spinning Solutions and Fiber Preparation (69) A total of 3 g of the undoped polymer powder, as prepared above, and 120 g 1,4-diaminocyclohexane are place in a 100-mL resin-kettle equipped with an air-driven 44-mm-dia three-bladed impeller, a nitrogen capillary bleed, and a vacuum port. Stirring is initiated at a 20 rpm under nitrogen at a 20 mm Hg vacuum for 2 h. A smooth, viscous solution results. When this concentration of undoped polymer is used with NMP alone as solvent, the mixture gels in about 10 min even when the mixture is stirred only intermittently with a spatula. When the polymer slurry in NMP is exposed to gaseous ammonia while being stirred as above under nitrogen and vacuum, the polymer dissolves immediately and forms a viscous solution of 17% solids. Likewise, a solution of undoped polymer (1 g in 5 g NMP and 3 g pyrrolidine) remains viscous even after warming to 70°C.

The polyaniline solution in 1,4-diaminocyclohexane (prepared above) is transferred under vacuum into a stainless-steel spinning cell. The solution is pumped through a 325-mesh screen and through a 0.004-in-dia by 0.008-in-long single-hole spinneret, through a 2-mm air gap into a water bath at 25°C. The immersion length in the bath is 20 in. The monofilament is wound up on a motor-driven spool at 2.42 m/min. The collected spools are stored in tap water at room temperature for 24 h, then air dried. The as-spun fiber is drawn two to three times over a hot pin at 215°C. The drawn fibers are 6 denier/filament and have a tenacity at break of 3.9 gpd; they have an elongation of 9.3% and a tensile modulus of 83 gpd.

The heat-drawn fibers are then acid doped as follows to impart significant conductivity. The drawn, undoped fibers are immersed relaxed in 9.8% aqueous sulfuric acid for 24 h. The fibers are then washed with water for 30 min and dried at room temperature for 2 days. Electrical conductivity, measured at ambient conditions on the fibers is 320.5 S/cm. Other drawn, undoped fiber is immersed in 3.6% aqueous HCl solution for 24 h. The fibers are air-dried for 6 days before testing. Electrical conductivity at ambient conditions is 157.8 S/cm. Average tenacity of the HCl-treated fiber is 1.4 gpd (148 MPa).

12.11.2. Polyphenylenevinylenes (70)

The parent polyphenylenevinylene is not very soluble and is difficult to process; however, ring substitution (e.g., with hexyloxy groups) has been used successively to improve solubility.

339. Preparation of Poly(2,5-di-n-hexyloxy-p-phenylenevinylene) (70)

$H_{13}C_6O$ ClH_2C CH_2Cl OC_6H_{13} I — 1 eq KOtBu, THF 25°C → $H_{13}C_6O$ ClH_2C $\overset{\ominus}{C}HCl$ OC_6H_{13} II

II — $-Cl^{\ominus}$ → $H_{13}C_6O$ Cl OC_6H_{13} III

III — II → $C_6H_{13}O$ Cl OC_6H_{13} Soluble Precursor IV

IV — 100°C, O,1 mm, −HCl → $H_{13}C_6O$ OC_6H_{13} DHPPV V

Schematic 12.2

339.A. Synthesis of 1,4-Di-n-hexyloxybenzene

In a 1-L three-necked flask equipped with an air condenser, stirrer, and nitrogen inlet is added 55 g (0.1 mol) hydroquinone and 500 mL dimethyl sulfoxide. The mixture is stirred under nitrogen at room temperature, and a solution is soon obtained. Then 56 g (1 mol) finely ground potassium hydroxide is added with rapid stirring and ice bath cooling, followed after 5 min by the addition of 165 g (1 mol) hexyl bromide over 5 min. The product begins to precipitate from solution, and stirring is continued for 12 h. The product is filtered, sucked dry, and recrystallized from isopropanol.

339.B. Synthesis of 2,5-Di-n-hexyloxy-α,α′,dichloro-p-xylene

In a three-necked 500-mL flask equipped with a magnetic stirrer, a fritted inlet tube (immersable below the surface of the liquid), and a water condenser with a lead at the top to a sodium hydroxide trap to adsorb excess hydrogen chloride is placed 40 g (0.14 mol) 1,4-dihexyloxybenzene, 22 g (0.73 mol) paraformaldehyde, 50 mL concentrated hydrochloric acid, 50 mL of 85% polyphosphoric acid, and 70 mL acetic acid. The flask is heated to 90°C and HCl gas is bubbled through at about 30 mL/min for 5 h. The passage of HCl is then stopped, the frit removed, and heating continued at 90°C for another 12 h. On cooling, the crude product precipitates and is filtered, sucked dry under a dental dam, and then recrystallized from hexane.

339.C. Dehydrohalogenation Polymerization of 2,5-Di-n-hexyloxy-α,α'-dichloro-p-xylene

Into a flamed dry, septum-capped flask equipped with a magnetic stirrer is added under nitrogen 3.75 g (0.01 mol) of the above dichloro product, and (75 mL) THF. Then 1.12 g (0.001 mol) of a solution of potassium *t*-butoxide in 75 mL THF is injected all at once, via a syringe, into the solution of the dichloride with stirring at ambient temperature. This removes 1 mol of HCl (from the α,α' positions) and forms the di-hexyloxyphenylene-α-chloroethylene polymer. A viscous, fluorescent yellow solution results and is stirred for 1 h. The polymer is precipitated by pouring into methanol. It is filtered, washed well with methanol, and dried under vacuum at room temperature. This precursor polymer is then dissolved in THF and cast into thin films. The films, after drying, are heated at 100°C under 0.1 torr vacuum for 10 to 12 h to eliminate HCl and convert the chloroethylene polymer to poly(phenylenevinylene). The final film is exposed in iodine vapor to saturation, resulting in a conductivity of 1×10^{-2} S/cm.

12.11.3. Polythiophenes

Polythiophenes are also conductive when doped. As with many of the other aromatic polymers, ring substitution is required to improve solubility for fabrication into fibers and films.

Schematic 12.3

340. Preparation of Poly(3-methyl-2,5-Thienylene) (71)

A 250-mL three-necked glass flask equipped with a nitrogen bleed, a reflux condenser with drying tube, a stoppered port, and a magnetic stirrer bar is flamed dry under nitrogen. Then 9.61 g (37.5 mol) 3-methyl-2,5-dibromothiophene in 20 mL calcium hydride–dried THF is added, followed by 0.911 g (37.5 mmol) magnesium. The mixture is kept at room temperature with stirring until a homogeneous solution results (about 30 min). Then, with stirring, is added 80 mL dibutyl ether, 0.045 g (0.16 mmol) magnesium, and 35 mg (0.125 mol) nickel(II) chloride 2,2′-bipyridine complex. The stirred reaction mixture is refluxed for 5 h, over which time a brown polymer separates. The mixture is cooled, and the solid is filtered through a glass frit. The solid is washed well with methanol and dried. Yield is 1.52 g, (40%).

The crude polymer is extracted in a Soxhlet apparatus with chloroform; 90% is soluble. The polymer on isolation is reddish brown. It has a high affinity for iodine and absorbes close to 100% by weight when exposed to iodine vapor at room temperature. The polymer alone is an insulator but is conductive when doped with iodine, having an electrical conductivity of about 2.8×10^{-2} S/cm at room temperature.

12.11.4. Electrically Conducting Polypyrroles

There has been considerable interest in doped conducting polypyrrole because of its ease of preparation and environmental stability, especially of samples prepared via electrochemical oxidative coupling. Conductivity of samples prepared by ferric chloride oxidation usually decreases on exposure to atmospheric moisture and oxygen. However, it has been found that stability can be significantly improved if ions of large volume, such as tosylates or dodecylbenzenesulfonates, are used as dopants (72).

341. Preparation of Conducting Polypyrroles (72)

Pyrrole monomer (Aldrich) is freshly distilled and stored under nitrogen. Ferric chloride hexahydrate (Baker) is used as received. The sulfonic acids used are either commercially available or are prepared via hydrolysis of the sulfonyl chlorides. Ferric hydroxide is prepared by slowly adding a solution of 44.1 g (1.11 mol) NaOH in 300 mL water to a vigorously stirred solution of 100 g (0.370 mol) ferric chloride hexahydrate. The precipitate is allowed to settle, and the supernatant liquor is siphoned off. Water is added, the mixture stirred, and the rinsing procedure repeated twice. The brown ferric hydroxide is then filtered through a 2-L medium-frit sintered funnel and washed once with water. The moist precipitate is transferred to a 2-L flask containing 207 g (1.05 mol) ethylbenzenesulfonic acid (95%) dissolved in 1-L methanol. The mixture is stirred for 3 h at 45°C, the solution is filtered, and the solvent is removed in a rotary evaporator. The resulting orange oil is transferred to a crystallizing dish and further dried in a vacuum oven (0.1 mm) at 50°C. The solid residue is then ground to a fine powder and stored under nitrogen in a sealed container.

To a rapidly stirred solution of 20 g (about 33 mmol) of ferric ethylbenzenesulfonate in 20 mL methanol at 0°C is added via a syringe 1.0 g (14.9 mmol) of precooled (0°C) pyrrole. The reaction mixture darkened rapidly, and a thick black suspension separated. After 30 min, the polypyrrole precipitate is filtered, and rinsed thoroughly with methanol and then acetone; it is then sucked dry. The polymer is dried in a vacuum at room temperature to yield 1.08 g polypyrrole ethylbenzenesulfonate (about 65% based on the pyrrole used). The polymer prepared at 0°C has a conductivity of 52 ohm-cm^{-1} (2×10^{-4} S/cm) compared to 26.9 for one made at ambient temperature. Testing is run on pellets.

12.11.5. Polyacetylenes

Undoped polyacetylene is not a particularly good conductor, and even the doped product is less conductive and much less oxidatively stable than polyaniline.

Considering the hazards in dealing with acetylene and its polymers we have not provided its preparation here. However, the technology is chemically significant, so we have included the preparation of *m*-diethynylbenzene. This gives a reasonably high polymer, capable—in the case of *m*-diethynylbenzene—of giving clear, flexible yellow films. The films are interesting in that when ignited in air at room temperature they explosively pyrolyze, with a weight loss of 5 to 6%. This accounts for all the hydrogen content (2.7%) and some of the carbon (59).

342. Preparation of Diethynylbenzenes (60)

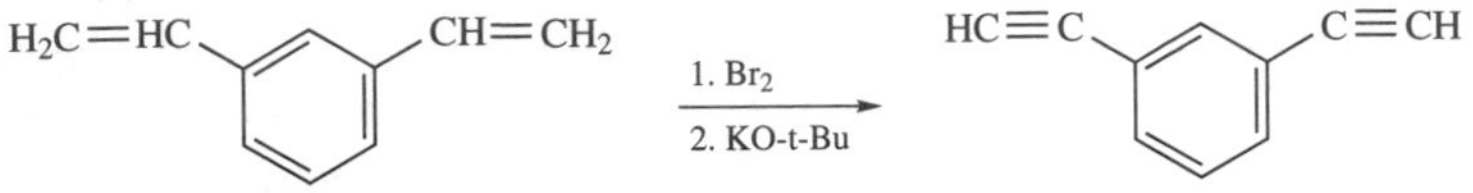

A total of 1300 g (8.13 mol) bromine is added over 2 h with stirring to a cooled solution of 750 g mixed divinylbenzene (40% = 2.3 mol *m*- and *p*-divinylbenzene) in 1.2 L chloroform. The mixture is cooled to 5°C and a voluminous precipitate separates and is filtered. Crystallization from chloroform yields 264 g (0.59 mol) 1,4-bis(1,2-dibromoethyl)benzene (mp 155 to 157°C). The two filtrates are combined, and the chloroform is removed in a rotating evaporator at 100°C at 3 mm. The residue is then fractionated in a molecular still. Distillation at 50°C (40 to 70 μ) and then at 80°C (20 to 50 μ) separates most of the dibromodiethylbenzenes. The residue, a viscous syrup, is distilled at 150°C (12 to 30 μ). The distillate crystallizes when triturated with cold alcohol and, after recrystallization from alcohol, yields about 420 g 1,3-bis(1,2-dibromoethyl)benzene (mp 65 to 66.5°C).

To a solution of 18 g (0.46 mol) potassium in 1-L *t*-butanol at reflux is added 50 g (0.11 mol) 1,3-bis(1,2-dibromoethyl)benzene. After 1 h, heating is stopped, and the reaction mixture is made up to 4-L with ice water. The product is isolated by ether extraction, the ether is removed, and the residue is distilled to give *m*-diethynylbenzene (bp 78°C/14 mm) in about 70% yield (9 g).

343. Polymerization of Diethynylbenzene (59, 60)

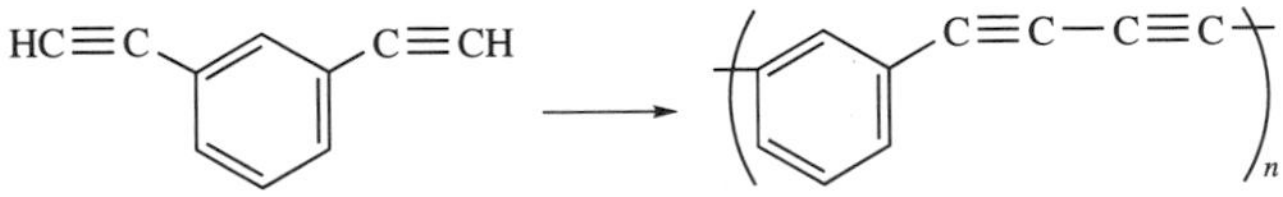

To a 125-mL flask is added 65 mL dry pyridine, 0.5 g Cu(I) chloride, and 12.5 g *m*-diethynylbenzene. The solution is kept in a 30°C bath and is vigorously stirred; oxygen is bubbled through it. The temperature rises as reaction takes place. At the end of the reaction, a pale yellow polymer separates in quantitative yield. The polymer is soluble in chlorobenzene and in nitrobenzene above 100°C. Clear, flexible, yellow films can be cast at 170°C under nitrogen from a nitrobenzene solution at about 30% concentration. The intrinsic viscosity may be no higher than 0.25 dL/g in nitrobenzene at 150°C. Molecular weight estimates based on ir determination of acetylene ends are 7000 or more.

12.12. POLY(ETHER KETONES)

Among thermoplastic resins used in composites, poly(ether ketones) and polyimides have probably evoked the highest interest. Initial work at DuPont on the synthesis of polyketones was based on the Friedel-Crafts reaction of acid chlorides with phenyl ethers. This route, however, unless carefully controlled, gives branched and/or colored polymer (73). Another route based on reaction of 4,4′-difluorobenzophenone with hydroquinone and other diphenols has been explored (80) and a product was commercialized as (poly ether ether ketone) (PEEK). Other syntheses of poly(ether ketones) based on modifications of this approach and on metal-catalyzed halide couplings have been published (75–78). Alternating poly(ether ketones) from phenyl ether and terephthaloyl/isophthaloyl chlorides are more crystalline and crystalize faster than the corresponding random copolymers and provide better property retention at elevated temperatures. Comparison syntheses are given below.

344. Preparations of Poly(ether Ketones) from Phenyl Ether and Terephthaloyl/Isophthaloyl Chlorides (73)

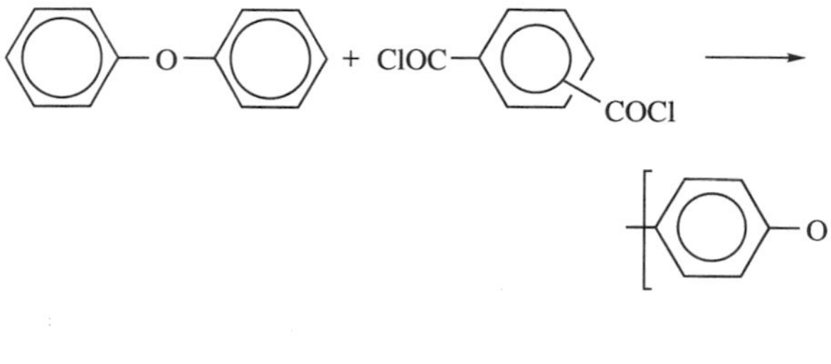

344.A. Random Copolymer (50/50; I/T)

To 228 g *o*-dichloro-benzene contained in a 2-L resin kettle with an outer jacked suitable for heating and cooling are added 16.92 g (0.083 mol) terephthaloyl chloride, 16.92 g (0.083 mol) isophthaloyl chloride, and 28.37 g (0.17 mol) phenyl ether. The resin kettle is also equipped with a high-speed resin-clad tantalum stirrer, a dry nitrogen inlet, a gas outlet to a scrubber, and a port for introducing the reagents. Stirring is started and the reactants cooled to −10°C. Then 68 g (0.51 mol) aluminum chloride is added over a 1-h period at −10°C. The reaction mixture is then warmed to 0°C and held at that temperature for 20 min. The contents are transferred (**Caution! Gloves and face shield.**) rapidly to another flask containing 625 mL dichlorobenzene at 107°C. The temperature falls to about 80°C and the system is reheated to 100°C and held at that temperature for 30 min. The heating source is removed, and the reaction is quenched by adding 1 L methanol cooled to −40°C. Stirring is continued until all the color is gone; about 1 h. The polymer is filtered and washed twice with 1-L methanol; it is then transferred to a beaker, and 1-L water is added and heated to about 100°C. The polymer is filtered and soaked in a solution of

100 mL concentrated. HCl in 750 mL water for 30 min, filtered, and soaked in 750 mL glacial formic acid for 30 min. After filtering, the polymer is washed with distilled water until the filtrate is neutral. The polymer is dried in a vacuum oven at 120°C under nitrogen. Inherent viscosity in concentrated sulfuric acid at 0.5% concentration is about 0.73.

344.B. Alternating Copolymer

The equipment used is the same as described above. Into the resin kettle is added 41.99 g (0.0893 mol) 1,4-bis(4-phenoxybenzoyl)benzene, 18.12 g (0.0893 mol) isophthaloyl chloride, and 350 mL *o*-dichlorobenzene. The system is cooled to −8°C. while 72 g aluminum chloride is added. The system is warmed to and held at 0°C for 20 min, then transferred rapidly to a similarly equipped resin kettle containing 625 mL *o*-dichlorobenzene at 100°C. The flask and contents are then kept at 100°C for 1 h. The heater is removed, and the system is quenched by addition of 1-L methanol precooled to −40°C. Stirring is continued until all the color is gone. The polymer is filtered and rewashed twice with 1-L methanol. The polymer is transferred to a beaker and boiled in 1-L water for 15 min. The polymer is refiltered and then soaked in 500 mL glacial formic acid for 30 min, filtered, washed with water until neutral, and then dried in a vacuum oven at 120°C under a slight nitrogen bleed. The polymer has a melt index at 390°C of 3 and an inherent viscosity of 1.7 at 0.5% concentration in concentrated sulfuric acid.

The 1,4-bis(4-phenoxybenzoyl)benzene used as starting material is prepared as follows: 65.3 g terephthaloyl chloride and 137.4 g diphenyl ether are dissolved in 0.65 L *o*-dichlorobenzene in a 0.5-gal glass-lined reactor. The system is cooled to −5 to 0°C with stirring, and 197.14 g aluminum chloride is added while keeping the temperature below 5°C. The mixture is warmed to and held at 20°C for 15 min. Then 0.5 gal methanol at −50°C is added over 15 to 20 min, while keeping the pot temperature below 70°C. The system is cooled to 20 to 30°C and stirred for 30 min. The slurry is filtered, reslurried in 0.5 gal methanol, and filtered. The product is dried in a vacuum oven at 120°C. A sample recrystallized from boiling dimethylacetamide melts at 215°C.

345. Preparation of Poly(ether-ether Ketone) (PEEK) from 4,4'Difluorobenzophenone and Hydroquinone (74)

$$F-C_6H_4-C(=O)-C_6H_4-F + HO-C_6H_4-OH \longrightarrow \left[C_6H_4-C(=O)-C_6H_4-O-C_6H_4-O \right]_n$$

Into a 200-mL four-necked resin kettle equipped with a stirrer, a nitrogen inlet, an air condenser, and a stoppered addition port is added 21.82 g (0.1 mol) 4,4′-difluorobenzophenone and 60 g diphenylsulfone. The mixture is heated with stirring to 180°C under nitrogen. Then 14.0 g (0.101 mol) anhydrous potassium carbonate (sieved under nitrogen through a 300-μm sieve and maintained under nitrogen to avoid air contamination) is added through the stoppered port. The temperature is raised to and maintained at 200°C for 1 h. The temperature is then raised to 320°C and kept at that temperature for 1 h. The polymer is in solution at this stage. The mixture is cooled, and the solid reaction product is filtered and milled to pass through a 500-μm sieve. The product is washed successively with acetone (two times), water (three times), and acetone/methanol (two times). The polymer is then dried at 140°C under vacuum. The inherent viscosity is about 1.4 at 0.5% concentration in 98% sulfuric acid and contains no gel. The PMT is 334°C and the T_g is 140°C. Tough film can be made by pressing at 400°C. The polymer is colorless and is stable at 400°C for over 1 h.

346. Preparation of an AB Poly(ether Ketone) from 4-Fluoro-4′-hydroxybenzophenone (75)

In a 250-mL round-bottom flask equipped with a stirrer, nitrogen inlet, and air condenser with drying tube is placed 108 g (0.5 mol) 4-fluoro-4′-hydroxybenzophenone (Aldrich), 71.8 g (0.52 mol) potassium carbonate and 40 g benzophenone. After purging with nitrogen, and maintaining a slow stream of nitrogen, the flask is immersed in a Woods metal bath at 300°C. The contents are stirred for 6 h. Then 5.45 g (0.025 mol) difluorobenzophenone is added as a stabilizer and heating is continued for another 30 min. The flask is cooled by removing it from the bath and immersing it in liquid nitrogen. The flask is carefully broken, and the contents are separated from the glass and crushed to a powder. The powder is then repeatedly extracted with acetone and with water to remove benzophenone and potassium fluoride. The polymer residue is dried under vacuum at 60°C. Yield is 85 g. Relative viscosity number of a 0.1% concentrated solution in 98% sulfuric acid is 1.02 dL/g.

A facile synthesis of poly(ether ketones) has been developed using a nickel-catalyzed coupling of dichlorobenzophenones in the presence of zinc, triphenylphosphine, and bipyridine. Polymerizations are smooth, and polymer inherent viscosity numbers up to 0.87dL/g.

347. Aromatic Poly(ether Ketones) by Nickel Catalyzed Coupling of Aromatic Dichlorobenzophenones (76)

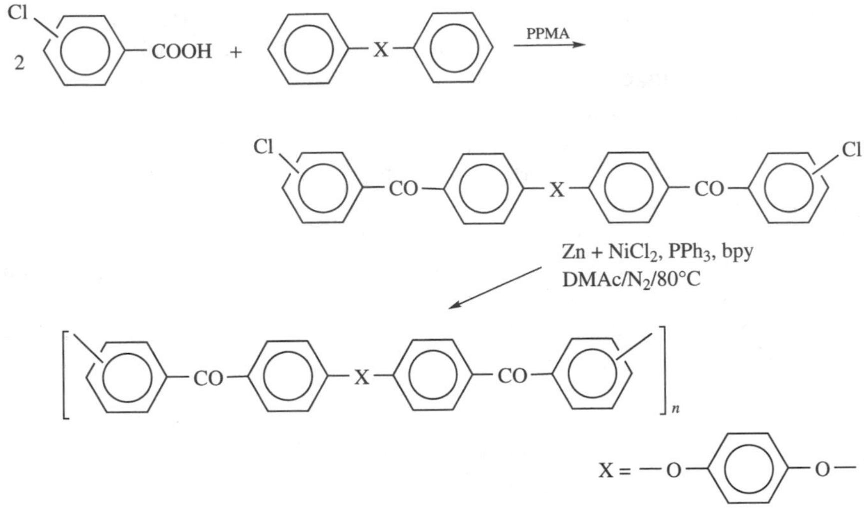

Cl, 2, COOH + X, PPMA

Cl, CO, X, CO, Cl

$Zn + NiCl_2$, PPh_3, bpy
DMAc/N_2/80°C

CO, X, CO, n

X = —O—C₆H₄—O—

347.A. Materials

Reagent-grade Ni(II)Cl is dried at 220°C under vacuum. Triphenylphosphine is purified by recrystallization from hexane. Powdered 400-mesh zinc is purified by stirring with acetic acid, filtering, washing thoroughly with ether, and drying under vacuum. All solvents (DMF, DMAc, NMP, and HMPA) are stirred with calcium hydride overnight and then distilled under reduced pressure, and stored over 4Åmolecular sieves. Phosphorus pentoxide/methanesulfonic acid (PPMA) is prepared by dissolving 1 part by weight of phosphorus pentoxide in 10 parts by weight of methanesulfonic acid.

347.B. Preparation of Bis[1,4-(3-chlorobenzoyl-p-phenoxy)]benzene

In a three-necked round-bottom flask equipped with a magnetic stirrer bar and a drying tube is placed 2.62 g (0.01 mol) 1,4-diphenoxybenzene, 3.13 g (0.02 mol) 3-chlorobenzoic acid, and 50 g PPMA. The mixture is stirred at room temperature for 60 h, then added to 200 mL water, and the product filtered. Recrystallization from ethyl acetate gives the product as white needles (mp 199 to 200°C).

347.C. Polymerization

In a 50-mL round-bottom flask equipped with a nitrogen inlet, an inlet port sealed with a rubber septum, a stoppered port for ingredients addition, and a magnetic stirrer, are added 16.2 mg (0.125 mmol), nickel(II) chloride, 65.6 mg (0.25 mmol) triphenylphosphine, 19.5 mg (0.125 mmol) 2,2′-bipyridine, 506 mg (7.75 mmol) zinc and 1.35 g (2.5 mmol) bis-1,4-[(3-chlorobenzoyl-p-phenoxy)]benzene. The flask is evacuated to remove air and is refilled with nitrogen. This process is repeated three times. Then 3.0 mL dry DMAc is added via a syringe through the septum cap and with continuous flushing with a slow steam of nitrogen. The flask is stirred and heated to 90°C. The mixture becomes reddish brown after about 30 min. Stirring is continued for 2.5 h, and the resultant viscous mixture is then diluted with 25 mL DMAc and poured into 250 mL of a 50/50 mixture of methanol/concentrated aqueous HCl. The precipitated polymer is filtered, washed well with methanol, and dried at 80°C at house vacuum for 10 h. Polymer yield is 100%. The inherent viscosity in NMP at 0.5% concentration is 0.87 dL/g.

A novel method of synthesizing poly(ether ketones) (PEK) was developed by reaction of 4,4'-dichlorobenzophenone with sodium carbonate in the presence of silica and cupric choride. The silica acts not only as a catalyst but also as an agent to control molecular weight. When too much silica is used, low molecular weight polymer with hydroxyl termination at both ends results. Molecular weight is also inversely proportional to the amount of silica used. The optimum amount of silica (e.g., Aerosil 300) used to obtain high molecular weight polymer is found to be 5 to 20 wt % based on the dihalide. Cuprous and cupric salts work as cocatalysts with the silica and promote the polymerization reaction; alone they have no catalytic activity. The optimum amount of copper salt is 0.5 to 3 wt % as copper atom based on the silica; higher amounts produce gel.

348. Preparation of Aromatic Poly(ether Ketones) from Aromatic Dihalides (79)

$$Cl-C_6H_4-\overset{O}{\overset{\|}{C}}-C_6H_4-Cl \xrightarrow[Na_2CO_3]{SiO_2/Cu} \left[C_6H_4-\overset{O}{\overset{\|}{C}}-C_6H_4-O \right]_n$$

348.A. Materials

Diphenylsulfone and 1,4-dichlorobenzophenone are obtained from Aldrich. Sodium carbonate and silica gel (Aerosil 300) or 400-mesh silica gel are dried at 200°C for 8 h before use.

348.B. Polymerization

A 1-L resin kettle equipped with a stirrer, nitrogen inlet, air condenser with drying tube, and feed-port is flamed out under a stream of nitrogen, and a constant low flow of nitrogen is maintained throughout the reaction. To the kettle is added 59.2 g (0.548 mol) finely divided sodium carbonate, 125 g (0.498 mol) 4,4′-dichlorobenzophenone, 10 g finely divided silica, 0.27 g (0.002 mol) cupric chloride, and 300 g diphenyl sulfone. The mixture is stirred, flushed with nitrogen for 30 min, and heated to 200°C with stirring under a constant stream of nitrogen. After 30 min, the temperature is raised to 280°C and held for 2 h; then it is raised to 300°C for 1 h and to 320°C for a further 2 h. Polymerization progress is monitored by evolution of carbon dioxide. When this has ceased, the flask is cooled, and the product removed and pulverized in water. The powdered product is then extracted thoroughly with acetone and with water and is finally soaked in 4% aqueous sodium hydroxide to remove the silica. The residue is filtered, washed well with acetone, and dried at 100°C under vacuum. An off-white polymer results. Inherent viscosity at 0.1% concentration in NMP is 1.0 to 1.2.

High molecular weight polyarylether sulfones are produced by nickel catalyzed coupling of 4,4′-bis(p-chlorophenoxy) diphenyl sulfone. Metallic zinc is used to drive the polymerization reaction, which takes place under mild conditions in the presence of triphenylphosphine and a dipolar aprotic solvent.

349. Preparation of 4,4′-Bis(p-chlorophenoxy)diphenyl Sulfone Monomer (78)

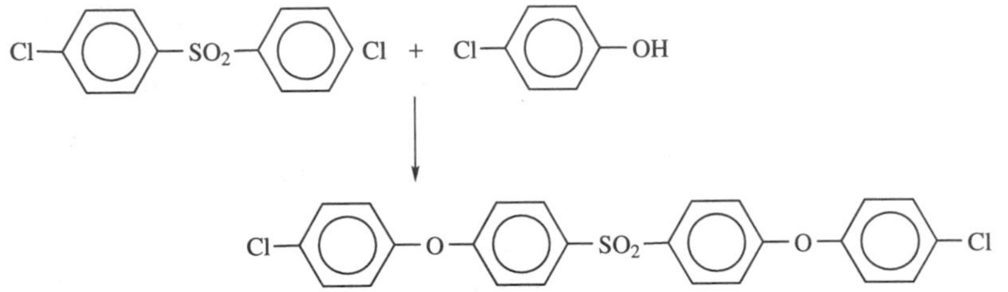

Cl—C6H4—SO_2—C6H4—Cl + Cl—C6H4—OH

↓

Cl—C6H4—O—C6H4—SO_2—C6H4—O—C6H4—Cl

Reagents are obtained from Aldrich and *p*-Chlorophenol is purified by distillation. Into a 1-L three-necked flask equipped with a thermometer, a mechanical stirrer , a Claisen adapter fitted with a nitrogen purge line, and a Dean-Stark trap with a condenser is added 180 g 4,4′-dichlorodiphenylsulfone, 161 g *p*-chlorophenol, 113 g potassium carbonate, 550 mL dimethylacetamide, and 350 mL toluene. The mixture is stirred and purged with nitrogen for 30 min and then heated to reflux (112 to 115°C). Toluene and water is condensed in the Dean-Stark trap. As the toluene is removed, the temperature of the solution increases until 160°C is reached. The temperature is then held at that level for 4 h. After cooling, the reaction mixture is filtered to remove the excess potassium carbonate and is coagulated in a methanol/water mixture. The product is recrystallized from isopropanol/water, filtered, and dried in an oven.

350. Polymerization of 4,4′-Bis(p-chlorophenoxy)diphenyl Sulfone by Reductive Halogen Coupling (78)

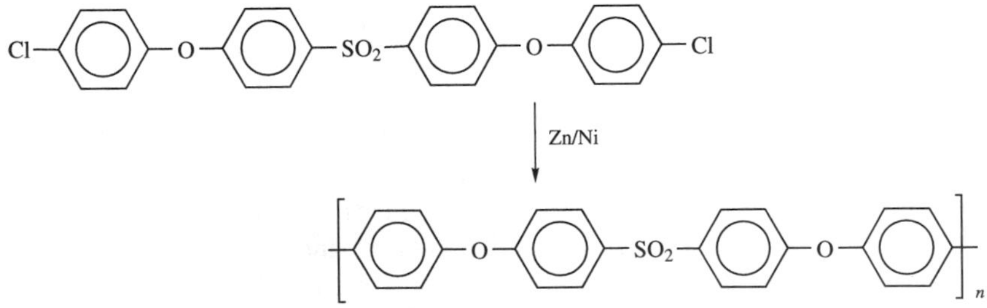

Cl—C6H4—O—C6H4—SO_2—C6H4—O—C6H4—Cl

↓ Zn/Ni

$[$—C6H4—O—C6H4—SO_2—C6H4—O—C6H4—$]_n$

Into a three-necked 250-mL flask equipped with a magnetic stirring bar are placed 8.0 g zinc dust, 6.0 g triphenylphosphine, 0.15 g 2,2′-bipyridine, and 0.1 g nickel(II) chloride. (The zinc must be pure, have a low level of oxide, and a high surface area. Poor-quality zinc is usually the cause of failure to attain high molecular weight. The proper color of the reaction mixture is reddish brown, which turns greenish on addition of monomer, but displays brownish streaks within a short time. A persistent deep green indicates that the nickel is not being reduced quickly, and a grayish color indicates total catalyst deactivation.) One neck of the flask is sealed with a serum cap. A side arm flask containing 20 g crystalline monomer is attached to the second neck, and a stopcock is attached to the third. The stopcock is connected to a double-manifold vacuum line by pressure tubing, and the apparatus is then evacuated and filled with nitrogen or argon several times. (It is essential to remove all traces of oxygen and maintain this state throughout the polymerization.) Then 60 mL dry, nitrogen-sparged DMAc is added via a syringe to the reaction flask, and the reaction apparatus is placed in an oil bath at 70°C. Once the red-brown catalyst forms, the monomer is added from the side-arm flask to the catalyst solution. The polymerization is run at 70°C for 10 to 15 h under nitrogen or argon. At the end of this period, 0.5 mL chlorobenzene is added and reacted for 30 min to end cap the polymer and remove any residual nickel attached to the polymer. The reaction mixture is diluted with an equal volume of DMAc and filtered through a medium frit sintered-glass funnel to remove any excess zinc. The polymer is coagulated in 1 L methanol in a blender. The polymer is then slurried at 0°C for 1 h in 1 L. deionized water. This is followed by filtration, washing the polymer cake several times with an equal volume of methanol, and drying in a vacuum oven. The polymer should have a relative viscosity of about 0.8 dL/g in *N*-methylpyrrolidone at 25°C.

12.13. POLYAMIDES

Polyamides have been prepared from amines, aryl dibromides, and carbon monoxide in the presence of palladium catalysts. The mechanism involves palladium complex formation with the halide, insertion of CO, followed by displacement of the halide by amine (79).

351. Aliphatic-Aromatic Polyamides from Aliphatic Diamines, Aromatic Dibromides and Carbon Monoxide (79)

$$H_2N-R-NH_2 + Br-Ar-Br + 2\,CO \longrightarrow -[-NH-R-NH-\overset{O}{\overset{\|}{C}}-Ar-\overset{O}{\overset{\|}{C}}-]_n-$$

351.A. Materials

Bis(bromophenyl ether) is crystallized from ethanol. Hexamethylenediamine is purified by vacuum distillation. Triphenylphosphine (PPh_3) is crystallized from hexane. 1,8-Diazabicyclo-[5.4. 0] undecene (DBU) is purified by vaccum distillation. 1,4-Diazabicyclo[2.2.2] octane (DABCO) is recrystallized from ethanol.

351.B. Polymerization

In a three-necked flask equipped with a stirrer, a carbon monoxide inlet and a reflux condenser is added 0.2095 g (2.5 mmol) hexamethylenediamine, 0.8200 g (2.5 mmol) 4,4′-bisbromophenyl ether, 0.1053 g (0.15 mmol) $PdCl_2(PPh_3)_2$, 0.0787 g (0.3 mmol) triphenylphosphine, and 5 mL dimethylacetamide. The flask and reactants are purged several times with carbon monoxide and heated with stirring at 115°C in an oil bath. To the mixture is added via a syringe, 0.97 mL (6.5 mmol) DBU. The reaction mixture is stirred at 115°C until the takeup of carbon monoxide ceases, about 1.5 h. The reaction solution is diluted with 25 mL DMAc and poured into 400 mL methanol. The polymer precipitates and is filtered, washed copiously with hot methanol, and dried at room temperature in vacuo. The yield is about 0.74 g. The inherent viscosity is 1 to 1.2 dL/g at 0.5% concentration in 98% sulfuric acid.

In place of DBU, 0.73 g (0.112 mmol) DABCO can be used, but the reaction should then be run for 48 h. Inherent viscosity is lower, about 0.7.

An unusual example of a hydrogen addition to a double bond to form a polymer is the Michael-type addition of acryamide to itself to form poly(β-alanine), or 3-nylon (80). It represents the unusual case of an AB monomer in an anionic, proton transfer addition reaction. Although the polymerization gives a relatively low molecular weight product, it is included here because of the unusual character of the polymerization and the fact that 3-nylon is one of the lower possible homologs of the polyamide series derived from ω-amino acids. Because of the short chain separation between amide groups, the PMT (320 to 330°C) of 3-nylon is high. Hydrolysis of the polymer provides a short, high-yield synthesis of β-alanine from the commercially available acrylamide.

Acrylamide can also be polymerized by a typical free-radical route to vinyl polymer with pendant amide groups. Polyacrylamide prepared this way is a water soluble, high molecular weight material that forms polyacrylic acid on hydrolysis.

352. Preparation of Poly(β-alanine) [3-Nylon] (80)

$$CH_2{=}CH{-}\overset{\displaystyle O}{\overset{\|}{C}}{-}NH_2 \xrightarrow{Na} \left[CH_2{-}CH_2{-}\overset{\displaystyle O}{\overset{\|}{C}}{-}\underset{\displaystyle H}{N} \right]_n \xrightarrow{H_2O} H_2N{-}CH_2{-}CH_2{-}CO_2H$$

The acrylamide used in this polymerization can be purified by recrystallization from ethyl acetate and sublimation at <1 mm; the T_m is 85°C. (**Care in handling! Acrylamide is toxic.**) The DMF used must be dry; it can be distilled from a small quantity of phosphorus pentoxide at reduced pressure.

A solution of 4.4 g acrylamide (0.062 mol) in 4 mL DMF is prepared in a suitable small reaction vessel that is protected from the atmosphere by a nitrogen inlet tube attached to a mercury bubbler to vent nitrogen when the vessel is closed. The solution is heated to 100°C in an oil bath and 2 drops of 50% dispersion of sodium in xylene (about 0.018 g) is added. Polymerization occurs and is completed in 3 to 5 min. The reaction mixture is quenched in water and the polymer filtered and washed with water several times in a blender. The polymer, dried at 70°C for 24 h in a vacuum, weighs 3.5 to 4.0 g (80 to 90%), has a T_m of 320 to 330°C, and has an inherent viscosity of 0.33 at 0.5% concentration in 90% formic acid at 25°C. It is soluble in strong acids, such as formic acid, from which brittle films can be cast.

This polymer is refluxed with an excess of 50% aqueous sulfuric acid for 4 h. The solution is then neutralized to pH 7 with hot aqueous barium hydroxide. The barium sulfate is filtered and washed twice by trituration with 50 mL water. The combined water filtrates are evaporated to dryness by heating at water aspirator pressure on a steam bath. The residual syrup crystallizes on cooling to give 80 to 90% yield of β-alanine, which may be recrystallized from hot methanol. The product melts at 195 to 196°C.

12.14. POLYANHYDRIDES

Elimination of water between carboxyl groups can give polyanhydrides when the spacial requirements favor linear rather than cyclic anhydride molecules. The behavior of diacids toward anhydride formation has been studied (81). Malonic acid forms only polymeric anhydride, whereas succinic and glutaric acids form only cyclic anhydrides. Adipic acid forms both cyclic and linear anhydrides, but higher acids form only linear products. Studies found that sebacic acid initially gave an anhydride of about 5,000 molecular weight (α-form). When this was subjected to molecular distillation, a cyclic dimer (β-anhydride) distilled, and a higher molecular weight (about 20,000) ω-polyanhydride remained. The β-anhydride was converted to a γ-form on standing and remelting. The latter was practically identical with the α-anhydride and was thought to contain very large ring structures, while the α-form was linear.

The ω-anhydride polymers are melt-spinnable to strong, lustrous, crystalline fibers. However, the aliphatic anhydrides are all so hydrolytically unstable that the polymers and fibers degrade rapidly on standing.

A series of polyanhydrides based on aromatic diacids was prepared (82), and these are much more stable than their aliphatic counterparts. Preparation is best accomplished by forming the mixed anhydride with acetic acid and heating that intermediate under vacuum to eliminate acetic anhydride. The products are crystalline, high melting, and thermally stable enough to be melt spun. Other polyanhydrides have been reported using essentially the same reaction as above (83).

353. Preparation of 1,3-Bis(p-carboxyphenoxy)propane (82)

$$2\ HO{-}C_6H_4{-}CO_2H + Br(CH_2)_3Br \xrightarrow[2.\ NaOH/H_2SO_4]{1.\ NaOH} HOC(=O){-}C_6H_4{-}O(CH_2)_3O{-}C_6H_4{-}C(=O)OH + 2NaBr$$

In a 1-L three-necked flask fitted with a reflux condenser, stirrer, and dropping funnel 138 g (1 mol) *p*-hydroxybenzoic acid is dissolved in a solution of 80 g (2 mol) of sodium hydroxide in 400 mL water. Through the funnel, 102 g (0.5 mol) 1,3-dibromopropane is added over 1 h, while the contents of the flask are stirred and kept at reflux temperature. The reaction mixture is then heated a further 3.5 h at reflux. After this period, 20 g (0.5 mol) solid sodium hydroxide is added to the mixture, which is heated for 2 h more at reflux. The reaction mixture is left to cool overnight. The fine, powdery, white precipitate of the disodium salt is filtered and washed with 200 mL methanol. The wet precipitate is dissolved in 1-L distilled water and acidified with 6 *N* sulfuric acid. The dibasic acid is filtered and dried in a vacuum oven at 80°C. The yield is 79 g (50%). The neutralization equivalent is 157; (158 calculated). The dibromopropane can be replaced by dichloro using the same procedure, but a longer reaction time (6 h) is required.

354. Preparation of the Mixed Anhydride of 1,3-Bis(p-caboxyphenoxy)propane and Acetic Acid

$$HOC(=O){-}C_6H_4{-}O(CH_2)_3O{-}C_6H_4{-}C(=O)OH + 2(CH_3C(=O))_2O \longrightarrow$$

$$CH_3C(=O){-}O{-}C(=O){-}C_6H_4{-}O(CH_2)_3O{-}C_6H_4{-}C(=O){-}O{-}C(=O)CH_3 + 2CH_3C(=O){-}OH$$

A total of 60 g (0.19 mol) 1,3-bis(carboxyphenoxy)propane is refluxed with 650 mL acetic anhydride while a slow stream of dry nitrogen is bubbled through the solution. After 30 min, almost all the dibasic acid is dissolved. The mixture is filtered while hot. The filtrate, slightly yellow, is concentrated to a volume of 60 mL by distilling acetic anhydride/acetic acid under vacuum at 65°C. The concentrated reaction mixture is stored overnight in a refrigerator. The white, needle-like crystals are separated by filtration and washed with dry ether. Yield is 66 g (87%), and the T_m is 102 to 103°C.

355. Preparation of Poly(1,3bis[p-carboxyphenoxy]propane Anhydride (82)

$$CH_3C(=O)-O-C(=O)-C_6H_4-O(CH_2)_3O-C_6H_4-C(=O)-O-C(=O)CH_3 \longrightarrow$$

$$\left[C(=O)-C_6H_4-O(CH_2)_3O-C_6H_4-C(=O)-O \right]_n + (CH_3C(=O))_2O$$

In a polymer tube equipped with a side arm leading to a receiving flask is placed 20 g of the mixed anhydride of 1,3-bis(p-carboxyphenoxy)propane with acetic acid. A capillary reaching to the bottom of the tube is inserted. The polymer tube is heated in a 280°C (dimethyl phthalate) vapor bath that is brought to that temperature after the tube is in position. Nitrogen is passed through the mixture and acetic anhydride distils. After 30 min, at 280°C, a vacuum of about 1 mm is applied. A stream of nitrogen is continually passed through the increasingly viscous melt. Periodically, the vacuum may be released and a strong current of nitrogen flushed through the viscous melt for additional mixing. At the end of 30 min, the polycondensation is terminated. On cooling the tube, adhesion of the polymer to the walls of the tube and shrinkage during crystallization may cause the tube to shatter (**Caution!**). The polymer is obtained as a yellowish, opaque hard block, which can be crystallized by further annealing at 130°C in an oven for about 30 min. The T_m is about 275°C. From the melt, yellowish lustrous fibers can be spun and cold drawn.

12.15. DENSE STAR POLYMERS (DENDRITES)

Organic polymers are generally classified as being linear or branched, the latter resulting from some of the polymerizing units, in the case of vinyl monomers, having a valency greater than two, and in the case of AA-BB step-addition polymers, one of the units having at least three reacting groups. Branching in this sense excludes regularly repeating side groups, such as methyl, ethyl, and other nonreacting groups in the polymerizing process. Carried to the extreme, e.g., in the case of AA-BB polymers, when the branch ends are one third of the total ends, branching leads to cross-linking. However, un-cross-linked multibranched polymers with star-structured branching (dendritic = tree-like) have been developed, wherein the individual branches radiate from a nucleus, and there are three branches per nucleus. An elegant example of star-branched polymers is shown in **12.4** (84, 85).

Star-branched polymers have terminal group densities greater than conventional polymers and are often more sensitive to degradation by shearing than

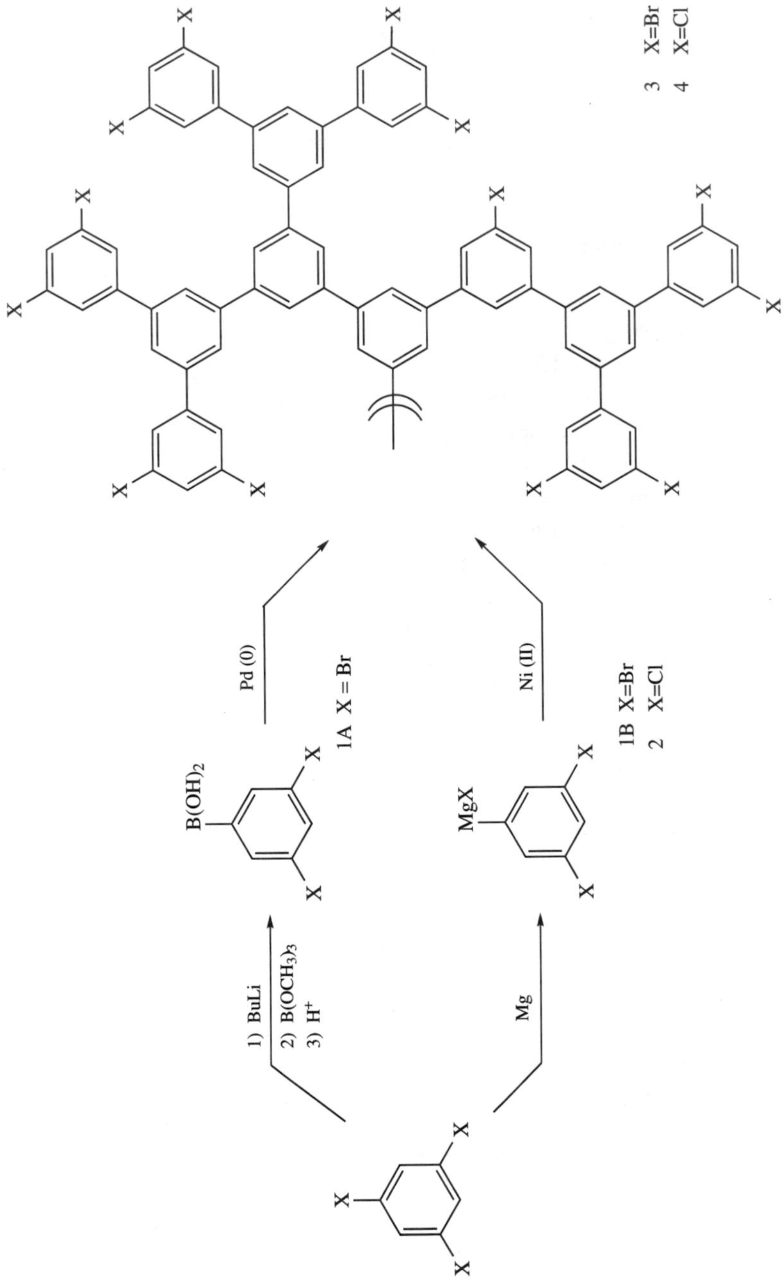

1) BuLi
2) $B(OCH_3)_3$
3) H^+

$B(OH)_2$

1A X = Br

Pd (0)

Mg

MgX

1B X=Br
2 X=Cl

Ni (II)

3 X=Br
4 X=Cl

Schematic 12.4

conventional polymers and are useful in paints and as wet-strength agents in the manufacture of paper. As expected, the star-branched polymers have very low intrinsic viscosities because of their compact structures.

356. Preparation of Polybiphenyl Star Polymer (85)

To 6.30 g (20.0 mmol) tribromobenzene in 30 mL ether is added 12.9 mL (20.0 mmol) n-butyl lithium at $-78°C$. The mixture is stirred for 10 min. To the lithiate is added 6.06 g (25 mmol) of M_gBr_2 etherate at $-78°C$. The solution is warmed to room temperature and is added to 500 mL THF and 6.45 mg (2.5 mmol) of a solution of Ni(acac)$_2$. The mixture is refluxed for 24 h, and then 5 mL water is added carefully at room temperature. The solution is concentrated under reduced pressure, and the polymer is precipitated from petroleum ether. After washing with methanol, water, and a dilute HCl solution and drying, 1.82 (59%) of a white powder is obtained. GPC; $\overline{M}_n = 3900$, $\overline{M}_w = 7080$.

REFERENCES

1. G. B. Butler and R. J. Angelo, *J. Am. Chem. Soc.*, **79**, 3128 (1957).
2. C. S. Marvel and R. D. Vest, *J. Am. Chem. Soc.*, **79**, 5711 (1957).
3. J. F. Jones, *J. Poly. Sci.*, **33**, 15 (1958).
4. J. F. Walker, *Formaldehyde*, Reinhold, New York, 1953.
5. H. Staudinger and A. Gaule, *Berichte chemic*, **49**, 1897 (1916).
6. H. Staudinger and R. Signer, *Helv. Chim. Acta.*, **11**, 1847 (1858).
7. U.S. Pat. 2,795,571 (June 11, 1957), A. K. Schneider to DuPont.
8. U.S. Pat. 2,768,994 (October 30, 1956), R. N. MacDonald to DuPont.
9. C. D. Kennedy et al., per presented at the 150th American Chemistry Society Meeting, Atlantic City, 1965.
10. F. M. Berardinelli et al., *J. Appl. Poly. Sci.*, **9**, 1419 (1965).
11. S. Okamura et al., *J. Poly. Sci. C*, **4**, 827 (1963).
12. D. B. Miller, *Am. Chem. Soc. Poly. Preprints*, **6**, 613 (1965).
13. S. Okamura et al., *J. Poly. Sci.*, **58**, 925 (1962).
14. E. Gipstein et al., *J. Poly. Sci. B*, **1**, 237 (1963).
15. J. B. Lando and V. J. Stannett, *J. Poly. Sci. B*, **2**, 375 (1965).
16. D. B. Miller, Stanford Research Institute, personal communication.
17. O. Vogl, *J. Poly. Sci. A*, **2**, 4591 (1964).
18. I. Rosen and C. L. Sturm, *J. Poly. Sci. A*, **3**, 3741 (1965).
19. I. Rosen et al., *J. Poly. Sci. A*, **3**, 1545 (1965).
20. W. J. Middleton et al., *J. Org. Chem.*, **30**, 1375 (1965).
21. A. Schonberg and A. Stephenson, *Ber.*, **66B**, 567 (1933).
22. A. L. Barney et al., *J. Poly. Sci. A*, **4**, 2617 (1966).
23. W. Sweeny, *J. Appl. Poly. Sci.*, **7**, 1983 (1963).
24. V. E. Shashoua et al., *J. Am. Chem. Soc.*, **82**, 866 (1960).
25. C. King, *J. Am. Chem. Soc.*, **86**, 437 (1964).
26. T. W. Campbell et al., *J. Am. Chem. Soc.*, **84**, 3673 (1962).

27. J. J. Monagle et al., *J. Am. Chem. Soc.*, **84**, 4288 (1962).
28. T. W. Campbell and K. C. Smeltz, *J. Org. Chem.*, **28**, 2069 (1963).
29. U.S. Pat. 2,941,966 (June 21, 1960) T. W. Campbell (Dec 22, 1953) to DuPont.
30. U.S. Pats., 2,663,736 and 2,663,739; W. B. McCormack. W. B. McCormack, *Org. Syn.*, **43**, 73 (1963).
31. C. E. H. Bawn and A. Ledwith, *Chem. Ind.*, **1957**, 1180 (1957).
32. G. D. Buckley, and N. H. Ray, *J. Chem. Soc.*, **1952**, 3701 (1952).
33. A. Nasini et al., *J. Poly. Sci.*, **34**, 106 (1959).
34. C. E. H. Bawn et al., *J. Poly. Sci.*, **34**, 93 (1959).
35. A. Ledwith, *Chem. Ind.*, **1956**, 1310 (1956).
36. G. Saini et al., *Gazz. Chim. Ital.*, **87**, 342 (1957).
37. M. Szwarc, *J. Poly. Sci.*, **6**, 319 (1951).
38. M. Szwarc, *J. Chem. Phys.*, **16**, 128 (1948).
39. U.S. Pat. 2,757,146 (July 31 1956), F. S. Fawcett to DuPont.
40. H. E. Winberg et al., *J. Am. Chem. Soc.*, **82**, 1428 (1960).
41. Brit. Pat. 807,196 (Jan 7, 1959) to DuPont, T. E. Young.
42. W. F. Gorham, *Poly. Preprints*, **6**, 73 1965.
43. H. G. Gilch and W. L Wheelwright, *J. Poly. Sci. A*, **4**, 1337 (1966).
44. H. Gilch, *Angew. Chem. Intern. Educ.*, **4**, 598 (1965) and *J. Poly. Sci. A*, **4**, 43 (1966).
45. H. Gilch, *J. Poly. Sci. A*, **4**, 1351 (1966).
46. L. A. Auspos et al., *J. Poly. Sci.*, **15**, 9 (1955).
47. J. R. Schaefgen, *J. Poly. Sci.*, **15**, 203 (1955).
48. H. E. Winberg and F. S. Fawcett in V. Boekelheide, ed., *Organic Synthesis*, vol. 42, John Wiley & Sons, New York, 1962.
49. U.S. Pat. 2,721,189 (October 1, 1955), A. W. Anderson and N. G. Merckling.
50. Ger. Pat. 1,037,103 (1958), A. W. Anderson et al.
51. W. L. Truett et al., *J. Am. Chem. Soc.*, **82**, 2337 (1960).
52. E. E. Magat et al., *J. Am. Chem. Soc.*, **73**, 1028 (1952).
53. E. E. Magat et al., *J. Am. Chem. Soc.*, **73**, 1031 (1951).
54. E. E. Magat and L. F. Salisbury, *J. Am. Chem. Soc.*, **73**, 1035 (1951).
55. E. E. Magat, *J. Am. Chem. Soc.*, **73**, 1367 (1951).
56. W. H. Hunter and M J. Morse, *J. Am. Chem. Soc.*, **55**, 3701 (1933).
57. G. Staffin and C. C. Price, *Rubber World*, **139**, 408 (1958).
58. A. S. Hay et al., *J. Am. Chem. Soc.*, **81**, 6335 (1959).
59. A. S. Hay, *J. Org. Chem.*, **25**, 1275 (1960).
60. A. S. Hay, *J. Org. Chem.*, **25**, 637 (1960).
61. G. Natta et al., *J. Am. Chem. Soc.*, **82**, 4742 (1960).
62. G. Natta et al., *Makromol. Chem.*, **51**, 148 (1962).
63. G. Natta et al., *Makromol. Chem.*, **44–46**, 537 (1961).
64. Belg. Pat. 623,181 (1963), G. Natta et al.
65. W. E. Hanford and J. C. Sauer, in R. Adams; ed., *Organic Reactions*, vol. 3, John Wiley & Sons, New York, 1947.
66. E. U. Elam, *Org. Chem. Bull.*, **36**, 1 (1964).
67. U.S. Pat. 2,672,480 (March 16, 1954), A. S. Matlack to Hercules Powder Corp.
68. H. Naarman, *Adv. Mater.*, **2**, 354 (1990). S. Roth and M. Filzmoser, *Adv. Mater.* **2**, 356 (1990).
69. C.-H. Hsu et al., *Synthetic Metals*, **59**, 37 (1993). A. G. Macdairmid et al., *Mol. Cryst. Liq. Cryst.*, **121**, 173 (1985).

70. W. J. Swatos and B. Gordon III, *Polym. Preprints*, **256**, (1990). S. Machida and S. Miyata, *Synthetic Metals*, **31**, 311 (1989). I. Murase et al., *Synthetic Metals*, **17**, 639 (1987). S. Tokito, P. Smith, and A. J. Heeger, *Synthetic Metals*, **36**, 183 (1990).
71. T. Yamamoto and K. Sanechika, *Chem. Ind.*, 301 (1982).
72. J. A. Walker, L. F. Warren, and E. F. Witucki, *J. Poly. Science* **26** 1285 (1988). L. F. Warren and D. P. Anderson, *J. Electrochem. Soc.*, **134**, 101 (1987).
73. U.S. Pat. 4,816,556 (March 28, 1989), F. P. Gay to DuPont.
74. U.S. Pat. 4,320,224 (1982), J. B. Rose and P. A. Staniland to ICI Ltd.
75. I. Fukawa and F. Tanabe, *J. Poly. Sci. A*, **31**, 535 (1993). U.S. Pat., 3,401,0147 (1977), J. B. Rose to ICI Ltd.
76. M. Ueda and F. Ichikawa, *Macromolecules*, **23**, 926 (1990).
77. I. Fukawa et al., *Macromolecules*, **24**, 3838 (1991).
78. I. Colon and G. T. Kwiatkowski, *J. Poly. Sci. A*, **28**, 367 (1990).
79. M. Yoneyama et al., *J. Poly. Sci. A*, **27**, 1985 (1989).
80. U.S. Pat. 2,672,480 (March 16, 1954), A. S. Matlack.
81. H. F. Mark and S. B. Whitby, (eds.), *Collected Papers of Wallace Hume Carothers*, Wiley-Interscience, NewYork, 1940.
82. A. Conix, *J. Poly. Sci.*, **29**, 343 (1958).
83. N. Yoda, *Macromol. Chem.*, **32**, 1 (1959).
84. U.S Pat. 4,507,466 (March 26, 1985), D. A. Tomalia and J. R. Dewald.
85. Y. H. Kim and O. W. Webster, *Macromolecules*, **1992**, 25 (1992).

AUTHOR INDEX

SUBJECT INDEX

T